Plants Secondary Metabolites and Pigments

The Author

Dr. U.D. Chavan obtained his B. Sc. and M. Sc. (Agri. in Biochemistry) degrees from Mahatma Phule Krishi Vidyapeeth Rahuri in 1985 and 1987 respectively. He received his Ph.D. degree in Food Science from Memorial University of Newfoundland St. John's Canada in 1999. He has done International training on "Global Nutrition 2002" at Uppsala University Uppsala, Sweden in 2002. He also attended follow-up International workshop on "Global Nutrition 2002" at Hanoi, Vietnam in 2002. Dr. Chavan visited Denmark, Finland, Ireland, France, Switzerland, Poland, Spain, Vietnam, Thailand, England and U.S.A. under "Global Nutrition 2002" Programme sponsored by Swedish International Development Agency (SIDA). During this programme he worked on human nutritional diet and disorders as well as on genetically modified organisms (GMO).

Dr. Chavan worked as Senior Research Assistant in the Department of Biochemistry and Food Science and Technology at Mahatma Phule Krishi Vidyapeeth, Rahuri from 1988 to 2000. During his Ph. D., he worked as Technician/Research Associate at Atlantic Cool Climate Crop Research Center and Agriculture and Agri-Food, Canada. He received D.Sc. degree in 2006 from USA. He has guided 22 students for M.Sc. (Agri.) in Biochemistry and Food Science and Technology. From 2000 to 2004 he worked as Assistant Professor of Biochemistry at Mahatma Phule Krishi Vidyapeeth Rahuri.

Dr. Chavan received Korgavkar Trust Fellowship, under-graduate and post-graduate merit scholarship as well as Senior Research Fellowship from ICAR, New Delhi and United States Department of Agriculture U.S.A. He received University Graduate Fellowship from Memorial University of Newfoundland St. John's Canada for his Ph. D. programme. Dr. Chavan also received International Scholar Award and Excellentiam Award for his Ph. D. research work from Memorial University of Newfoundland St. John's Canada. He has received a "Certificate of Appreciation" from the U.S. Department of Agriculture in 1992 for his work on processing of groundnut under the guidance of Dr. S. S. Kadam (Principal Investigator). He has written 129 research papers and 135 popular articles. He has authored 19 books in Marathi, 10 books in English and six book chapters in English. He has been awarded "Literary Award" for his best book on "Growth Regular" on agriculture during 1997. Dr. Chavan was selected as a best group leader and best presentation for "Global Nutrition 2002" by SIDA. He has been awarded "Life Time Achievement Award 2004" for his outstanding contribution in post-harvest technology of fruits and vegetables and allied fields by United Writers' Association of India. He has been elected as a Fellow of United Writers' Association of India in 2004. Dr. Chavan received pride of nation and Maharashtra Gunigan Ratna Award for 2006. He also received Jewels of India Award for 2006 for his contribution in the field of Food Science and Technology. He has been awarded "Literary Awards" for his best book on "Sorghum Grain Processing" on agriculture during 2009-2010 and Pulses cultivation to processing and value addition for 2012. He is recipient of "Krishi Goorav Award-2011" from Bhartiya Krishak Samaj, Maharashtra State, Nasik. He also contributed in the development of crop varieties in wheat NIAW-917 and in Sorghum Phule Anuradha (RSV 423), Phule Chitra (RSV 1546), Phule Suchitra (RSV 1098), Phule Vasudha (RSV 423), Phule Revati (RSV 1006), SPV-1626, Phule Panchami (RSPOV-3) for Sorghum pops, Phule Godhan (SPV-2057 for forage single cut, CSH-50 (Sweet Sorghum), Phule Madhur (RSSGV – 46) for sorghum *hurda* and Phule Rohini (RPASV-3) for sorghum *papad*. Now he is working as a Professor in the Department of Food Science and Technology, Senior Cereal Food Technologist in Sorghum Improvement Project and Foreign Student Advisor at Mahatma Phule Krishi Vidyapeeth Rahuri, Dist., Ahmednagar, Maharashtra, India.

Plants Secondary Metabolites and Pigments

— *Author* —

U.D. Chavan

Professor
Department of Food Science and Technology
Mahatma Phule Krishi Vidyapeeth, Rahuri, Maharashtra, India

2016

Daya Publishing House®
A Division of

Astral International Pvt. Ltd.
New Delhi – 110 002

Cataloging in Publication Data--DK
Courtesy: D.K. Agencies (P) Ltd. <docinfo@dkagencies.com>

Chavan, U. D., author.
Plants secondary metabolites and pigments / author, U.D. Chavan.
    pages cm
    Includes index.

    ISBN 9789351309185 (International Edition)

    1. Plant metabolites. 2. Plant pigments. I. Title.

QK881.C43 2016   DDC 572.42   23

Published by   :   **Daya Publishing House®**
                            A Division of
                            Astral International Pvt. Ltd.
                            – ISO 9001:2008 Certified Company –
                            4760-61/23, Ansari Road, Darya Ganj
                            New Delhi-110 002
                            Ph. 011-43549197, 23278134
                            E-mail: info@astralint.com
                            Website: www.astralint.com

Laser Typesetting   :   **Classic Computer Services**, Delhi - 110 035

Printed at   :   **Thomson Press India Limited**

Acknowledgement

Author is thankful to the Hon. Dr. S. S. Kadam, Ex. Vice-Chancellor, Marathwada Agricultural University, Parbhani for his critical and fruitful suggestions while the preparation of **"Plants Secondary Metabolites and Pigments"** book. I am thankful to Dr. J. K. Chavan, Dr. R. N. Adsule, Dr. V. D. Pawar, Dr. U. Y. Bhoite, Dr. K. D. Chavan, Dr. C. A. Nimbalkar, for helping during this manuscript preparation, as well as those scientists who worked on plant secondary metabolites and plant pigments identification, utilization and usefulness for human beings and their contribution is acknowledged in this book as well their sources are cited in the reference section. Author is also thankful to all scientists and friends those helped directly or indirectly while preparation of this manuscript. We would like to express over profound thanks to our family members for their constant support, patience and devotion, which inspired me to complete this manuscript.

U.D. Chavan

Preface

Plant species produce a variety of secondary metabolites in their cells. According to Kossel, who introduces the term 'secondary metabolites'. "Primary metabolites are present in every plant cell that is capable of division, where as secondary metabolites are present only accidentally and are not essential for plant life." However, many secondary plant products are known to have important roles in growth, development and adaptation of plant species. Some provide defence against micro-organisms, insect parasites and herbivores. Several classes of secondary products are 'induced' by infection, wounding, or herbivory. Genetic variation in the speed and extent of such induction may account at least in part for the difference between resistant and susceptible varieties. The biosynthesis of several plant secondary compounds occurs in specialized organs and plant cells at precise developmental stages. They are important in plant metabolism either due to their own biological activities or as precursors to active compounds. The two types of metabolisms are strongly interlinked. So it is often difficult to decide whether a metabolite is a primary or secondary one. A major distinction is that primary metabolites are few in number and are common to all plant species. Many secondary metabolites are derived from the products of primary metabolism although many have their own independent metabolic pathways.

Secondary products are generally small molecules when compared to the size of macro molecules such as proteins and nucleic acids. Their molecular structure is often based on a multiple ring hydrocarbon skeleton to which atoms of other elements (oxygen and nitrogen) are joined but some compounds have a straight chain molecular structure. Secondary metabolic products from plants are classified in many different ways - by their function, by the species from which it is derived; or by chemical composition.

Secondary metabolites have economic value in modern industry. They are natural sources to chemical compounds important in the pharmaceutical industry, precursors to important vitamins and raw material to many industrial products.

It has been estimated that about 30 per cent of the world's plant species have been used for medicinal purpose. Some products act on the autonomic and central nervous systems. Morphine, the principal alkaloid of the opium poppy is used for the relief of severe pain. Compounds like cannabis from hemp, caffeine and theobromine, from coffee and cocoa respectively and cocaine from *Erythroxylim coca*, stimulate mental activity, where as reserpine from *Rauwolfia* spp. depresses mental activity. The alkaloid atropine from solanaceous plants is used to dilate the pupil of the eye during surgery. Steroidal glycosides of *Digitalis* spp. are mainly used to impart a slowing and strengthening effect of heart muscles. Quinidine from cinchona bark is used to control atrial fibrillation.

Many non pharmaceutical products important in industry are also derived from plants [dyes, green chlorophyll, saffron yellow dye, purple or red colours from the roots of madder (*Rubia tinctorum*)]. Many secondary products provide protection from pests and diseases. The seed extract of neem (*Azadirachta indica*), root powder of *Derris* spp., flowers of *Chrysanthemum cinerariaefolium* and leaves and stalks of *Nicotiana* have insecticidal properties. *Schoenocaulon officionale* (Sabadilla) is used as a rodent poison. Therefore, the information given in this book will help to the teachers, students, researchers and pharmaceuticals also.

Pigments are chemical compounds which reflect only certain wavelengths of visible light. This makes them appear "colourful". Flowers, corals and even animal skin contain pigments which give them their colours. More important than their reflection of light is the ability of pigments to absorb certain wavelengths. Because they interact with light to absorb only certain wavelengths, pigments are useful to plants and other autotrophs organisms which make their own food using photosynthesis. In plants, algae and cyanobacteria, pigments are the means by which the energy of sunlight is captured for photosynthesis. However, since each pigment reacts with only a narrow range of the spectrum, there is usually a need to produce several kinds of pigments, each of a different colour, to capture more of the sun's energy. There are three basic classes of pigments.

1. **Chlorophylls** are greenish pigments which contain a porphyrin ring. This is a stable ring-shaped molecule around which electrons are free to migrate. Because the electrons move freely, the ring has the potential to gain or lose electrons easily and thus the potential to provide energized electrons to other molecules. This is the fundamental process by which chlorophyll "captures" the energy of sunlight. There are several kinds of chlorophyll, the most important being chlorophyll "a". This is the molecule which makes photosynthesis possible, by passing its energized electrons on to molecules which will manufacture sugars. All plants, algae and cyanobacteria which photosynthesize contain chlorophyll "a". A second kind of chlorophyll is chlorophyll "b", which occurs only in "green algae" and in the plants. A third form of chlorophyll which is common is (not surprisingly) called chlorophyll "c" and is found only in the photosynthetic members of the Chromista as well as the dinoflagellates. The difference between the chlorophylls of these major groups was one of the first clues that they were not as closely related as previously thought.

2. **Carotenoids** are usually red, orange, or yellow pigments and include the familiar compound carotene, which gives carrots their colour. These compounds are composed of two small six-carbon rings connected by a "chain" of carbon atoms. As a result, they do not dissolve in water and must be attached to membranes within the cell. Carotenoids cannot transfer sunlight energy directly to the photosynthetic pathway, but must pass their absorbed energy to chlorophyll. For this reason, they are called accessory pigments. One very visible accessory pigment is fucoxanthin the brown pigment which colours kelps and other brown algae as well as the diatoms.

3. **Phycobilins** are water-soluble pigments and are therefore found in the cytoplasm, or in the stroma of the chloroplast. They occur only in Cyanobacteria and Rhodophyta. The picture at the right shows the two classes of phycobilins which may be extracted from these "algae". The vial on the left contains the bluish pigment phycocyanin, which gives the Cyanobacteria their name. The vial on the right contains the reddish pigment phycoerythrin, which gives the red algae their common name. Phycobilins are not only useful to the organisms which use them for soaking up light energy; they have also found use as research tools. Both pycocyanin and phycoerythrin fluoresce at a particular wavelength. That is, when they are exposed to strong light, they absorb the light energy and release it by emitting light of a very narrow range of wavelengths. The light produced by this fluorescence is so distinctive and reliable, that phycobilins may be used as chemical "tags". The pigments are chemically bonded to antibodies, which are then put into a solution of cells. When the solution is sprayed as a stream of fine droplets past a laser and computer sensor, a machine can identify whether the cells in the droplets have been "tagged" by the antibodies. This has found extensive use in cancer research, for "tagging" tumor cells.

Therefore, I hope this book will be useful to under-graduate as well as post-graduate students in Science Faculty, in Agriculture, Food Technology, Biotechnology, Biochemistry, Horticulture, Home Science and Industrial Sectors.

U.D. Chavan

Contents

Acknowledgement   *v*

Preface   *vii*

Part I: Plants Secondary Metabolites

1.   Introduction   3

2.   Alkaloids   15

3.   Alpha Amylase Inhibitor   33

4.   BOAA (Lathyrus)   67

5.   Carotenoids   85

6.   Chymotrypsin   95

7.   Flavonoids   103

8.   Glycosides   117

9.   Gossypol   123

10. Non-protein Amino Acids   127

11. Oxalic Acid   145

12. Phytic Acid   151

13. Phytoestrogens   165

14. Polyphenols   175

15. Saponin   189

16. Steroids   217

17. Tannins   233

18. Terpenoids   321

19. Trypsin Inhibitor   325

20. Potential Health Benefits and Problems   339

Part II: Plant Pigments

21. General Biological Pigment   377

22. Adaptive Colouration   387

23. Anthocyanins   405

24. Betalain   419

25. Bioluminescence   427

26. Carotenoids   437

27. Chlorophyll   447

28. Chromatophore   465

29. Melanin   477

30. Photo-Protective Pigments   491

31. Lighting and Physiology   499

32. Tetrapyrroles   523

    Index   537

Part–I

Plants Secondary Metabolites

Chapter 1

Introduction

Plant species produce a variety of secondary metabolites in their cells. According to Kossel (1891) who introduces the term 'secondary metabolites' "primary metabolites are present in every plant cell that is capable of division, where as secondary metabolites are present only accidentally and are not essential for plant life." However, many secondary plant products are known to have important roles in growth, development and adaptation of plant species. Some provide defence against micro-organisms, insect parasites and herbivores. Several classes of secondary products are 'induced' by infection, wounding, or herbivory. Genetic variation in the speed and extent of such induction may account at least in part for the difference between resistant and susceptible varieties. The biosynthesis of several plant secondary compounds occurs in specialised organs and plant cells at precise developmental stages. They are important in plant metabolism either due to their own biological activities or as precursors to active compounds. The two types of metabolism are strongly interlinked. So it is often difficult to decide whether a metabolite is a primary or secondary one. A major distinction is that primary metabolites are few in number and are common to all plant species. Many secondary metabolites are derived from the products of primary metabolism although many have their own independent metabolic pathways.

Secondary products are generally small molecules when compared to the size of macro molecules such as proteins and nucleic acids. Their molecular structure is often based on a multiple ring hydrocarbon skeleton to which atoms of other elements (oxygen and nitrogen) are joined but some compounds have a straight chain molecular structure. Secondary metabolic products from plants are classified in many different ways - by their function, by the species from which it is derived; or by chemical composition.

Economic Significance of Secondary Metabolites in Plants

Secondary metabolites have economic value in modern industry. They are natural sources to chemical compounds important in the pharmaceutical industry, precursors to important vitamins and raw material to many industrial products.

It has been estimated that about 30 per cent of the world's plant species have been used for medicinal purpose. Some products act on the autonomic and central nervous systems. Morphine, the principal alkaloid of the opium poppy is used for the relief of severe pain. Compounds like cannabis from hemp, caffeine and theobromine, from coffee and cocoa respectively and cocaine from *Erythroxylim coca*, stimulate mental activity, whereas reserpine from *Rauwolfia* spp. depresses mental activity. The alkaloid atropine from solanaceous plants is used to dilate the pupil of the eye during surgery. Steroidal glycosides of *Digitalis* spp. are mainly used to impart a slowing and strengthening effect of heart muscles. Quinidine from cinchona bark is used to control atrial fibrillation.

Many non-pharmaceutical products important in industry are also derived from plants [dyes, green chlorophyll, saffron yellow dye, purple or red colours from the roots of madder (*Rubia tinctorum*)]. Many secondary products provide protection from pests and diseases. The seed extract of neem (*Azadirachta indica*), root powder of *Derris* spp., flowers of *Chrysanthemum cinerariaefolium* and leaves and stalks of *Nicotiana* have insecticidal properties. *Schoenocaulon officionale* (Sabadilla) is used as a rodent poison.

Some provides defence against micro-organisms and insect parasites. The phytoalexins, for example, provides defence against fungi. Many fungal pathogens have evolved ability to metabolise phytoalexins. Nonetheless, the complex variety of phytoalexins produced by plants in response to fungal attack effectively counteracts the ability of fungi for detoxification. Some phytoalexins with anti microbial properties are specific to a taxonomic group, sesqiterpenes for example are specific to solanaceae and isoflavonoids to Leguminosae (Rhodes, 1994) Secondary metabolites act as repellents to predatory animals and are also known to affect plant-plant relationships. Secondary products like β-propionitrile in legume species inhibits the growth of other species in the near vicinity, a phenomenon known as allelopathy.

Tobacco contains the highly toxic alkaloid nicotine, plant parts are used to make cigarettes, pipe, tobacco, cigars, snuff and chewing tobacco. The opium poppy (*Papaver somniferum*) is used for oil and narcotics; hemp (*Canabis sativa*) for fibre, oil and cannabis drugs.

On the debit side there are several secondary metabolites that are toxic to man and domestic animals. Their presence in cultivated crop plants limits the use of crop products which could be of excellent nutritional value. Such a toxic compound (β-ODAP) is present in *Lathyrus sativus*, besides its high nutritional value and good agronomical characters. Excessive consumption of the seeds of *Lathyrus sativus* causes the upper motor neurone disease known as lathyrism.

Secondary metabolites are organic compounds that are not directly involved in the normal growth, development, or reproduction of an organism. Unlike primary metabolites, absence of secondary metabolites does not result in immediate death, but rather in long-term impairment of the organism's survivability, fecundity, or aesthetics, or perhaps in no significant change at all. Secondary metabolites are often restricted to a narrow set of species within a phylogenetic group. Secondary metabolites often play an important role in plant defense against herbivory and other interspecies defenses. Humans use secondary metabolites as medicines, flavorings and recreational drugs.

Categories

Most of the secondary metabolites of interest to humankind fit into categories which classify secondary metabolites based on their biosynthetic origin. Since secondary metabolites are often created by modified primary metabolite syntheses, or "borrow" substrates of primary metabolite origin, these categories should not be interpreted as saying that all molecules in the category are secondary metabolites (for example the steroid category), but rather that there are secondary metabolites in these categories.

Small "Small Molecules"

Alkaloids (usually a small, heavily derivatized amino acid)

- ☆ Hyoscyamine, present in *Datura stramonium*
- ☆ Atropine, present in Atropa belladonna, Deadly nightshade
- ☆ Cocaine, present in Erythroxylon coca the Coca plant
- ☆ Scopolamine, present in the Solanaceae (nightshade) plant family
- ☆ Codeine and Morphine, present in Papaver somniferum, the opium poppy
- ☆ Tetrodotoxin, a microbial product in Fugu and some salamanders
- ☆ Vincristine and Vinblastine, mitotic inhibitors found in the Rosy Periwinkle

Terpenoids (come from semiterpene oligomerization)

- ☆ Azadirachtin, (Neem tree)
- ☆ Artemisinin, present in Artemisia annua Chinese wormwood
- ☆ tetrahydrocannabinol, present in cannabis
- ☆ Steroids (Terpenes with a particular ring structure)
- ☆ Saponins (plant steroids, often glycosylated)

Glycosides (heavily modified sugar molecules)

- ☆ Nojirimycin
- ☆ Glucosinolates

Natural Phenols

- ☆ Resveratrol

Phenazines

- ☆ Pyocyanin
- ☆ Phenazine-1-carboxylic acid (and derivatives)

Biphenyls and Dibenzofurans

- ☆ These are phytoalexins of the Pyrinae

Beta-lactams

☆ Penicillin, present in Penicillium fungi.

Big "Small Molecules", produced by large, modular, "molecular factories"

Polyketides

☆ Erythromycin

☆ Lovastatin and other statins

☆ Discodermolide

☆ Aflatoxin B1

☆ Avermectins

☆ Nystatin

☆ Rifamycin

Fatty Acid Synthase Products

☆ FR-900848

☆ U-106305

☆ phloroglucinols

☆ Nonribosomal peptides:

☆ Vancomycin

☆ Thiostrepton

☆ Ramoplanin

☆ Teicoplanin

☆ Gramicidin

☆ Bacitracin

☆ Ciclosporin

Hybrids of the above Three

☆ Epothilone

☆ Polyphenols

Non-"small Molecules"–DNA, RNA, Ribosome, or Polysaccharide "Classical" Biopolymers

Ribosomal Peptides

☆ Microcin-J25

Plant secondary metabolism produces products that aid in the growth and development of plants but are not required for the plant to survive. Secondary metabolism facilitates the primary metabolism in plants. This primary metabolism consists of chemical reactions that allow the plant to live. In order for the plants to stay healthy, secondary metabolism plays a pinnacle role in keeping all the of plants' systems working properly. A common role of secondary metabolites in plants is defense mechanisms. They are used to fight off herbivores, pests and pathogens. Although researchers know that this trait is common in many plants it is still difficult to determine the precise role each secondary metabolite. Secondary metabolites are used in anti-feeding activity, toxicity or acting as precursors to physical defense systems.

History

Research into secondary plant metabolism primarily took off in the latter half of the 19th century; however, there was still much confusion over what the exact function and usefulness of these compounds were. All that was known was that secondary plant metabolites were "by-products" of the primary metabolism and were not crucial to the plant's survival. Early research only succeeded as far as categorizing the secondary plant metabolites but did not give real insight into the actual function of the secondary plant metabolites. The study of plant metabolites is thought to have started in the early 1800s when Friedrich Willhelm Serturner isolated morphine from opium poppy and after that new discoveries were made rapidly. In the early half of the 1900s, the main research around secondary plant metabolism was dedicated to the formation of secondary metabolites in plants and this research was compounded by the use of tracer techniques which made deducing metabolic pathways much easier. However, there was still not much research being conducted into the functions of secondary plant metabolites until around the 1980s. Before then, secondary plant metabolites were thought of as simply waste products. In the 1970s, however, new research showed that secondary plant metabolites play an indispensable role in the survival of the plant in its environment. One of the most ground breaking ideas of this time argued that plant secondary metabolites evolved in relation to environmental conditions and this indicated the high gene plasticity of secondary metabolites, but this theory was ignored for about half a century before gaining acceptance. Recently, the research around secondary plant metabolites is focused around the gene level and the genetic diversity of plant metabolites. Biologists are now trying to trace back genes to their origin and re-construct evolutionary pathways.

Primary vs. Secondary Plant Metabolism

Primary metabolism in a plant comprises all metabolic pathways that are essential to the plant's survival. Primary metabolites are compounds that are directly involved in the growth and development of a plant whereas secondary metabolites are compounds produced in other metabolic pathways that, although important, are not essential to the functioning of the plant. However, secondary plant metabolites are useful in the long term, often for defense purposes and give plants characteristics such as colour. Secondary plant metabolites are also used in signaling and regulation of primary metabolic pathways. Plant hormones, which are secondary metabolites, are often used to regulate the metabolic activity within cells and oversee the overall development of the plant. As mentioned above in the History tab, secondary plant metabolites help the plant maintain an intricate balance with the environment, often adapting to match the environmental needs.

Plant metabolites that colour the plant are a good example of this, as the colouring of a plant can attract pollinators and also defend against attack by animals.

Types of Secondary Metabolites

There is no fixed, commonly agreed upon system for classifying secondary metabolites. Based on their biosynthetic origins, plant secondary metabolites can be divided into three major groups:

1. Flavonoids and allied phenolic and polyphenolic compounds,
2. Terpenoids and
3. Nitrogen-containing alkaloids and sulphur-containing compounds.

Other researchers have classified secondary metabolites into following, more specific types.

Class	Type	Number of known Metabolites	Examples
Alkaloids	Nitrogen-containing	21000	Cocaine, Psilocin, Caffeine, Nicotine, Morphine, Berberine, Vincristine, Reserpine, Galantamine, Atropine, Vincamine, Quinidine, Ephedrine, Quinine
Non-protein amino acids (NPAAs)	Nitrogen-containing	700	NPAAs are produced by specific plant families such as Leguminosae, Cucurbitaceae, Sapindaceae, Aceraceae and Hippocastanaceae. Examples: Azatyrosine, Canavanine
Amines	Nitrogen-containing	100	
Cyanogenic glycosides	Nitrogen-containing	60	Amygdalin, Dhurrin, Linamarin, Lotaustralin, Prunasin
Glucosinolates	Nitrogen-containing	100	
Alkamides	Nitrogen-containing	150	
Lectins, peptides and polypeptides	Nitrogen-containing	2000	
Terpenes	Without nitrogen	>15,000	Azadirachtin, Artemisinin, Tetrahydro-cannabinol
Steroids and saponins	Without nitrogen	NA	These are terpenoids with a particular ring structure. Cycloartenol
Flavonoids and Tannins	Without nitrogen	5000	
Phenylpropanoids, lignins, coumarins and lignans	Without nitrogen	2000	
Polyacetylenes, fatty acids and waxes	Without nitrogen	1500	
Polyketides	Without nitrogen	750	
Carbohydrates and organic acids	Without nitrogen	200	

Atropine

Atropine is a type of secondary metabolite called a tropane alkaloid. Alkaloids contain nitrogens, frequently in a ring structure and are derived from amino acids. Tropane is an organic compound containing nitrogen and it is from tropane that atropine is derived. Atropine is synthesized by a reaction between tropine and tropate, catalyzed by atropinase. Both of the substrates involved in this reaction are derived from amino acids, tropine from pyridine (through several steps) and tropate directly from phenylalanine. Within Atropa belladonna atropine synthesis has been found to take place primarily in the root of the plant. The concentration of synthetic sites within the plant is indicative of the nature of secondary metabolites. Typically, secondary metabolites are not necessary for normal functioning of cells within the organism meaning the synthetic sites are not required throughout the organism. As atropine is not a primary metabolite, it does not interact specifically with any part of the organism, allowing it to travel throughout the plant.

Flavonoids

Flavonoids are one class of secondary plant metabolites that are also known as Vitamin P or citrin. These metabolites are mostly used in plants to produce yellow and other pigments which play a big role in colouring the plants. In addition, Flavonoids are readily ingested by humans and they seem to display important anti-inflammatory, anti-allergic and anti-cancer activities. Flavonoids are also found to be powerful anti-oxidants and researchers are looking into their ability to prevent cancer and cardiovascular diseases. Flavonoids help prevent cancer by inducing certain mechanisms that may help to kill cancer cells and researches believe that when the body processes extra flavonoid compounds, it triggers specific enzymes that fight carcinogens. Good dietary sources of Flavonoids are all citrus fruits, which contain the specific flavanoids hesperidins,quercitrin,and rutin, berries, tea, dark chocolate and red wine and many of the health benefits attributed to these foods come from the Flavonoids they contain. Flavonoids are synthesized by the phenylpropanoid metabolic pathway where the amino acid phenylalanine is used to produce 4-coumaryol-CoA and this is then combined with malonyl-CoA to produce chalcones which are backbones of Flavonoids Chalcones are aromatic ketones with two phenyl rings that are important in many biological compounds. The closure of chalcones causes the formation of the flavonoid structure. Flavonoids are also closely related to flavones which are actually a sub class of flavonoids and are the yellow pigments in plants. In addition to flavones, 11 other subclasses of Flavonoids including, isoflavones, flavans, flavanones, flavanols, flavanolols, anthocyanidins, catechins (including proanthocyanidins), leukoanthocyanidins, dihydrochalcones and aurones.

Cyanogenic Glycoside

Many plants have adapted to iodine-deficient terrestrial environment by removing iodine from their metabolism, in fact iodine is essential only for animal cells. An important antiparasitic action is caused by the block of the transport of iodide of animal cells inhibiting sodium-iodide symporter (NIS). Many plant pesticides are cyanogenic glycoside which liberates cyanide, which, blocking cytochrome c oxidase and NIS, is poisonous only for a large part of parasites and herbivores and not for the plant cells in which it seems useful inseed dormancy phase. To get a better understanding of how secondary metabolites play a big role in plant defense mechanisms we can focus on the recognizable defense-related secondary metabolites, cyanogenic glycosides. The compounds of these secondary metabolites are found in over 2000 plant species. Its structure allows the release of cyanide, a poison produced by certain bacteria, fungi and algae that is found in numerous plants. Animals and humans possess the ability to detoxify cyanide from their systems naturally. Therefore cyanogenic glycosides can be used for positive benefits in animal systems always. For example, the larvae of the southern armyworm consumes plants that contain this certain metabolite and have shown a better growth rate with this metabolite in their diet, as opposed to other secondary metabolite-containing plants. Although this example shows cyanogenic glycosides being beneficial to the larvae many still argue that this metabolite can do harm. To help in determining whether cyanogenic glycosides are harmful or helpful researchers look closer at its biosynthetic pathway. Past research suggests that cyanogenic glucosides stored in the seed of the plant are metabolized during germination to release nitrogen for seedling to grow. With this, it can be inferred that cyanogenic glycosides play various roles in plant metabolism. Though subject to change with future research, there is no evidence showing that cyanogenic glycosides are responsible for infections in plants.

Phytic Acid

Phytic acid is the main method of phosphorus storage in plant seeds, but is not readily absorbed by many animals (only absorbed by ruminant animals). Not only is phytic acid a phosphorus storage unit, but it also is a source of energy and cations, a natural antioxidant for plants and can be a source of myoinositol which is one of the preliminary pieces for cell walls.

Phytic acid is also known to bond with many different minerals and by doing so prevents those minerals from being absorbed; making phytic acid an anti-nutrient. There is a lot of concern with phytic acids in nuts and seeds because of its anti-nutrient characteristics. In preparing foods with high phytic acid concentrations, it is recommended they be soaked in after being ground to increase the surface area. Soaking allows the seed to undergo germination which increases the availability of vitamins and nutrient, while reducing phytic acid and protease inhibitors, ultimately increasing the nutritional value. Cooking can also reduce the amount of phytic acid in food but soaking is much more effective.

Phytic acid is an antioxidant found in plant cells that most likely serves the purpose of preservation. This preservation is removed when soaked, reducing the phytic acid and allowing the germination and growth of the seed. When added to foods it can help prevent discolouration by inhibiting lipid peroxidation. There is also some belief that the chelating of phytic acid may have potential use in the treatment of cancer.

Gossypol

Gossypol has a yellow pigment and is found in cotton plants. It occurs mainly in the root and/or seeds of different species of cotton plants. Gossypol can have various chemical structures. It can exist in three forms: gossypol, gossypol acetic acid and gossypol formic acid. All of these forms have very similar biological properties. Gossypol is a type of aldehyde, meaning that it has a formyl group. The formation of gossypol occurs through an isoprenoid pathway. Isoprenoid pathways are common among secondary metabolites. Gossypol's main function in the cotton plant is to act as an enzyme inhibitor. An example of gossypol's enzyme inhibition is its ability to inhibit nicotinamide adenine dinucleotide-linked enzymes of *Trypanosoma cruzi*. *Trypanosoma cruzi* is a parasite which causes Chaga's disease.

For some time it was believed that gossypol was merely a waste product produced during the processing of cottonseed products. Extensive studies have shown that gossypol has other functions. Many of the more popular studies on gossypol discuss how it can act as a male contraceptive. Gossypol has also been linked to causing hypokalemic paralysis. Hypokalemic paralysis is a disease characterized by muscle weakness or paralysis with a matching fall in potassium levels in the blood. Hypokalemic paralysis associated with gossypol in-take usually occurs in March, when vegetables are in short supply and in September, when people are sweating a lot. This side effect of gossypol in-take is very rare however. Gossypol induced hypokalemic paralysis is easily treatable with potassium repletion.

Phytoestrogens

Plants synthesize certain compounds called secondary metabolites which are not naturally produced by humans but can play vital roles in protection or destruction of human health. One such group of metabolites is phytoestrogens, found in nuts, oilseeds, soy and other foods. Phytoestrogens are chemicals which act like the hormone estrogen.Estrogen is important for women's bone and heart health, but high amounts of it has been linked to breast cancer. In the plant, the phytoestrogens are involved in the defense system against fungi. Phytoestrogens can do two different things in a human body. At low doses it mimics estrogen, but at high doses it actually blocks the body's natural estrogen. The estrogen receptors in the body which are stimulated by estrogen will acknowledge the phytoestrogen, thus the body may reduce its own production of the hormone. This has a negative result, because there are various abilities of the phytoestrogen which estrogen does not do. Its effects the communication pathways between cells and has effects on other parts of the body where estrogen normally does not play a role. It has also been found to induce tumor growth of the estrogen receptor cells in the breast. But, one role of estrogens which phytoestrogens mimic is its protective behavior for the heart. So, an intake of phytoestrogens has also been seen to reduce the risk of cardiovascular disease. Resveratrol, a phytoestrogen found in grapes is responsible for this. For example, the French suffer relatively little heart disease despite the average French diet being relatively high in fat. One proposed reason for this is the resveratrol found in red wine, which has been linked to decreased risk of cardiovascular disease.

Carotenoids

Carotenoids are organic pigments found in the chloroplasts and chromoplasts of plants. They are also found in some organisms such as algae, fungi, some bacteria and certain species of aphids. There are over 600 known carotenoids. They are split into two classes, xanthophylls and carotenes. Xanthophylls are carotenoids with molecules containing oxygen, such as lutein and zeaxanthin. Carotenes are carotenoids with molecules that are unoxygenated, such as α-carotene, β-carotene and lycopene. In plants, carotenoids can occur in roots, stems, leaves, flowers and fruits. Carotenoids have two important functions in plants. First, they can contribute to photosynthesis. They do this by transferring some of the light energy they absorb to chlorophylls, which then uses this energy for photosynthesis. Second, they can protect plants which are over-exposed to sunlight. They do this by harmlessly dissipating excess light energy which they absorb as heat. In the absence of carotenoids, this excess light energy could destroy proteins, membranes and other molecules. Some plant physiologists believe that carotenoids may have an additional function as regulators of certain developmental responses in plants. Tetraterpenes are synthesized from DOXP precursors in plants and some bacteria. Carotenoids involved in photosynthesis are formed in chloroplasts; others are formed in plastids. Carotenoids formed in fungi are presumably formed from mevalonic acid precursors. Carotenoids are formed by a head-to-head condensation of geranylgeranyl pyrophosphate or diphosphate (GGPP) and there is no NADPH requirement.

References

Chizzali, Cornelia and Beerhues, Ludger (2012). "Phytoalexins of the Pyrinae: Biphenyls and dibenzofurans". *Beilstein J. Org. Chem.* 8: 613–620.

Crozier, Alan and Hiroshi Ashihara. Plant Secondary Metabolites: Occurrence, Structure and Role in the Human Diet. Ames, IA: Blackwell Publishing Professional, 2006. Print.

Fraenkel, Gottfried S. (1959). "The raison d'Etre of secondary plant substances". *Science* 129 (3361): 1466–1470.

Fred R. West, Jr. and Edward S. Mika. "Synthesis of Atropine by Isolated Roots and Root-Callus Cultures of Belladonna." *Botanical Gazette*: Vol. 119, No. 1 (Sep., 1957), pp. 50-54 http: //www.jstor.org/stable/2473263

Graf, E, Eaton, JW (1990). "Antioxidant functions of phytic acid". *Free radical biology and medicine* 8 (1): 61–9. doi: 10.1016/0891-5849(90)90146-A. PMID 2182395.

Hartmann, Thomas (2007). "From waste products to ecochemicals: fifty years research of plant secondary metabolism." *Phytochemistry* 68: 22-24: 2831-2846. Web. 31 Mar 2011.

Heinstein, PF, P. F. Heinstein, D. L. Herman, S. B. Tove and F. H. Smith (1970). "Biosynthesis of Gossypol". *The Journal of Biological Chemistry* **245** (18): 4658–4665. PMID 4318479. Retrieved 2011-03-31.

http: //naturalbias.com/a-hidden-danger-with-nuts-grains-and-seeds/

http://www.genome.jp/dbget-bin/www_bget?3.1.1.10+R03563

http://www.phyticacid.org/nuts/phytic-acid-in-nuts/

Introduction: Biochemistry, Physiology and Ecological Functions of Secondary Metabolites". *Annual Plant Reviews Volume 40: Biochemistry of Plant Secondary Metabolism, Second Edition*. 26 Mar 2010. doi: 10.1002/9781444320503.ch1.

Kopp, Peter (1998). "Resveratrol, a phytoestrogen found in red wine. A possible." *European Journal of Endocrinology* 138: 619–620. Web. 1 Apr 2011. <http://www.eje.org/cgi/reprint/138/6/619.pdf>.

Montamat, EE, C Burgos, NM Gerez de Burgos, LE Rovai, A Blanco, EL Segura (1982). "Inhibitory action of gossypol on enzymes and growth of Trypanosoma cruzi". *Science*. (4569) **218** (4569): 288–289. doi: 10.1126/science.6750791.PMID 6750791.

Phenols, Polyphenols and Tannins (2007). An Overview". *Plant Secondary Metabolites: Occurrence, Structure and Role in the Human Diet*. Nov 12. doi: 10.1002/9780470988558.

Qian, Shao-Zhen and, Wang, Zhen-Gang, Wang, Z (1984). "GOSSYPOL: A POTENTIAL ANTIFERTILITY AGENT FOR MALES". *Annual Reviews* **24**: 329 360.doi: 10.1146/annurev pa 24.040184.001553. Retrieved 2011-03-27.

Richard C. Leegood, Per Lea (1998). *Plant Biochemistry and Molecular Biology*. John Wiley and Sons. p. 211. ISBN 978-0-471-97683-7.

Samuni-Blank, M, Izhaki, I, Dearing, MD, Gerchman, Y, Trabelcy, B, Lotan, A, Karasov, WH, Arad, Z (2012). Intraspecific directed deterrence by the mustard oil bomb in a desert plant.*Current Biology*. 22: 1-3.

Schultz, Jack. "Secondary Metabolites in Plants". Biology Reference. Retrieved 2011-03-27.

Stamp, Nancy (2003). "Out of the quagmire of plant defense hypotheses". *The Quarterly Review of Biology* 78 (1): 23–55.

Thompson LU, Boucher BA, Liu Z, Cotterchio M, Kreiger N (2006). "Phytoestrogen content of foods consumed in Canada, including isoflavones, lignans and coumestan". *Nutrition and Cancer* **54** (2): 184- 201. doi: 10.1207/s15327914nc 5402_5. PMID 16898863.

Urbano, G, López-Jurado, M, Aranda, P, Vidal-Valverde, C, Tenorio, E, Porres, J (2000). "The role of phytic acid in legumes: antinutrient or beneficial function?". *Journal of physiology and biochemistry* **56** (3): 283–94. doi: 10.1007/bf03179796. PMID 11198165.

Venturi, S., Donati, F.M., Venturi, A., Venturi, M. (2000). "Environmental Iodine Deficiency: A Challenge to the Evolution of Terrestrial Life?". *Thyroid* **10** (8): 727–9.doi: 10.1089/10507250050137851. PMID 11014322.

Venturi, Sebastiano (2011). "Evolutionary Significance of Iodine". *Current Chemical Biology-* **5** (3): 155–162. doi: 10.2174/187231311796765012. ISSN 1872-3136.

Warren, Barbour S. and Carol Devine (2010). "Phytoestrogens and Breast Cancer." Cornell University. Cornell University, 31/03/2010. Web. 1 Apr 2011. <http://envirocancer.cornell.edu/factsheet/diet/fs1.phyto.cfm>.

Warren, Barbour S. and Carol Devine (2011). "Phytoestrogens and Breast Cancer." Cornell University. Cornell University, 31/03/2010. Web. 1 Apr 2011. <http://envirocancer.cornell.edu/factsheet/diet/fs1.phyto.cfm>.

Chapter 2

Alkaloids

Alkaloids are a group of naturally occurring chemical compounds that contain mostly basic nitrogen atoms. This group also includes some related compounds with neutra and even weakly acidic properties. Some synthetic compounds of similar structure are also termed alkaloids. In addition to carbon, hydrogen and nitrogen, alkaloids may also contain oxygen, sulfur and more rarely other elements such as chlorine, bromine and phosphorus.

Alkaloids are produced by a large variety of organisms including bacteria, fungi, plants and animals. They can be purified from crude extracts of these organisms by acid-base extraction. Alkaloids have a wide range of pharmacological activities ncluding antimalarial (*e.g.* quinine), antiasthma (*e.g.* ephedrine), anticancer (*e.g.* homoharringtonine), cholinomimetic (*e.g.* galantamine), vasodilatory(*e.g.* vincamine), antiarrhythmic (*e.g.* quinidine), analgesic (*e.g.* morphine), antibacterial (*e.g.* chelerythrine) and antihyperglycemic activities (*e.g.* piperine). Many have found use in traditional or modern medicine, or as starting points for drug discovery. Other alkaloids possess psychotropic (*e.g.* psilocin) and stimulant activities (*e.g.* cocaine, caffeine, nicotine) and have been used in entheogenic rituals or as recreational drugs. Alkaloids can be toxic too (*e.g.* atropine, tubocurarine). Although alkaloids act on a diversity of metabolic systems in humans and other animals, they almost uniformly invoke a bitter taste.

The boundary between alkaloids and other nitrogen-containing natural compounds is not clear-cut. Compounds like amino acid peptides, proteins, nucleotides, nucleic acid, amines and antibiotics are usually not called alkaloids. Natural compounds containing nitrogen in the exocyclic position (mescaline, serotonin, dopamine, etc.) are usually attributed to amines rather than alkaloids. Some authors, however, consider alkaloids a special case of amines.

History

Alkaloid-containing plants have been used by humans since ancient times for therapeutic and recreational purposes. For example, medicinal plants have been known in the Mesopotamia at least around 2000 BC. The *Odyssey* of Homer referred to a gift given to Helen by the Egyptian queen, a drug bringing oblivion. It is believed that the gift was an opium-containing drug. A Chinese book on houseplants written in 1st–3rd centuries BC mentioned a medical use of Ephedra and opium poppies. Also, coca leaves have been used by South American Indians since ancient times. Extracts from plants containing toxic alkaloids, such as aconitine and tubocurarine, were used since antiquity for poisoning arrows. Studies of alkaloids began in the 19th century. In 1804, the German chemist Friedrich Sertürner isolated from opium a "soporific principle" (Latin: *principium somniferum*), which he called "morphium" in honor of Morpheus, the Greek god of dreams; in German and some other Central-European languages, this is still the name of the drug. The term "morphine", used in English and French, was given by the French physicist Joseph Louis Gay-Lussac. A significant contribution to the chemistry of alkaloids in the early years of its development was made by the French researchers Pierre Joseph Pelletier and Joseph Bienaimé Caventou, who discovered quinine (1820) and strychnine (1818). Several other alkaloids were discovered around that time, including xanthine (1817), atropine (1819), caffeine (1820), coniine (1827), nicotine (1828), colchicine (1833), sparteine (1851) and cocaine (1860). The first complete synthesis of an alkaloid was achieved in 1886 by the German chemist Albert Ladenburg. He produced coniine by reacting 2-methylpyridine with acetaldehydeand reducing the resulting 2-propenyl pyridine with sodium. The development of the chemistry of alkaloids was accelerated by the emergence of spectroscopic and chromatographic methods in the 20th century, so that by 2008 more than 12,000 alkaloids had been identified.

Classifications

Compared with most other classes of natural compounds, alkaloids are characterized by a great structural diversity and there is no uniform classification of alkaloids.[32] First classification methods have historically combined alkaloids by the common natural source, *e.g.*, a certain type of plants. This classification was justified by the lack of knowledge about the chemical structure of alkaloids and is now considered obsolete. More recent classifications are based on similarity of the carbon skeleton (*e.g.*, indole-, isoquinoline- and pyridine-like) or biochemical precursor (ornithine, lysine, tyrosine, tryptophan, etc.). However, they require compromises in borderline cases; for example, nicotine contains a pyridine fragment from nicotinamide and pyrrolidine part from ornithine and therefore can be assigned to both classes.

Alkaloids are often divided into the following major groups:

1. "True alkaloids", which contain nitrogen in the heterocycle and originate from amino acids. Their characteristic examples are atropine,nicotine and morphine. This group also includes some alkaloids that besides nitrogen heterocycle contain terpene (*e.g.*, evonine) or peptide fragments (*e.g.* ergotamine). This group also includes piperidine alkaloids coniine and coniceine although they do not originate from amino acids.

2. "Protoalkaloids", which contain nitrogen and also originate from amino acids. Examples include mescaline, adrenaline and ephedrine.

3. Polyamine alkaloids – derivatives of putrescine, spermidine and spermine.

4. Peptide and cyclopeptide alkaloids.

5. Pseudalkaloids – alkaloid-like compounds that do not originate from amino acids. This group includes, terpene-like and steroid-like alkaloids, as well as purine-like alkaloids such as caffeine, theobromine, theacrine and theophylline. Some authors classify as pseudoalkaloids such compounds such as ephedrine and cathinone. Those originate from the amino acid phenylalanine, but acquire their nitrogen atom not from the amino acid but through transamination.

Some alkaloids do not have the carbon skeleton characteristic of their group. So, galantamine and homoaporphines do not contain isoquinolinefragment, but are, in general, attributed to isoquinoline alkaloids.

Main classes of monomeric alkaloids are listed in the Table 2.1.

Properties

Most alkaloids contain oxygen in their molecular structure; those compounds are usually colourless crystals at ambient conditions. Oxygen-free alkaloids, such as nicotine or coniine, are typically volatile, colourless, oily liquids. Some alkaloids are coloured, like berberine (yellow) and sanguinarine (orange). Most alkaloids are weak bases, but some, such as theobromine and theophylline, are amphoteric. Many alkaloids dissolve poorly in water but readily dissolve in organic solvents, such as diethyl ether, chloroform or 1,2-dichloroethane. Caffeine, cocaine, codeine and nicotine are water soluble (with a solubility of $\geq 1g/L$), whereas others, including morphine and yohimbine are highly water soluble (0.1–1 g/L). Alkaloids and acids form salts of various strengths. These salts are usually soluble in water and ethanoland poorly soluble in most organic solvents. Exceptions include scopolamine hydrobromide, which is soluble in organic solvents and the water-soluble quinine sulfate. Most alkaloids have a bitter taste or are poisonous when ingested. Alkaloid production in plants appeared to have evolved in response to feeding by herbivorous animals; however, some animals have evolved the ability to detoxify alkaloids. Some alkaloids can produce developmental defects in the offspring of animals that consume but cannot detoxify the alkaloids. One example is the alkaloid cyclopamine, produced in the leaves of corn lily. During the 1950s, up to 25 per cent of lambs born by sheep that had grazed on corn lily had serious facial deformations. These ranged from deformed jaws to cyclopia (see picture). After decades of research, in the 1980s, the compound responsible for these deformities was identified as the alkaloid 11-deoxyjervine, later renamed to cyclopamine.

Distribution in Nature

Alkaloids are generated by various living organisms, especially by higher plants – about 10 to 25 per cent of those contain alkaloids. Therefore, in the past the term "alkaloid" was associated with plants. The alkaloids content in plants is usually within a few percent and is inhomogeneous over the plant tissues. Depending on the type of plants, the maximum concentration is observed in the leaves (black henbane),

Table 2.1: Main Classes of Monomeric Alkaloids

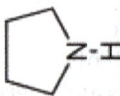

Class	Major Groups	Main Synthesis Steps	Examples
Alkaloids with nitrogen heterocycles (true alkaloids)			
Pyrrolidine derivatives		Ornithine or arginine → putrescine → N-methylputrescine → N-methyl-Δ¹-pyrroline	Cuscohygrine, hygrine, hygroline, stachydrine
Tropane derivatives	Atropine group Substitution in positions 3, 6 or 7	Ornithine or arginine → putrescine → N-methylputrescine → N-methyl-Δ¹-pyrroline	Atropine, scopolamine, hyoscyamine
	Cocaine group Substitution in positions 2 and 3		Cocaine, ecgonine
Pyrrolizidine derivatives	Non-esters	In plants: ornithine or arginine → putrescine → homospermidine →retronecine	Retronecine, heliotridine, laburnine
	Complex esters of monocarboxylic acids		Indicine, lindelophin, sarracine
	Macrocyclic diesters		Platyphylline, trichodesmine
	1-aminopyrrolizidines (lolines)	In fungi: L-proline + L-homoserine → N-(3-amino-3-carboxypropyl)proline → norloline	Loline, N-formylloline, N-acetylloline
Piperidine derivatives		Lysine → cadaverine → Δ¹-piperideine	Sedamine, lobeline, anaferine, piperine
		Octanoic acid → coniceine → coniine	Coniine, coniceine

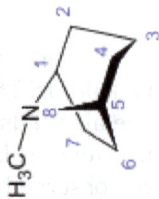

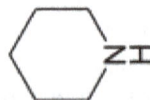

The $Main Synthesis Steps$ column for Tropane derivatives applies where the Main synthesis steps use: Ornithine or arginine → putrescine → N-methylputrescine → N-methyl-Δ¹-pyrroline

H_3C-N (tropane numbering 1, 2, 3, 4, 5, 6, 7, 8)

Table 2.1–Contd...

Class	Major Groups	Main Synthesis Steps	Examples
Quinolizidine derivatives 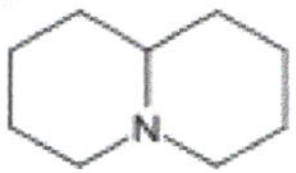	Lupinine group	Lysine → cadaverine → Δ¹-piperideine	Lupinine, nupharidin
	Cytisine group		Cytisine
	Sparteine group		Sparteine, lupanine, anahygrine
	Matrine group		Matrine, oxymatrine, allomatridine
	Ormosanine group		Ormosanine, piptantine
Indolizidine derivatives 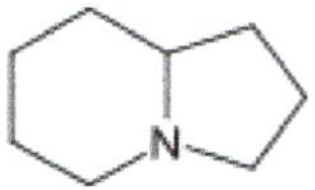		Lysine → Δ-semialdehyde of α-aminoadipic acid → pipecolic acid → 1 indolizidinone	Swainsonine, castanospermine
Pyridine derivatives 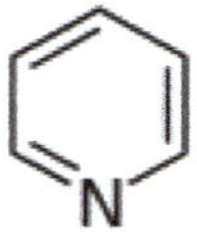	Simple derivatives of pyridine	Nicotinic acid → dihydronicotinic acid → 1,2-dihydropyridine	Trigonelline, ricinine, arecoline
	Polycyclic noncondensing pyridine derivatives		Nicotine, nornicotine, anabasine, anatabine
	Polycyclic condensed pyridine derivatives		Actinidine, gentianine, pediculinine
	Sesquiterpene pyridine derivatives	Nicotinic acid, isoleucine	Evonine, hippocrateine, triptonine
Isoquinoline derivatives and related alkaloids 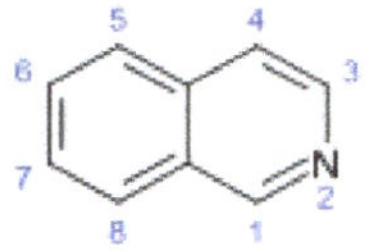	Simple derivatives of isoquinoline	Tyrosine or phenylalanine → dopamine or tyramine (for alkaloids Ama-illis)	Salsoline, lophocerine
	Derivatives of 1- and 3-isoquinolines		N-methylcoridaldine, noroxyhydrastinine
	Derivatives of 1- and 4-phenyltetrahydroisoquinolines		Cryptostilin
	Derivatives of 5-naftil-isoquinoline		Ancistrocladine

Contd...

Table 2.1–Contd...

Class	Major Groups	Main Synthesis Steps	Examples
	Derivatives of 1- and 2-benzyl-izoquinolines		Papaverine, laudanosine, sendaverine
	Cularine group		Cularine, yagonine
	Pavines and isopavines		Argemonine, amurensine
	Benzopyrrocolines		Cryptaustoline
	Protoberberines		Berberine, canadine, ophiocarpine, mecambridine, corydaline
	Phthalidisoquinolines		Hydrastine, narcotine (Noscapine)
	Spirobenzylisoquinolines		Fumaricine
	Ipecacuanha alkaloids		Emetine, protoemetine, ipecoside
	Benzophenanthridines		Sanguinarine, oxynitidine, corynoloxine
	Aporphines		Glaucine, coridine, liriodenine
	Proaporphines		Pronuciferine, glaziovine
	Homoaporphines		Kreysiginine, multifloramine
	Homoproaporphines		Bulbocodine
	Morphines		Morphine, codeine, thebaine, sinomenine
	Homomorphines		Kreysiginine androcymbine
	Tropoloisoquinolines		Imerubrine
	Azofluoranthenes		Rufescine, imeluteine
	Amaryllis alkaloids		Lycorine, ambelline, tazettine, galantamine, montanine
	Erythrina alkaloids		Erysodine, erythroidine
	Phenanthrene derivatives		Atherosperminine
	Protopins		Protopine, oxomuramine, corycavidine
	Aristolactam		Doriflavin

Contd...

Table 2.1–Contd...

Class	Major Groups	Main Synthesis Steps	Examples
Oxazole derivatives 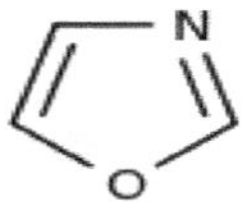		Tyrosine → tyramine	Annuloline, halfordinol, texaline, texamine
Isoxazole derivatives 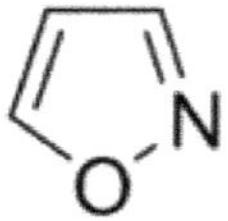		Ibotenic acid → Muscimol	Ibotenic acid, Muscimol
Thiazole derivatives 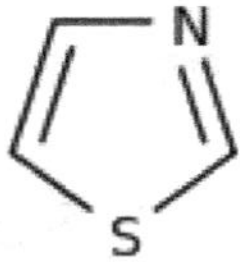		1-Deoxy-D-xylulose 5-phosphate (DOXP), tyrosine, cysteine	Nostocyclamide, thiostreptone
Quinazoline derivatives 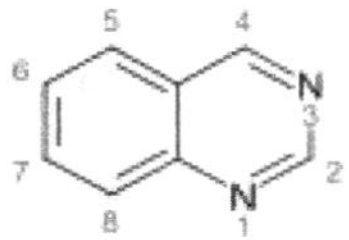	3,4-Dihydro-4-quinazolone derivatives	Anthranilic acid or phenylalanine or ornithine	Febrifugine
	1,4-Dihydro-4-quinazolone derivatives		Glycorine, arborine, glycosminine
	Pyrrolidine and piperidine quinazoline derivatives		Vazicine (peganine)
Acridine derivatives 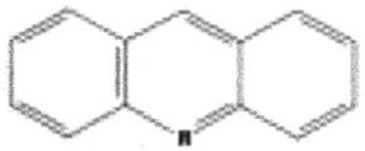		Anthranilic acid	Rutacridone, acronicine

Contd...

Table 2.1—Contd...

Class	Major Groups	Main Synthesis Steps	Examples
Quinoline derivatives 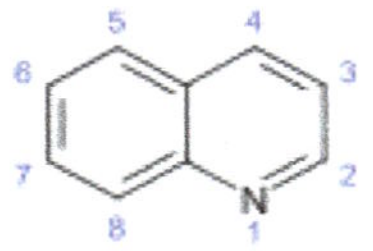	Simple derivatives of quinoline derivatives of 2-quinolones and 4-quinolone	Anthranilic acid → 3-carboxyquinoline	Cusparine, echinopsine, evocarpine
	Tricyclic terpenoids		Flindersine
	Furanoquinoline derivatives		Dictamnine, fagarine, skimmianine
	Quinines	Tryptophan → tryptamine → strictosidine (with secologanin) → korinanteal → cinhoninon	Quinine, quinidine, cinchonine, cinhonidine
Indole derivatives 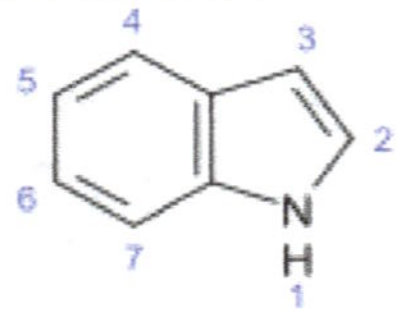 *See also: indole alkaloids*	*Non-isoprene indole alkaloids*		
	Simple indole derivatives	Tryptophan → tryptamine or 5-hydroxitriptofan	Serotonin, psilocybin, dimethyltryptamine (DMT), bufotenin
	Simple derivatives of β-carboline		Harman, harmine, harmaline, eleagnine
	Pyrroloindole alkaloids		Physostigmine (eserine), etheramine, physovenine, eptastigmine
	Semiterpenoid indole alkaloids		
	Ergot alkaloids	Tryptophan → chanoclavine → agroclavine → elimoclavine → paspalic acid → lysergic acid	Ergotamine, ergobasine, ergosine
	Monoterpenoid indole alkaloids		
	Corynanthe type alkaloids	Tryptophan → tryptamine → strictosidine (with secologanin)	Ajmalicine, sarpagine, vobasine, ajmaline, yohimbine, reserpine, mitragynine, group strychnine and (Strychnine brucine, aquamicine, vomicine)
	Iboga-type alkaloids		Ibogamine, ibogaine, voacangine
	Aspidosperma-type alkaloids		Vincamine, vinca alkaloids, vincotine, aspidospermine

Contd...

Table 2.1–*Contd*...

Class	Major Groups	Main Synthesis Steps	Examples
Imidazole derivatives 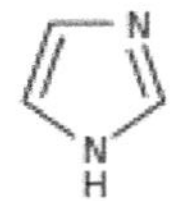		Directly from histidine	Histamine, pilocarpine, pilosine, stevensine
Purine derivatives 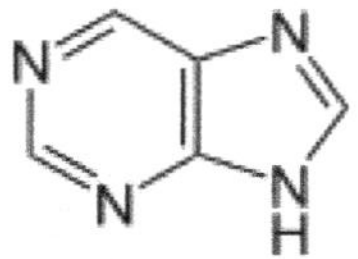		Xanthosine (formed in purine biosynthesis) → 7 methylxantosine → 7-methyl xanthine → theobromine → caffeine	Caffeine, theobromine, theophylline, saxitoxin
Alkaloids with nitrogen in the side chain (protoalkaloids)			
β-Phenylethylamine derivatives 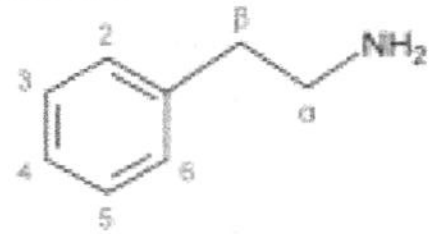		Tyrosine or phenylalanine → dioxyphenilalanine → dopamine → adrenaline and mescaline tyrosine → tyramine phenylalanine → 1-phenylpropane-1,2-dione → cathinone → ephedrine and pseudoephedrine	Tyramine, ephedrine, pseudoephedrine, mescaline, cathinone, catecholamines (adrenaline, noradrenaline, dopamine)
Colchicine alkaloids 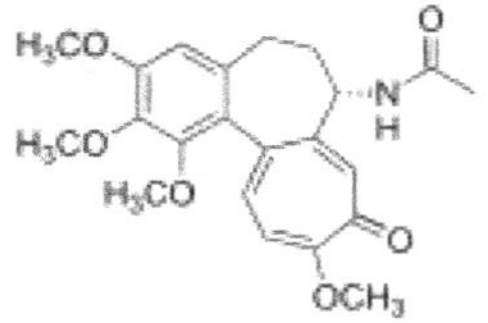		Tyrosine or phenylalanine → dopamine → autumnaline → colchicine	Colchicine, colchamine

Contd...

Table 2.1–Contd...

Class	Major Groups	Main Synthesis Steps	Examples
Muscarine 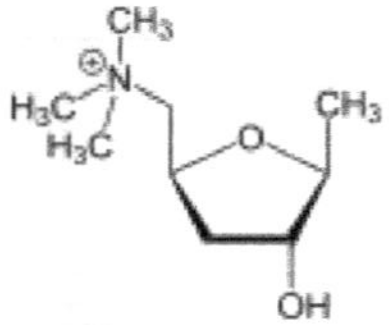		Glutamic acid → 3-ketoglutamic acid → muscarine (with pyruvic acid)	Muscarine, allomuscarine, epimuscarine, epiallomuscarine
Benzylamine 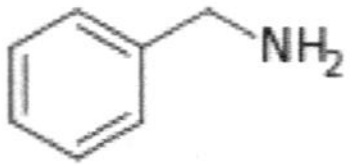		Phenylalanine with valine, leucine or isoleucine	Capsaicin, dihydrocapsaicin, nordihydrocapsaicin, vanillylamine
Polyamines alkaloids			
Putrescine derivatives 		Ornithine → putrescine → spermidine → spermine	Paucine
Spermidine derivatives 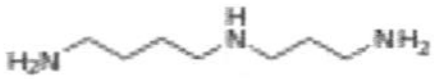			Lunarine, codonocarpine
Spermine derivatives 			Verbascenine, aphelandrine
Peptide (cyclopeptide) alkaloids			
Peptide alkaloids with a 13-membered cycle	Nummularine C type	From different amino acids	Nummularine C, Nummularine S
	Ziziphine type		Ziziphine A, sativanine H

Contd...

Table 2.1–Contd...

Class	Major Groups	Main Synthesis Steps	Examples
Peptide alkaloids with a 14-membered cycle	Frangulanine type		Frangulanine, scutianine J
	Scutianine A type		Scutianine A
	Integerrine type		Integerrine, discarine D
	Amphibine F type		Amphibine F, spinanine A
	Amfibine B type		Amphibine B, lotusine C
Peptide alkaloids with a 15-membered cycle	Mucronine A type		Mucronine A
Pseudoalkaloids (terpenes and steroids)			
Diterpenes	Lycoctonine type	Mevalonic acid → izopenterilpyrophosfate → geranyl pyrophosphate	Aconitine, delphinine
Steroids		Cholesterol, arginine	Solasodine, solanidine, veralkamine, batrachotoxin

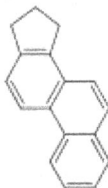

fruits or seeds (Strychnine tree), root (*Rauwolfia serpentina*) or bark (cinchona). Furthermore, different tissues of the same plants may contain different alkaloids. Beside plants, alkaloids are found in certain types of fungi, such as psilocybin in the fungus of the genus Psilocybe and in animals, such as bufotenin in the skin of some toads. Many marine organisms also contain alkaloids. Some amines, such as adrenaline andserotonin, which play an important role in higher animals, are similar to alkaloids in their structure and biosynthesis and are sometimes called alkaloids.

Extraction

Because of the structural diversity of alkaloids, there is no single method of their extraction from natural raw materials. Most methods exploit the property of most alkaloids to be soluble in organic solvents but not in water and the opposite tendency of their salts. Most plants contain several alkaloids. Their mixture is extracted first and then individual alkaloids are separated. Plants are thoroughly ground before extraction. Most alkaloids are present in the raw plants in the form of salts of organic acids. The extracted alkaloids may remain salts or change into bases. Base extraction is achieved by processing the raw material with alkaline solutions and extracting the alkaloid bases with organic solvents, such as 1,2-dichloroethane, chloroform, diethyl ether or benzene. Then, the impurities are dissolved by weak acids; this converts alkaloid bases into salts that are washed away with water. If necessary, an aqueous solution of alkaloid salts is again made alkaline and treated with an organic solvent. The process is repeated until the desired purity is achieved. In the acidic extraction, the raw plant material is processed by a weak acidic solution (*e.g.*, acetic acid in water, ethanol, or methanol). A base is then added to convert alkaloids to basic forms that are extracted with organic solvent (if the extraction was performed with alcohol, it is removed first and the remainder is dissolved in water). The solution is purified as described above. Alkaloids are separated from their mixture using their different solubility in certain solvents and different reactivity with certain reagents or by distillation.

Biosynthesis

Biological precursors of most alkaloids are amino acids, such as ornithine, lysine, phenylalanine, tyrosine, tryptophan, histidine, aspartic acid and anthranilic acid. Nicotinic acid can be synthesized from tryptophan or aspartic acid. Ways of alkaloid biosynthesis are too numerous and cannot be easily classified. However, there are a few typical reactions involved in the biosynthesis of various classes of alkaloids, including synthesis of Schiff bases and Mannich reaction.

Schiff bases can be obtained by reacting amines with ketones or aldehydes. These reactions are a common method of producing C=N bonds.

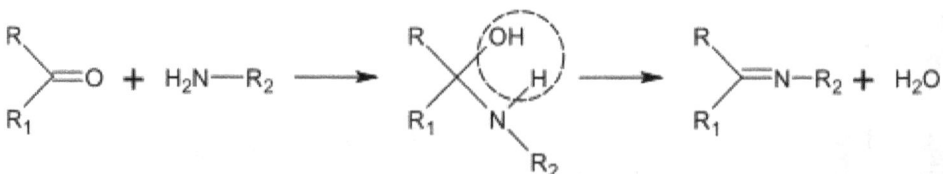

In the biosynthesis of alkaloids, such reactions may take place within a molecule, such as in the synthesis of piperidine:

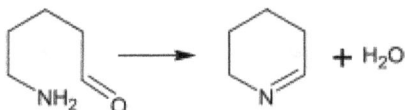

An integral component of the Mannich reaction, in addition to an amine and a carbonyl compound, is a carbanion, which plays the role of the nucleophile in the nucleophilic addition to the ion formed by the reaction of the amine and the carbonyl.

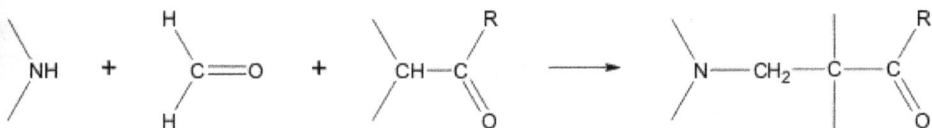

The Mannich reaction can proceed both intermolecularly and intramolecularly:

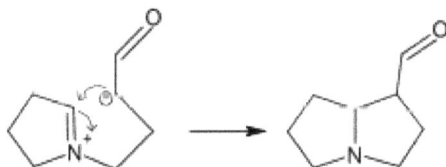

Dimer Alkaloids

In addition to the described above monomeric alkaloids, there are also dimeric and even trimeric and tetrameric alkaloids formed upon condensation of two, three and four monomeric alkaloids. Dimeric alkaloids are usually formed from monomers of the same type through the following mechanisms:

☆ Mannich reaction, resulting in, e.g., voacamine
☆ Michael reaction (villalstonine)
☆ Condensation of aldehydes with amines (toxiferine)
☆ Oxidative addition of phenols (dauricine, tubocurarine)
☆ Lactonization (carpaine).

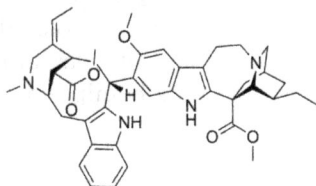

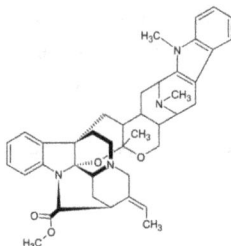

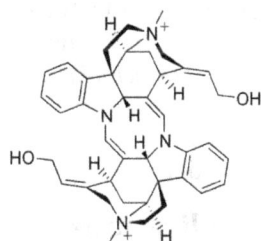

Voacamine   **Villalstonine**   **Toxiferine**

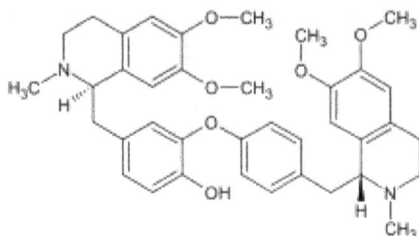

Dauricine

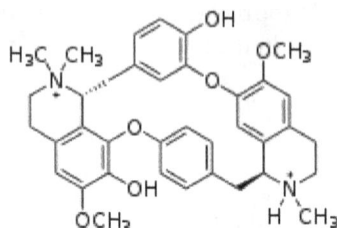

Tubocurarine

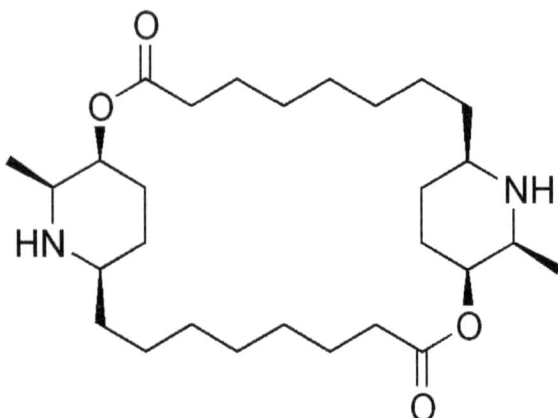

Carpaine

Biological Role

The role of alkaloids for living organisms that produce them is still unclear. It was initially assumed that the alkaloids are the final products of nitrogen metabolism in plants, asurea in mammals. It was later shown that alkaloid a concentration varies over time and this hypothesis was refuted. Most of the known functions of alkaloids are related to protection. For example, aporphine alkaloid liriodenine produced by the tulip tree protects it from parasitic mushrooms. In addition, the presence of alkaloids in the plant prevents insects and chordate animals from eating it. However, some animals are adapted to alkaloids and even use them in their own metabolism. Such alkaloid-related substances as serotonin, dopamine and histamine are important neurotransmitters in animals. Alkaloids are also known to regulate plant growth. Another example of an organism that uses alkaloids for protection is the *Utetheisa ornatrix*, more commonly known as the ornate moth. Pyrrolizidine alkaloids render these larvae and adult moths unpalatable to many of their natural enemies like coccinelid beetles, green lacewings, insectivorous hemiptera and insectivorous bats.

Applications

In Medicine

Medical use of alkaloid-containing plants has a long history and, thus, when the first alkaloids were isolated in the 19th century, they immediately found application in clinical practice. Many alkaloids are still used in medicine, usually in the form of salts, including the following:

Alkaloid	Action
Ajmaline	Antiarrhythmic
Atropine, scopolamine, hyoscyamine	Anticholinergic
Caffeine	Stimulant, adenosine receptor antagonist
Codeine	Cough medicine, analgesic
Colchicine	Remedy for gout
Emetine	Antiprotozoal agent
Ergot alkaloids	Sympathomimetic, vasodilator, antihypertensive
Morphine	Analgesic
Nicotine	Stimulant, nicotinic acetylcholine receptor agonist
Physostigmine	Inhibitor of acetylcholinesterase
Quinidine	Antiarrhythmic
Quinine	Antipyretics, antimalarial
Reserpine	Antihypertensive
Tubocurarine	Muscle relaxant
Vinblastine, vincristine	Antitumor
Vincamine	Vasodilating, antihypertensive
Yohimbine	Stimulant, aphrodisiac

Many synthetic and semisynthetic drugs are structural modifications of the alkaloids, which were designed to enhance or change the primary effect of the drug and reduce unwanted side-effects. For example, naloxone, an opioid receptor antagonist, is a derivative of thebaine that is present in opium.

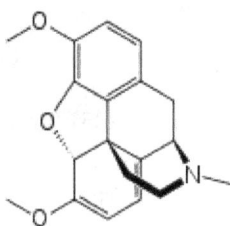

Thebaine

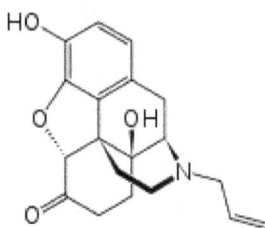

Naloxone

In Agriculture

Prior to the development of a wide range of relatively low-toxic synthetic pesticides, some alkaloids, such as salts of nicotine and anabasine, were used as insecticides. Their use was limited by their high toxicity to humans.

Use as Psychoactive Drugs

Preparations of plants containing alkaloids and their extracts and later pure alkaloids, have long been used as psychoactive substances. Cocaine, caffeine and cathinone are stimulants of the central nervous system. Mescaline and many of indole alkaloids (such as psilocybin, dimethyltryptamine and ibogaine) have hallucinogenic effect. Morphine and codeine are strong narcotic pain killers. There are alkaloids that do not have strong psychoactive effect themselves, but are precursors for semi-synthetic psychoactive drugs. For example, ephedrine andpseudoephedrine are used to produce methcathinone and methamphetamine. The baine is used in the synthesis of many painkillers such as oxycodone.

References

Andreas Luch (2009). *Molecular, clinical and environmental toxicology*. Springer. p. 20. ISBN 3-7643-8335-6.

Atta-ur-Rahman and M. Iqbal Choudhary (1997)."Diterpenoid and steroidal alkaloids". *Nat. Prod. Rep*14 (2): 191–203. doi: 10.1039/np9971400191.

Blankenship JD, Houseknecht JB, Pal S, Bush LP, Grossman RB, Schardl CL (2005). "Biosynthetic precursors of fungal pyrrolizidines, the loline alkaloids". *Chembiochem* 6 (6): 1016–1022. doi: 10.1002/cbic.200400327. PMID 15861432.

Cushnie T, Cushnie B, Lamb A (2014). "Alkaloids: An overview of their antibacterial, antibiotic-enhancing and antivirulence activities". *Int J Antimicrob Agents* 44 (5): 377–386. doi: 10.1016/j.ijantimicag.2014.06.001.PMID 25130096.

Dimitris C. Gournelif, Gregory G. Laskarisb and Robert Verpoorte (1997). "Cyclopeptide alkaloids". *Nat. Prod. Rep.* 14 (1): 75–82. doi: 10.1039/NP9971400075.

Faulkner JR, Hussaini SR, Blankenship JD, Pal S, Branan BM, Grossman RB, Schardl CL (2006). "On the sequence of bond formation in loline alkaloid biosynthesis". *Chembiochem* 7 (7): 1078–1088. doi: 10.1002/cbic.200600066.

IUPAC. Compendium of Chemical Terminology, 2nd ed. (The "Gold Book"). Compiled by A. D. McNaught and A. Wilkinson. Blackwell Scientific Publications, Oxford (1997) ISBN 0-9678550-9-8doi: 10.1351/goldbook

John R. Lewis (2000). "Amaryllidaceae, muscarine, imidazole, oxazole, thiazole and peptide alkaloids and other miscellaneous alkaloids". *Nat. Prod. Rep* 17 (1): 57–84. doi: 10.1039/a809403i. PMID 10714899.

Joseph P. Michael (2002). "Indolizidine and quinolizidine alkaloids". *Nat. Prod. Rep* 19: 458–475.doi: 10.1039/b208137g.

Kenneth W. Bentley (1997). "β-Phenylethylamines and the isoquinoline alkaloids". *Nat. Prod. Rep* 14 (4): 387–411. doi: 10.1039/NP9971400387. PMID 9281839.

Kittakoop P, Mahidol C, Ruchirawat S (2014). "Alkaloids as important scaffolds in therapeutic drugs for the treatments of cancer, tuberculosis and smoking cessation". *Curr Top Med Chem* 14 (2): 239–252. doi: 10.2174/1568026613666131216105049

Leland J. Cseke (2006). Natural Products from Plants Second Edition. – CRC. p. 30 ISBN 0-8493-2976-0

Manske, R. H. F. (1965). *The Alkaloids. Chemistry and Physiology.* Volume VIII. – New York: Academic Press, 1965, p. 673

Oscar Jacobsen (1882)., "Alkaloide" in: Ladenburg,*Handwörterbuch der Chemie* (Breslau, Germany: Eduard Trewendt), vol. 1, pp. 213–422.

Qiu S, Sun H, Zhang A, Xu H, Yan G, Han Y (2014). "Natural alkaloids: basic aspects, biological roles and future perspectives". *Chin J Nat Med* 12 (6): 401–406. doi: 10.1016/S1875-5364(14)60063-7.PMID 24969519.

Raj K Bansal (2004). A Text Book of Organic Chemistry. 4th Edition, New Age International, p. 644 ISBN 81-224-1459-1

Raymond S. Sinatra, Jonathan S. Jahr, J. Michael Watkins-Pitchford (2010). *The Essence of Analgesia and Analgesics.* Cambridge University Press. pp. 82–90. ISBN 1139491989.

Rhoades, David F (1979). "Evolution of Plant Chemical Defense against Herbivores". In Rosenthal, Gerald A. and Janzen, Daniel H. *Herbivores: Their Interaction with Secondary Plant Metabolites.* New York: Academic Press. p. 41. ISBN 0-12-597180-X.

Richard B. Herbert, Herbert, Richard B., Herbert, Richard B. (1999). "The biosynthesis of plant alkaloids and nitrogenous microbial metabolites". *Nat. Prod. Rep*16: 199–208. doi: 10.1039/a705734b.

Robbers JE, Speedie MK and Tyler VE (1996). "Chapter 9: Alkaloids". *Pharmacognosy and Pharmacobiotechnology.* Philadelphia: Lippincott, Williams and Wilkins. p. 143-185. ISBN 068308500X.

Robert A. Meyers (1998). *Encyclopedia of Physical Science and Technology* – Alkaloids, 3rd edition. ISBN 0-12-227411-3

Robert Alan Lewis (1998). *Lewis' dictionary of toxicology.* CRC Press, p. 51 ISBN 1-56670-223-2

Russo P, Frustaci A, Del Bufalo A, Fini M, Cesario A (2013). "Multitarget drugs of plants origin acting on Alzheimer's disease". *Curr Med Chem* 20 (13): 1686–93. doi: 10.2174/0929867311320130008.PMID 23410167.

Schardl CL, Grossman RB, Nagabhyru P, Faulkner JR, Mallik UP (2007). "Loline alkaloids: currencies of mutualism". *Phytochemistry* 68 (7): 980–996. doi: 10.1016/j. phytochem. 2007.01.010.PMID 17346759.

The suffix "ine" is a Greek feminine patronymic suffix and means "daughter of", hence, for example, "atropine" means "daughter of Atropa (belladonna)": Development of Systematic Names for the Simple Alkanes. yale.edu

William Johnson, A. (1999). Invitation to Organic Chemistry, Jones and Bartlett, p. 433 ISBN 0-7637-0432-6

Chapter 3

Alpha Amylase Inhibitor

In molecular biology, alpha-amylase inhibitor is a protein family which inhibits mammalian alpha-amylases specifically, by forming a tight stoichiometric 1:1 complex with alpha-amylase. This family of inhibitors has no action on plant and microbial alpha amylases. A crystal structure has been determined for tendamistat, the 74-amino acid inhibitor produced by *Streptomyces tendae* that targets a wide range of mammalian alpha-amylases. The binding of tendamistat to alpha-amylase leads to the steric blockage of the active site of the enzyme. The crystal structure of tendamistat revealed an immunoglobulin-like fold that could potentially adopt multiple conformations. Such molecular flexibility could enable an induced-fit type of binding that would both optimise binding and allow broad target specificity.

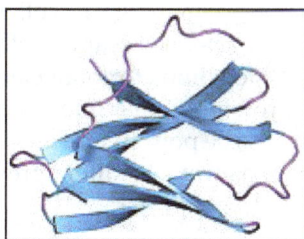

Amylase Inhibitor

Crystal structure determination, refinement and the molecular model of the alpha-amylase inhibitor hoe-467a

Identifiers

Symbol	A_amylase_inhib
Pfam	PF01356
InterPro	IPR000833
SCOP	1hoe
SUPERFAMILY	1hoe

An enzyme inhibitor is a molecule that binds to an enzyme and decreases its activity. Since blocking an enzyme's activity can kill a pathogen or correct a metabolic imbalance, many drugs are enzyme inhibitors. They are also used as herbicides and pesticides. Not all molecules that bind to enzymes are inhibitors; *enzyme activators* bind to enzymes and increase their enzymatic activity, while enzyme substrates bind and are converted to products in the normal catalytic cycle of the enzyme.

The binding of an inhibitor can stop a substrate from entering the enzyme's active site and/or hinder the enzyme from catalyzing its reaction. Inhibitor binding is either reversible or irreversible. Irreversible inhibitors usually react with the enzyme and change it chemically (*e.g.* via covalent bond formation). These inhibitors modify key amino acid residues needed for enzymatic activity. In contrast, reversible inhibitors bind non-covalently and different types of inhibition are produced depending on whether these inhibitors bind to the enzyme, the enzyme-substrate complex, or both.

Many drug molecules are enzyme inhibitors, so their discovery and improvement is an active area of research in biochemistry and pharmacology. A medicinal enzyme inhibitor is often judged by its specificity (its lack of binding to other proteins) and its potency (its dissociation constant, which indicates the concentration needed to inhibit the enzyme). A high specificity and potency ensure that a drug will have few side effects and thus low toxicity.

Enzyme inhibitors also occur naturally and are involved in the regulation of metabolism. For example, enzymes in a metabolic pathway can be inhibited by downstream products. This type of negative feedback slows the production line when products begin to build up and is an important way to maintain homeostasis in a cell. Other cellular enzyme inhibitors are proteins that specifically bind to and inhibit an enzyme target. This can help control enzymes that may be damaging to a cell, like proteases or nucleases. A well-characterised example of this is the ribonuclease inhibitor, which binds toribonucleases in one of the tightest known protein–protein interactions. Natural enzyme inhibitors can also be poisons and are used as defences against predators or as ways of killing prey.

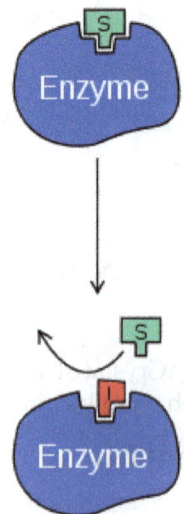

Reversible Inhibitors

Types of Reversible Inhibitors

Reversible inhibitors attach to enzymes with non-covalent interactions such as hydrogen bonds, hydrophobic interactions and ionic bonds. Multiple weak bonds between the inhibitor and the active site combine to produce strong and specific binding. In contrast to substrates and irreversible inhibitors, reversible inhibitors generally do not undergo chemical reactions when bound to the enzyme and can be easily removed by dilution or dialysis.

Competitive Inhibition: Substrate (S) and Inhibitor (I) Compete for the Active Site.

There are four kinds of reversible enzyme inhibitors. They are classified according to the effect of varying the concentration of the enzyme's substrate on the inhibitor.

☆ In competitive inhibition, the substrate and inhibitor cannot bind to the enzyme at the same time, as shown in the figure on the left. This usually results from the inhibitor having an affinity for the active site of an enzyme where the substrate also binds; the substrate and inhibitor compete for access to the enzyme's active site. This type of inhibition can be overcome by sufficiently high concentrations of substrate (Vmax remains constant), i.e., by out-competing the inhibitor. However, the apparent Km will increase as it takes a higher concentration of the substrate to reach the Km point, or half the Vmax. Competitive inhibitors are often similar in structure to the real substrate (see examples below).

☆ In uncompetitive inhibition, the inhibitor binds only to the substrate-enzyme complex, it should not be confused with non-competitive inhibitors. This type of inhibition causes Vmax to decrease (maximum velocity decreases as a result of removing activated complex) and Km to decrease (due to better binding efficiency as a result of Le Chatelier's principle and the effective elimination of the ES complex thus decreasing the Km which indicates a higher binding affinity).

☆ In mixed inhibition, the inhibitor can bind to the enzyme at the same time as the enzyme's substrate. However, the binding of the inhibitor affects the binding of the substrate and vice versa. This type of inhibition can be reduced, but not overcome by increasing concentrations of substrate. Although it is possible for mixed-type inhibitors to bind in the active site, this type of inhibition generally results from an allosteric effect where the inhibitor binds to a different site on an enzyme. Inhibitor binding to this allosteric site changes the conformation (i.e., tertiary structure or three-dimensional shape) of the enzyme so that the affinity of the substrate for the active site is reduced.

☆ Non-competitive inhibition is a form of mixed inhibition where the binding of the inhibitor to the enzyme reduces its activity but does not affect the binding of substrate. As a result, the extent of inhibition depends only on the concentration of the inhibitor. Vmax will decrease due to the inability for the reaction to proceed as efficiently, but Km will remain the same as the actual binding of the substrate, by definition, will still function properly.

Quantitative Description of Reversible Inhibition

Reversible inhibition can be described quantitatively in terms of the inhibitor's binding to the enzyme and to the enzyme-substrate complex and its effects on the kinetic constantsof the enzyme. In the classic Michaelis-Menten scheme below, an enzyme (E) binds to its substrate (S) to form the enzyme–substrate complex ES. Upon catalysis, this complex breaks down to release product P and free enzyme.

The inhibitor (I) can bind to either E or ES with the dissociation constants K_i or K_i', respectively.

☆ Competitive inhibitors can bind to E, but not to ES. Competitive inhibition increases Km (i.e., the inhibitor interferes with substrate binding), but does not affect Vmax (the inhibitor does not hamper catalysis in ES because it cannot bind to ES).

☆ Non-competitive inhibitors have identical affinities for E and ES (Ki = Ki'). Non-competitive inhibition does not change Km (*i.e.*, it does not affect substrate binding) but decreases Vmax (*i.e.*, inhibitor binding hampers catalysis).

☆ Mixed-type inhibitors bind to both E and ES, but their affinities for these two forms of the enzyme are different (Ki ≠ Ki'). Thus, mixed-type inhibitors interfere with substrate binding (increase Km) and hamper catalysis in the ES complex (decrease Vmax).

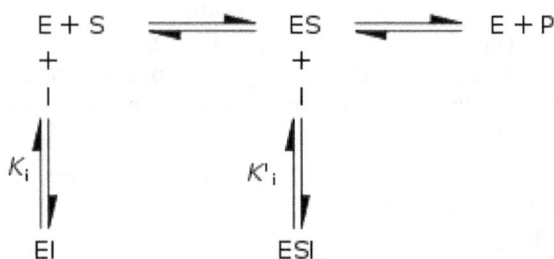

$$E + S \rightleftharpoons ES \rightleftharpoons E + P$$

$$+ \qquad\qquad +$$

$$I \qquad\qquad I$$

$$K_i \downarrow \qquad\qquad K_i' \downarrow$$

$$EI \qquad\qquad ESI$$

Kinetic Scheme for Reversible Enzyme Inhibitors

When an enzyme has multiple substrates, inhibitors can show different types of inhibition depending on which substrate is considered. This results from the active site containing two different binding sites within the active site, one for each substrate. For example, an inhibitor might compete with substrate A for the first binding site, but be a non-competitive inhibitor with respect to substrate B in the second binding site.

Measuring the Dissociation Constants of a Reversible Inhibitor

As noted above, an enzyme inhibitor is characterised by its two dissociation constants, K_i and K_i', to the enzyme and to the enzyme-substrate complex, respectively. The enzyme-inhibitor constant K_i can be measured directly by various methods; one extremely accurate method isisothermal titration calorimetry, in which the inhibitor is titrated into a solution of enzyme and the heat released or absorbed is measured.[4]However, the other dissociation constant K_i' is difficult to measure directly, since the enzyme-substrate complex is short-lived and undergoing a chemical reaction to form the product. Hence, K_i' is usually measured indirectly, by observing the enzyme activity under various substrate and inhibitor concentrations and fitting the data to a modified Michaelis – Menten equation.

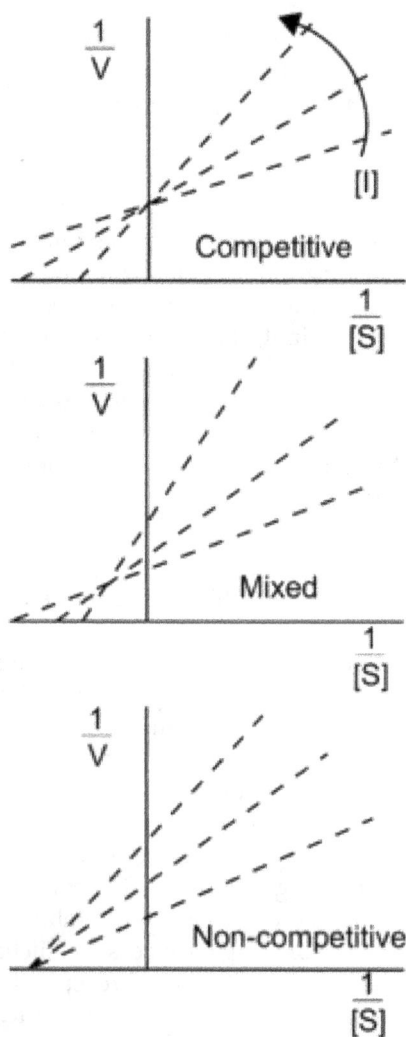

Lineweaver–Burk Plots of different Types of Reversible Enzyme Inhibitors. The arrow shows the effect of increasing concentrations of inhibitor.

$$V = \frac{V_{max}[S]}{\alpha K_m + \alpha'[S]} = \frac{(1/\alpha')V_{max}[S]}{(\alpha/\alpha')K_m + [S]}$$

where, the modifying factors α and α' are defined by the inhibitor concentration and its two dissociation constants

$$\alpha = 1 + \frac{[I]}{K_i} \qquad \alpha' = 1 + \frac{[I]}{K_i'}.$$

Thus, in the presence of the inhibitor, the enzyme's effective K_m and V_{max} become $(\alpha/\alpha')K_m$ and $(1/\alpha')V_{max}$, respectively. However, the modified Michaelis - Menten equation assumes that binding of the inhibitor to the enzyme has reached equilibrium, which may be a very slow process for inhibitors with sub-nanomolar dissociation constants. In these cases, it is usually more practical to treat the tight-binding inhibitor as an irreversible inhibitor (see below); however, it can still be possible to estimate K_i' kinetically if K_i is measured independently.

The effects of different types of reversible enzyme inhibitors on enzymatic activity can be visualized using graphical representations of the Michaelis – Menten equation, such as Lineweaver–Burk and Eadie-Hofstee plots. For example, in the Lineweaver–Burk plots at the right, the competitive inhibition lines intersect on the y-axis, illustrating that such inhibitors do not affect V_{max}. Similarly, the non-competitive inhibition lines intersect on the x-axis, showing these inhibitors do not affect K_m. However, it can be difficult to estimate K_i and K_i' accurately from such plots, so it is advisable to estimate these constants using more reliable nonlinear regression methods, as described above.

Reversible Inhibitors

Traditionally reversible enzyme inhibitors have been classified as competitive, uncompetitive, or non-competitive, according to their effects on K_m and V_{max}. These different effects result from the inhibitor binding to the enzyme E, to the enzyme–substrate complex ES, or to both, respectively. The division of these classes arises from a problem in their derivation and results in the need to use two different binding constants for one binding event. The binding of an inhibitor and its effect on the enzymatic activity are two distinctly different things, another problem the traditional equations fail to acknowledge. In non-competitive inhibition the binding of the inhibitor results in 100 per cent inhibition of the enzyme only and fails to consider the possibility of anything in between. The common form of the inhibitory term also obscures the relationship between the inhibitor binding to the enzyme and its relationship to any other binding term be is the Michaelis – Menten equation or a dose response curve associated with legend receptor binding. To demonstrate the relationship the following rearrangement can be made:

$$\frac{V_{max}}{1 + \dfrac{[I]}{K_i}}$$

$$\frac{V_{max}}{\dfrac{[I] + K_i}{K_i}}$$

Adding zero to the bottom ([I]-[I])

$$\frac{\dfrac{V_{max}}{[I] + K_i}}{[I] + K_i - [I]}$$

Dividing by [I]+K$_i$

$$\frac{\dfrac{V_{max}}{1}}{1 - \dfrac{[I]}{[I] + K_i}}$$

$$V_{max} - V_{max}\frac{[I]}{[I] + K_i}$$

This notation demonstrates that similar to the Michaelis – Menten equation, where the rate of reaction depends on the percent of the enzyme population interacting with substrate.

Fraction of the enzyme population bound by substrate

$$\frac{[S]}{[S] + K_m}$$

Fraction of the enzyme population bound by inhibitor

$$\frac{[I]}{[I] + K_i}$$

The effect of the inhibitor is a result of the percent of the enzyme population interacting with inhibitor. The only problem with this equation in its present form is that it assumes absolute inhibition of the enzyme with inhibitor binding, when in fact there can be a wide range of effects anywhere from 100 per cent inhibition of substrate turn over to just >0 per cent. To account for this the equation can be easily modified to allow for different degrees of inhibition by including a delta V_{max} term.

$$V_{max} - \Delta V_{max}\frac{[I]}{[I] + K_i}$$

or

$$V_{max}1 - (V_{max}1 - V_{max}2)\frac{[I]}{[I] + K_i}$$

This term can then define the residual enzymatic activity present when the inhibitor is interacting with individual enzymes in the population. However the inclusion of this term has the added value of allowing for the possibility of activation if the secondary V_{max} term turns out to be higher than the initial term. To account for the possibly of activation as well the notation can then be rewritten replacing the inhibitor "I" with a modifier term denoted here as "X".

$$V_{\max}1 - (V_{\max}1 - V_{\max}2)\frac{[X]}{[X] + K_x}$$

While this terminology results in a simplified way of dealing with kinetic effects relating to the maximum velocity of the Michaelis – Menten equation, it highlights potential problems with the term used to describe effects relating to the K_m. The K_m relating to the affinity of the enzyme for the substrate should in most cases relate to potential changes in the binding site of the enzyme which would directly result from enzyme inhibitor interactions. As such a term similar to the one proposed above to modulate V_{max} should be appropriate in most situations.

$$K_m1 - (K_m1 - K_m2)\frac{[X]}{[X] + K_x}$$

Special Cases

☆ The mechanism of partially competitive inhibition is similar to that of non-competitive, except that the EIS complex has catalytic activity, which may be lower or even higher (partially competitive activation) than that of the enzyme–substrate (ES) complex. This inhibition typically displays a lower Vmax, but an unaffected Km value.

☆ Non-competitive inhibition occurs when the inhibitor binds only to the enzyme–substrate complex, not to the free enzyme; the EIS complex is catalytically inactive. This mode of inhibition is rare and causes a decrease in both Vmax and the Km value.

☆ Substrate and product inhibition is where either the substrate or product of an enzyme reaction inhibits the enzyme's activity. This inhibition may follow the competitive, uncompetitive or mixed patterns. In substrate inhibition there is a progressive decrease in activity at high substrate concentrations. This may indicate the existence of two substrate-binding sites in the enzyme. At low substrate, the high-affinity site is occupied and normal kinetics are followed. However, at higher concentrations, the second inhibitory site becomes occupied, inhibiting the enzyme. Product inhibition is often a regulatory feature in metabolism and can be a form of negative feedback.

☆ Slow-tight inhibition occurs when the initial enzyme–inhibitor complex EI undergoes isomerisation to a second more tightly held complex, EI*, but the overall inhibition process is reversible. This manifests itself as slowly increasing enzyme inhibition. Under these conditions, traditional Michaelis–Menten kinetics gives a false value for Ki, which is time–dependent. The true value of Ki can be obtained through more complex analysis of the on (kon) and off (koff) rate constants for inhibitor association. See irreversible inhibition below for more information.

Examples of Reversible Inhibitors

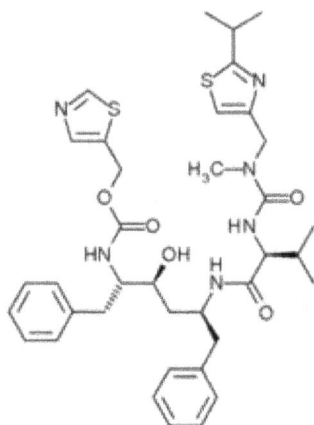

Peptide-Based HIV-1 Protease Inhibitor ritonavir

As enzymes have evolved to bind their substrates tightly and most reversible inhibitors bind in the active site of enzymes, it is unsurprising that some of these inhibitors are strikingly similar in structure to the substrates of their targets. An example of these substrate mimics are the protease inhibitors, a very successful class of antiretroviral drugs used to treat HIV. The structure of ritonavir, a protease inhibitor based on a peptide and containing three peptide bonds, is shown on the right. As this drug resembles the protein that is the substrate of the HIV protease, it competes with this substrate in the enzyme's active site.

Enzyme inhibitors are often designed to mimic the transition state or intermediate of an enzyme-catalyzed reaction. This ensures that the inhibitor exploits the transition state stabilising effect of the enzyme, resulting in a better binding affinity (lower K_i) than substrate-based designs. An example of such a transition state inhibitor is the antiviral drug oseltamivir; this drug mimics the planar nature of the ring oxonium ion in the reaction of the viral enzyme neuraminidase.

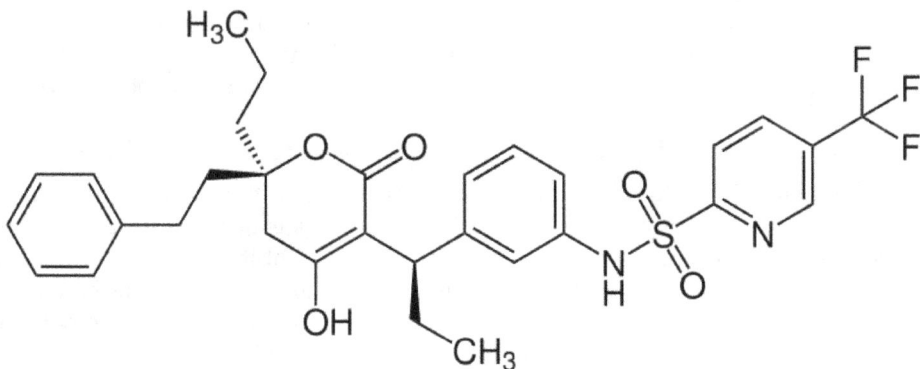

Nonpeptidic HIV-1 Protease Inhibitor tipranavir

However, not all inhibitors are based on the structures of substrates. For example, the structure of another HIV protease inhibitor tipranavir is shown on the left. This molecule is not based on a peptide and has no obvious structural similarity to a protein substrate. These non-peptide inhibitors can be more stable than inhibitors containing peptide bonds, because they will not be substrates for peptidases and are less likely to be degraded.[14]

In drug design it is important to consider the concentrations of substrates to which the target enzymes are exposed. For example, someprotein kinase inhibitors have chemical structures that are similar to adenosine triphosphate, one of the substrates of these enzymes. However, drugs that are simple competitive inhibitors will have to compete with the high concentrations of ATP in the cell. Protein kinases can also be inhibited by competition at the binding sites where the kinases interact with their substrate proteins and most proteins are present inside cells at concentrations much lower than the concentration of ATP. As a consequence, if two protein kinase inhibitors both bind in the active site with similar affinity, but only one has to compete with ATP and then the competitive inhibitor at the protein-binding site will inhibit the enzyme more effectively.

Irreversible Inhibitors

Types of Irreversible Inhibition

Irreversible inhibitors usually covalently modify an enzyme and inhibition can therefore not be reversed. Irreversible inhibitors often contain reactive functional groups such as nitrogen mustards, aldehydes, haloalkanes, alkenes, Michael acceptors, phenyl sulfonates, or fluorophosphonates. These electrophilic groups react with amino acid side chains to form covalent adducts. The residues modified are those with side chains containing nucleophiles such as hydroxyl or sulfhydryl groups; these include the amino acids serine (as inDFP, right), cysteine, threonine, or tyrosine.

Irreversible inhibition is different from irreversible enzyme inactivation. Irreversible inhibitors are generally specific for one class of enzyme and do not inactivate all proteins; they do not function by destroying protein structure but by specifically altering the active site of their target. For example, extremes of pH or temperature usually cause denaturation of all protein structure, but this is a non-specific effect. Similarly, some non-specific chemical treatments destroy protein structure: for example, heating in concentrated hydrochloric acid will hydrolyse the peptide bonds holding proteins together, releasing free amino acids.

Irreversible inhibitors display time-dependent inhibition and their potency therefore cannot be characterised by an IC_{50} value. This is because the amount of active enzyme at a given concentration of irreversible inhibitor will be different depending on how long the inhibitor is pre-incubated with the enzyme. Instead, $k_{obs}/[I]$ values are used, where k_{obs} is the observed pseudo-first order rate of inactivation (obtained by plotting the log of per cent activity vs. time) and $[I]$ is the concentration of inhibitor. The $k_{obs}/[I]$ parameter is valid as long as the inhibitor does not saturate binding with the enzyme (in which case $k_{obs} = k_{inact}$).

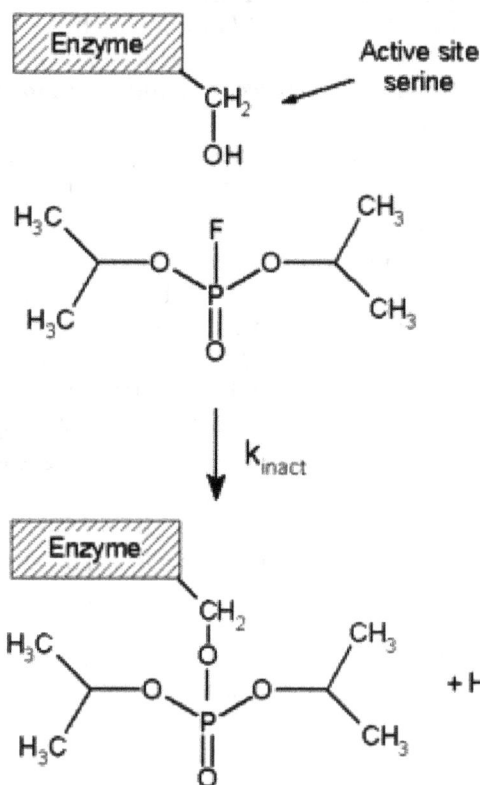

Reaction of the Irreversible Inhibitor Diisopropylfluorophosphate (DFP) with a Serine Protease.

Analysis of Irreversible Inhibition

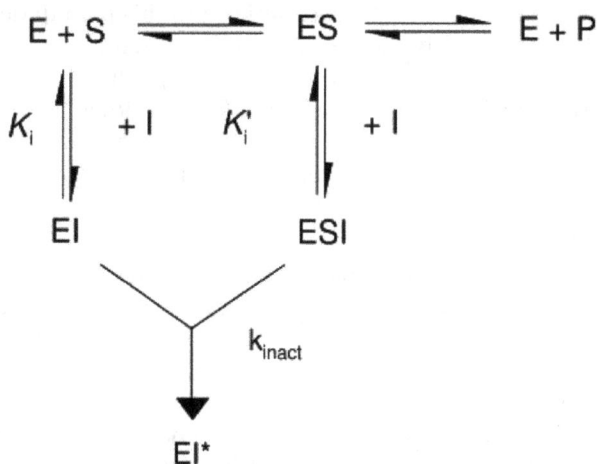

Kinetic Scheme for Irreversible Inhibitors

As shown in the figure to the left, irreversible inhibitors form a reversible non-covalent complex with the enzyme (EI or ESI) and this then reacts to produce the covalently modified "dead-end complex" EI*. The rate at which EI* is formed is called the inactivation rate or k_{inact}. Since formation of EI may compete with ES, binding of irreversible inhibitors can be prevented by competition either with substrate or with a second, reversible inhibitor. This protection effect is good evidence of a specific reaction of the irreversible inhibitor with the active site.

The binding and inactivation steps of this reaction are investigated by incubating the enzyme with inhibitor and assaying the amount of activity remaining over time. The activity will be decreased in a time-dependent manner, usually following exponential decay. Fitting these data to a rate equation gives the rate of inactivation at this concentration of inhibitor. This is done at several different concentrations of inhibitor. If a reversible EI complex is involved the inactivation rate will be saturable and fitting this curve will give k_{inact} and K_i.

Another method that is widely used in these analyses is mass spectrometry. Here, accurate measurement of the mass of the unmodified native enzyme and the inactivated enzyme gives the increase in mass caused by reaction with the inhibitor and shows the stoichiometry of the reaction. This is usually done using a MALDI-TOF mass spectrometer. In a complementary technique, peptide mass fingerprinting involves digestion of the native and modified protein with a protease such as trypsin. This will produce a set of peptides that can be analysed using a mass spectrometer. The peptide that changes in mass after reaction with the inhibitor will be the one that contains the site of modification.

Special Cases

Not all irreversible inhibitors form covalent adducts with their enzyme targets. Some reversible inhibitors bind so tightly to their target enzyme that they are essentially irreversible. These tight-binding inhibitors may show kinetics similar to covalent irreversible inhibitors. In these cases, some of these inhibitors rapidly bind to the enzyme in a low-affinity EI complex and this then undergoes a slower rearrangement to a very tightly bound EI* complex (see figure above). This kinetic behaviour is called slow-binding. This slow rearrangement after binding often involves a conformational changeas the enzyme "clamps down" around the inhibitor molecule. Examples of slow-binding inhibitors include some important drugs, such methotrexate, allopurinol and the activated form of acyclovir.

Uses of Inhibitors

Enzyme inhibitors are found in nature and are also designed and produced as part of pharmacology and biochemistry. Natural poisons are often enzyme inhibitors that have evolved to defend a plant or animal against predators. These natural toxins include some of the most poisonous compounds known. Artificial inhibitors are often used as drugs, but can also be insecticides such as malathion, herbicides such as glyphosate, or disinfectants such as triclosan. Other artificial enzyme inhibitors block acetylcholinesterase, an enzyme which breaks down acetylcholine and are used as nerve agents in chemical warfare.

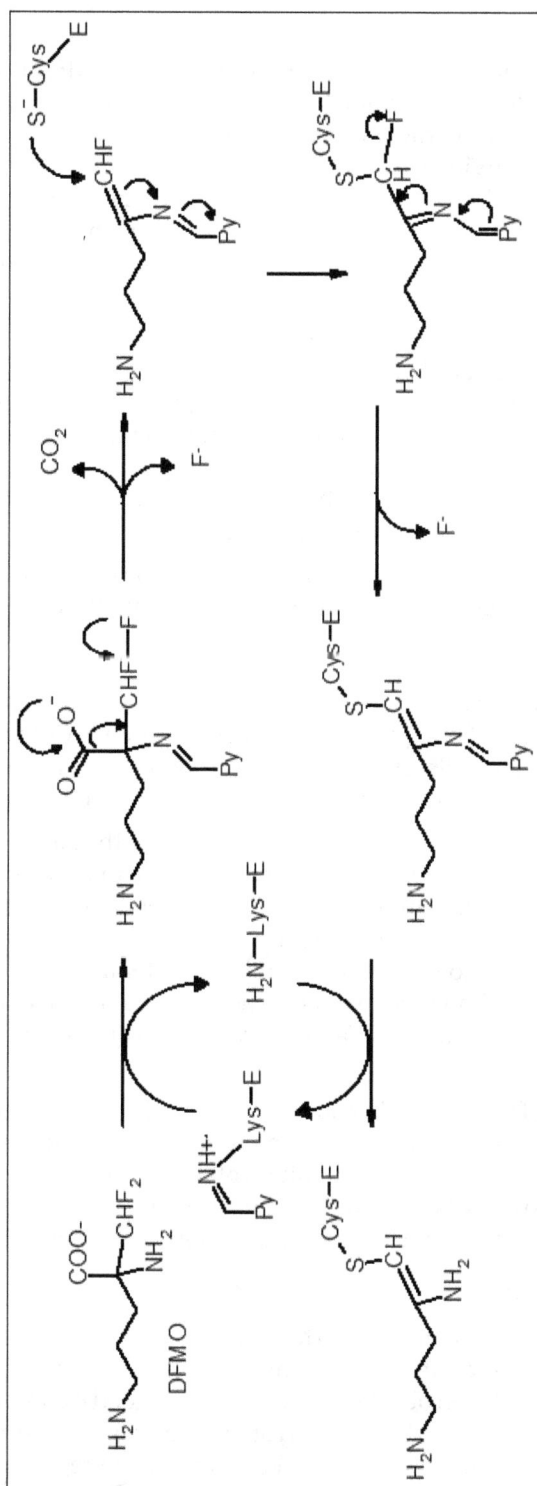

Chemical Mechanism for Irreversible inhibition of Crnithine Decarboxylase by DFMO.
Pyridoxal 5'-phosphate (Py) and enzyme (E) are not shown.

Chemotherapy

The most common uses for enzyme inhibitors are as drugs to treat disease. Many of these inhibitors target a human enzyme and aim to correct a pathological condition. However, not all drugs are enzyme inhibitors. Some, such as anti-epileptic drugs, alter enzyme activity by causing more or less of the enzyme to be produced. These effects are called enzyme induction and inhibition and are alterations in gene expression, which is unrelated to the type of enzyme inhibition discussed here. Other drugs interact with cellular targets that are not enzymes, such as ion channels or membrane receptors.

An example of a medicinal enzyme inhibitor is sildenafil (Viagra), a common treatment for male erectile dysfunction. This compound is a potent inhibitor of cGMP specific phosphodiesterase type 5, the enzyme that degrades the signalling molecule cyclic guanosine monophosphate. This signalling molecule triggers smooth muscle relaxation and allows blood flow into the corpus cavernosum, which causes an erection. Since the drug decreases the activity of the enzyme that halts the signal, it makes this signal last for a longer period of time.

Another example of the structural similarity of some inhibitors to the substrates of the enzymes they target is seen in the figure comparing the drug methotrexate to folic acid. Folic acid is a substrate of dihydrofolate reductase, an enzyme involved in making nucleotides that is potently inhibited by methotrexate. Methotrexate blocks the action of dihydrofolate reductase and thereby halts the production of nucleotides. This block of nucleotide biosynthesis is more toxic to rapidly growing cells than non-dividing cells, since a rapidly growing cell has to carry out DNA replication, therefore methotrexate is often used in cancer chemotherapy.

Drugs also are used to inhibit enzymes needed for the survival of pathogens. For example, bacteria are surrounded by a thick cell wall made of a net-like polymer called peptidoglycan. Many antibiotics such as penicillin and vancomycin inhibit the enzymes that produce and then cross-link the strands of this polymer together. This causes the cell wall to lose strength and the bacteria to burst. In the figure, a molecule of penicillin (shown in a ball-and-stick form) is shown bound to its target, the transpeptidase from the bacteria *Streptomyces* R61 (the protein is shown as a ribbon-diagram).

Drug design is facilitated when an enzyme that is essential to the pathogen's survival is absent or very different in humans. In the example above, humans do not make peptidoglycan, therefore inhibitors of this process are selectively toxic to bacteria. Selective toxicity is also produced in antibiotics by exploiting differences in the structure of the ribosomes in bacteria, or how they make fatty acids.

Metabolic Control

Enzyme inhibitors are also important in metabolic control. Many metabolic pathways in the cell are inhibited by metabolitesthat control enzyme activity through allosteric regulation or substrate inhibition. A good example is the allosteric regulation of the glycolytic pathway. This catabolic pathway consumes glucose and produces ATP, NADH and pyruvate. A key step for the regulation of glycolysis is an early reaction in the pathway catalysed by phosphofructokinase-1 (PFK1). When ATP levels rise, ATP binds an allosteric site in PFK1 to decrease the rate of the enzyme reaction; glycolysis is inhibited and ATP production falls. This negative feedback control helps maintain a steady concentration of ATP in the cell. However, metabolic pathways are not just regulated through inhibition since enzyme activation is equally important. With respect to PFK1, fructose 2,6-bisphosphate and ADP are examples of metabolites that are allosteric activators.

Physiological enzyme inhibition can also be produced by specific protein inhibitors. This mechanism occurs in the pancreas, which synthesises many digestive precursor enzymes known as zymogens. Many of these are activated by the trypsin protease, so it is important to inhibit the activity of trypsin in the pancreas to prevent the organ from digesting itself. One way in which the activity of trypsin is controlled is the production of a specific and potent trypsin inhibitor protein in the pancreas. This inhibitor binds tightly to trypsin, preventing the trypsin activity that would otherwise be detrimental to the organ. Although the trypsin inhibitor is a protein, it avoids being hydrolysed as a substrate by the protease by excluding water from trypsin's active site and destabilising the transition state. Other examples of physiological enzyme inhibitor proteins include the barstarinhibitor of the bacterial ribonuclease barnase and the inhibitors of protein phosphatases.

Pesticides and Herbicides

Many herbicides and pesticides are enzyme inhibitors. Acetylcholinesterase (AChE) is an enzyme found in animals from insects to humans. It is essential to nerve cell function through its mechanism of breaking down the neurotransmitter acetylcholine into its constituents, acetate and choline. This is somewhat unique among neurotransmitters as most, including serotonin, dopamine and norepinephrine, are absorbed from the synaptic cleft rather than cleaved. A large number of AChE inhibitors are used in both medicine and agriculture. Reversible competitive inhibitors, such as edrophonium, physostigmine and neostigmine, are used in the treatment of myasthenia gravis and in anaesthesia. The carbamate pesticides are also examples of reversible AChE inhibitors. The organophosphate insecticides such as malathion, parathion and chlorpyrifos irreversibly inhibit acetylcholinesterase. The herbicide glyphosate is an inhibitor of 3-phosphoshikimate 1-carboxyvinyltransferase, other herbicides, such as the sulfonylureas inhibit the enzyme acetolactate synthase. Both these enzymes are needed for plants to make branched-chain amino acids. Many other enzymess are inhibited by herbicides, including enzymes needed for the biosynthesis of lipids and carotenoids and the processes of photosynthesis and oxidative phosphorylation.

Natural Poisons

Animals and plants have evolved to synthesise a vast array of poisonous products including secondary metabolites, peptides and proteins that can act as inhibitors. Natural toxins are usually small organic molecules and are so diverse that there are probably natural inhibitors for most metabolic processes. The metabolic processes targeted by natural poisons encompass more than enzymes in metabolic pathways and can also include the inhibition of receptor, channel and structural protein functions in a cell. For example, paclitaxel (taxol), an organic molecule found in the Pacific yew tree, binds tightly to tubulindimers and inhibits their assembly into microtubules in the cytoskeleton.

Many natural poisons act as neurotoxins that can cause paralysis leading to death and have functions for defence against predators or in hunting and capturing prey. Some of these natural inhibitors, despite their toxic attributes, are valuable for therapeutic uses at lower doses. An example of a neurotoxin are the glycoalkaloids, from the plant species in the *Solanaceae* family (includes potato, tomato and eggplant), that are acetylcholinesterase inhibitors. Inhibition of this enzyme causes an uncontrolled increase in the acetylcholine neurotransmitter, muscular paralysis and then death. Neurotoxicity can also result from the inhibition of receptors; for example, atropine from deadly nightshade (*Atropa belladonna*) that functions as a competitive antagonist of the muscarinic acetylcholine receptors.

Although many natural toxins are secondary metabolites, these poisons also include peptides and proteins. An example of a toxic peptide is alpha-amanitin, which is found in relatives of the death cap mushroom. This is a potent enzyme inhibitor, in this case preventing the RNA polymerase II enzyme from transcribing DNA. The algal toxin microcystinis also a peptide and is an inhibitor of protein phosphatases. This toxin can contaminate water supplies after algal blooms and is a known carcinogen that can also cause acute liver hemorrhage and death at higher doses.

Proteins can also be natural poisons or antinutrients, such as the trypsin inhibitors (discussed above) that are found in some legumes, as shown in the figure above. A less common class of toxins are toxic enzymes: these act as irreversible inhibitors of their target enzymes and work by chemically modifying their substrate enzymes. An example isricin, an extremely potent protein toxin found in castor oil beans. This enzyme is a glycosidase that inactivates ribosomes. Since ricin is a catalytic irreversible inhibitor, this allows just a single molecule of ricin to kill a cell.

Multiprotein Complex

A multiprotein complex (or protein complex) is a group of two or more associated polypeptide chains. If the different polypeptide chains contain different protein domain, the resulting multiprotein complex can have multiple catalytic functions. This is distinct from a multienzyme polypeptide, in which multiple catalytic domains are found in a single polypeptide chain. Protein complexes are a form of quaternary structure. Proteins in a protein complex are linked by non-covalent protein–protein interactions and different protein complexes have different degrees of stability over time. These complexes are a cornerstone of many (if not most) biological processes and together they form various types of molecular machinery that perform a vast array of biological functions. Increasingly, scientists view the cell as composed of modular supramolecular complexes, each of which performs an independent, discrete biological function.

Through proximity, the speed and selectivity of binding interactions between enzymatic complex and substrates can be vastly improved, leading to higher cellular efficiency. Unfortunately, many of the techniques used to break open cells and isolate proteins are inherently disruptive to such large complexes, so their protein complexes within the cell may be even more widespread than can be detected. Examples include the proteasome for molecular degradation, the metabolon for oxidative energy generation and the ribosome for protein synthesis. In stable complexes, large hydrophobic interfaces between proteins typically bury surface areas larger than 2500 square angstroms.

Function

Protein complex formation sometimes serves to activate or inhibit one or more of the complex members and in this way, protein complex formation can be similar to phosphorylation. Individual proteins can participate in the formation of a variety of different protein complexes. Different complexes perform different functions and the same complex can perform very different functions that depend on a variety of factors. Some of these factors are:

- ☆ Which cellular compartment the complex exists in when it is contained
- ☆ Which stage in the cell cycle the complexes are present
- ☆ The nutritional status of the cell
- ☆ Others

Many protein complexes are well understood, particularly in the model organism *Saccharomyces cerevisiae* (a strain of yeast). For this relatively simple organism, the study of protein complexes is now being performed genome wide and the elucidation of most protein complexes of the yeast is undergoing.

Types of Protein Complexes

Obligate vs Non-obligate Protein Complex

If a protein can form stable crystal structure of its own (without any other associated protein) *in vivo*, then the complexes formed by such proteins are called "non-obligate protein complex". On the other hand, some proteins can't be found to create a crystal structure alone, but can be found as a part of a protein complex which creates a stable crystal structure. Such protein complexes are called "obligate protein complex".

Transient vs Permanent/Stable Protein Complex

Transient protein complexes form and break down transiently *in vivo*, whereas permanent complexes have a relatively long half-life. Typically, the obligate interactions (protein-protein interactions in an obligate complex) are permanent, whereas non-obligate interactions have been found to be either permanent or transient. Note that there is no clear distinction between obligate and non-obligate interaction, rather there exist a continuum between them which depends on various conditions *e.g.* pH, protein concentration etc.[5]However, there are important distinctions between the properties of transient and permanent/stable interactions: stable interactions are highly conserved but transient interactions are far less conserved, interacting proteins on the two sides of a stable interaction have more tendency of being co-expressed than those of a transient interaction (in fact, co-expression probability between two transiently interacting proteins is not higher than two random proteins) and transient interactions are much less co-localized than stable interactions. Though, transient by nature, transient interactions are very important for cell biology: human interactome is enriched in such interactions, these interactions are the dominating players of gene regulation and signal transduction and proteins with *intrinsically disordered regions* (IDR: regions in protein that show dynamic inter-converting structures in the native state) are found to be enriched in transient regulatory and signaling interactions.

Fuzzy Complex

Fuzzy protein complexes have more than one structural forms or dynamic structural disorder in the bound state. This means that proteins may not fold completely in either transient or permanent complexes. Consequently, specific complexes can have ambiguous interactions, which vary according to the environmental signals. Hence different ensemble of structures result in different (even opposite) biological functions. Post-translational modifications, protein interactions or alternative splicing modulate the conformational ensembles of fuzzy complexes, to fine-tune affinity or specificity of interactions. These mechanisms are often used for regulation within the eukaryotic transcription machinery.

Essential Proteins in Protein Complexes

Although some early studies suggested a strong correlation between essentiality and protein interaction degree (the "centrality-lethality" rule) subsequent analyses have shown that this correlation is weak for binary or transient interactions (*e.g.,* yeast two-hybrid). However, the correlation is robust for networks of stable cocomplex interactions. In fact, a disproportionate number of essential belong to protein complexes. This led to the conclusion that essentiality is a property of molecular machines (*i.e.* complexes) rather than individual components. Wang *et al.* (2009) noted that larger protein complexes are more likely to be essential, explaining why essential genes are more likely to have high cocomplex interaction degree. Ryan *et al.* (2013) referred to the observation that entire complexes appear essential as "modular essentiality". These authors also showed that complexes tend to be composed of either essential or non-essential proteins rather than showing a random distribution. However, this not an all or nothing phenomenon: only about 26 per cent (105/401) of yeast complexes consist of solely essential or solely nonessential subunits. In humans, genes whose protein products belong to the same complex are more likely to result in the same disease phenotype.

Homomultimeric and Heteromultimeric Proteins

The subunits of a multimeric protein may be identical as in a homomultimeric (homooligomeric) protein or different as in a heteromultimeric protein. Many soluble and membrane proteins form homomultimeric complexes in a cell, majority of proteins in the Protein Data Bank are homomultimeric. Homooligomers are responsible for the diversity and specificity of many pathways may mediate and regulate gene expression, activity of enzymes, ion channels, receptors and cell adhesion processes.

The voltage-gated potassium channels in the plasma membrane of a neuron are heteromultimeric proteins composed of four of forty known alpha subunits. Subunits must be of the same subfamily to form the multimeric protein channel. The tertiary structure of the channel allows ions to flow through the hydrophobic plasma membrane. Connexons are an example of a homomultimeric protein composed of six identical connexins. A cluster of connexons forms the gap-junction in two neurons that transmit signals through an electrical synapse.

Structure Determination

The molecular structure of protein complexes can be determined by experimental techniques such as X-ray crystallography or nuclear magnetic resonance. Increasingly the theoretical option of protein–protein docking is also becoming available. One method that is commonly used for identifying the meomplexes is immunoprecipitation. Recently, Raicu and coworkers developed a method to determine the quaternary structure of protein complexes in living cells. This method is based on the determination of pixel-level Forster resonance energy transfer (FRET) efficiency in conjunction with spectrally resolved two-photon microscope. The distribution of FRET efficiencies are simulated against different models to get the geometry and stoichiometry of the complexes.

Molecular Binding

Molecular binding is an attractive interaction between two molecules that results in a stable association in which the molecules are close to each other. The result of molecular binding is the formation of a molecular complex. A molecular complex in turn is a loose association involving two or more molecules. The attractive bonding between the components of a complex is normally weaker than in a covalent bond.

Types

Molecular binding can be classified into the following types:

- ☆ Non-covalent – no chemical bonds are formed between the two interacting molecules hence the association is fully reversible
- ☆ Reversible covalent – a chemical bond is formed, however the free energy difference separating the noncovalently-bonded reactants from bonded product is near equilibrium and the activation barrier is relatively low such that the reverse reaction which cleaves the chemical bond easily occurs
- ☆ Irreversible covalent – a chemical bond is formed in which the product is thermodynamically much more stable than the reactants such that the reverse reaction does not take place.

Driving Force

In order for the complex to be stable, the free energy of complex by definition must be lower than the solvent separated molecules. The binding may be primarily entropy-driven (release of ordered solvent molecules around the isolated molecule that results in a net increase of entropy of the system). When the solvent is water, this is known as the hydrophobic effect. Alternatively the binding may be enthalpy-driven where non-covalent attractive forces such as electrostatic attraction, hydrogen bonding and van der Waals/London dispersion forces are primarily responsible for the formation of a stable complex. Complexes that have a strong entropy contribution to formation tend to have weak enthalpy contributions. Conversely complexes that have strong enthalpy component tend to have a weak entropy component. This phenomenon is known as enthalpy-entropy compensation.

Measurement

The association constant (K_I) also known as the binding constant (K_A) between the components of complex is the ratio of the concentration of the complex divided by the product of the concentrations of the isolated components at equilibrium:

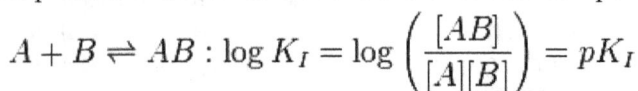

$$A + B \rightleftharpoons AB : \log K_I = \log \left(\frac{[AB]}{[A][B]} \right) = pK_I$$

Examples

Molecules that can participate in molecular binding include proteins, nucleic acids, carbohydrates, lipids and small organic molecules such as drugs. Hence the types of complexes that form as a result of molecular binding include:

- ☆ Protein – protein
- ☆ Protein – DNA
- ☆ Protein – hormone
- ☆ Protein – drug

Proteins that form stable complexes with other molecules are often referred to as receptors while their binding partners are called ligands.

Sensitivity and Specificity

Sensitivity and specificity are statistical measures of the performance of a binary classification test, also known in statistics as classification function. Specificity (also called the true positive rate, or the recall rate in some fields) measures the proportion of actual positives which are correctly identified as such (*e.g.*, the percentage of sick people who are correctly identified as having the condition) and is complementary to the false negative rate. Sensitivity (sometimes called the true negative rate) measures the proportion of negatives which are correctly identified as such (*e.g.*, the percentage of healthy people who are correctly identified as not having the condition) and is complementary to the false positive rate.

A perfect predictor would be described as 100 per cent sensitive (*e.g.*, all sick are identified as sick) and 100 per cent specific (*e.g.*, all healthy are identified as healthy); however, theoretically any predictor will possess a minimum error bound known as the Bayes error rate. For any test, there is usually a trade-off between the measures. For instance, in an airport security setting in which one is testing for potential threats to safety, scanners may be set to trigger on low-risk items like belt buckles and keys (low specificity), in order to reduce the risk of missing objects that do pose a threat to the aircraft and those aboard (high sensitivity). This trade-off can be represented graphically as a receiver operating characteristic curve.

Definitions

Imagine a study evaluating a new test that screens people for a disease. Each person taking the test either has or does not have the disease. The test outcome can be positive (predicting that the person has the disease) or negative (predicting that the person does not have the disease). The test results for each subject may or may not match the subject's actual status. In that setting:

☆ True positive: Sick people correctly diagnosed as sick

☆ False positive: Healthy people incorrectly identified as sick

☆ True negative: Healthy people correctly identified as healthy

☆ False negative: Sick people incorrectly identified as healthy

In general, Positive = identified and negative = rejected. Therefore:

☆ True positive = correctly identified

☆ False positive = incorrectly identified

☆ True negative = correctly rejected

☆ False negative = incorrectly rejected

Let us define an experiment from **P** positive instances and **N** negative instances for some condition.

Sensitivity

Sensitivity relates to the test's ability to identify a condition correctly. Consider the example of a medical test used to identify a disease. Sensitivity of the test is the proportion of people known to have the disease, who test positive for it. Mathematically, this can be expressed as:

$$\text{sensitivity} = \frac{\text{number of true positives}}{\text{number of true positives} + \text{number of false negatives}}$$

$$= \frac{\text{number of true positives}}{\text{total number of sick individuals in population}}$$

$$= \text{probability of a positive test, given that the patient is ill}$$

A negative result in a test with high sensitivity is useful for ruling out disease. A high sensitivity test is reliable when its result is negative, since it rarely misdiagnoses those who have the disease. A test with 100 per cent sensitivity will recognize all patients with the disease by testing positive. A negative test result would definitively *rule out* presence of the disease in a patient.

A positive result in a test with high sensitivity is not useful for ruling in disease. Suppose a 'bogus' test kit is designed to show only one reading, positive. When used on diseased patients, all patients test positive, giving the test 100 per cent sensitivity. However, sensitivity by definition does not take into account false positives. The bogus test also returns positive on all healthy patients, giving it a false positive rate of 100 per cent, rendering it useless for diagnosing or "ruling in" the disease.

Sensitivity is not the same as the precision or positive predictive value (ratio of true positives to combined true and false positives), which is as much a statement about the proportion of actual positives in the population being tested as it is about the test.

The calculation of sensitivity does not take into account indeterminate test results. If a test cannot be repeated, indeterminate samples either should be excluded from the analysis (the number of exclusions should be stated when quoting sensitivity) or can be treated as false negatives (which gives the worst-case value for sensitivity and may therefore underestimate it). A test with high sensitivity has a low type II error rate. In non-medical contexts, sensitivity is sometimes called recall.

Specificity

Specificity relates to the test's ability to exclude a condition correctly. Consider the example of a medical test for diagnosing a disease. Specificity of a test is the proportion of healthy patients known not to have the disease, who will test negative for it. Mathematically, this can also be written as:

$$specificity = \frac{number\ of\ true\ negatives}{number\ of\ true\ negatives + number\ of\ false\ positives}$$

$$= \frac{number\ of\ true\ negatives}{total\ number\ of\ well\ individuals\ in\ population}$$

$$= probability\ of\ a\ negative\ test\ given\ that\ the\ patient\ is\ well$$

Positive result in a test with high specificity is useful for ruling in disease. The test rarely gives positive results in healthy patients. A test with 100 per cent specificity will read negative and accurately exclude disease from all healthy patients. A positive result will highlight a high probability of the presence of disease.

Negative result in a test with high specificity is not useful for ruling out disease. Assume a 'bogus' test is designed to read only negative. This is administered to healthy patients and reads negative on all of them. This will give the test a specificity of 100 per cent. Specificity by definition does not take into account false negatives. The same test will also read negative on diseased patients; therefore it has a false negative rate of 100 per cent and will be useless for ruling out disease. A test with a high specificity has a low type I error rate.

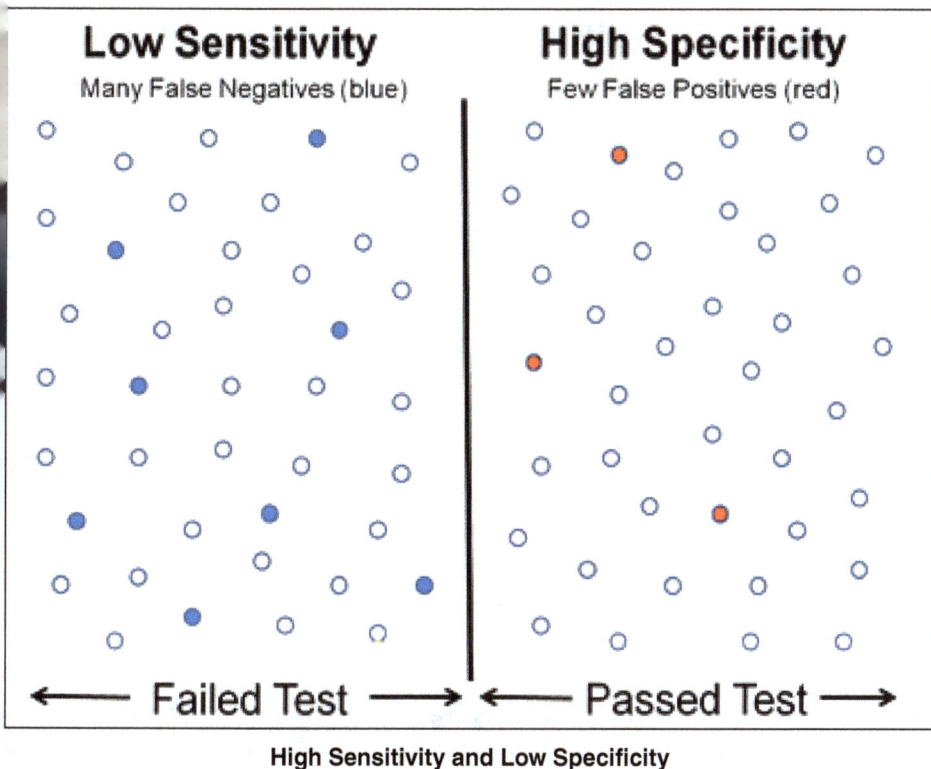

Low Sensitivity
Many False Negatives (blue)

High Specificity
Few False Positives (red)

← Failed Test → ← Passed Test →

High Sensitivity and Low Specificity

Medical Examples

In medical diagnostics, test sensitivity is the ability of a test to correctly identify those with the disease (true positive rate), whereas test specificity is the ability of the test to correctly identify those without the disease (true negative rate). If 100 patients known to have a disease were tested and 43 test positive, then the test has 43 per cent sensitivity. If 100 with no disease are tested and 96 return a negative result, then the test has 96 per cent specificity. Sensitivity and specificity are prevalence-independent test characteristics, as their values are intrinsic to the test and do not depend on the disease prevalence in the population of interest. Positive and negative predictive values, but not sensitivity or specificity, are values influenced by the prevalence of disease in the population that is being tested.

Misconceptions

It is often claimed that a highly specific test is effective at ruling in a disease when positive, while a highly sensitive test is deemed effective at ruling out a disease when negative. This has led to the widely used mnemonics SPIN and SNOUT, according to which a highly SPecific test, when Positive, rules IN disease (SP-P-IN) and a highly 'SeNsitive' test, when Negative rules OUT disease (SN-N-OUT). Both rules of thumb are, however, inferentially misleading, as the diagnostic power of any test is determined by both its sensitivity and its specificity.

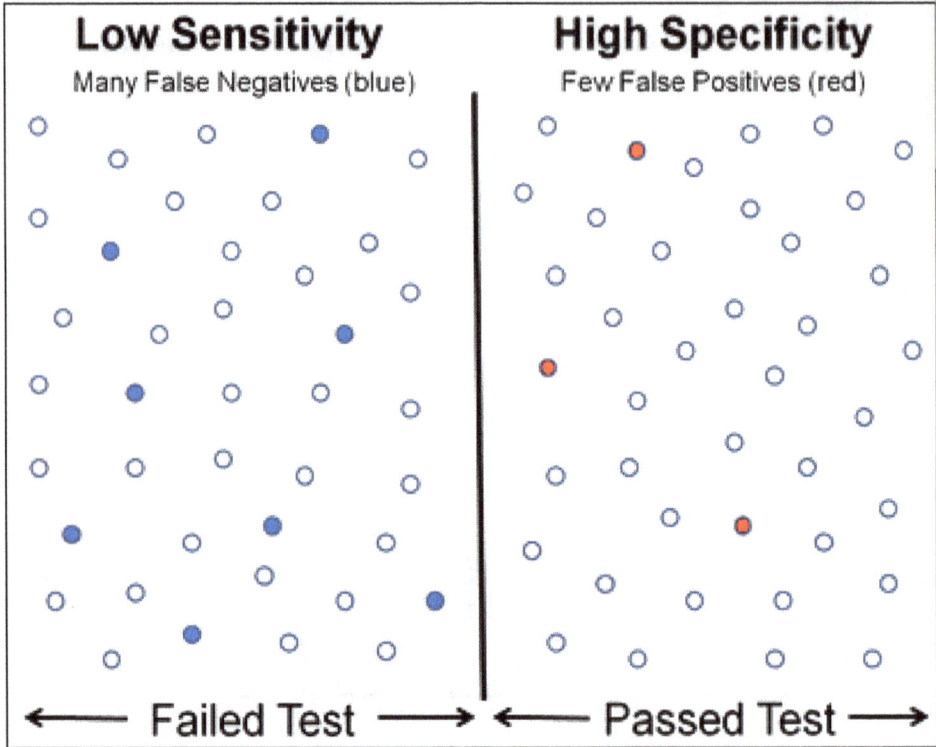

Low Sensitivity and High Specificity

Sensitivity Index

The sensitivity index or d' (pronounced 'dee-prime') is a statistic used in signal detection theory. It provides the separation between the means of the signal and the noise distributions, compared against the standard deviation of the noise distribution. For normally distributed signal and noise with mean and standard deviations μ_s and σ_s and μN and σ_N, respectively, d' is defined as:

$$d' = \frac{\mu_S - \mu_N}{\sqrt{\frac{1}{2}(\sigma_S^2 + \sigma_N^2)}}$$

An estimate of d' can be also found from measurements of the hit rate and false-alarm rate. It is calculated as:

$d' = Z(\text{hit rate}) - Z(\text{false alarm rate})$,

where function $Z(p)$, $p \in [0,1]$, is the inverse of the cumulative Gaussian distribution.

d' is a dimensionless statistic. A higher d' indicates that the signal can be more readily detected.

A Worked Example

A diagnostic test with sensitivity 67 per cent and specificity 91 per cent is applied to 2030 people to look for a disorder with a population prevalence of 1.48 per cent.

Patients with Bowel Cancer (As confirmed on endoscopy)

		Condition Positive	Condition Negative	
Fecal occult blood screen test outcome	Test outcome positive	**True positive** (TP) = 20	**False positive** (FP) = 180	Positive predictive value = TP/(TP + FP) = 20/(20 + 180) = **10 per cent**
	Test outcome negative	**False negative** (FN) = 10	**True negative** (TN) = 1820	Negative predictive value = TN/(FN + TN) = 1820/(10 + 1820) ≈ **99.5 per cent**
		Sensitivity = TP/(TP + FN) = 20/(20 + 10) ≈ **67 per cent**	**Specificity** = TN/(FP + TN) = 1820/(180 + 1820) = **91 per cent**	

Related Calculations

☆ False positive rate (α) = type I error = 1 – specificity = FP/(FP + TN) = 180/(180 + 1820) = 9 per cent

☆ False negative rate (β) = type II error = 1 – sensitivity = FN/(TP + FN) = 10/(20 + 10) = 33 per cent

☆ Power = sensitivity = 1 – β

☆ Likelihood ratio positive = sensitivity/(1 – specificity) = 66.67 per cent/ (1 – 91 per cent) = 7.4

☆ Likelihood ratio negative = (1 – sensitivity)/specificity = (1 – 66.67 per cent)/91 per cent = 0.37

Hence with large numbers of false positives and few false negatives, a positive screen test is in itself poor at confirming the disorder (PPV = 10 per cent) and further investigations must be undertaken; it did, however, correctly identify 66.7 per cent of all cases (the sensitivity). However as a screening test, a negative result is very good at reassuring that a patient does not have the disorder (NPV = 99.5 per cent) and at this initial screen correctly identifies 91 per cent of those who do not have cancer (the specificity).

Practical Aspects of Transgenic Plants Expressing α-amylase Inhibitors

The transgenic plant approach provides an attractive alternative to the use of chemical pesticides and insecticides and could contribute to the production of crop varieties that are inherently tolerant/resistant to their major target insect pests. Besides the benefit on agricultural crop production, the use of genes that encode insecticidal proteins in transgenic crops also has the potential to benefit the environment. The first reports of transgenic plants appeared in 1984 and since then there has been rapid progress using this new technology for crop improvement. Several different classes of plant proteins have been shown to be insecticidal towards a range of economically important insect pests when tested in artificial diets or transgenic plants. α-Amylase inhibitors show particular promise against bruchids of stored grains that depend to a large extent on α-amylase activity for survival.

The first practical demonstration involving α-amylase inhibitors used α-AI1, which specifically inhibits the α-amylases of the three Old World bruchids; the pea weevil *Bruchus pisorum*, the cowpea weevil and the azuki bean weevil. In transgenic pea plants, complete resistance against these bruchids was observed for α-AI1 levels in the range 0.8–1.0 per cent, with complete larval mortality of the first or second instars. Similar observations were made under field conditions. Similarly, azuki bean plants expressing α-AI1 were completely resistant to the azuki bean weevil. As α-AI2 is a much less effective inhibitor of pea weevil α-amylase, this inhibitor was only partially effective in protecting field-grown transgenic peas against pea weevils. Nevertheless, feeding tests carried out with artificial diets containing proteinaceous extracts of this α-AI2-expressing transgenic pea showed complete effectiveness against *Z. subfasciatus*. At an earlier stage of development, wheat inhibitors, as potent *in#$vitro*inhibitors of the gut hydrolytic enzymes from the larvae of seed storage weevils, are promising future weapons against these pests.

Just as important as the proof of protection of transgenic crops against pests is the demonstration that the new crops present no health risk to consumers. This appears to be less of a problem in the case of human consumption as such crops would be cooked before human consumption, with concomitant protein denaturation and inactivation. However, such crops may also be used as animal feed so that potential differences between uncooked normal and transgenic crops should be evaluated. A recent study has addressed this issue by feeding rats with transgenic peas expressing high levels of α-AI1 and monitoring possible effects on intestinal metabolism, growth and starch and protein digestibility. The minimal nutritional differences seen up to a dietary level of 300 g per kg of transgenic pea should encourage the use of transgenic crops as animal feed.

A further important factor affecting the practical development of transgenic plants expressing α-amylase inhibitors lies outside the purely scientific arena. The degree of social acceptance of transgenic crops depends on consumer reactions, which in turn depend on the social context of the transgenic technology. Although rigorous specifications have been established to ensure the safety of the transgenic products for human health and for the environment, severe consumer doubts remain about whether the transgenic technology is necessary or if it could causes human diseases. For example, if transgenic development is believed to be for the benefit of industry and not the consumer, public acceptance is likely to be lower. In addition, the genetic modification of animals is more acceptable if it is applied within a medical context than a food-related context. Bearing in mind these factors, it is essential to create an informed consumer who is able to make rational decisions regarding consumption of products and provide input into the strategic development of the science.

Different Structural Classes of α-amylase Inhibitors, Based on a Classification by Richardson

Structural Class	Source and References	Residue Numbers	Disulfide Bonds	CATH[b] Code and Family	SCOP Fold	Names
Legume lectin type	Common beans	240–250	5	2.60.120.60 [c]Lectin	Concanavalin A-like lectins/glucanases	α-A11 and α-A12
Knottin type	Amaranth	32	3	ND[a]	Knottins	AA1
Cereal type	Wheat barley and Indian finger millet	124–160	5	1.10.120.10 Cereal inhibitor	Bifunctional inhibitor/ lipid-transfer protein/ seed storage 2S albumin	0.19, 0.53, 0.28 WRP25, WRP26, WRP27 and RBI
Kunitz type	Barley, wheat and rice	176–181	1–2	2.80.10.50.6 Proteinase inhibitor	β-Trefoil	BASI, WASI and RASI
Thaumatin type	Maize	173–235	5–8	2.60.110.10 Sweet tasting protein	Osmotin Thaumatin-like proteins	Zeamatin
γ-Purothionin type	Sorghum	47–48	5	3.30.30.10 Antibacterial protein	Knottins	Slα1, Slα2 and Slα3

a: This inhibitor class was not classified by CATH program. b: Orengo *et al.* and 3 from CATH code represents mainly α helix, mainly β sheets and mixed α helix and β sheets, respectively.

Abbreviations

AAI: *Amaranthus*α-amylase inhibitor

α-AI1 and α-AI2: α-amylase inhibitors 1 and 2 from the common bean

AMY1 and AMY2: α-amylases from barley seeds

BASI: barley α-amylase subtilisin inhibitor

BLA: *Bacillus licheniformis*α-amylase

CAI: cowpea α-amylase inhibitor

CHFI: corn Hageman factor inhibitor

HAS: human salivary α-amylase

LCAI: *Lachrima jobi* chitinase/α-amylase inhibitor

PAI: pigeonpea α-amylase inhibitor

PPA: porcine pancreatic α-amylase

RASI: rice α-amylase/subtilisin inhibitor

RBI: ragi bifunctional inhibitor

SIα1: SIα2 and SIα3, *Sorghum* α-amylase inhibitors 1–3

TASI: triticale α-amylase/subtilisin inhibitor

TMA: *Tenebrio molitor* α-amylase

WASI: wheat α-amylase subtilisin inhibitor

ZSA: *Zabrotes subfasciatus* α-amylase.

References

Abal, M andreu, JM, Barasoain, I (2003). "Taxanes: microtubule and centrosome targets and cell cycle dependent mechanisms of action". *Current cancer drug targets* 3(3): 193–203.

Adam, GC, Cravatt, BF, Sorensen, EJ (2001). "Profiling the specific reactivity of the proteome with non-directed activity-based probes". *Chemistry and biology* 8 (1): 81–95.

Amoutzias G, Van de Peer Y (2010). "Single-Gene and Whole-Genome Duplications and the Evolution of Protein–Protein Interaction Networks. Evolutionary genomics and systems biology". pp. 413–429.

Baron, JA (1994). "Too bad it isn't true". *Medical decision making: an international journal of the Society for Medical Decision Making* 14 (2): 107.

Berg J., Tymoczko J. and Stryer L. (2002) *Biochemistry.* W. H. Freeman and Company, ISBN 0-7167-4955-6.

Bischoff, K (2001). "The toxicology of microcystin-LR: occurrence, toxicokinetics, toxicodynamics, diagnosis and treatment". *Veterinary and human toxicology* 43 (5): 294–7.

Bogoyevitch, MA, Barr, RK, Ketterman, AJ (2005). "Peptide inhibitors of protein kinases-discovery, characterisation and use". *Biochimica et Biophysica Acta* 1754 (1–2): 79–99.

Boyko, EJ (1994). "Ruling out or ruling in disease with the most sensitive or specific diagnostic test: short cut or wrong turn?". *Medical decision making : an international journal of the Society for Medical Decision Making* 14 (2): 175–179.

Brenner, G. M. (2000): *Pharmacology*. Philadelphia, PA: W.B. Saunders Company.

Brown KR, Jurisica I (2007). "Unequal evolutionary conservation of human protein interactions in interologous networks". *Genome Biol.* 8 (5): R95.

Cohen, J.A., Oosterbaan, R.A., Berends, F. (1967). "[81] Organophosphorus compounds". *Enzyme Structure*. Methods in Enzymology 11. p. 686.

Defrates, LJ, Hoehns, JD, Sakornbut, EL, Glascock, DG, Tew, AR (2005). "Antimuscarinic intoxication resulting from the ingestion of moonflower seeds". *The Annals of pharmacotherapy* 39 (1): 173–6.

Dixon, M. Webb, E.C., Thorne, C.J.R. and Tipton K.F., *Enzymes* (3rd edition) Longman, London (1979) p. 126

Duke SO (1990). "Overview of herbicide mechanisms of action". *Environ. Health Perspect.* (Brogan and #38) 87: 263 71.

Fawcelt, Tom (2006). "An Introduction to ROC Analysis". *Pattern Recognition Letters* 27 (8): 861–874.

Fischer PM (2003). "The design, synthesis and application of stereochemical and directional peptide isomers: a critical review". *Curr. Protein Pept. Sci.* 4 (5): 339–56.

Fraser, H. B., Plotkin, J. B. (2007). "Using protein complexes to predict phenotypic effects of gene mutation". *Genome Biology* 8 (11): R252.

Fuxreiter M (2012). "Fuzziness: linking regulation to protein dynamics". *Mol Biosyst* 8 (1): 168–77.

Fuxreiter M, Simon I, Bondos S (2011). "Dynamic protein-DNA recognition: beyond what can be seen". *Trends Biochem. Sci.* 36 (8): 415–23.

Gale, SD, Perkel, DJ (2010). "A basal ganglia pathway drives selective auditory responses in songbird dopaminergic neurons via disinhibition". *The Journal of neuroscience : the official journal of the Society for Neuroscience* 30 (3): 1027–1037.

Glen RC, Allen SC (2003). "Ligand-protein docking: cancer research at the interface between biology and chemistry". *Curr. Med. Chem.* 10 (9): 763–7.

Gohlke H, Klebe G (2002). "Approaches to the description and prediction of the binding affinity of small-molecule ligands to macromolecular receptors". *Angew. Chem. Int. Ed. Engl.* 41 (15): 2644–76.

Hart, G. T., Lee, I, Marcotte, E. R. (2007). "A high-accuracy consensus map of yeast protein complexes reveals modular nature of gene essentiality". *BMC Bioinformatics* 8: 236.

Hartley, MR, Lord, JM (2004). "Cytotoxic ribosome-inactivating lectins from plants". *Biochimica et Biophysica Acta* 1701 (1–2): 1–14.

Hartley, RW (1989). "Barnase and barstar: two small proteins to fold and fit together".*Trends in Biochemical Sciences* 14 (11): 450–4.

Hartwell LH, Hopfield JJ, Leibler S, Murray AW (1999). "From molecular to modular cell biology". *Nature* 402 (6761 Suppl): C47–52.

Hashimoto K, Nishi H, Bryant S, Panchenko AR (2011). "Caught in self-interaction: evolutionary and functional mechanisms of protein homooligomerization". *Phys Biol* 8 (3): 035007.

Holdgate, GA (2001). "Making cool drugs hot: isothermal titration calorimetry as a tool to study binding energetics". *BioTechniques* 31 (1): 164–6, 168, 170.

Holmes, CF, Maynes, JT, Perreault, KR, Dawson, JF, James, MN (2002). "Molecular enzymology underlying regulation of protein phosphatase-1 by natural toxins". *Current medicinal chemistry* 9 (22): 1981–9.

Hostettmann, K., Borloz, A., Urbain, A., Marston, A. (2006). "Natural Product Inhibitors of Acetylcholinesterase". *Current Organic Chemistry* 10 (8): 825.

Hsu, JT, Wang, HC, Chen, GW, Shih, SR (2006). "Antiviral drug discovery targeting to viral proteases". *Current pharmaceutical design* 12 (11): 1301–14.

Irwin H. Segel, *Enzyme Kinetics : Behavior and Analysis of Rapid Equilibrium and Steady-State Enzyme Systems.* Wiley–Interscience, New edition (1993), ISBN 0-471-30309-7.

Jeong, H, Mason, S. P., Barabási, A. L., Oltvai, Z. N. (2001). "Lethality and centrality in protein networks". *Nature* 411 (6833): 41–2.

Katz, AH, Caufield, CE (2003). "Structure-based design approaches to cell wall biosynthesis inhibitors". *Current pharmaceutical design* 9 (11): 857–66.

König V, Vértesy L, Schneider TR (2003). "Structure of the alpha-amylase inhibitor tendamistat at 0.93 A". *Acta Crystallogr. D Biol. Crystallogr.* 59 (Pt 10): 1737–43.

Koppitz M, Eis K (2006). "Automated medicinal chemistry". *Drug Discov. Today* 11(11–12): 561–8.

Lage, K, Karlberg, E. O., Størling, Z. M., Olason, P. I., Pedersen, A. G., Rigina, O, Hinsby, A. M., Tümer, Z, Pociot, F, Tommerup, N, Moreau, Y, Brunak, S (2007). "A human phenome-interactome network of protein complexes implicated in genetic disorders".*Nature Biotechnology* 25 (3): 309–16.

Leatherbarrow, RJ (1990). "Using linear and non-linear regression to fit biochemical data". *Trends in Biochemical Sciences* 15 (12): 455–8.

Lew W, Chen X, Kim CU (2000). "Discovery and development of GS 4104 (oseltamivir): an orally active influenza neuraminidase inhibitor". *Curr. Med. Chem.* 7 (6): 663–72.

Loo JA, DeJohn DE, Du P, Stevenson TI, Ogorzalek Loo RR (1999). "Application of mass spectrometry for target identification and characterization". *Med Res Rev* 19 (4): 307–19.

Lundblad R. L. (2004). *Chemical Reagents for Protein Modification* CRC Press Inc. ISBN 0-8493-1983-8.

Macmillan, Neil A., Creelman, C. Douglas (2004). *Detection Theory: A User's Guide.* Psychology Press. p. 7.

Maggi, M, Filippi, S, Ledda, F, Magini, A, Forti, G (2000). "Erectile dysfunction: from biochemical pharmacology to advances in medical therapy". *European Journal of Endocrinology* 143 (2): 143–54.

Maurer, T, Fung, HL (2000). "Comparison of methods for analyzing kinetic data from mechanism-based enzyme inactivation: application to nitric oxide synthase". *AAPS pharmSci* 2 (1): 68–77.

McGuire, JJ (2003). "Anticancer antifolates: current status and future directions". *Current pharmaceutical design* 9 (31): 2593–613.

Nooren IM, Thornton JM (2003). "Diversity of protein-protein interactions". *EMBO J.* 22 (14): 3486–92.

Okar, DA, Lange, AJ (1999). "Fructose-2,6-bisphosphate and control of carbohydrate metabolism in eukaryotes". *BioFactors (Oxford, England)* 10 (1): 1–14.

Oliver, CJ, Shenolikar, S (1998). "Physiologic importance of protein phosphatase inhibitors". *Frontiers in bioscience: a journal and virtual library* 3: D961–72.

Oti, M, Brunner, H. G. (2007). "The modular nature of genetic diseases". *Clinical Genetics* 71 (1): 1–11.

Pereira-Leal JB, Levy ED, Teichmann SA (2006). "The origins and evolution of functional modules: lessons from protein complexes". *Philos. Trans. R. Soc. Lond., B, Biol. Sci.* 361 (1467): 507–17.

Pewsner, D, Battaglia, M, Minder, C, Marx, A, Bucher, HC, Egger, M (2004)."Ruling a diagnosis in or out with "SpPIn" and "SnNOut": a note of caution". *BMJ (Clinical research ed.)* 329 (7459): 209–13.

Pick, FM, McGartoll, MA, Bray, RC (1971). "Reaction of formaldehyde and of methanol with xanthine oxidase". *European journal of biochemistry/FEBS* 18 (1): 65–72.

Poulin, R, Lu, L, Ackermann, B, Bey, P, Pegg, AE (1992). "Mechanism of the irreversible inactivation of mouse ornithine decarboxylase by alpha-difluoromethylornithine. Characterization of sequences at the inhibitor and coenzyme binding sites". *The Journal of Biological Chemistry* 267 (1): 150–8.

Price NC, Stevens L (1999). *Fundamentals of enzymology: The cell and molecular biology of catalytic protein.* Oxford, New York: Oxford University Press.

Price, N., Hames, B. and Rickwood, D. (eds.) (1996) *Proteins LabFax* Academic Press.

Price, Nicholas and Stevens, Lewis (1999) *Fundamentals of Enzymology*, Oxford University Press.

Raicu V, Stoneman MR, Fung R, Melnichuk M, Jansma DB, Pisterzi LF, Rath S, Fox, M, Wells, JW, Saldin DK (2008). "Determination of supramolecular structure and spatial distribution of protein complexes in living cells.". *Nature Photonics* 3: 107–113.

Reardon, JE (1989). "Herpes simplex virus type 1 and human DNA polymerase interactions with 2'-deoxyguanosine 5'-triphosphate analogues. Kinetics of incorporation into DNA and induction of inhibition". *The Journal of Biological Chemistry* 264 (32): 19039–44.

Ryan, C. J., Krogan, N. J., Cunningham, P, Cagney, G (2013). "All or nothing: Protein complexes flip essentiality between distantly related eukaryotes". *Genome Biology and Evolution* 5 (6): 1049–59.

Saravanamuthu, A, Vickers, TJ, Bond, CS, Peterson, MR, Hunter, WN, Fairlamb, AH (2004). "Two interacting binding sites for quinacrine derivatives in the active site of trypanothione reductase: a template for drug design". *The Journal of Biological Chemistry* 279 (28): 29493–500.

Scapin G (2006). "Structural biology and drug discovery". *Curr. Pharm. Des.* 12 (17): 2087–97.

Segel, Irwin H. (1993) *Enzyme Kinetics : Behavior and Analysis of Rapid Equilibrium and Steady-State Enzyme Systems*. Wiley-Interscience, New edition, ISBN 0-471-30309-7.

Shapiro, R, Vallee, BL (1991). "Interaction of human placental ribonuclease with placental ribonuclease inhibitor". *Biochemistry* 30 (8): 2246–55.

Smyth, TP (2004). "Substrate variants versus transition state analogues as noncovalent reversible enzyme inhibitors". *Bioorganic and Medicinal Chemistry* 12 (15): 4081–8.

Stone, SR, Morrison, JF (1986). "Mechanism of inhibition of dihydrofolate reductases from bacterial and vertebrate sources by various classes of folate analogues". *Biochimica et Biophysica Acta* 869 (3): 275–85.

Szedlacsek, SE, Duggleby, RG (1995). "Kinetics of slow and tight-binding inhibitors". *Enzyme Kinetics and Mechanism Part D: Developments in Enzyme Dynamics. Methods in Enzymology* 249. pp. 144–80.

Tan S, Evans R, Singh B (2006). "Herbicidal inhibitors of amino acid biosynthesis and herbicide-tolerant crops". *Amino Acids* 30 (2): 195–204.

Tan, G, Gyllenhaal, C, Soejarto, DD (2006). "Biodiversity as a source of anticancer drugs". *Current drug targets* 7 (3): 265–77.

Tompa P, Fuxreiter M (2008). "Fuzzy complexes: polymorphism and structural disorder in protein-protein interactions". *Trends Biochem. Sci.* 33 (1): 2–8.

Tseng, SJ, Hsu, JP (1990). "A comparison of the parameter estimating procedures for the Michaelis-Menten model". *Journal of Theoretical Biology* 145 (4): 457–64.

Vetter, J (1998). "Toxins of Amanita phalloides". *Toxicon* 36 (1): 13–24.

Walsh, R., Martin, E., Darvesh, S. (2007). "A versatile equation to describe reversible enzyme inhibition and activation kinetics: Modeling β-galactosidase and butyrylcholinesterase". *Biochimica et Biophysica Acta (BBA) - General Subjects* 1770(5): 733–746.

Walsh, R., Martin, E., Darvesh, S. (2011). "Limitations of conventional inhibitor classifications". *Integrative Biology* (Royal Society of Chemistry) 3 (12): 1197–1201.

Walsh, Ryan (2012). "Ch. 17. Alternative Perspectives of Enzyme Kinetic Modeling". In Ekinci, Deniz. *Medicinal Chemistry and Drug Design*. InTech. pp. 357–371. ISBN 978-953-51-0513-8.

Wang, H, Kakaradov, B, Collins, S. R., Karotki, L, Fiedler, D, Shales, M, Shokat, K. M., Walther, T. C., Krogan, N. J., Koller, D (2009). "A complex-based reconstruction of the Saccharomyces cerevisiae interactome". *Molecular and Cellular Proteomics* 8 (6): 1361–81.

Yu, H, Braun, P, Yildirim, M. A., Lemmens, I, Venkatesan, K, Sahalie, J, Hirozane-Kishikawa, T, Gebreab, F, Li, N, Simonis, N, Hao, T, Rual, J. F., Dricot, A, Vazquez, A, Murray, R. R., Simon, C, Tardivo, L, Tam, S, Svrzikapa, N, Fan, C, De Smet, A. S., Motyl, A, Hudson, M. E., Park, J, Xin, X, Cusick, M. E., Moore, T, Boone, C, Snyder, M, Roth, F. P. (2008). "High-quality binary protein interaction map of the yeast interactome network".*Science* 322 (5898): 104–10.

Zotenko, E, Mestre, J, O'Leary, D. P., Przytycka, T. M. (2008). "Why do hubs in the yeast protein interaction network tend to be essential: Reexamining the connection between the network topology and essentiality". *PLoS Computational Biology* 4 (8): e1000140.

Chapter 4
BOAA (Lathyrus)

Lathyrism or neurolathyrism is a neurological disease of humans and domestic animals, caused by eating certain legumes of thegenus *Lathyrus*. This problem is mainly associated with *Lathyrus sativus* (also known as *Grass pea, Kesari Dhal, Khesari Dhal* or *Almorta*) and to a lesser degree with *Lathyrus cicera, Lathyrus ochrus* and *Lathyrus clymenum* containing the toxin ODAP. The lathyrism resulting from the ingestion of *Lathyrus odoratus* seeds (*sweet peas*) is often referred to as odoratism or osteolathyrism, which is caused by a different toxin (beta-aminopropionitrile) that affects the linking of collagen, a protein of connective tissues.

Symptoms

The consumption of large quantities of *Lathyrus* grain containing high concentrations of the glutamate analogue neurotoxin β-oxalyl-L-α,β-diaminopropionic acid (ODAP, also known as β-N-oxalyl-amino-L-alanine, or BOAA) causes paralysis, characterized by lack of strength in or inability to move the lower limbs and may involve pyramidal tracts producing signs of upper motor neuron damage. The toxin may also cause aortic aneurysm. A unique symptom of lathyrism is the atrophy ofgluteal muscles (buttocks). ODAP is a poison of mitochondria leading to excess cell death, especially in motor neurons.

Prevalence

This disease is prevalent in some areas of Bangladesh, Ethiopia, India and Nepal and affects more men than women, because the gene for G6PD deficiency is carried on the X chromosome, the condition is much more common in men.

Causes

The toxicological cause of the disease has been attributed to the neurotoxin ODAP which acts as a structural analogue of theneurotransmitter glutamate. Ingestion of legumes containing the toxin results mostly from ignorance of their

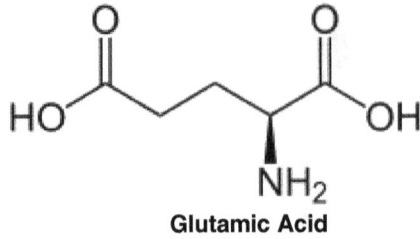

Glutamic Acid

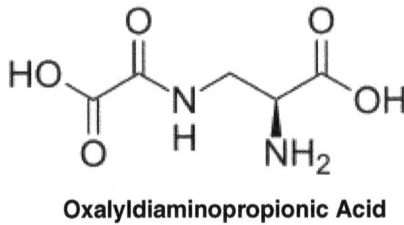

Oxalyldiaminopropionic Acid

toxicity and usually occurs where the despair of poverty and malnutrition leaves few other food options. Lathyrism can also be caused by food adulteration.

Prevention

Recent research suggests that sulfur amino acids have a protective effect against the toxicity of ODAP. Food preparation is also an important factor. Toxic amino acids are readily soluble in water and can be leached. Bacterial (lactic acid) and fungal (tempeh) fermentation is useful to reduce ODAP content. Moist heat (boiling, steaming) denatures protease inhibitors which otherwise add to the toxic effect of raw grasspea through depletion of protective sulfur amino acids. The underlying cause for excessive consumption of grasspea is a lack of alternative food sources. This is a consequence of poverty and political conflict. The prevention of lathyrism is therefore a socio-economic challenge.

Historical Occurrence

The first mentioned intoxication goes back to ancient India and also Hippocrates mentions a neurological disorder 46 B.C. in Greece caused by Lathyrus seed. Lathyrism was occurring on a regular basis. During the Spanish War of Independence against Napoleon, grasspea served as a famine food. This was the subject of one of Francisco de Goya's famous aquatint prints titled *Gracias a la Almorta* ("Thanks to the Grasspea"), depicting poor people surviving on a porridge made from grasspea flour, one of them lying on the floor, already crippled by it.

During WWII, on the order of Colonel I. Murgescu, commandant of the Vapniarka camp in Transnistria, the detainees - most of them Jews - were fed nearly exclusively with fodder pea; consequently they got ill of "lathyrisms". In the film *Ashes* [English title] by Andrzej Wajda based on the novel *Popioly* [Polish title] translated as *Lost army* [English title] by Stefan eromski spanning the period 1798–1812, a horse is poisoned by grain from a Spanish village. The footage of the horse losing control of its hind legs suggests that it was fed with Almortas.

Modern Occurrence

During the post-war period in Spain, there were several outbreaks of lathyrism, caused by the shortage of food, which led people to consume excessive amounts of Almorta flour. In Spain, a seed mixture known as comuña consisting of *Lathyrus sativus*, *L. cicera*, *Vicia sativa* and *V. ervilia* provides a potent mixture of toxic amino acids to poison monogastric (single stomached) animals. Particularly the toxin beta-cyanoalanine from seeds of *V. sativa* enhances the toxicity of such a mixture through its inhibition of sulfur amino acid metabolism [conversion of methionine to cysteine leading to excretion of cystathionine in urine] and hence depletion of protective reduced thiols. Its use for sheep does not pose any lathyrism problems if doses do not exceed 50 percent of the ration. Ronald Hamilton suggested in his paper *The Silent Fire: ODAP and the death of Christopher McCandless* that Christopher McCandless may have died from starvation after being unable to hunt or gather food due to lathyrism induced paralysis of his legs caused by eating the seeds of *Hedysarum alpinum*. In 2014, a lab analysis indicated that the seeds did contain ODAP.

Related Conditions

A related disease has been identified and named osteolathyrism, because it affects the bones and connecting tissues, instead of the nervous system. It is a skeletal disorder, caused by the toxin beta-aminopropionitrile (BAPN) and characterized by hernias, aortic dissection, exostoses and kyphoscoliosis and other skeletal deformities, apparently as the result of defective aging of collagen tissue. The cause of this disease is attributed to beta-aminopropionitrile, which inhibits the copper-containing enzyme lysyl oxidase, responsible for cross-linking procollagen and proelastin. BAPN is also a metabolic product of a compound present in sprouts of grasspea, pea and lentils. A disorder that is clinically similar is konzo.

Human Neurolathyrism

The term lathyrism was first introduced by a Neapolitan physician, Arnolda Cantani in 1874. The term 'neurolathyrism' was first coined by Selye (1957) who distinguished between neurolathyrism and osteolathyrism. Neurolathyrism is defined as an upper motor-neuron disease producing cortico-spinal dysfunction following excessive consumption of the grass pea (*L. sativus*) seeds. Lathyrism is an irreversible spastic paraparesis which results when two-third of the diet is made up of the seeds of *L.sativus*.

Kessler (1947) gave the best documented evidence of Lathyrism in human beings. He observed the outbreak of lathyrism in about 800 political prisoners in a World War II camp in a Ukrainian town near the Russian Roumanian border. They were fed with cooked *L. sativus* seeds as the staple diet. Within two months they developed leg weakness, muscle spasm and signs of vascular insufficiency; and finally individuals were bed ridden. From this he concluded that the crippling disease was due to the excessive consumption of the seeds of *L. sativus*.

Lathyrism is characterised by signs of muscular rigidity, weakness of the leg muscles and death in extreme cases. It causes irreversible paralysis of the lower limbs which may appear suddenly or insidiously. Symptoms range from inability to walk to complete paralysis of the lower limbs and in extreme cases death. In mild cases there is a slight bending of the knees and difficulty in running. In some cases the victim walks with bent knees. In advanced cases a crossed gait develops with a tendency to walk on the toes. Such patients are forced to use sticks for support. These symptoms are indicative of the degeneration of cortico-spinal nerve fibers which under normal conditions regulate leg movement.

The disease affects more males than females in a population. People under 40 and the low income sections of the population are reported to be more susceptible to the disease. Roldon *et al.* (1994) observed that the patients aged between 49-70 showed a characteristic pattern of neurological deficit and deterioration of motor performance (increased leg stiffness shortening of steps and tendency to scrape the floor with toes) which gradually developed over 2-10 years. This was often accompanied by painful nocturnal calf muscle cramping. Cohn (1994) has believed that lathyrism is not a longetive effect. Haque (1990) has reported that after reaching its peak the symptoms of lathyrism remain static.The neurotoxic action of β-ODAP has been demonstrated in animal species such as young rats and pigeons, young squirrel monkeys, (*Saimiri sciureus*) and mice. β -ODAP produces flaccid paraplegia in Rhesus monkeys (*Macaca mullata*) when administered intrathecally. Prolonged feeding of *L. sativus* seeds to undernourished Rhesus monkey combined with daily oral doses of *L. sativus* extract proved to be an effective method of inducing paralytic disorders.

Neurological Disorder

A neurological disorder is any disorder of the body nervous system. Structural, biochemical or electrical abnormalities in the brain, spinal cord or other nerves can result in a range of symptoms. Examples of symptoms include paralysis, muscle weakness, poor coordination, loss of sensation, seizures, confusion, pain and altered levels of consciousness. There are many recognized neurological disorders, some relatively common, but many rare. They may be assessed by neurological examination and studied and treated within the specialties of neurology and clinical neuropsychology. Interventions for neurological disorders include preventative measures, lifestyle changes, physiotherapy or other therapy, neurorehabilitation, pain management, medication, or operations performed by neurosurgeons. The World Health Organization estimated in 2006 that neurological disorders and their sequelae (direct consequences) affect as many as one billion people worldwide and identified health inequalities and social stigma/discrimination as major factors contributing to the associated disability and suffering.

Causes

Although the brain and spinal cord are surrounded by tough membranes, enclosed in the bones of the skull and spinal vertebrae and chemically isolated by the so-called blood–brain barrier, they are very susceptible if compromised. Nerves tend to lie deep under the skin but can still become exposed to damage. Individual neurons and the neural networks and nerves into which they form, are susceptible to electrochemical and structural disruption. Neuroregeneration may

```
        ┌─────────────┐
        │     APP     │
        └──────┬──────┘
               │
               ▼
   ┌──────────────────────────┐
   │  Errors in APP Processing │
   └──────────┬───────────────┘
              │
              ▼
   ┌───────────────┐      ┌──────────────────────┐
   │  Soluble Aβ   │─────▶│  Fibrillar Aβ plaques │
   └───────┬───────┘      └──────────┬───────────┘
           │                         │
           ▼                         ▼
   ┌───────────────┐      ┌──────────────────────┐
   │     ↓LTP      │─────▶│   Neurodegeneration   │
   └───────┬───────┘      └──────────┬───────────┘
           │                         │
           ▼                         ▼
┌──────────────────────┐  ┌──────────────────────┐
│ Late cognitive decline│  │ Early cognitive decline│
└──────────────────────┘  └──────────────────────┘
```

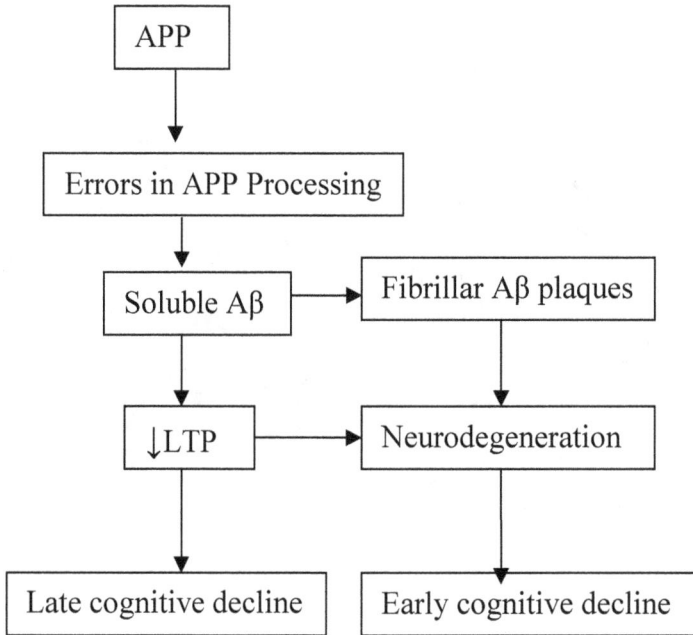

Part of the Causal Chain Leading to Alzheimer's Disease

occur in the peripheral nervous system and thus overcome or work around injuries to some extent; it is thought to be rare in the brain and spinal cord. The specific causes of neurological problems vary, but can include genetic disorders, congenital abnormalities or disorders, infections, lifestyle or environmental health problems including malnutrition and brain injury, spinal cord injury or nerve injury. The problem may start in another body system that interacts with the nervous system. For example, cerebrovascular disorders involve brain injury due to problems with the blood vessels(cardiovascular system) supplying the brain; autoimmune disorders involve damage caused by the body's own immune system; lysosomal storage diseases such as Niemann-Pick disease can lead to neurological deterioration. In a substantial minority of cases of neurological symptoms, no neural cause can be identified using current testing procedures and such "idiopathic" conditions can invite different theories about what is occurring.

Classification

Neurological disorders can be categorized according to the primary location affected, the primary type of dysfunction involved, or the primary type of cause. The broadest division is between central nervous system disorders and peripheral nervous system disorders. TheMerck Manual lists brain, spinal cord and nerve disorders in the following overlapping categories:

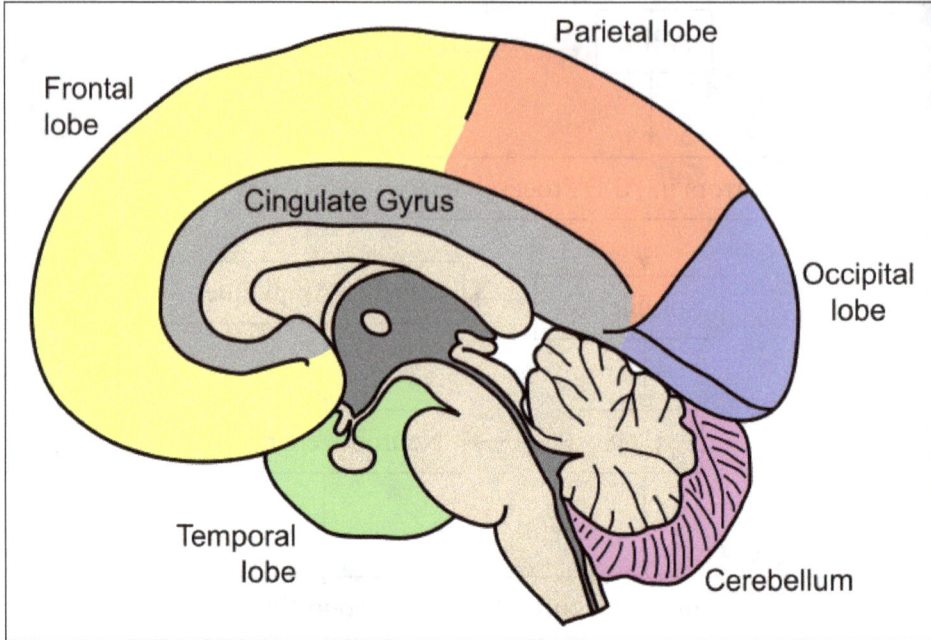

Anatomy of the Human Brain

Anatomical Terminology

☆ Brain:

- ☆ Brain damage according to cerebral lobe (see also 'lower' brain areas such as basal ganglia, cerebellum, brainstem):
  - ☆ Frontal lobe damage
  - ☆ Parietal lobe damage
  - ☆ Temporal lobe damage
  - ☆ Occipital lobe damage
- ☆ Brain dysfunction according to type:
  - ☆ Aphasia (language)
  - ☆ Dysgraphia (writing)
  - ☆ Dysarthria (speech)
  - ☆ Apraxia (patterns or sequences of movements)
  - ☆ Agnosia (identifying things or people)
  - ☆ Amnesia (memory)

☆ Spinal cord disorders (see spinal pathology, injury, inflammation)

☆ Peripheral neuropathy and other Peripheral nervous system disorders

☆ Cranial nerve disorder such as Trigeminal neuralgia

☆ Autonomic nervous system disorders such as dysautonomia, Multiple System Atrophy

Nervous System

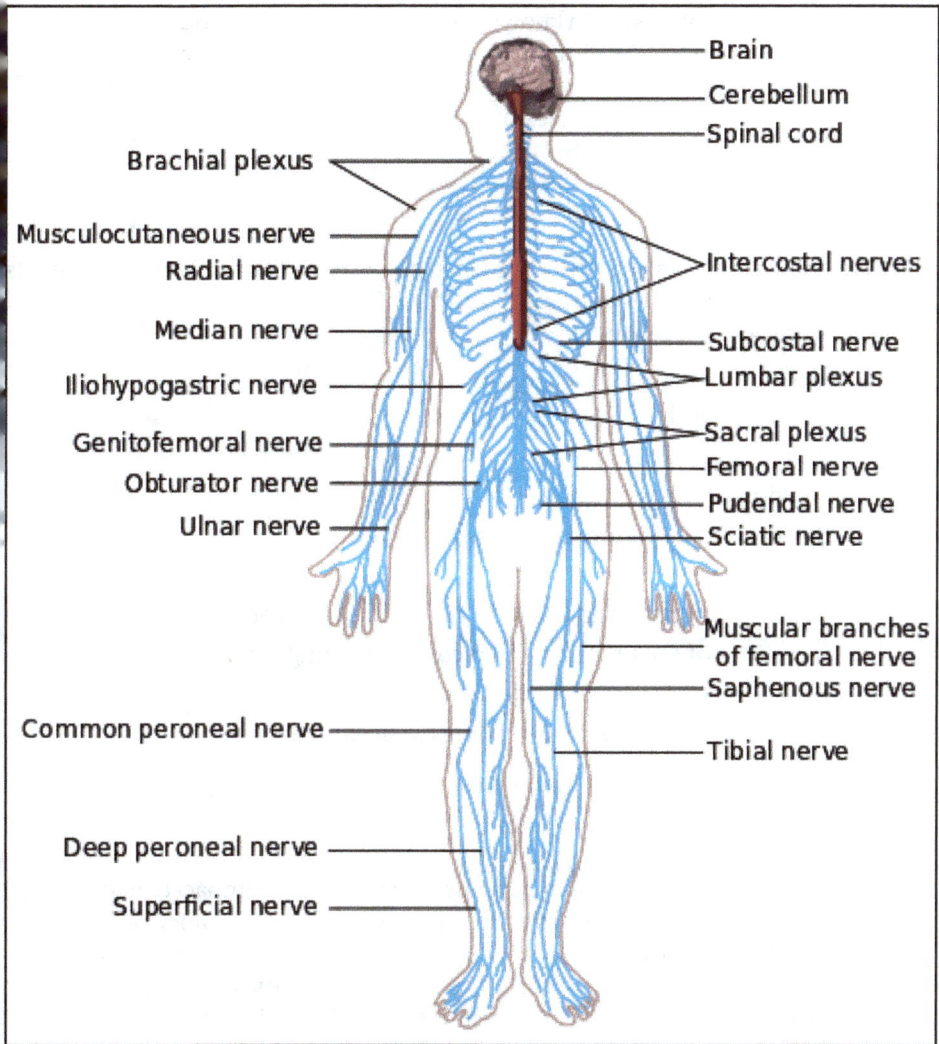

The Human Nervous System

☆ Seizure disorders such as epilepsy

☆ Movement disorders of the central and peripheral nervous system such as Parkinson's disease, Essential tremor, Amyotrophic lateral sclerosis, Tourette's Syndrome, Multiple Sclerosis and various types of Peripheral Neuropathy

☆ Sleep disorders such as Narcolepsy

☆ Migraines and other types of Headache such as Cluster Headache and Tension Headache

☆ Lower back and neck pain (see Back pain)

☆ Central Neuropathy (see Neuropathic pain)

☆ Neuropsychiatric illnesses (diseases and/or disorders with psychiatric features associated with known nervous system injury, underdevelopment, biochemical, anatomical, or electrical malfunction and/or disease pathology e.g. Attention deficit hyperactivity disorder, Autism, Tourette's Syndrome and some cases of Obsessive compulsive disorder as well as the neurobehavioral associated symptoms of degeneratives of the nervous system such as Parkinson's disease, Essential tremor, Huntington's disease, Alzheimer's disease, Multiple sclerosis and organic psychosis.)

Many of the diseases and disorders listed above have neurosurgical treatments available (*e.g.* Tourette's Syndrome, Parkinson's disease, Essential tremor and Obsessive compulsive disorder).

☆ Delirium and dementia such as Alzheimer's disease

☆ Dizziness and vertigo

☆ Stupor and coma

☆ Head injury

☆ Stroke (CVA, cerebrovascular attack)

☆ Tumors of the nervous system (e.g. cancer)

☆ Multiple sclerosis and other demyelinating diseases

☆ Infections of the brain or spinal cord (including meningitis)

☆ Prion diseases (a type of infectious agent)

☆ Complex regional pain syndrome (a chronic pain condition)

Neurological disorders in non-human animals are treated by veterinarians.

Mental Functioning

A neurological examination can to some extent assess the impact of neurological damage and disease on brain function in terms of behavior, memory or cognition. Behavioral neurology specializes in this area. In addition, clinical neuropsychology uses neuropsychological assessment to precisely identify and track problems in mental functioning, usually after some sort of brain injury or neurological impairment. Alternatively, a condition might first be detected through the presence of abnormalities in mental functioning, and further assessment may indicate an underlying neurological disorder. There are sometimes unclear boundaries in the distinction between disorders treated within neurology and mental disorders treated within the other medical specialty of psychiatry, or other mental health professions such as clinical psychology. In practice, cases may present as one type but be assessed as more appropriate to the other. Neuropsychiatry deals with mental disorders arising from specific identified diseases of the nervous system.

One area that can be contested is in cases of idiopathic neurological symptoms - conditions where the cause cannot be established. It can be decided in some cases, perhaps by exclusion of any accepted diagnosis, that higher-level brain/mental

Some of the Fields that Contribute to Understanding Mental Functioning

activity is causing symptoms, rather than the symptoms originating in the area of the nervous system from which they may appear to originate. Classic examples are "functional" seizures, sensory numbness, "functional" limb weakness and functional neurological deficit ("functional" in this context is usually contrasted with the old term "organic disease"). Such cases may be contentiously interpreted as being "psychological" rather than "neurological". Some cases may be classified as mental disorders, for example as conversion disorder, if the symptoms appear to be causally linked to emotional states or responses to social stress or social contexts. On the other hand, dissociation refer to partial or complete disruption of the integration of a person's conscious functioning, such that a person may feel detached from one's emotions, body and/or immediate surroundings. At one extreme this may be diagnosed as depersonalization disorder. There are also conditions viewed as neurological where a person appears to consciously register neurological stimuli that cannot possibly be coming from the part of the nervous system to which they would normally be attributed, such as phantom pain or synesthesia, or where limbs act without conscious direction, as in alien hand syndrome. Theories and assumptions about consciousness, free will, moral responsibility and social stigma can play a part in this, whether from the perspective of the clinician or the patient.

Conditions that are classed as mental disorders, or learning disabilities and forms of Intellectual disability, are not themselves usually dealt with as neurological disorders. Biological psychiatry seeks to understand mental disorders in terms of their basis in the nervous system, however. In clinical practice, mental disorders are usually indicated by a mental state examination, or other type of structured interview or questionnaire process. At the present time, neuroimaging (brain scans) alone cannot accurately diagnose a mental disorder or tell the risk of developing one; however, it can be used to rule out other medical conditions such as a brain tumor. In research, neuroimaging and other neurological tests can show correlations between reported and observed mental difficulties and certain aspects of neural function or differences in brain structure. In general, numerous fields intersect to try and understand the basic processes involved in mental functioning, many of which are brought together in cognitive science. The distinction between neurological and mental disorders can be a matter of some debate, either in regard to specific facts about the cause of a condition or in regard to the general understanding of brain and mind. Moveover, the definition of disorder in medicine or psychology is sometimes contested in terms of what is considered abnormal, dysfunctional, harmful or unnatural in neurological, evolutionary, psychometric or social terms.

Osteolathyrism

Osteolathyrism is a collagen cross-linking deficiency brought on by dietary over-reliance on the seeds of *Lathyrus sativus* or grass pea, a legume often grown as a famine crop in Asia and East Africa. Other members of the genus are also known to cause the disease, including *L. sylvestris*, *L. cicera* and *L. clymenum*. *L. sativus* grows well under famine conditions, often severe drought, where it is cultivated. The condition results in damage to bone and mesenchymal connective tissues. It is seen in people in combination with neurolathyrism and angiolathyrism in areas where famine demands reliance on a crop with known detrimental effects. It occurs in cattle and horses with diets over reliant upon the grass pea. Osteolathyrism is caused by a variety of osteolathyrogenic compounds, specifically excitatory amino-compounds. The most widely-studied of these compounds is beta-aminopropionitrile (BAPN), which exerts its deleterious effect by an unknown yet potently irreversible mechanism. Other instigators are ureides, semicarbazides andthiosemicarbazides, which are believed to chelate the prosthetic Cu(II)-bipyridine cofactor complex in the enzyme lysyl oxidase.

Cause

Lysyl oxidase is an important enzyme for the creation of crosslinks between collagen triple-helices in connective tissue. By oxidizing the terminal amino group of lysine, analdehyde is created. This aldehyde can undergo several reactions with neighboring aldehydes or amines to create strong covalent cross-links between collagen tertiary structuresin bone and cartilage. The main product of these reactions is the aldimine compound dehydrohydroxylysinonorleucine. This unique crosslink can be formed by the Schiff basemechanism in which the lone pair of electrons on a primary amine reacts with the carbonyl carbon of an aldehyde. Other crosslinks include the formation of α, β-unsaturated ketone via aldol condensation and hydroxylysinonorleucine. If these crosslinks are not formed, as in the case of osteolathyrism, the synthesis of strong mesenchymal and mesodermal tissue is inhibited. Symptoms of osteolathyrism include weakness and fragility of connective tissue (*i.e.*, skin, bones and blood vessels (angiolathyrism) and the paralysis of the lower extremities associated with neurolathyrism. For these reasons, compounds containing lathyrogens should be avoided during pregnancy and growth of a child. Aminopropionitrile, also known as β-aminopropionitrile (BAPN), is an organic compound with both amine and nitrile functional groups. It is an antirheumatic agent in veterinary medicine. It can cause osteolathyrism, neurolathyrism and/or angiolathyrism. Aminopropionitrile is prepared by the reaction of ammonia with acrylonitrile. BAPN is a component of *lathyrus odoratus*.

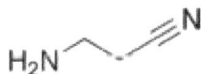

3-Aminopropanenitrile

Lathyrus is a genus of flowering plant species known as sweet peas and vetchlings. *Lathyrus* is in the legume family, Fabaceae and contains approximately 160 species. They are native to temperate areas, with a breakdown of 52 species in Europe, 30 species in North America, 78 in Asia, 24 in tropical East Africa and 24 in temperate South America. There are annual and perennialspecies which may be climbing or bushy. This genus has numerous sections, including *Orobus*, which was once a separate genus.

Uses

Many species are cultivated as garden plants. The genus includes the garden sweet pea (*Lathyrus odoratus*) and the perennial everlasting pea (*Lathyrus latifolius*). Flowers on these cultivated species may be rose, red, maroon, pink and white, yellow, purple or blue and some are bicoloured. They are also grown for their fragrance. Cultivated species are susceptible to fungal infections including downy and powdery mildew. Other species are grown for food, including the Indian pea (*L. sativus*) and the red pea (*L. cicera*) and less commonly Cyprus-vetch (*L. ochrus*) and Spanish vetchling (*L. clymenum*). The tuberous pea (*L. tuberosus*) is grown as a root vegetable for its starchy edible tuber. The seeds of some *Lathyrus* species contain the toxic amino acid oxalyldiaminopropionic acid and if eaten in large quantities can causelathyrism, a serious disease.

Oxalyldiaminopropionic Acid

Oxalyldiaminopropionic acid (ODAP), a structural analogue of the neurotransmitter glutamate, is the neurotoxin responsible for lathyrism.

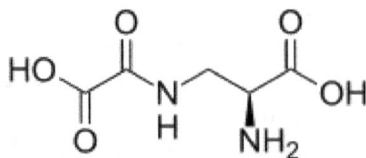

3-[(Carboxycarbonyl) amino] Alanine

Plants belonging to the *Lathyrus* species are known to be sometimes toxic to man and animal due to the presence of large amounts of lathyrogenic compounds, especially under severe drought conditions. Both the vegetative parts and seeds of *Lathyrus* species contain unusual ninhydrin-reactive toxic amino acids; these are nitrogenous secondary metabolites responsible for osteolathyrism and neurolathyrism. The toxic γ-glutamyl derivative of β-aminopropionitrile (BAPN) and β-N-oxalylamino-L-alanine (BOAA) are present in *Lathyrus* species. It has been shown that β-(γ-glutamyl) aminopropionitrile is mostly responsible for the inhibition of synthesis of desmosine and isodesmosine which results in failure of cross-linking between the polypeptide chains in elastin and presumably in collagen and ultimately osteolathyrism (O'Dell *et al.*, 1966). Bell and O'Donovan (1966) and Roy and Narasinga Rao (1968) have shown that BOAA (β-isomer) exists naturally in an isomeric mixture with the α-isomer. This formation could be due to the migration of oxalyl moiety from the β-amino to the α-amino group of BOAA. β-isomer is the main neurotoxic component of *Lathyrus sativus* seed, while α-isomer has been shown to be less toxic to experimental animals. The proportion of α- to β-isomers in *Lathyrus sativus* seeds is approximately 5:95. There are different nomenclatures suggested for the β-isomer of the neurotoxin from *Lathyrus sativus*. These are β-N-oxalylamino-L-alanine (BOAA), β-N-oxalylamino-α, β-diaminopropionic acid (Ox-dapro or ODAP) and L-3-oxalylamino-2-aminopropionic acid (OAP). The α-isomer of the neurotoxin has been referred to as the α-isomer of Noxalylamino-α, β-diaminopropionic acid and as L-2-oxalylamino-3- aminopropionic acid. The α-isomer of the neurotoxin has been shown to be produced in the plant by a nontoxic rearrangement. Rao *et al.* (1964) have reported that β-form of N-oxalylamino-L-alanine (BOAA) is responsible for the neurotoxicity (Neurolathyrism) in humans. Neurolathyrism is characterized by such symptoms as muscular rigidity, weakness and paralysis of the leg muscles and death in extreme cases upon excessive consumption of *Lathyrus* seeds especially those grown under drought conditions. In most recorded cases, the onset of the disease is sudden. It has generally been concluded from the nature of the sysmptoms that the disease primarily affects the central nervous system. The content of these lathyrogenic compounds in *Lathyrus* species depends on cultivar, geographical area and climatic conditions, with cultivar having the most important effect. The content of β-N-oxalylamino-L-alanine in *Lathyrus sativus* seeds ranges from 0.22 to 11.00 g/kg.

Biosynthesis

Naturally-occurring neurotoxic amino acids (Neurolathyrogens) such as β-cyanoalanine, β-N-oxalylamino-Lalanine and osteolathyrogens, β-aminopropionitrile and β-(γ-glutamyl) aminopropionitrile are biosynthetically

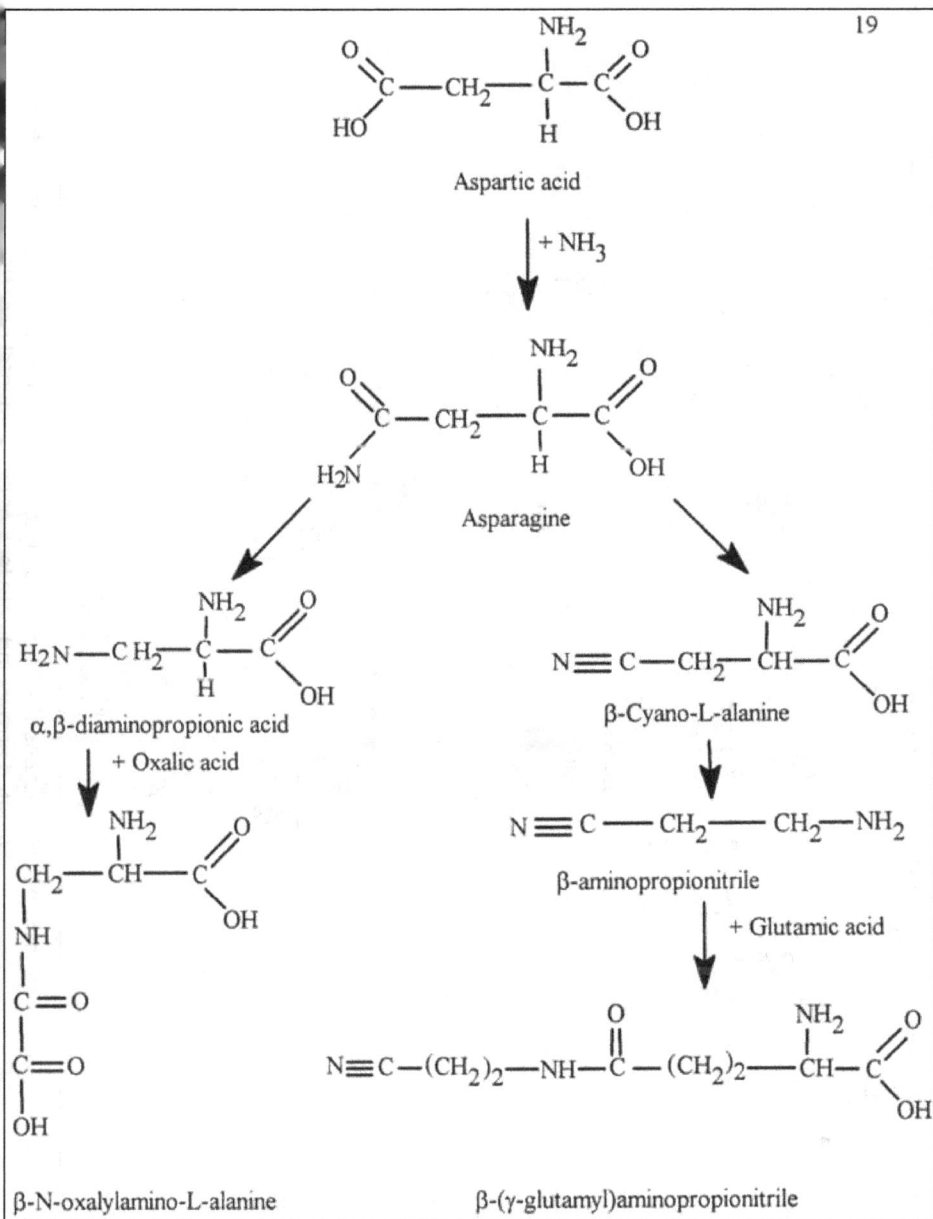

Possible Pathways for Biosynthesis of β–N-oxalylamino-L-alanine and β–(γ-glutamyl) Aminopropionitrile

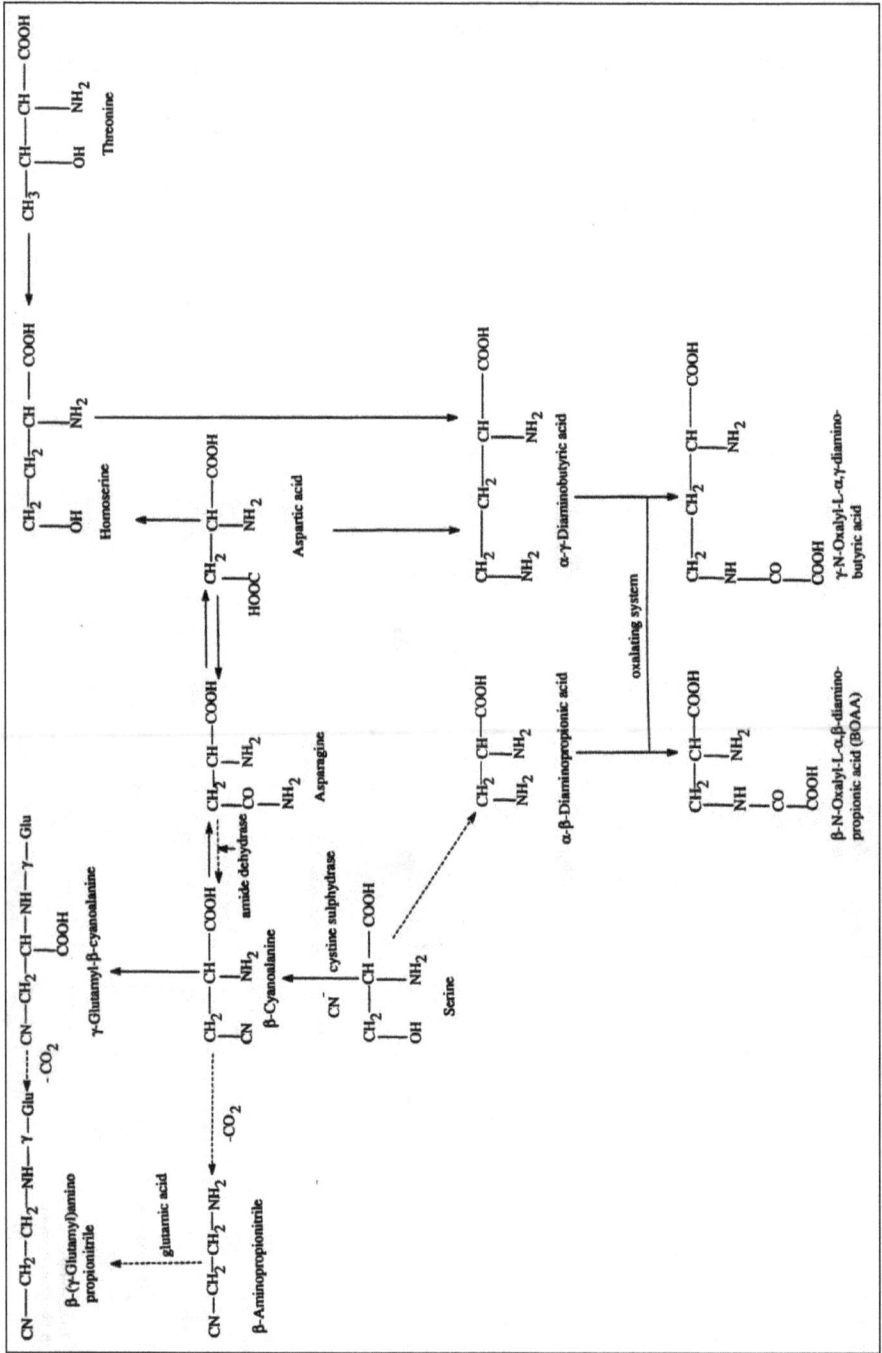

Biosynthesis of Lathyrogens [Solid arrows indicate experimentally confirmed pathways; dashed arrows indicate lack of experimental proof]

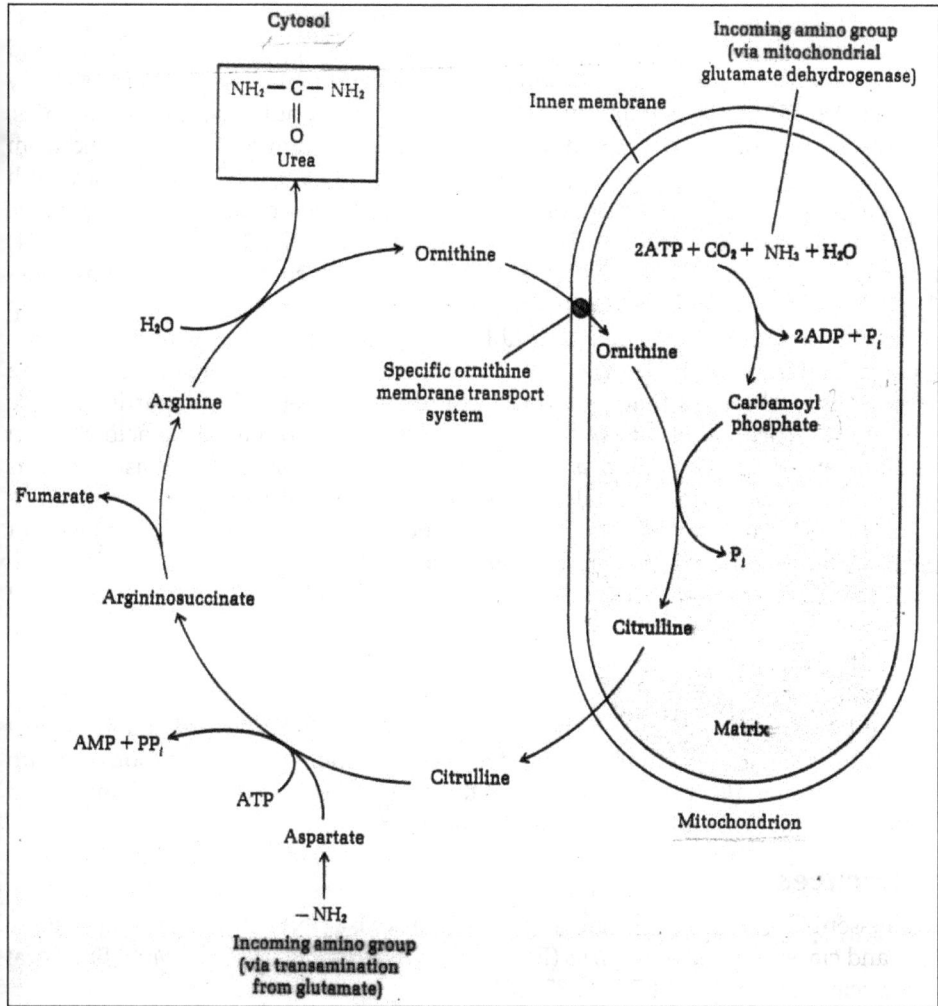

Exchange of Amino Acids between Cytosol and Mitochondria.

inter-related. The biosynthetic pathways of some toxic amino acids and nitriles in *Lathyrus* plants have been postulated by Murti *et al.* (1964) and Sarma and Padmanaban (1969). Nigam and Ressler (1964) have shown that labelled serine incorporation into the dipeptide [β-(γ-glutamyl) aminopropionitrile] in *Vicia sativa* and *Lathyrus* species is possible only in the presence of cyanide. The incorporation of cyanide into serine is catalyzed by an enzyme prepared by *Vicia sativa* and also from *Lathyrus* species. It has been suggested that β-cyanoalanine may be formed through non-specific enzyme reactions catalyzed by cysteine sulphydrase. Ressler *et al.* (1961) proposed that β-cyanoalanine itself may be derived from asparagine by a hypothetical enzyme, "amide dehyrase", although the evidence obtained so far has confirmed only the reverse pathway. Ressler *et al.* (1961) suggested that α, β-diaminobutyric acid may be derived from β-cyanoalanine by another hypothetical enzyme, "nitrile reductase". The pathways from asparagine to β-cyano-L-alanine and the formation of α, γ-diaminobutyric acid from β-cyano-L-alanine were suggested by Ressler *et al.* (1961); subsequently this was supported by autoradiographic studies. It is resonable to assume that α, β-diaminopropionic acid may also be formed from the same source, the –CONH2 group being replaced by a –NH2 by analogy with the well-known Hofmann reaction. The subsequent step of acylation with oxalic acid is an acceptable biochemical process (Figure, Murti *et al.*, 1964). Kuo *et al.* (1994) reported that the incorporation of [^{14}C]-label from the precursor β-(isoxazolin-5-one-2-yl)-L-alanine into β-BOAA occurs in intact fruits as well as in separated pericarp and immature seeds, in both high and low toxic varieties of *Lathyrus sativus*. This incorporation gradually decreases in the pericarp while increasing in the maturing seeds. It was observed that L-(3H) homoserine and DL-(1-14C) aspartic acid are efficiently incorporated into α, γ-diaminobutyric acid from aspartic acid (Nigam and Ressler, 1966). Ikegami *et al.* (1993) reported that cysteine synthase which is present in *Lathyrus sativus* catalyses the formation of β-(isoxazolin-5-one-2-yl)-L-alanine (BIA), the biosynthetic precursor of the neurotoxin BOAA and some other heterocyclic β-substituted alanines from *O*-acetyl-L-serine (OAS) as an additional catalytic activity.

Biological Significance

Non-protein amino acids (such as neurotoxic and osteotoxic amino acids) can be potent toxicants able to benefit the plant by affording protection against predation and disease and by improving its competition for habitat resources with other plants. Simola (1967) reported that the presence of an uncommon amino acid in the genus *Lathyrus* may prevent hybridization.

References

Asmussen, C. B and A. Liston. (1998). "Chloroplast DNA characters, phylogeny and classification of *Lathyrus* (Fabaceae)". *American Journal of Botany* (Botanical Society of America) 85 (3): 387–401.

AZCOYTIA, Carlos (2006). "Historia de la Almorta or el veneno que llegó con el hambre tras la Guerra Civil Española". "HistoriaCocina"". Historiacocina.com. Retrieved 2013-09-23.

Bailey AJ, Peach CM (1971). "The Chemistry of the Collagen Cross-Links: The Absence of Reduction of Dehydrolysinonorleucine and Dehydrohydroxylysinonorleucine *in Vivo*". *The Biochemical Journal* 121: 257-259.

Barrow, M. V. (1974). "Lathyrism: A Review". *The Quarterly Review of Biology* 49 (2): 101–128.

Bell, E. A. and O'Donovan, P. (1966). The isolation of a and âoxalyl derivatives of a, b-iaminobutyric acid from seeds of Lathyrus latifolius and the detection of the a-oxalyl isomer of the neurotoxin a-amino-â-oxalyl amino propionic acid which occurs together with the neurotoxin in this and other species. *Phytochemistry.* 5: 1211–9.

COHN, D.F. (1995) "Are other systems apart from the nervous system involved in human lathyrism?" in *Lathyrus sativus and Human Lathyrism: Progress and Prospects*. Ed. Yusuf H, Lambein F. University of Dhaka. Dhaka pp. 101-2.

Dawson DA, Rinaldi AC, Poch G (2002). "Biochemical and toxicological evaluation of agent-cofactor reactivity as a mechanism of action for osteolathyrism". *Toxicology* 177 (2-3): 267-284.

Fred, Edwin Broun, Baldwin, Ira Lawrence, McCoy, Elizabeth (1932). *Root Nodule Bacteria and Leguminous Plants*. UW-Madison Libraries Parallel Press. p. 142. ISBN 1-893311-28-7.

GRIN Species Records of *Lathyrus*. Germplasm Resources Information Network (GRIN).

Hernández Bermejo, J. E. and J. León (1994). *Neglected crops 1492 from a different perspective*. Rome: Food and Agriculture Organization of the United Nations. ISBN 92-5-103217-3.

http://www.newyorker.com/books/page-turner/how-chris-mccandless-died

Ikegami, F., Ongena, G., Sakai, R., Itagaki, S., Kobori, M. Ishikawa, T., Kuo, Y.H., Lambein, F. and Murakoshi, I.1993. Biosynthesis of β-(Isoxazolin-5-on-2-yl)-L-alanine by cysteine synthase in *Lathyrus sativus. Phytochemistry.* 33: 93-98.

Kuo, Y.H., Khan, J.K. and Lambein, F. (1994). Biosynthesis of the neurotoxin b-ODAP in developing pods of *Lathyrus sativus. Phytochemistry.* 35: 911-913.

Lathyrism. Egton Medical Information Systems Limited.

Mark V. Barrow, Charles F. Simpson, Edward J. Miller (1974). "Lathyrism: A Review".*The Quarterly Review of Biology* 49 (2): 101–128.

Medical problems caused by plants: Lathyrism" at *Prince Leopold Institute of Tropical Medicine* online database

Murti, V.V.S., Seshadri, T.R. and Vankitasubramanian, T.A. (1964). Neurotoxic compounds of the seeds of *Lathyrus sativus*. Phytochemistry. 3: 73-78.

Nigam, S.N. and Ressler, C. (1964). Biosynthesis in *Vicia sativa* (common vetch) of g-glutamyl-b-cyanoalanine from (b-^{14}C) serine and its relation to cyanide metabolism. *Biochem. Biophys. Acta.* 93: 339-345.

O'Dell, B.L., Elsden, D.F., Thomas, J., Partridge, S.M., Smith, R.H. and Palmer, R. (1966). Inhibition of the biosynthesis of the cross-links in elastin by a Lathyrogen. Nature. 209: 401-404.

Rao, S.L.N., Adiga, P.R. and Sarma, P.S. (1964). Isolation and characterisation of b-oxalyl-L-b-diaminopropionic acid: a neurotoxin from the seeds of *Lthyrus sativus*. Biochem. 3: 432-436.

Ravindranath, V. (2002). "Neurolathyrism: Mitochondrial dysfunction in excitotoxicity mediated by L-β-oxalyl aminoalanine".*Neurochemistry International* 40 (6): 505–509.

Ressler, C., Redstone, P.A. and Erenberg, R.H. (1961). Isolation and identification of neuroactive factor from *Lathyrus latifolius*. Sci. 134: 188-190.

Ronald Hamilton (2013). "ODAP and the death of Christopher McCandless".

Rosenthal, Gerald (2003). "Toxic Constituents and their Related Metabolites". *Plant Nonprotein Amino and Imino Acids: Biological, Biochemical and Toxicological Properties*. Elsevier. ISBN 9780323157742.

Roy, D.N. and Narasinga Rao, B.S. (1968). Distribution of a- and b-isomer of N-oxalyl-a,-diaminopropionic acid in some varieties of *Lathyrus sativus*. *Curr. Sci.* 37: 395-399.

Sarma, P.S. and Padmanaban, G. (1969). *In Toxic Constituents of Plant Foodstuffs*. I.E. Liener (Ed.), p. 267-291. Academic Press. New York, NY.

Simola, L. K. (1967). The effect of some non-protein amino acids on pollen germination and pollen-tube growth in five species of the Vicieae. *Planta*. 77: 287-290.

Spencer P. S., Ludolph A. C., Kisby G. E. (1993). "Neurologic diseases associated with use of plant components with toxic potential". *Environmental Research* 62 (1): 106–113.

Sriram K., Shankar S.K., Boyd M.R., Ravindranath V. (1998). "Thiol Oxidation and Loss of Mitochondrial Complex I Precede Excitatory Amino Acid-Mediated Neurodegeneration". *The Journal of Neuroscience* 18 (24): 10287–10296.

Sunset Western Garden Book (1995), p. 606–607

The etymological origin of this name is from "común" (*common*) in its meaning of mixture, referring to the mix of seeds obtained when cleaning the grain and which contaminate the main grain, generally wheat.

The Holocaust in Romania Under the Antonescu Government by Marcu Rozen -http://isurvived.org/

William Howlett (2012). *Neurology in Africa*. pp. 248–249.

Wilmarth KR, Froines JR (1992). "*In vitro* and *In vivo* Inhibition of Lysyl Oxidase by Aminopropionitriles". *Journal of Toxicology and Environmental Health* 37 (3): 411-423.

Chapter 5

Carotenoids

Introduction

Carotenoids are organic pigments that are found in the chloroplasts and chromoplasts of plants and some other photosyntheticorganisms, including some bacteria and some fungi. Carotenoids can be produced from fats and other basic organic metabolic building blocks by all these organisms. Carotenoids can also be sourced from various animal tissues, as is often the case involving food grade additives and within the dietary supplement industry in the United States. Carniverous animal life sources Carotenoids from the consumption of animal tissue. The only animal known to produce Carotenoids is the aphid. There are over 600 known carotenoids; they are split into two classes, xanthophylls (which contain oxygen) and carotenes (which are purely hydrocarbons and contain no oxygen). All carotenoids are tetraterpenoids, meaning that they are produced from 8 isoprenemolecules and contain 40 carbon atoms. In general, carotenoids absorb wavelengths ranging from 400-550 nanometers (violet to green light). They serve two key roles in plants and algae: they absorb light energy for use in photosynthesis and they protect chlorophyll from photodamage. In humans, three carotenoids (beta-carotene, alpha-carotene and beta-cryptoxanthin) have vitamin A activity (meaning that they can be converted to retinal) and these and other carotenoids can also act as antioxidants. In the eye, certain other carotenoids (lutein, astaxanthin and zeaxanthin) apparently act directly to absorb damaging blue and near-ultraviolet light, in order to protect themacula of the retina, the part of the eye with the sharpest vision. People consuming diets rich in carotenoids from natural foods, such as fruits and vegetables, are healthier and have lower mortality from a number of chronic illnesses. Although a recent meta-analysis of 68 reliable antioxidant supplementation experiments involving a total of 232,606 individuals concluded additional β-carotene from supplements is unlikely to be beneficial and may actually be harmful, this may be due to the inclusion of studies involving smokers - β-carotene under intense oxidative stress (*e.g.* induced by heavy smoking) gives breakdown products that reduce plasma vitamin A and worsen the lung cell proliferation induced by smoke. With the notable exception of the gac fruit and crude palm oil, most carotenoid-rich fruits and vegetables are low in lipids. Since dietary lipids have been hypothesized to be an important factor for carotenoid bioavailability, a 2005 study investigated whether addition of avocado fruit or oil, as lipid sources, would enhance carotenoid absorption in humans. The study found that the addition of either avocado fruit or oil significantly enhanced the subjects' absorption of all carotenoids tested (α-carotene, β-carotene, lycopene and lutein).

Biosynthesis

Properties

Carotenoids belong to the category of tetraterpenoids (*i.e.*, they contain 40 carbon atoms, being built from four terpene units each containing 10 carbon atoms). Structurally, carotenoids take the form of a polyene hydrocarbon chain which is sometimes terminated by rings and may or may not have additional oxygen atoms attached.

☆ Carotenoids with molecules containing oxygen, such as lutein and zeaxanthin, are known as xanthophylls.

☆ The unoxygenated (oxygen free) carotenoids such as α-carotene, β-carotene and lycopene, are known as carotenes. Carotenes typically contain only carbon and hydrogen (*i.e.*, are hydrocarbons) and are in the subclass of unsaturated hydrocarbons.

Probably the most well-known carotenoid is the one that gives this second group its name, carotene, found in carrots (also apricots) and are responsible for their bright orange colour. Crude palm oil, however, is the richest source of carotenoids in nature in terms of retinol (provitamin A) equivalent. Vietnamese Gac fruit contains the highest known concentration of the carotenoid lycopene. Their colour, ranging from pale yellow through bright orange to deep red, is directly linked to their structure. Xanthophylls are often yellow, hence their class name. The double carbon-carbon bonds interact with each other in a process called conjugation, which allows electrons in the molecule to move freely across these areas of the molecule. As the number of conjugated double bonds increases, electrons associated with conjugated systems have more room to move and require less energy to change states. This causes the range of energies of light absorbed by the molecule to decrease. As more frequencies of light are absorbed from the short end of the visible spectrum, the compounds acquire an increasingly red appearance. Carotenoids are usually lipophilic due to the presence of long unsaturated aliphatic chains as in some fatty acids. The physiological absorption of these fat-soluble vitamins in humans and other organisms depends directly on the presence of fats and bile salts.

Physiological Effects

In oxygenic photosynthetic organisms, specifically flora and cyanobacteria, the carotenoid β-carotene plays a vital role in the photosynthetic reaction centre where, due to quantum mechanical reasons arising from the symmetry of the molecule, it provides a mechanism for photoprotection against auto-oxidation. They also participate in the energy-transfer process. In non-photosynthesizing organisms, such as humans, carotenoids have been linked to oxidation-preventing mechanisms. Carotenoids have many physiological functions. Given their structure, carotenoids are efficient free-radical scavengers and they enhance the vertebrate immune system. There are several dozen carotenoids in foods people consume and most carotenoids have antioxidant activity. Epidemiological studies have shown that people with high β-carotene intake and high plasma levels of β-carotene have a significantly reduced risk of lung cancer. However, studies of supplementation with large doses of β-carotene in smokers have shown an increase in cancer risk (possibly because β-carotene under intense oxidative stress (*e.g.* induced by heavy smoking) gives breakdown products that reduce plasmavitamin A and worsen the lung cell proliferation induced by smoke). Similar results have been found in other animals. Humans and animals are mostly incapable of synthesizing carotenoids and must obtain them through their diet. The notable exception is the red pea aphid, which has the genesnecessary for synthesizing carotenoids, thought to have been acquired from fungi via horizontal gene transfer. Carotenoids are a common and often ornamental feature in animals. For example, the pink colour of flamingos and salmon and the red colouring of cooked lobsters are due to carotenoids. It has been proposed that carotenoids are used in ornamental traits (for extreme examples see puffin birds) because, given their physiological and chemical properties, they can be used as honest indicators of individual health and hence they can be used by animals when selecting potential mates. In the macula lutea of the human eye, certain carotenoids are actively concentrated to the point that they cause a yellow

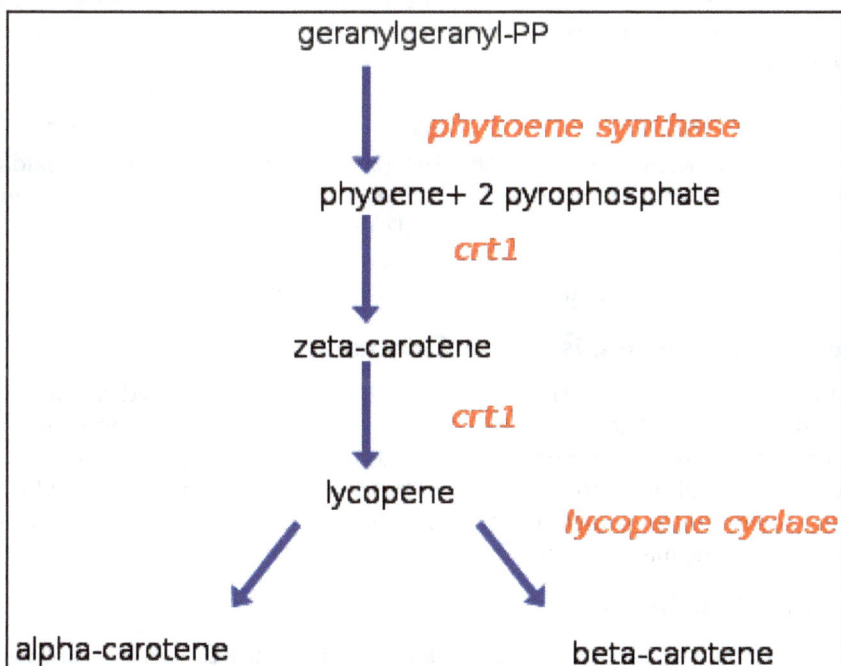

Simplified Carotenoid Synthesis Pathway

colouring and this may help to protect the retina from blue and actinic light, in the same way that carotenoids protect the photosystems of plants. Carotenoids are also actively concentrated in the corpus luteum of the ovaries, where they impart the characteristic colour and may act as general antioxidants.

The most common carotenoids include lycopene and the vitamin A precursor β-carotene. In plants, the xanthophyll lutein is the most abundant carotenoid and its role in preventing age-related eye disease is currently under investigation. Lutein and the other carotenoid pigments found in mature leaves are often not obvious because of the masking presence of chlorophyll. When chlorophyll is not present, as in young foliage and also dying deciduous foliage (such as autumn leaves), the yellows, reds and oranges of the carotenoids are predominant. For the same reason, carotenoid colours often predominate in ripe fruit (*e.g.*, oranges, tomatoes, bananas), after being unmasked by the disappearance of chlorophyll. However, the reds, the purples and their blended combinations that decorate autumn foliage usually come from another group of pigments in the cells called anthocyanins. Unlike the carotenoids, these pigments are not present in the leaf throughout the growing season, but are actively produced towards the end of summer.

Aroma Chemicals

Products of carotenoid degradation such as ionones, damascones and damascenones are also important fragrance chemicals that are used extensively in the perfumes and fragrance industry. Both β-damascenone and β-ionone although low in concentration in rose distillates are the key odour-contributing compounds in flowers. In fact, the sweet floral smells present in black tea, aged tobacco, grape and many fruits are due to the aromatic compounds resulting from carotenoid breakdown.

Disease

Some carotenoids are produced by bacteria to protect themselves from oxidative immune attack. The golden pigment that gives some strains of *Staphylococcus aureus* their name (*aureusis* = golden) is a carotenoid called staphyloxanthin. This carotenoid is a virulence factor with an antioxidant action that helps the microbe evade death by reactive oxygen species used by the host immune system.

Question of Synthesis in the Corpus Luteum

Following a 1968 report that beta-carotene was synthesized in laboratory conditions in slices of corpus luteum from cows, an organ known to concentrate beta-carotene (hence its colour and name), attempts have been made to replicate these findings, but have not succeeded. The idea is not presently accepted by the scientific community. Rather, the mammalian corpus luteum, like the macula lutea in the retina of the mammalian eye, merely concentrates carotenoids from the diet.

Artificial Synthesis

Microorganisms can be genetically modified to produce certain C40 carotenoids, including lycopene and beta carotene.

Naturally Occurring Carotenoids

Hydrocarbons

☆ Lycopersene 7,8,11,12,15,7',8',11',12',15'-Decahydro-γ,γ-carotene

☆ Phytofluene

☆ Hexahydrolycopene 15-*cis*-7,8,11,12,7',8'-Hexahydro-γ,γ-carotene

☆ Torulene 3',4'-Didehydro-β,γ-carotene

☆ α-Zeacarotene 7',8'-Dihydro-ε,γ-carotene

Alcohols

☆ Alloxanthin

☆ Cynthiaxanthin

☆ Pectenoxanthin

☆ Cryptomonaxanthin (3R,3'R)-7,8,7',8'-Tetradehydro-β,β-carotene-3,3'-diol

☆ Crustaxanthin β,-Carotene-3,4,3',4'-tetrol

☆ Gazaniaxanthin (3R)-5'-cis-β,γ-Carotene-3-ol

☆ OH-Chlorobactene 1',2'-Dihydro-f,γ-carotene-1'-ol

☆ Loroxanthin β,ε-Carotene-3,19,3'-triol

☆ Lutein (3R,32R,62R)-β,ε-carotene-3,3-diol

☆ Lycoxanthin γ,γ-Carotene-16-ol

☆ Rhodopin 1,2-Dihydro-γ,γ-carotene-l-ol

☆ Rhodopinol aka Warmingol 13-*cis*-1,2-Dihydro-γ,γ-carotene-1,20-diol

☆ Saproxanthin 3',4'-Didehydro-1',2'-dihydro-β,γ-carotene-3,1'-diol

☆ Zeaxanthin

Glycosides

☆ Oscillaxanthin 2,2'-Bis(β-L-rhamnopyranosyloxy)-3,4,3',4'-tetradehydro-1,2,1',2'-tetrahydro-γ,γ-carotene-1,1'-diol

☆ Phleixanthophyll 1'-(β-D-Glucopyranosyloxy)-3',4'-didehydro-1',2'-dihydro-β,γ-carotene-2'-ol

Ethers

☆ Rhodovibrin 1'-Methoxy-3',4'-didehydro-1,2,1',2'-tetrahydro-γ,γ-carotene-1-ol

☆ Spheroidene 1-Methoxy-3,4-didehydro-1,2,7',8'-tetrahydro-γ,γ-carotene

Epoxides

☆ Diadinoxanthin 5,6-Epoxy-7',8'-didehydro-5,6-dihydro–carotene-3,3-diol

☆ Luteoxanthin 5,6: 5',8'-Diepoxy-5,6,5',8'-tetrahydro-β,β-carotene-3,3-diol

☆ Mutatoxanthin

☆ Citroxanthin

☆ Zeaxanthin furanoxide 5,8-Epoxy-5,8-dihydro-β,β-carotene-3,3'-diol

☆ Neochrome 5',8'-Epoxy-6,7-didehydro-5,6,5',8'-tetrahydro-β,β-carotene-3,5,3'-triol

☆ Foliachrome

☆ Trollichrome

☆ Vaucheriaxanthin 5',6'-Epoxy-6,7-didehydro-5,6,5',6'-tetrahydro-β,β-carotene-3,5,19,3'-tetrol

Aldehydes

☆ Rhodopinal

☆ Warmingone 13-cis-1-Hydroxy-1,2-dihydro-γ,γ-carotene-20-al

☆ Torularhodinaldehyde 3',4'-Didehydro-β,γ-carotene-16'-al

Acids and Acid Esters

☆ Torularhodin 3',4'-Didehydro-β,γ-carotene-16'-oic acid

☆ Torularhodin methyl ester Methyl 3',4'-didehydro-β,γ-carotene-16'-oate

Ketones

☆ Astacene

☆ Astaxanthin

☆ Canthaxanthin aka Aphanicin, Chlorellaxanthin β,β-Carotene-4,4'-dione

☆ Capsanthin (3R,3'S,5'R)-3,3'-Dihydroxy-β,chachuu-carotene-6'-one

☆ Capsorubin (3S,5R,3'S,5'R)-3,3'-Dihydroxy-chachuu,chachuu-carotene-6,6'-dione

☆ Cryptocapsin (3'R,5'R)-3'-Hydroxy-β,chachuu-carotene-6'-one

☆ 2,2'-Diketospirilloxanthin 1,1'-Dimethoxy-3,4,3',4'-tetradehydro-1,2,1',2'-tetrahydro-γ,γ-carotene-2,2'-dione

☆ Flexixanthin 3,1'-Dihydroxy-3',4'-didehydro-1',2'-dihydro-β,γ-carotene-4-one

☆ 3-OH-Canthaxanthin aka Adonirubin aka Phoenicoxanthin 3-Hydroxy-β,β-carotene-4,4'-dione

☆ Hydroxyspheriodenone 1'-Hydroxy-1-methoxy-3,4-didehydro-1,2,1',2',7',8'-hexahydro-γ,γ-carotene-2-one

☆ Okenone 1'-Methoxy-1',2'-dihydro-c,γ-carotene-4'-one

☆ Pectenolone 3,3'-Dihydroxy-7',8'-didehydro-β,β-carotene-4-one

☆ Phoeniconone aka Dehydroadonirubin 3-Hydroxy-2,3-didehydro-β,β-carotene-4,4'-dione

☆ Phoenicopterone β,ε-carotene-4-one

☆ Rubixanthone 3-Hydroxy-β,γ-carotene-4'-one

☆ Siphonaxanthin 3,19,3'-Trihydroxy-7,8-dihydro-β,ε-carotene-8-one

Esters of Alcohols

☆ Astacein 3,3'-Bispalmitoyloxy-2,3,2',3'-tetradehydro-β,β-carotene-4,4'-dione or 3,3'-dihydroxy-2,3,2',3'-tetradehydro-β,β-carotene-4,4'-dione dipalmitate

☆ Fucoxanthin 3'-Acetoxy-5,6-epoxy-3,5'-dihydroxy-6',7'-didehydro-5,6,7,8,5',6'-hexahydro-β,β-carotene-8-one

☆ Isofucoxanthin 3'-Acetoxy-3,5,5'-trihydroxy-6',7'-didehydro-5,8,5',6'-tetrahydro-β,β-carotene-8-one

☆ Physalien

☆ Zeaxanthin (3R,3'R)-3,3'-Bispalmitoyloxy-β,β-carotene or (3R,3'R)-β,β-carotene-3,3'-diol

☆ Siphonein 3,3'-Dihydroxy-19-lauroyloxy-7,8-dihydro-β,ε-carotene-8-one or 3,19,3'-trihydroxy-7,8-dihydro-β,å-carotene-8-one 19-laurate

Apocarotenoids

☆ β-Apo-2'-carotenal 3',4'-Didehydro-2'-apo-β-carotene-2'-al

☆ Apo-2-lycopenal

☆ Apo-6'-lycopenal 6'-Apo-y-carotene-6'-al

☆ Azafrinaldehyde 5,6-Dihydroxy-5,6-dihydro-10'-apo-β-carotene-10'-al

☆ Bixin 6'-Methyl hydrogen 9'-cis-6,6'-diapocarotene-6,6'-dioate

☆ Citranaxanthin 5',6'-Dihydro-5'-apo-β-carotene-6'-one or 5',6'-dihydro-5'-apo-18'-nor-β-carotene-6'-one or 6'-methyl-6'-apo-β-carotene-6'-one

☆ Crocetin 8,8'-Diapo-8,8'-carotenedioic acid

☆ Crocetinsemialdehyde 8'-Oxo-8,8'-diapo-8-carotenoic acid

☆ Crocin Digentiobiosyl 8,8'-diapo-8,8'-carotenedioate

☆ Hopkinsiaxanthin 3-Hydroxy-7,8-didehydro-7',8'-dihydro-7'-apo-β-carotene-4,8'-dione or 3-hydroxy-8'-methyl-7,8-didehydro-8'-apo-β-carotene-4,8'-dione

☆ Methyl apo-6'-lycopenoate Methyl 6'-apo-y-carotene-6'-oate

☆ Paracentrone 3,5-Dihydroxy-6,7-didehydro-5,6,7',8'-tetrahydro-7'-apo-b-carotene-8'-one or 3,5-dihydroxy-8'-methyl-6,7-didehydro-5,6-dihydro-8'-apo-β-carotene-8'-one

☆ Sintaxanthin 7',8'-Dihydro-7'-apo-β-carotene-8'-one or 8'-methyl-8'-apo-b-carotene-8'-one

Nor- and Seco-carotenoids

☆ Actinioerythrin 3,3'-Bisacyloxy-2,2'-dinor-b,b-carotene-4,4'-dione

☆ β-Carotenone 5,6:5',6'-Diseco-b,b-carotene-5,6,5',6'-tetrone

☆ Peridinin 3'-Acetoxy-5,6-epoxy-3,5'-dihydroxy-6',7'-didehydro-5,6,5',6'-tetrahydro-12',13',20'-trinor-β,β-carotene-19,11-olide

☆ Pyrrhoxanthininol 5,6-epoxy-3,3'-dihydroxy-7',8'-didehydro-5,6-dihydro-12',13',20'-trinor-β,β-carotene-19,11-olide

☆ Semi-α-carotenone 5,6-Seco-b,e-carotene-5,6-dione

☆ Semi-β-carotenone 5,6-seco-b,b-carotene-5,6-dione or 5',6'-seco-β,β-carotene-5',6'-dione

☆ Triphasiaxanthin 3-Hydroxysemi-b-carotenone 3'-Hydroxy-5,6-seco-β,β-carotene-5,6-dione or 3-hydroxy-5',6'-seco-b,b-carotene-5',6'-dione

Retro-Carotenoids and Retro-apo-Carotenoids

☆ Eschscholtzxanthin 4',5'-Didehydro-4,5'-retro-b,b-carotene-3,3'-diol

☆ Eschscholtzxanthone 3'-Hydroxy-4',5'-didehydro-4,5'-retro-β,β-carotene-3-one

☆ Rhodoxanthin 4',5'-Didehydro-4,5'-retro-β,β-carotene-3,3'-dione

☆ Tangeraxanthin 3-Hydroxy-5'-methyl-4,5'-retro-5'-apo-β-carotene-5'-one or 3-hydroxy-4,5'-retro-5'-apo-β-carotene-5'-one

Higher Carotenoids

☆ Nonaprenoxanthin 2-(4-Hydroxy-3-methyl-2-butenyl)-7',8',11',12'-tetrahydro-e,y-carotene

☆ Decaprenoxanthin 2,2'-Bis(4-hydroxy-3-methyl-2-butenyl)-e,e-carotene

☆ C.p. 450 2-[4-Hydroxy-3-(hydroxymethyl)-2-butenyl]-2'-(3-methyl-2-butenyl)-β,β-carotene

☆ C.p. 473 2'-(4-Hydroxy-3-methyl-2-butenyl)-2-(3-methyl-2-butenyl)-3',4'-didehydro-1',2'-dihydro-b,y-carotene-1'-ol

☆ Bacterioruberin 2,2'-Bis(3-hydroxy-3-methylbutyl)-3,4,3',4'-tetradehydro-1,2,1',2'-tetrahydro-y,y-carotene-1,1'-dio

References

Alija AJ, Bresgen N, Sommerburg O, Siems W, Eckl PM (2004). "Cytotoxic and genotoxic effects of β-carotene breakdown products on primary rat hepatocytes". *Carcinogenesis* 25(5): 827–31. doi: 10.1093/carcin/bgh056. PMID 14688018.

Alija, A. J, Bresgen N, Sommerburg O, Siems W, Eckl PM (2004). "Cytotoxic and genotoxic effects of β-carotene breakdown products on primary rat hepatocytes". *Carcinogenesis* 25(5): 827–31. doi: 10.1093/carcin/bgh056. PMID 14688018.

Armstrong GA, Hearst JE (1996). "Carotenoids 2: Genetics and molecular biology of carotenoid pigment biosynthesis". *FASEB J.* 10 (2): 228–37. PMID 8641556.

β-Carotene and other carotenoids as antioxidants. From U.S. National Library of Medicine. November, 2008.

Bjelakovic G, *et al.*, Nikolova, D, Gluud, LL, Simonetti, RG, Gluud, C (2007). "Mortality in randomized trials of antioxidant supplements for primary and secondary prevention: systematic review and meta-analysis". *JAMA* 297 (8): 842– 57. doi: 10.1001/jama. 297.8.842.

Brian H. Davies (19991). Carotenoid metabolism as a preparation for function. Pure and Applied Chemistry, Vol. 63, No. 1, pp. 131-140, available online. Accessed April 30, 2010.

Daviesþ, Kevin M. (2004). *Plant pigments and their manipulation*. Wiley-Blackwell. p. 6. ISBN 1-4051-1737-0.

Diplock1, A. T., J.-L. Charleux, G. Crozier-Willi, F. J. Kok, C. Rice-Evans, M. Roberfroid, W. Stahl, J. Vina-Ribes (1998). Functional food science and defence against reactive oxidative species, British Journal of Nutrition, 80, Suppl. 1, S77–S112

Kidd, Parris (2011). "Astaxanthin, Cell Membrane Nutrient with Diverse Clinical Benefits and Anti-Aging Potential". *Alternative Medicine Review* 16 (4): 335–364.

Linus Pauling Institute. "Micronutrient Information Center-Carotenoids". Retrieved 3 August 2013.

Liu GY, Essex A, Buchanan JT (2005). "*Staphylococcus aureus* golden pigment impairs neutrophil killing and promotes virulence through its antioxidant activity". *J. Exp. Med.*202 (2): 209–15. doi: 10.1084/jem.20050846. PMC 2213009, PMID 16009720.

Lozano, G. A. (1994). Carotenoids, parasites and sexual selection. *Oikos* 70: 309-311.

Patent Pending: US Application Number 11/817,120

Seyoung Choi and Sangho Koo (2005). Efficient Syntheses of the Keto-carotenoids Canthaxanthin, Astaxanthin and Astacene. *J. Org. Chem.*, 70 (8): 3328–3331, doi: 10.1021/jo050101l.

Unlu N, *et al.,* Bohn, T, Clinton, SK, Schwartz, SJ (2005). "Carotenoid Absorption from Salad and Salsa by Humans Is Enhanced by the Addition of Avocado or Avocado Oil".*Human Nutrition and Metabolism* 135 (3): 431–6. PMID 15735074.

Chapter 6

Chymotrypsin

Chymotrypsin (EC 3.4.21.1, *chymotrypsins A and B, alpha-chymar ophth, avazyme, chymar, chymotest, enzeon, quimar, quimotrase, alpha-chymar, alpha-chymotrypsin A, alpha-chymotrypsin*) is a digestive enzyme component of pancreatic juice acting in the duodenum where it performs proteolysis, the breakdown of proteins and polypeptides. Chymotrypsin preferentially cleaves peptide amide bonds where the carboxyl side of the amide bond (the P_1 position) is a large hydrophobic amino acid (tyrosine, tryptophan and phenylalanine). These amino acids contain an aromatic ring in their side chain that fits into a 'hydrophobic pocket' (the S_1 position) of the enzyme. It is activated in the presence of trypsin. The hydrophobic and shape complementarily between the peptide substrate P_1 side chain and the enzyme S_1 binding cavity accounts for the substrate specificity of this enzyme. Chymotrypsin also hydrolyzes other amide bonds in peptides at slower rates, particularly those containing leucine and methionine at the P_1 position. Structurally, it is the archetypal structure for its super family, the PA clan of proteases.

Activation

Chymotrypsin is synthesized in the pancreas by protein biosynthesis as precursor called chymotrypsinogen that is enzymatically inactive. On cleavage by trypsin into two parts that are still connected via an S-S bond, cleaved chymotrypsinogen molecules can activate each other by removing two small peptides in a *trans*-proteolysis. The resulting molecule is active chymotrypsin, a three-polypeptide molecule interconnected via disulfide bonds.

Mechanism of Action and Kinetics

In vivo, chymotrypsin is a proteolytic enzyme (Serine protease) acting in the digestive systems of many organisms. It facilitates the cleavage of peptide bonds by a hydrolysis reaction, which despite being thermodynamically favorable occurs extremely slowly in the absence of a catalyst. The main substrates of chymotrypsin include tryptophan, tyrosine, phenylalanine, leucine and methionine, which are

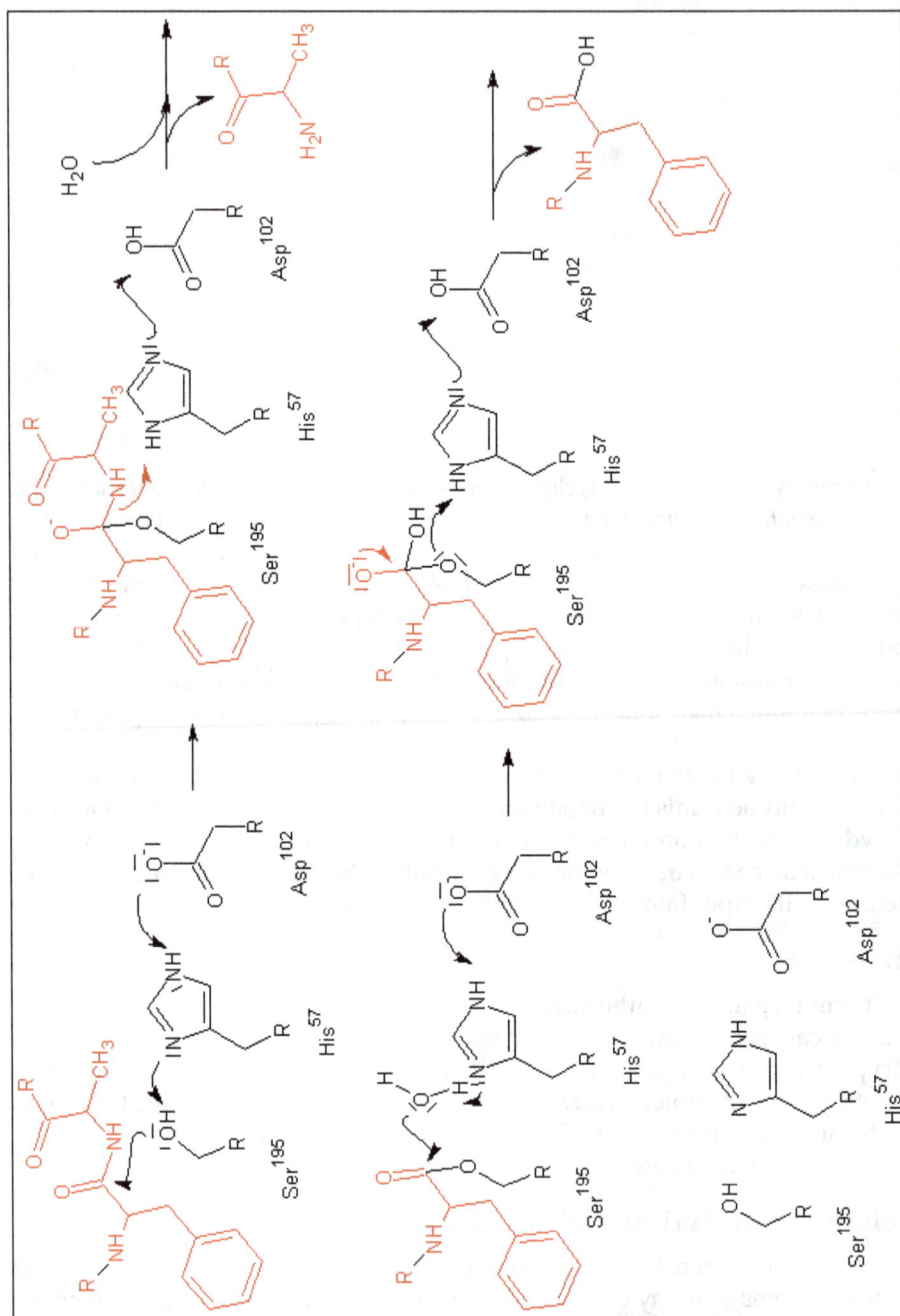

Mechanism of Peptide Bond Cleavage in α-chymotrypsin.

cleaved at the carboxyl terminal. Like many proteases, chymotrypsin will also hydrolyse amide bonds *in vitro*, a virtue that enabled the use of substrate analogs such as N-acetyl-L-phenylalanine p-nitrophenyl amide for enzyme assays.

Chymotrypsin cleaves peptide bonds by attacking the unreactive carbonyl group with a powerful nucleophile, the serine 195 residue located in the active site of the enzyme, which briefly becomes covalently bonded to the substrate, forming an enzyme-substrate intermediate. Along with histidine 57 andaspartic acid 102, this serine residue constitutes the catalytic triad of the active site. These findings rely on inhibition assays and the study of the kinetics of cleavage of the aforementioned substrate, exploiting the fact that the enzyme-substrate intermediate p-nitrophenolate has a yellow colour, enabling us to measure its concentration by measuring light absorbance at 410 nm. It was found that the reaction of chymotrypsin with its substrate takes place in two stages, an initial "burst" phase at the beginning of the reaction and a steady-state phase following Michaelis-Menten kinetics. It is also called "ping-pong" mechanism. The mode of action of chymotrypsin explains this as hydrolysis takes place in two steps. First acylation of the substrate to form an acyl-enzyme intermediate and then deacylation in order to return the enzyme to its original state. This occurs via the concerted action of the three amino acid residues in the catalytic triad.[5] Aspartate hydrogen bonds to the N-Δ hydrogen of histidine, increasing the pKa of its å nitrogen and thus making it able to deprotonate serine. It is this deprotonation that allows the serine side chain to act as a nucleophile and bind to the electron-deficient carbonyl carbon of the protein main chain. Ionization of the carbonyl oxygen is stabilized by formation of two hydrogen bonds to adjacent main chain N-hydrogens. This occurs in the oxyanion hole. This forms a tetrahedral adduct and breakage of the peptide bond. An acyl-enzyme intermediate, bound to the serine, is formed and the newly formed amino terminus of the cleaved protein can dissociate. In the second reaction step, a water molecule is activated by the basic histidine and acts as a nucleophile. The oxygen of water attacks the carbonyl carbon of the serine-bound acyl group, resulting in formation of a second tetrahedral adduct, regeneration of the serine -OH group and release of a proton, as well as the protein fragment with the newly formed carboxyl terminus.

Digestive enzymes synthesized and secreted by pancreatic acinar cells, breakdown the foods that we eat: starches (amylases), fats (lipases) and proteins (proteases). Pancreatic proteases have long been used in medicine both diagnostically and therapeutically for pancreatic disease and more recently for their involvement in cancer metastases. Proteases represent a diverse array of enzymes that act on the peptide bonds within proteins. They can be divided into discreet protein families that differ with respect to structure and catalytic type, including serine, threonine, cysteine, aspartic and metallo proteases. The two proteases trypsin and chymotrypsin are grouped into the serine protease family.

Trypsin and Chymotrypsin, Serine Protease Digestive Enzymes

Trypsin and chymotrypsin, like most proleotytic enzymes, are synthesized as inactive zymogen precursors (trypsinogen and chymotrypsinogen) to prevent unwanted destruction of cellular proteins and to regulate when and where enzyme activity occurs. The inactive zymogens are secreted into the duodenum, where they travel the small and large intestines prior to excretion. Zymogens also enter the bloodstream, where they can be detected in serum prior to excretion in urine. Zymogens are converted to the mature, active enzyme by proteolysis to split off a pro-peptide, either in a subcellular compartment or in an extracellular space where they are required for digestion.

Trypsin and chymotrypsin are structurally very similar, although they recognise different substrates. Trypsin acts on lysine and arginine residues, while chymotrypsin acts on large hydrophobic residues such as tryptophan, tyrosine and phenylalanine, both with extraordinary catalytic efficiency. Both enzymes have a catalytic triad of serine, histidine and aspartate within the S1 binding pocket, although the hydrophobic nature of this pocket varies between the two, as do other structural interactions beyond the S1 pocket.

The human pancreas secretes three isoforms of trypsinogen: cationic (trypsinogen-1), anionic (trypsinogen-2) and mesotrypsinogen (trypsinogen-3). Cationic and anionic trypsins are the major isoforms responsible for digestive protein degradation, occurring in a ratio of 2:1, while mesotrypsinogen accounts for less than 5 per cent of pancreatic secretions. Mesotrypsin is a specialised protease known for its resistance to trypsin inhibitors. It is thought to play a special role in the degradation of trypsin inhibitors, possibly to aid in the digestion of inhibitor-rich foods such as soybeans and lima beans. An alternatively spliced mesotrypsinogen in which the signal peptide is replaced with a different exon 1 is expressed in the human brain; the function of this brain trypsinogen is unknown.

There are two isoforms of pancreatic chymotrypsin, A and B, which are known to cleave proteins selectively at specific peptide bonds formed by the hydrophobic residues tryptophan, phenylalanine and tyrosine.

The Need for Inhibitors

Proteases perform many beneficial functions that are essential to life, but uncontrolled they can be dangerous. Protease inhibitors are used as the major form of control once the protease has been activated. In higher organisms, there is a delicate balance between proteases and their natural inhibitors to help control the activation and catabolism of many intra- and extra-cellular proteins. In mammals, the bloodstream is a major carrier for many glycoproteins that act as protease inhibitors. Protease inhibitors use a reactive site peptide bond to serve as a substrate for various proteases, forming very stable complexes where the inhibitor peptide bond is hydrolysed by the protease extremely slowly, thereby effectively removing the protease from circulation. There are at least eighteen families of protease inhibitors, all of which share a common conformation surrounding the reactive site peptide bond, even though they differ in their global structure. Some members only hydrolyse trypsin, such as chicken ovomucoid, while others hydrolyse both trypsin and chymotrypsin using different inhibitory domains, such as turkey ovomucoid.

Protease inhibitors are ubiquitous in nature. They are widely distributed in plant seeds, particularly in legumes. The presence of these inhibitors in seeds acts as a feeding deterrent, especially in insects where they inhibit midgut proteases. Inhibitors can deter other animals as well; isolated soybean inhibitors have been found to cause enlargement of the pancreas in certain species, such as rat and mouse. Many bacterial species produce protease inhibitors that help them to survive the digestive processes of the gut, such as ecotin in *Escherichia coli*, which is effective against several different pancreatic proteases because of its flexibility.

Protease inhibitors can have nutritional value as well. The Bowman-Birk inhibitor (BBI) from soybean may play a role in the prevention of tumourigenesis. BBI is also an effective inhibitor of nephrotoxicity induced by the antibiotic gentamicin.

Pancreatitis, when Control Fails

Any disturbance of the balance between proteolytic enzymes and their inhibitors, or of the activation process, can result in pancreatits, where premature, intracellular activation of zymogens cause the autodigestion of the pancreas. During pancreatits, anionic trypsinogen becomes the predominant isoforms secreted from the pancreas and there is an increased level of zymogen in serum. Anionic trypsinogen exhibits increased autocatalytic degradation over the cationic isoforms. Pancreatitis can occur as a hereditary, autodominant condition, as part of the pathogenesis of cystic fibrosis or as a result of alcohol toxicity.

Hereditary pancreatitis results in recurrent attacks of acute pancreatitis, progressing to chronic pancreatitis at a young age. Mutations in the cationinc trypsinogen I gene have been characterised from several patients. One of these mutations was shown to alter a trypsin recognition site, preventing the deactivation of trypsin within the pancreas, resulting in autodigestion. Hereditary pancreatitis can also result in patients with mutations in the trypsin/trypsinogen inhibitor SPINK1, which leads to the inappropriate activation of pancreatic zymogens. SPINK1 mutations appear to cause a more rapid progression of chronic pancreatitis than cationic trypsinogen mutations.

Cystic fibrosis (CF) is an autosomal recessive disorder caused by mutations in the CF transmembrane conductance regulator (CFTR) gene. Patients with CF often suffer from recurrent pancreatitis. CF is thought ot interfere with the negative feedback loop that helps regulate pancreatic zymogen secretion. A high concentration of protease in the duodenum inhibits cholecystokinin (CCK) secretion, thereby reducing pancreatic secretion of zymogens; a high concentration of inhibitors induces pancreatic zymogen secretion through CCK. In CF, it has been suggested that obstruction to pancreatic zymogen secretion could induce CCK release, which in turn stimulates pancreatic zymogen production. The blockage of zymogen secretion together with its continued synthesis could result in the built-up of digestive enzymes within lysosomes, which could then rupture into the cytoplasm.

Chronic alcohol consumption is known to increase the synthesis of several digestive enzymes in the pancreas, including trypsinogen. Progression to chronic pancreatitis can be rapid. As fewer than 10 per cent of alcoholics develop chronic pancreatits, other genetic or environmental factors must be involved.

Trypsin, Chymotrypsin and Cancer

Hereditary pancreatitis carries a 40 per cent risk of pancreatic cancer (5 per cent risk for non-hereditary pancreatitis), a very invasive cancer with high mortality rates. Pancreatic inflammation promotes intensive cell proliferation to regenerate the damaged pancreas, during which the amplification of pathological changes in DNA can occur.

Elevated levels of trypsin have been found in a variety of other tumours, such as ovarian and colourectal carcinomas, where it may have a role in malignant tumour formation or metastasis. Trypsins appear to be necessary for cancer cells to invade normal tissue and to enter the bloodstream and lymphatic channels. A critical step in cancer metastasis involves breaking down the extracellular matrix surrounding the malignant tumour, which allows it to invade and spread. Type I collagen degradation involves the action of matrix metalloproteases (MMP-1, -8 and -13), which are activated by MMP-3. Trypsin-2 can directly activate all four pro-MMP enzymes and can degrade type I collagen, acting as a potent tumour-associated matrix serine protease. The inhibitor TATI (tumour-associated trypsin inhibitor) can inhibit trypsin activation of pro-MMPs and trypsin degradation of type I collagen. Not surprisingly, TATI has been detected at elevated levels in a variety of tumours and in the serum of cancer patients. Trypsin has also been shown to activate a G protein-coupled receptor PAR-2 (protease-activated receptor-2) in both colon and pancreatic cancer cell lines, where it appears to act as a potent mitogen *in vitro*, functioning as a growth factor.

Certain protease inhibitors appear to be capable of suppressing carcinogenesis in a variety of model systems, the most potent of which appear to inhibit chymotrypsin or chymotrypsin-like proteases. Both soybean-derived BBI (inhibits trypsin and chymotrypsin) and human alpha1-antichymotrypsin is potent inhibitors of chymotrypsin, is highly anti-inflammatory and may be involved in defence of their respective organisms. BBI was found to help suppress a variety of tumours in different organisms, both *in vitro* and *in vivo*. Extracts high in proteolytic enzymes, especially trypsin, chymotrypsin and papain, have also been used in cancer studies. Enzyme therapy appears to be anti-inflammatory by its induction of protease inhibitors.

References

Appel W (1986). "Chymotrypsin: molecular and catalytic properties". *Clin. Biochem.* 19 (6): 317–22.

Berger A, Schechter I (1970). "Mapping the active site of papain with the aid of peptide substrates and inhibitors". *Philos. Trans. R. Soc. Lond., B, Biol. Sci.* 257 (813): 249–64.

PDB 1CHG, Freer ST, Kraut J, Robertus JD, Wright HT, Xuong NH (1970). "Chymotrypsinogen: 2.5-angstrom crystal structure, comparison with alpha-chymotrypsin and implications for zymogen activation". *Biochemistry* 9 (9): 1997–2009.

Petsko, Gregory, Ringe, Dagmar (2009). *Protein Structure and Function*. Oxford: Oxford University Press. pp. 78–79.

Wilcox PE (1970). "Chymotrypsinogens chymotrypsins". *Methods in Enzymology*. Methods in Enzymology 19: 64–108.

Chapter 7

Flavonoids

Flavonoids (or bioflavonoids) (from the Latin word *flavus* meaning yellow, their colour in nature) are a class of plant secondary metabolites. Flavonoids were referred to as Vitamin P (probably because of the effect they had on the permeability of vascular capillaries) from the mid-1930s to early 50s, but the term has since fallen out of use.

Chemically, they have the general structure of a 15-carbon skeleton, which consists of two phenyl rings (A and B) and heterocyclic ring (C). This carbon structure can be abbreviated C6-C3-C6. According to the IUPAC nomenclature, they can be classified into:

- ☆ Flavonoids or bioflavonoids
- ☆ Isoflavonoids, derived from 3-phenylchromen-4-one (3-phenyl-1,4-benzopyrone) structure
- ☆ Neoflavonoids, derived from 4-phenylcoumarine (4-phenyl-1,2-benzopyrone) structure

The three flavonoid classes above are all ketone-containing compounds and as such, are anthoxanthins (flavones and flavonols). This class was the first to be termed *bioflavonoids*. The terms *flavonoid* and *bioflavonoid* have also been more loosely used to describe non-ketone polyhydroxy polyphenol compounds which are more specifically termed flavanoids. The three cycle or heterocycles in the flavonoid backbone are generally called ring A, B and C. Ring A usually shows a phloroglucinol substitution pattern.

Flavonoid Biosynthesis

Flavonoids are synthesized by the phenylpropanoid metabolic pathway in which the amino acid phenylalanine is used to produce 4-coumaroyl-CoA. This can be combined withmalonyl-CoA to yield the true backbone of flavonoids, a group of compounds called chalcones, which contain two phenyl rings. Conjugate

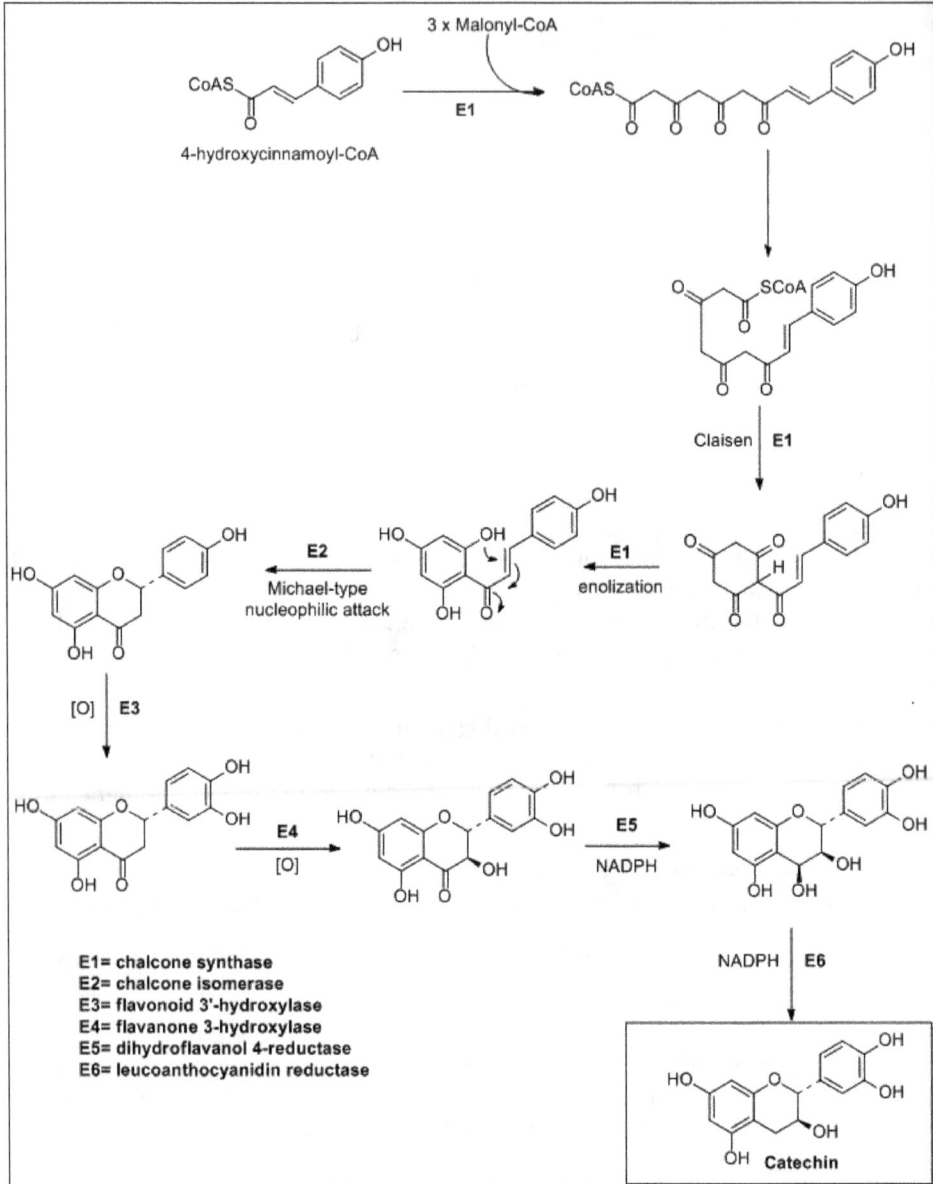

E1= chalcone synthase
E2= chalcone isomerase
E3= flavonoid 3'-hydroxylase
E4= flavanone 3-hydroxylase
E5= dihydroflavanol 4-reductase
E6= leucoanthocyanidin reductase

ring-closure of chalcones results in the familiar form of flavonoids, the three-ringed structure of a flavone. The metabolic pathway continues through a series of enzymatic modifications to yield flavanones →dihydroflavonols → anthocyanins. Along this pathway, many products can be formed, including the flavonols, flavan-3-ols, proanthocyanidins (tannins) and a host of other various polyphenolics.

Enzymes

The biosynthesis of flavonoids involves several enzymes.

☆ Anthocyanidin reductase

☆ Chalcone isomerase

☆ Dihydrokaempferol 4-reductase

☆ Flavone synthase

☆ Flavonoid 3'-monooxygenase

☆ Flavonol synthase

☆ Flavanone 3-dioxygenase

☆ Flavanone 4-reductase

☆ Leucoanthocyanidin reductase

☆ Leucocyanidin oxygenase

☆ Naringenin-chalcone synthase

Methylation

☆ Apigenin 4'-O-methyltransferase

☆ Luteolin O-methyltransferase

☆ Quercetin 3-O-methyltransferase

Glycosylation

☆ Anthocyanidin 3-O-glucosyltransferase

☆ Flavone 7-O-beta-glucosyltransferase

☆ Flavone apiosyltransferase

☆ Flavonol-3-O-glucoside L-rhamnosyltransferase

☆ Flavonol 3-O-glucosyltransferase

Further Acetylations

☆ Isoflavone-7-O-beta-glucoside 6"-O-malonyltransferase

Functions of Flavonoids in Plants

Flavonoids are widely distributed in plants, fulfilling many functions. Flavonoids are the most important plant pigments for flower colouration, producing yellow or red/blue pigmentation in petals designed to attract pollinator animals. In higher plants, flavonoids are involved in UV filtration, symbiotic nitrogen fixation and floral pigmentation. They may also act as chemical messengers, physiological regulators and cell cycle inhibitors. Flavonoids secreted by the root of their host plant help *Rhizobia* in the infection stage of their symbiotic relationship with legumes like peas, beans, clover and soy. Rhizobia living in soil are able to sense the flavonoids and this triggers the secretion of Nod factors, which in turn are recognized by the host plant and can lead to root hair deformation and several cellular responses such as ion fluxes and the formation of a root nodule. In addition, some flavonoids have inhibitory activity against organisms that cause plant diseases, *e.g. Fusarium oxysporum.*

Medical Research

Though there is ongoing research into the potential health benefits of individual flavonoids, neither the Food and Drug Administration (FDA) nor the European Food Safety Authority (EFSA) has approved any health claim for flavonoids or approved any flavonoids as pharmaceutical drugs. Moreover, several companies have been cautioned by the FDA over misleading health claims.

In vitro

Flavonoids have been shown to have a wide range of biological and pharmacological activities in *in vitro* studies. Examples include anti-allergic, anti-inflammatory, antioxidant, anti-microbial (antibacterial, antifungal and antiviral), anti-cancer and anti-diarrheal activities. Flavonoids have also been shown to inhibittopoisomerase enzymes and to induce DNA mutations in the mixed-lineage leukemia (*MLL*) gene *in vitro* studies. However, in most of the above cases no follow up *in vivo* or clinical research has been performed, leaving it impossible to say if these activities have any beneficial or detrimental effect on human health. Biological and pharmacological activities which have been investigated in greater depth are described below.

Antioxidant

Research at the Linus Pauling Institute and the European Food Safety Authority shows that flavonoids are poorly absorbed in the human body (less than 5 per cent), with most of what is absorbed being quickly metabolized and excreted. These findings suggest that flavonoids have negligible systemic antioxidant activity and that the increase in antioxidant capacity of blood seen after consumption of flavonoid-rich foods is not caused directly by flavonoids, but is due to production of uric acid resulting from flavonoid depolymerizationand excretion.

Inflammation

Inflammation has been implicated as a possible origin of numerous local and systemic diseases, such as cancer, cardiovascular disorders, diabetes mellitus and celiac disease. Preliminary studies indicate that flavonoids may affect anti-inflammatory mechanisms via their ability to inhibit reactive oxygen or nitrogen compounds. Flavonoids have also been proposed to inhibit the pro-inflammatory activity of enzymes involved in free radical production, such as cyclooxygenase, lipoxygenase or inducible nitric oxide synthase and to modify intracellular signaling pathways in immune cells. Procyanidins, a class of flavonoids, have been shown in preliminary research to have anti-inflammatory mechanisms including modulation of the arachidonic acid pathway, inhibition of gene transcription, protein expression and activity of inflammatory enzymes, as well as secretion of anti-inflammatory mediators.

Cancer

Clinical studies investigating the relationship between flavonoid consumption and cancer prevention/development are conflicting for most types of cancer, probably because most studies are retrospective in design and use a small sample size. Two apparent exceptions are gastric carcinoma and smoking-related cancers. Dietary flavonoid intake isassociated with reduced gastric carcinoma risk in women and reduced aerodigestive tract cancer risk in smokers.

Cardiovascular Diseases

Among the most intensively studied of general human disorders possibly affected by dietary flavonoids, preliminary cardiovascular disease research has revealed the following mechanisms under investigation in patients or normal subjects:

- ✰ Inhibit coagulation, thrombus formation or platelet aggregation
- ✰ Reduce risk of atherosclerosis
- ✰ Reduce arterial blood pressure and risk of hypertension
- ✰ Reduce oxidative stress and related signaling pathways in blood vessel cells
- ✰ Modify vascular inflammatory mechanisms
- ✰ Improve endothelial and capillary function
- ✰ Modify blood lipid levels
- ✰ Regulate carbohydrate and glucose metabolism
- ✰ Modify mechanisms of aging

Listed on the clinical trial registry of the US National Institutes of Health (November 2013) are 36 human studies completed or underway to study the dietary effects of plant flavonoids on cardiovascular diseases.

Antibacterial

Flavonoids have been shown to have (a) direct antibacterial activity, (b) synergistic activity with antibiotics and (c) the ability to suppress bacterial virulence factors in numerous *in vitro* and a limited number of *in vivo* studies. Noteworthy among the *in vivo* studies is the finding that oral quercetin protects guinea pigs against the Group 1 carcinogen *Helicobacter pylori*. Researchers from the European Prospective Investigation into Cancer and Nutrition have speculated this may be one reason why dietary flavonoid intake is associated with reduced gastric carcinoma risk in European women. Additional *in vivo* and clinical research is needed to determine if flavonoids could be used as pharmaceutical drugs for the treatment of bacterial infection, or whether dietary flavonoid intake offers any protection against infection.

Dietary Sources

Flavonoids (specifically flavanoids such as the catechins) are "the most common group of polyphenolic compounds in the human diet and are found ubiquitously in plants". Flavonols, the original bioflavonoids such as quercetin, are also found ubiquitously, but in lesser quantities. The widespread distribution of flavonoids, their variety and their relatively low toxicity compared to other active plant compounds(for instance alkaloids) mean that many animals, including humans, ingest significant quantities in their diet. Foods with a high flavonoid content include parsley, onions, blueberries and other berries, black tea, green tea and oolong tea, bananas, all citrus fruits, Ginkgo biloba, red wine, sea-buckthorns and dark chocolate (with a cocoa content of 70 per cent or greater). Further information on dietary sources of flavonoids can be obtained from the US Department of Agriculture flavonoid database.

Subgroups

Over 5000 naturally occurring flavonoids have been characterized from various plants. They have been classified according to their chemical structure and are usually subdivided into the following subgroups:

Anthoxanthins

Anthoxanthins are divided into two groups:

Group	Skeleton			Examples
	Description	Functional Groups	Structural Formula	
		3-hydroxyl 2,3-dihydro		
Flavone	2-phenyl chromen-4-one	✗ ✗		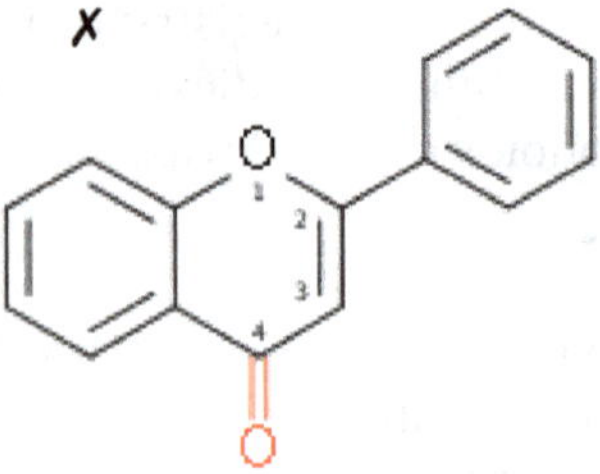 Luteolin, Apigenin, Tangeritin
Flavonol or 3-hydroxy- flavone	3-hydroxy- 2-phenyl- chromen-4-one	✔ ✗		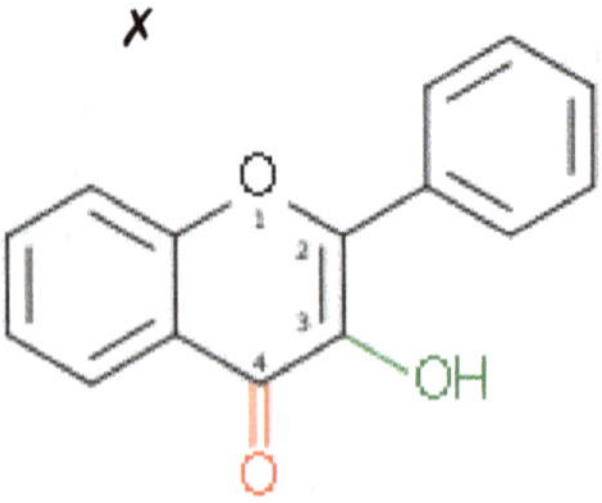 Quercetin, Kaempferol, Myricetin, Fisetin, Galangin, Isorhamnetin, Pachypodol, Rhamnazin, Pyranoflavonols, Furanoflavonols

Flavanones

Group	Skeleton			Examples
	Description	Functional Groups	Structural Formula	
		3-hydroxyl 2,3-dihydro		
Flavanone	2,3-dihydro-2-phenylchromen-4-one	✗ ✔		Hesperetin, Naringenin, Eriodictyol, Homoeriodictyol

Flavanonols

Group	Skeleton			Examples
	Description	Functional Groups	Structural Formula	
		3-hydroxyl 2,3-dihydro		
Flavanonol or 3-Hydroxy-flavanone or 2,3-dihydro-flavonol	3-hydroxy-2,3-dihydro-2-phenyl chromen-4-one	✔ ✔		Taxifolin (or Dihydroquercetin), Dihydrokaempferol

Flavans

Include flavan-3-ols (flavanols), flavan-4-ols and flavan-3,4-diols.

Skeleton	Name
	Flavan-3-ol (flavanol)
	Flavan-4-ol
	Flavan-3,4-diol (leucoanthocyanidin)

☆ Flavan-3-ols (flavanols)

  ☆ Flavan-3-ols use the 2-phenyl-3,4-dihydro-2H-chromen-3-ol skeleton

    Examples: Catechin (C), Gallocatechin (GC), Catechin 3-gallate (Cg), Gallocatechin 3-gallate (GCg)), Epicatechins (Epicatechin (EC)), Epigallocatechin (EGC),Epicatechin 3-gallate (ECg), Epigallocatechin 3-gallate (EGCg)

☆ Theaflavin

  Examples: Theaflavin-3-gallate, Theaflavin-3'-gallate, Theaflavin-3,3'-digallate

☆ Thearubigin

☆ Proanthocyanidins are dimers, trimers, oligomers, or polymers of the flavanols

☆ Anthocyanidins

  Anthocyanidins are the aglycones of anthocyanins; they use the **flavylium** (2-phenylchromenylium) ion skeleton

  Examples: Cyanidin, Delphinidin, Malvidin, Pelargonidin, Peonidin, Petunidin

Isoflavonoids

☆ Isoflavonoids

☆ Isoflavones use the 3-phenylchromen-4-one skeleton (with no hydroxyl group substitution on carbon at position 2)

  Examples: Genistein, Daidzein, Glycitein

  ☆ Isoflavanes

  ☆ Isoflavandiols

  ☆ Isoflavenes

  ☆ Coumestans

  ☆ Pterocarpans

Synthesis, Detection, Quantification and Semi-synthetic Alterations

Availability through Microorganisms

Several recent research articles have demonstrated the efficient production of flavonoid molecules from genetically engineered microorganisms.

Tests for Detection

Shinoda Test

Four pieces of magnesium fillings (ribbon) are added to the ethanolic extract followed by few drops of concentrated hydrochloric acid. A pink or red colour indicates the presence of flavonoid. Colours varying from orange to red indicated flavones, red to crimson indicated flavonoids, crimson to magenta indicated flavonones.

Sodium Hydroxide Test

About 5 mg of the compound is dissolved in water, warmed and filtered. 10 per cent aqueous sodium hydroxide is added to 2 ml of this solution. This produces a yellow colouration. A change in colour from yellow to colourless on addition of dilute hydrochloric acid is an indication for the presence of flavonoids.

p-Dimethylaminocinnamaldehyde Test

A colourimetric assay based upon the reaction of A-rings with the chromogen p-dimethylaminocinnamaldehyde (DMACA) has been developed for flavanoids in beer that can be compared with the vanillin procedure.

Quantification

Lamaison and Carnet have designed a test for the determination of the total flavonoid content of a sample ($AlCl_3$ method). After proper mixing of the sample and the reagent, the mixture is incubated for 10 minutes at ambient temperature and the absorbance of the solution is read at 440 nm. Flavonoid content is expressed in mg/g of quercetin.

Semi-synthetic Alterations

Immobilized *Candida antarctica* lipase can be used to catalyze the regioselective acylation of flavonoids.

References

A new colourimetric assay for flavonoids in pilsner beers. Jan A. Delcour and Didier Janssens de Varebeke, *Journal of the Institute of Brewing*, January–February 1985, Volume 91, Issue 1, pages 37–40, doi: 10.1002/j.2050-0416.1985.tb04303.x

A theoretical study of the conformational behavior and electronic structure of taxifolin correlated with the free radical-scavenging activity. Patrick Trouillas, Catherine Fagnère, Roberto Lazzaroni, Claude Calliste, Abdelghafour Marfak and Jean-Luc Duroux, *Food Chemistry*, Volume 88, Issue 4, December 2004, Pages 571-582

Babu P, Liu D, Gilbert E (2013). "Recent advances in understanding the anti-diabetic actions of dietary flavonoids". *The Journal of Nutritional Biochemistry* 24 (11): 1777–1789. doi: 10.1016/j.jnutbio.2013.06.003.

Bandele O, Clawson S, Osheroff N (2008). "Dietary polyphenols as topoisomerase II poisons: B-ring substituents determine the mechanism of enzyme-mediated DNA cleavage enhancement". *Chemical Research in Toxicology* 21 (6): 1253–1260. doi: 10.1021/tx8000785.

Barjesteh van Waalwijk van Doorn-Khosrovani S, Janssen J, Maas L, Godschalk R, Nijhuis J, van Schooten F (2007). "Dietary flavonoids induce MLL translocations in primary human CD34+ cells". *Carcinogenesis* 28 (8): 1703–9.doi: 10.1093/carcin/bgm102.

Bello I, Ndukwe G, Audu O, Habila J (2011). "A bioactive flavonoid from Pavetta crassipes K. Schum". *Organic and Medicinal Chemistry Letters* 1 (1): 14.doi: 10.1186/2191-2858-1-14. PMC 3305906.

Benthsath, A., Rusznyak, S. T., Szent-Györgyi, A. (1937). "Vitamin P". *Nature* 139(3512): 326–327. Bibcode: 1937Natur.139R.326B. doi: 10.1038/139326b0.

Cazarolli L, Zanatta L, Alberton E, Figueiredo M, Folador P, Damazio R. (2008). "Flavonoids: Prospective Drug Candidates". *Mini-Reviews in Medicinal Chemistry* 8 (13): 1429–1440. doi: 10.2174/138955708786369564.

Choi O, Yahiro K, Morinaga N, Miyazaki M, Noda M (2007). "Inhibitory effects of various plant polyphenols on the toxicity of Staphylococcal alpha-toxin". *Microbial Pathogenesis* 432 (5–6): 215–224. doi: 10.1016/j.micpath.2007.01.007.

Chukwumah Y, Walker L, Verghese M (2009). "Peanut skin colour: a biomarker for total polyphenolic content and antioxidative capacities of peanut cultivars". *Int J Mol Sci* 10(11): 4941–52. doi: 10.3390/ijms10114941.

Cushnie T, Lamb A (2005). "Antimicrobial activity of flavonoids". *International Journal of Antimicrobial Agents* 26 (5): 343–356.doi: 10.1016/j.ijantimicag.2005.09.002.

Cushnie T, Lamb A (2011). "Recent advances in understanding the antibacterial properties of flavonoids". *International Journal of Antimicrobial Agents* 38 (2): 99–107.doi: 10.1016/j.ijantimicag.2011.02.014.

de Sousa R, Queiroz K, Souza A, Gurgueira S, Augusto A, Miranda M (2007). "Phosphoprotein levels, MAPK activities and NFkappaB expression are affected by fisetin". *J Enzyme Inhib Med Chem* 22 (4): 439–444. doi: 10.1080/14756360601162063.

EFSA Panel on Dietetic Products, Nutrition and Allergies (NDA) 2, 3 European Food Safety Authority (EFSA), Parma, Italy (2010). "Scientific Opinion on the substantiation of health claims related to various food(s)/food constituent(s) and protection of cells from premature aging, antioxidant activity, antioxidant content and antioxidant properties and protection of DNA, proteins and lipids from oxidative damage pursuant to Article 13(1) of Regulation (EC) No. 1924/20061". *EFSA Journal* 8 (2): 1489.doi: 10.2903/j.efsa.2010.1489 (inactive 2015-02-01).

Esselen M, Fritz J, Hutter M, Marko D (2009). "Delphinidin Modulates the DNA-Damaging Properties of Topoisomerase II Poisons". *Chemical Research in Toxicology* 22(3): 554–64. doi: 10.1021/tx800293v.

F IUPAC. Compendium of Chemical Terminology, 2nd ed. (the "Gold Book"). Compiled by A. D. McNaught and A. Wilkinson. Blackwell Scientific Publications, Oxford (1997). XML on-line corrected version: http://goldbook.iupac.org (2006–) created by M. Nic, J. Jirat, B. Kosata, updates compiled by A. Jenkins. ISBN 0-9678550-9-8. doi: 10.1351/goldbook. Last update: 2012-08-19, version: 2.3.2. DOI of this term: doi: 10.1351/goldbook.F02424. (Original

PDF version: http: //goldbook.iupac. org/goldbook/F02424.html. The PDF version is out of date and is provided for reference purposes only.) Retrieved 16 September 2012.

FDA approved drug products. US Food and Drug Administration. Retrieved 8 November 2013.

Ferretti G, Bacchetti T, Masciangelo S, Saturni L (2012). "Celiac Disease, Inflammation and Oxidative Damage: A Nutrigenetic Approach". *Nutrients* 4 (12): 243–257. doi: 10.3390/nu4040243.

Flavonoids in cardiovascular disease clinical trials. *Clinicaltrials.gov*. US National Institutes of Health. November 2013. Retrieved November 24, 2013.

Friedman M (2007). "Overview of antibacterial, antitoxin, antiviral and antifungal activities of tea flavonoids and teas". *Molecular Nutrition and Food Research* 51 (1): 116–134. doi: 10.1002/mnfr.200600173.

Fruits Are Good for Your Health? Not So Fast: FDA Stops Companies From Making Health Claims About Foods. TheDailyGreen.com. Retrieved 25 October 2013.

Galeotti, F, Barile, E, Curir, P, Dolci, M, Lanzotti, V (2008). "Flavonoids from carnation (Dianthus caryophyllus) and their antifungal activity". *Phytochemistry Letters* 1: 44.doi: 10.1016/j.phytol.2007.10.001.

Gomes A, Couto D, Alves A, Dias I, Freitas M, Porto G (2012). "Trihydroxyflavones with antioxidant and anti-inflammatory efficacy". *BioFactors* 38 (5): 378–386. doi: 10.1002/biof.1033.

González C, Sala N, Rokkas T (2013). "Gastric cancer: epidemiologic aspects". *Helicobacter* 18 (Supplement 1): 34–38. doi: 10.1111/hel.12082.

González-Segovia R, Quintanar J, Salinas E, Ceballos-Salazar R, Aviles-Jiménez F, Torres-López J (2008). "Effect of the flavonoid quercetin on inflammation and lipid peroxidation induced by Helicobacter pylori in gastric mucosa of guinea pig". *Journal of Gastroenterology* 43 (6): 441–447. doi: 10.1007/s00535-008-2184-7.

Health Claims Meeting Significant Scientific Agreement. US Food and Drug Administration. Retrieved 8 November 2013.

Higdon, J, Drake, V, Frei, B (2009). "Non-Antioxidant Roles for Dietary Flavonoids: Reviewing the relevance to cancer and cardiovascular diseases". *Nutraceuticals World*. Rodman Media. Retrieved 24 November 2013.

Hwang E, Kaneko M, Ohnishi Y, Horinouchi S (2003). "Production of plant-specific flavanones by *Escherichia coli* containing an artificial gene cluster". *Appl. Environ. Microbiol.* 69 (5): 2699–706. doi: 10.1128/AEM.69.5.2699-2706.2003.PMC 154558.

Inspections, Compliance, Enforcement and Criminal Investigations (Flavonoid Sciences)". US Food and Drug Administration. Retrieved 8 November 2013.

Inspections, Compliance, Enforcement and Criminal Investigations (Unilever, Inc.). US Food and Drug Administration. Retrieved 25 October 2013.

Isolation of a UDP-glucose: Flavonoid 5-O-glucosyltransferase gene and expression analysis of anthocyanin biosynthetic genes in herbaceous peony (Paeonia lactiflora Pall.). Da Qiu Zhao, Chen Xia Han, Jin Tao Ge and Jun Tao, *Electronic Journal of Biotechnology*, 15 November 2012, Volume 15, Number 6, doi: 10.2225/vol15-issue6-fulltext-7

Izzi V, Masuelli L, Tresoldi I, Sacchetti P, Modesti A, Galvano F (2012). "The effects of dietary flavonoids on the regulation of redox inflammatory networks". *Frontiers in bioscience (Landmark edition)* 17 (7): 2396–2418. doi: 10.2741/4061.

Lamaison, JL and Carnet, A (1991). "Teneurs en principaux flavonoides des fleurs de Cratageus monogyna Jacq et de Cratageus Laevigata (Poiret D.C) en Fonction de la vegetation". *Plantes Medicinales Phytotherapie* 25: 12–16.

Lipton green tea is a drug. NutraIngredients-USA.com. Retrieved 25 October 2013.

Lotito S, Frei B (2006). "Consumption of flavonoid-rich foods and increased plasma antioxidant capacity in humans: cause, consequence, or epiphenomenon?". *Free Radic. Biol. Med.* 41 (12): 1727–46. doi: 10.1016/j.freeradbiomed.2006.04.033.

Manach C, Mazur A, Scalbert A (2005). "Polyphenols and prevention of cardiovascular diseases". *Current opinion in lipidology* 16 (1): 77–84. doi: 10.1097/00041433-200502000-00013.

Manner S, Skogman M, Goeres D, Vuorela P, Fallarero A (2013). "Systematic exploration of natural and synthetic flavonoids for the inhibition of *Staphylococcus aureus* biofilms". *International Journal of Molecular Sciences* 14 (10): 19434–19451.doi: 10.3390/ijms141019434.

Martinez-Micaelo N, González-Abuín N, Ardèvol A, Pinent M, Blay M (2012). "Procyanidins and inflammation: Molecular targets and health implications". *BioFactors*38 (4): 257–265. doi: 10.1002/biof.1019.

McNaught, Alan D, Wilkinson andrew, IUPAC (1997). "IUPAC Compendium of Chemical Terminology" (2 ed.). Oxford: Blackwell Scientific. Archived from the original on 29 June 2011.

Methods of analysis and separation of chiral flavonoids. Jaime A. Yáñeza, Preston K. Andrewsb and Neal M. *Journal of Chromatography* B, Volume 848, Issue 2, 1 April 2007, Pages 159-181

Mobh, Shiro (1938). "Research for Vitamin P". *The Journal of Biochemistry* 29 (3): 487–501.

Oh D, Kim J, Kim Y (2010). "Genistein inhibits Vibrio vulnificus adhesion and cytotoxicity to HeLa cells". *Archives of Pharmacal Research* 33 (5): 787–792.doi: 10.1007/s12272-010-0520-y.

Passicos E, Santarelli X, Coulon D (2004). "Regioselective acylation of flavonoids catalyzed by immobilized *Candida antarctica* lipase under reduced pressure". *Biotechnol Lett.* 26 (13): 1073–1076. doi: 10.1023/B: BILE.0000032967.23282.15.

Ravishankar D, Rajora A, Greco F, Osborn H (2013). "Flavonoids as prospective compounds for anti-cancer therapy". *The International Journal of Biochemistry and Cell Biology* 45 (12): 2821–2831. doi: 10.1016/j.biocel.2013.10.004.

Romagnolo D, Selmin O (2012). "Flavonoids and cancer prevention: a review of the evidence". *J Nutr Gerontol Geriatr* 31 (3): 206–38. doi: 10.1080/21551197.2012.702534.

Schuier M, Sies H, Illek B, Fischer H (2005). "Cocoa-related flavonoids inhibit CFTR-mediated chloride transport across T84 human colon epithelia". *J. Nutr.* 135 (10): 2320–5.

Serafini M, Bugianesi R, Maiani G, Valtuena S, De Santis S, Crozier A (2003). "Plasma antioxidants from chocolate". *Nature* 424 (6952): 1013.doi: 10.1038/4241013a.

Serafini M, Bugianesi R, Maiani G, Valtuena S, De Santis S, Crozier A (2003). "Nutrition: milk and absorption of dietary flavanols". *Nature* 424 (6952): 1013. doi: 10.1038/4241013a.

Siasos G, Tousoulis D, Tsigkou V, Kokkou E, Oikonomou E, Vavuranakis M (2013). "Flavonoids in atherosclerosis: An overview of their mechanisms of action". *Current medicinal chemistry* 20 (21): 2641–2660. doi: 10.2174/09298673311320210003.

Spencer J (2008). "Flavonoids: modulators of brain function?". *British Journal of Nutrition* 99: ES60–77. doi: 10.1017/S0007114508965776.

Stauth D (2007). "Studies force new view on biology of flavonoids". EurekAlert!, Adapted from a news release issued by Oregon State University.

Tangney C, Rasmussen H (2013). "Polyphenols, Inflammation and Cardiovascular Disease". *Current Atherosclerosis Reports* 15 (5): 324. doi: 10.1007/s11883-013-0324-x. PMC 3651847.

Taylor P, Hamilton-Miller J, Stapleton P (2005). "Antimicrobial properties of green tea catechins". *Food Science and Technology Bulletin* 2 (7): 71–81. doi: 10.1616/1476-2137.14184. PMC 2763290.

The devil in the dark chocolate. *Lancet* 370 (9605): 2070. 2007. doi: 10.1016/S0140-6736(07)61873-X.

Trantas E, Panopoulos N, Ververidis F (2009). "Metabolic engineering of the complete pathway leading to heterologous biosynthesis of various flavonoids and stilbenoids in *Saccharomyces cerevisiae*". *Metabolic Engineering* 11 (6): 355–366. doi: 10.1016/j.ymben.2009.07.004.

van Dam R, Naidoo N, Landberg R (2013). "Dietary flavonoids and the development of type 2 diabetes and cardiovascular diseases". *Current Opinion in Lipidology* 24 (1): 25–33. doi: 10.1097/MOL.0b013e32835bcdff.

Ververidis F, Trantas E, Douglas C, Vollmer G, Kretzschmar G, Panopoulos N (2007). "Biotechnology of flavonoids and other phenylpropanoid-derived natural products. Part I: Chemical diversity, impacts on plant biology and human health".*Biotechnology Journal* 2 (10): 1214–34. doi: 10.1002/biot.200700084.

Ververidis F, Trantas E, Douglas C, Vollmer G, Kretzschmar G, Panopoulos N (2007). "Biotechnology of flavonoids and other phenylpropanoid-derived natural products. Part II: Reconstruction of multienzyme pathways in plants and microbes". *Biotechnology Journal* 2 (10): 1235–49. doi: 10.1002/biot.200700184.

Ververidis Filippos, F, Trantas Emmanouil, Douglas Carl, Vollmer Guenter,

Kretzschmar Georg, Panopoulos Nickolas (2007). "Biotechnology of flavonoids and other phenylpropanoid-derived natural products. Part I: Chemical diversity, impacts on plant biology and human health". *Biotechnology Journal* 2 (10): 1214–34. doi: 10.1002/biot.200700084.PMID 17935117.

Williams R, Spencer J, Rice-Evans C (2004). "Flavonoids: antioxidants or signalling molecules?". *Free Radical Biology and Medicine* 36 (7): 838–49. doi: 10.1016/j.freeradbiomed.2004.01.001.

Woo H, Kim J (2013). "Dietary flavonoid intake and smoking-related cancer risk: a meta-analysis". *PLoS ONE* 8 (9): e75604. Bibcode: 2013PLoSO.875604W. doi: 10.1371/journal. pone. 0075604.

Yamamoto Y, Gaynor R (2001). "Therapeutic potential of inhibition of the NF-chachuuB pathway in the treatment of inflammation and cancer". *Journal of Clinical Investigation* 107 (2): 135–42. doi: 10.1172/JCI11914.

Yisa, Jonathan (2009). "Phytochemical Analysis and Antimicrobial Activity Of *Scoparia dulcis* and *Nymphaea lotus*". *Australian Journal of Basic and Applied Sciences* 3 (4): 3975–3979.

Zamora-Ros R, Agudo A, Luján-Barroso L, Romieu I, Ferrari P, Knaze V (2012). "Dietary flavonoid and lignan intake and gastric adenocarcinoma risk in the European Prospective Investigation into Cancer and Nutrition (EPIC) study". *American Journal of Clinical Nutrition* 96 (6): 1398–1408. doi: 10.3945/ajcn.112.037358.

Chapter 8

Glycosides

In chemistry, a glycoside is a molecule in which a sugar is bound to another functional group via a glycosidic bond. Glycosides play numerous important roles in living organisms. Many plants store chemicals in the form of inactive glycosides. These can be activated by enzyme hydrolysis, which causes the sugar part to be broken off, making the chemical available for use. Many such plant glycosides are used as medications. In animals and humans, poisons are often bound to sugar molecules as part of their elimination from the body. In formal terms, a glycoside is any molecule in which a sugar group is bonded through its anomeric carbon to another group via aglycosidic bond. Glycosides can be linked by an O- (an *O-glycoside*), N- (a *glycosylamine*), S-(a *thioglycoside*), or C- (a *C-glycoside*) glycosidic bond. According to the IUPAC, the name "C-glycoside" is a misnomer; the preferred term is "C-glycosyl compound". The given definition is the one used by IUPAC, which recommends the Haworth projection to correctly assign stereochemical configurations. Many authors require in addition that the sugar be bonded to a *non-sugar* for the molecule to qualify as a glycoside, thus excludingpolysaccharides. The sugar group is then known as the *glycone* and the non-sugar group as the *aglycone* or *genin* part of the glycoside. The glycone can consist of a single sugar group (monosaccharide) or several sugar groups (oligosaccharide).

The first glycoside ever identified was amygdalin, by the French chemists Pierre Robiquet and Antoine Boutron-Charlard, in 1830.

Related Compounds

Molecules containing an N-glycosidic bond are known as glycosylamines and are not discussed in this article. (Many authors in biochemistry call these compounds *N-glycosides* and group them with the glycosides; this is considered a misnomer and discouraged by IUPAC.) Glycosylamines and glycosides are grouped together as glycoconjugates; other glycoconjugates include glycoproteins, glycopeptides, peptidoglycans, glycolipids and lipopolysaccharides.

Chemistry

Much of the chemistry of glycosides is explained in the article on glycosidic bonds. For example, the glycone and aglycone portions can be chemically separated by hydrolysis in the presence of acid and can be hydrolyzed by alkali. There are also numerous enzymes that can form and break glycosidic bonds. The most important cleavage enzymes are theglycoside hydrolases and the most important synthetic enzymes in nature are glycosyltransferases. Genetically altered enzymes termed glycosynthases have been developed that can form glycosidic bonds in excellent yield. There are many ways to chemically synthesize glycosidic bonds. Fischer glycosidation refers to the synthesis of glycosides by the reaction of unprotected monosaccharides with alcohols (usually as solvent) in the presence of a strong acid catalyst. The Koenigs-Knorr reaction is the condensation of glycosyl halides and alcohols in the presence of metal salts such as silver carbonate or mercuric oxide.

Classification

Glycosides can be classified by the glycone, by the type of glycosidic bond and by the aglycone.

By Glycone/Presence of Sugar

If the glycone group of a glycoside is glucose, then the molecule is a glucoside; if it is fructose, then the molecule is a fructoside; if it is glucuronic acid, then the molecule is aglucuronide; etc. In the body, toxic substances are often bonded to glucuronic acid to increase their water solubility; the resulting glucuronides are then excreted.

By Type of Glycosidic Bond

Depending on whether the glycosidic bond lies "below" or "above" the plane of the cyclic sugar molecule, glycosides are classified as **α-glycosides** or **β-glycosides**. Some enzymes such as α-amylase can only hydrolyze α-linkages; others, such as emulsin, can only affect β-linkages.

There are four types of linkages present between glycone and aglycone:

☆ C-linkage/glycosidic bond, "nonhydrolysable by acids or enzymes"

☆ O-linkage/glycosidic bond

☆ N-linkage/glycosidic bond

☆ S-linkage/glycosidic bond

By Aglycone

Glycosides are also classified according to the chemical nature of the aglycone. For purposes of biochemistry and pharmacology, this is the most useful classification.

Alcoholic Glycosides

An example of an alcoholic glycoside is salicin, which is found in the genus *salix*. Salicin is converted in the body into salicylic acid, which is closely related to aspirin and hasanalgesic, antipyretic and antiinflammatory effects.

Anthraquinone Glycosides

These glycosides contain an aglycone group that is a derivative of anthraquinone. They have a laxative effect. They are mainly found in dicot plants except the Liliaceae family which are monocots. They are present in senna, rhubarb and *Aloe* species. Antron and anthranol are reduced forms of anthraquinone.

Coumarin Glycosides

Here, the aglycone is coumarin or a derivative. An example is apterin which is reported to dilate the coronary arteries as well as block calcium channels. Other coumarin glycosides are obtained from dried leaves of *Psoralea corylifolia*.

Chromone Glycosides

In this case, the aglycone is benzo-gamma-pyrone.

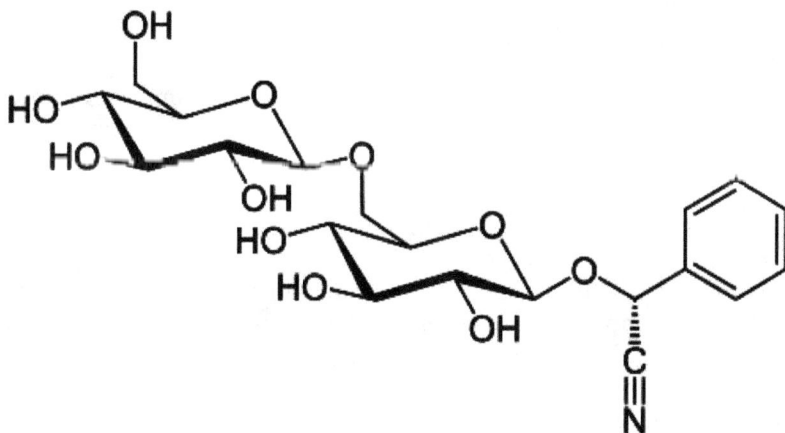

Amygdalin

In this case, the aglycone contains a cyanide group. All of these plants have these glycosides stored in the vacuole, but, if the plant is attacked, they are released and become activated by enzymes in the cytoplasm. These remove the sugar part of the molecule and release toxic hydrogen cyanide. Storing them in inactive forms in the vacuole prevents them from damaging the plant under normal conditions.

An example of these is amygdalin from almonds. They can also be found in the fruits (and wilting leaves) of the rose family (including cherries, apples, plums, almonds, peaches, apricots, raspberries and crabapples). Cassava, an important food plant in Africa and South America, contains cyanogenic glycosides and, therefore, has to be washed and ground under running water prior to consumption. Sorghum (*Sorghum bicolour*) expresses cyanogenic glycosides in its roots and thus is resistant to pests such as rootworms (*Diabrotica* spp.) that plague its cousin maize (*Zea mays* L.). It was once thought that cyanogenic glycosides might have anti-cancer properties, but this idea was disproven (see Amygdalin). A recent study may also show that increasing CO_2 levels may result in much higher levels of cyanogenic glycoside production in sorghum and cassava plants, making them highly toxic and inconsumable. A doubling of CO_2 concentration was found to double the concentration of cyanogenic glycosides in the leaves. Dhurrin, linamarin, lotaustralin and prunasin are also classified as cyanogenic glycosides.

Flavonoid Glycosides

Here, the aglycone is a flavonoid. Examples of this large group of glycosides include:

☆ Hesperidin (aglycone: Hesperetin, glycone: Rutinose)

☆ Naringin (aglycone: Naringenin, glycone: Rutinose)

☆ Rutin (aglycone: Quercetin, glycone: Rutinose)

☆ Quercitrin (aglycone: Quercetin, glycone: Rhamnose)

Among the important effects of flavonoids are their antioxidant effect. They are also known to decrease capillary fragility.

Phenolic Glycosides

Here, the aglycone is a simple phenolic structure. An example is arbutin found in the Common Bearberry *Arctostaphylos uva-ursi*. It has a urinary antiseptic effect.

Saponins

These compounds give a permanent froth when shaken with water. They also cause hemolysis of red blood cells. Saponin glycosides are found in liquorice. Their medicinal value is due to their expectorant and corticoid and anti-inflammatory effects. Steroid saponins, for example, in Dioscorea wild yam the sapogenin diosgenin–in form of its glycosidedioscin is an important starting material for production of semi-synthetic glucocorticoids and other steroid hormones such as progesterone. The ginsenosides are triterpeneglycosides and Ginseng saponins from *Panax Ginseng* C. A. Meyer, (Chinese ginseng) and *Panax quinquefolius* (American Ginseng). In general, the use of the term saponin in organic chemistry is discouraged, because many plant constituents can produce foam and many triterpene-glycosides are amphipolar under certain conditions, acting as asurfactant. More modern uses of saponins in biotechnology are as adjuvants in vaccines: Quil A and its derivative QS-21, isolated from the bark of Quillaja saponaria Molina, to stimulate both the Th1 immune response and the production of cytotoxic T-lymphocytes (CTLs) against exogenous antigens make them ideal for use in subunit vaccines and vaccines directed against intracellular pathogens as well as for therapeutic cancer vaccines but with the aforementioned side-effects of hemolysis.

Steroidal Glycosides or Cardiac Glycosides

Here the aglycone part is a steroidal nucleus. These glycosides are found in the plant genera *Digitalis*, *Scilla* and *Strophanthus*. They are used in the treatment of heart diseases, *e.g.*, congestive heart failure (historically as now recognised does not improve survivability; other agents are now preferred) and arrhythmia.

Steviol Glycosides

These sweet glycosides found in the stevia plant *Stevia rebaudiana* Bertoni have 40-300 times the sweetness of sucrose. The two primary glycosides, stevioside and rebaudioside A, are used as natural sweeteners in many countries. These glycosides have steviol as the aglycone part. Glucose or rhamnose-glucose combinations are bound to the ends of the aglycone to form the different compounds.

Thioglycosides

As the name implies (q.v. thio-), these compounds contain sulfur. Examples include sinigrin, found in black mustard and sinalbin, found in white mustard.

References

Brito-Arias, Marco (2007). *Synthesis and Characterization of Glycosides.* Springer. ISBN 978-0-387-26251-2.

Cassavas get cyanide hike from carbon emissions - environment". *New Scientist.* 13 July 2009. Retrieved 2009-07-13.

IUPAC Gold Book - Glycosides.

Lindhorst, T.K. (2007). *Essentials of Carbohydrate Chemistry and Biochemistry.* Wiley-VCH. ISBN 978-3-527-31528-4.

Robiquet, Boutron-Charlard (1830). "Nouvelles expériences sur les amandes amères et sur l'huile volatile qu'elles fournissent". *Annales de chimie et de physique* (in French) 44: 352–382.

Staples such as cassava on which millions of people depend become more toxic and produce much smaller yields in a world with higher carbon dioxide levels and more drought.*Yahoo.com.* July 2009.

Sun, Hong-Xiang, Xie, Yong, Ye, Yi-Ping (2009). "Advances in saponin-based adjuvants". *Vaccine* 27 (12): 1787–1796. doi: 10.1016/j.vaccine.2009.01.091.

Chapter 9

Gossypol

Gossypol is a natural phenol derived from the cotton plant (genus *Gossypium*). Gossypol is a phenolic aldehyde that permeates cells and acts as an inhibitor for several dehydrogenase enzymes. It is a yellow pigment. Among other things, it has been tested as a male oral contraceptive in China. In addition to its contraceptive properties, gossypol has also long been known to possess antimalarial properties. Other researchers are investigating the anticancer properties of gossypol.

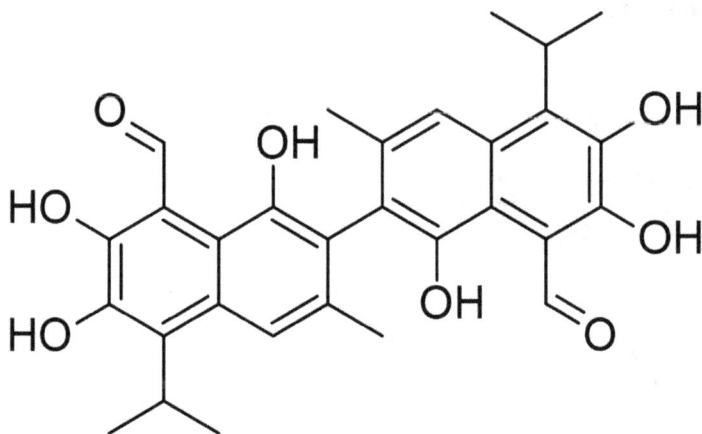

Gossypol

Biological Properties

It has proapoptotic properties, probably due to the regulation of the Bax and Bcl2. It also reversibly inhibits calcineurin and binds tocalmodulin. It inhibits replication of the HIV-1 virus. It is an effective protein kinase C inhibitor. It also causes low potassium levels and thus causes temporary paralysis.

Biosynthesis

Gossypol is a terpenoid aldehyde, which is formed metabolically through acetate via the isoprenoid pathway. Sesquiterpene dimer undergoes a radical coupling reaction to form gossypol. Geranyl pyrophosphate (GPP) and IPP make sesquiterpene precursor, farnesyl diphosphate (FPP), for gossypol. The biosynthesis of gossypol is summarized in figure below. The biosynthesis begins with the 0 compound derived from GPP and IPP. Cadinyl cation (1) is oxidized to 2 by (+)-d-cadinene synthase. The (+)-d-cadinene (2) is involved in making the basic aromatic sesquiterpene unit, homigossypol, by oxidation, which generates the 3 (8-hydroxy-d-cadinene) with the help of (+)-d-cadinene 8-hyroxylase. At 5, the 3 goes through various oxidative processes to make 4 (deoxyhemigossypol), which is oxidized by one electron into hemigossypol (5,6,7) and then undergoes a phenolic oxidative coupling, ortho to the phenol groups, to form 8 (gossypol). The coupling is catalyzed by a hydrogen peroxide-dependent peroxidise enzyme, which results in the final product.

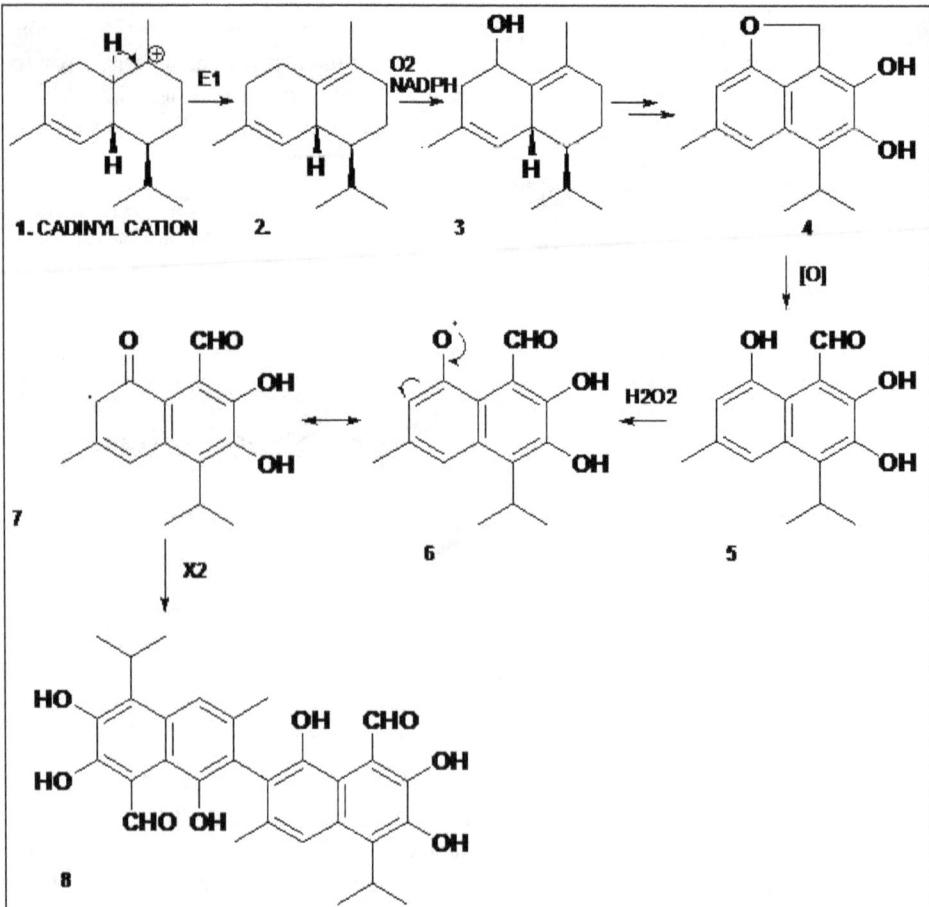

Research

Contraception

A 1929 investigation in Jiangxi showed correlation between low fertility in males and use of crude cottonseed oil for cooking. The compound causing the contraceptive effect was determined to be gossypol. In the 1970s, the Chinese government began researching the use of gossypol as a contraceptive. Their studies involved over 10,000 subjects and continued for over a decade. They concluded gossypol provided reliable contraception, could be taken orally as a tablet and did not upset men's balance of hormones. However, gossypol also had serious flaws. The studies also discovered an abnormally high rate of hypokalemia among subjects. Hypokalemia low blood potassium levels causes symptoms of fatigue, muscle weakness and at its most extreme, paralysis. In addition, about 7 per cent of subjects reported effects on their digestive systems and about 12 per cent had increased fatigue. Most subjects recovered after stopping treatment and taking potassium supplements. The same study showed taking potassium supplements during gossypol treatment did not prevent hypokalemia in primates. The potassium deficiency may also be a result of the Chinese diet or genetic predisposition. In the mid-1990s, the Brazilian pharmaceutical company Hebron announced plans to market a low-dose gossypol pill called Nofertil, but the pill never came to market. Its release was indefinitely postponed due to unacceptably high rates of permanent infertility. 5-25 per cent of the men remained azoospermic up to a year after stopping treatment. The longer the men had taken the drug and the higher their overall dosages, the more likely they were to have lowered fertility or to become completely infertile. Researchers have suggested gossypol might make a good noninvasive alternative to surgical vasectomy. In 1986, the Chinese stopped research because of these side effects. In 1998, the World Health Organization's Research Group on Methods for the Regulation of Male Fertility recommended the research should be abandoned. In addition to the other side effects, the WHO researchers were concerned about gossypol's toxicity: the toxic dose in primates is less than 10 times the contraceptive dose.[6] This report effectively ended further studies of gossypol as a temporary contraceptive, but research into using it as an alternative to vasectomy continues in Austria, Brazil, Chile, China, the Dominican Republic and Nigeria.

Cancer

Gossypol is also under investigation as a possible chemotherapy drug, especially in its R- state. It is currently believed that gossypol in itself will not kill cancerous cells; however, it changes the chemistry within the cancer cell and makes it more susceptible to traditional chemotherapy drugs. Phased trials have been done on resistant prostate and lung cancer.Few results have been published to date, so no conclusions can be drawn.

Toxicity and Potential Food Source

Food and animal agricultural industries must manage cotton-derivative product levels to avoid toxicity. For example, only ruminant microflora can digest gossypol, but only to a certain level and cottonseed oil must be refined. A research team at Texas A and M University has genetically engineered cotton plants that contain very little gossypol in the seed, but still contain the compound in the stems and leaves. This provides protection against pests and diseases, while allowing the seed to be used for oil and meal for human consumption. The plants are modified by RNA interference, shutting down the genes for gossypol production in the seed, while leaving them unaffected in the rest of the plant. The resulting gossypol-free cottonseed is then suitable as a high-quality protein source suitable for consumption not only by cattle, but also by humans. Protein makes up 23 per cent of the cottonseed.

Gossypol is toxic to erythrocytes *in vitro* by stimulating cell death contributing to the side effect of hemolytic anemia.

References

Burgos, M., Ito, S., Segal, J. S., Tran, T. P. (1997). "Effect of Gossypol on Ultrastructure of Spisula Sperm". *Biol. Bull.* 193 (2): 228–229.

Cottonseed Protein: From Farmers to Your Family Table. Medgadget.com (2006-11-22). Retrieved on 2012-06-09.

Coutinho, F. M. (2002). "Gossypol: a contraceptive for men". *Contraception* 65 (4): 259–263. doi: 10.1016/S0010-7824(02)00294-9.

Dewick, P. M. Medicinal Natural Product: A Biosynthetic approach. 3rd ed., 2008 ISBN 0-470-74167-8

Gossypol (Gossipol). Bioscreening.net (last updated 2008-07-09). Retrieved on 2012-06-09.

Gossypol. Malecontraceptives.org (2011-07-27). Retrieved on 2013-03-30.

Heinstein, P. F., Herman, L. D., Tove, B. S., Smith, H. F. (1970). "Biosynthesis of Gossypol". *J. Biol. Chem.* 245 (18): 4658–4665.

Polsky, B, Segal, SJ, Baron, PA, Gold, JW, Ueno, H, Armstrong, D (1989). "Inactivation of human immunodeficiency virus *in vitro* by gossypol". *Contraception* 39 (6): 579–87.doi: 10.1016/0010-7824(89)90034-6.

Walsh, Brian. Hungry? How About Some Protein-Rich Cotton., Time Magazine, September 14, 2009, p. 54

Zbidah, M, Lupescu, A, Shaik, N, Lang, F (2012). "Gossypol-induced suicidal erythrocyte death". *Toxicology* 302 (2–3): 101–5. doi: 10.1016/j.tox.2012.09.010.

Chapter 10

Non-protein Amino Acids

In biochemistry, non-coded, non-proteinogenic, or "unnatural" amino acids are those not naturally encoded or found in the genetic code of any organisms. Despite the use of only 23 amino acids (21 in eukaryotes) by the translational machinery to assemble proteins (the proteinogenic amino acids), over 140 natural amino acids are known and thousands of more combinations are possible. Several non-proteinogenic amino acids are noteworthy because they are:

☆ Intermediates in biosynthesis

☆ Post-translationally incorporated into protein

☆ Possess a physiological role (*e.g.* components of bacterial cell walls, neurotransmitters and toxins)

☆ Natural and man-made pharmacological compounds

☆ Present in meteorites and in prebiotic experiments (*e.g.* Miller–Urey experiment)

Definition by Negation

Technically, any organic compound with an amine (-NH$_2$) and a carboxylic acid (-COOH) functional group is an amino acid. The proteinogenic amino acids are small subset of this group that possess central carbon atom (α- or 2-) bearing an amino group, a carboxyl group, a side chain and an α-hydrogen levo conformation, with the exception of glycine, which is achiral and proline, whose amine group is a secondary amine and is consequently frequently

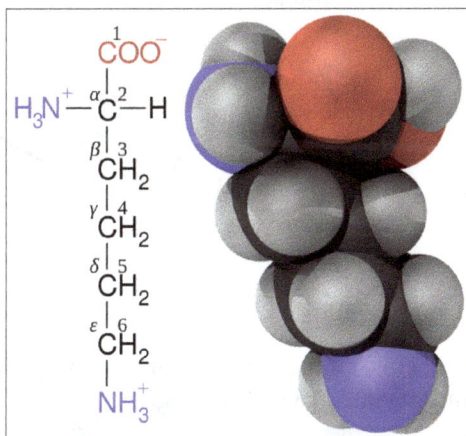

Lysine

referred to as an imino acid for traditional reasons, albeit not an imino. The genetic code encodes 20 standard amino acids. However, there are three extra proteinogenic amino acids: selenocysteine,pyrrolysine and N-formylmethionine. The former two do not have a dedicated codon, but are added in place of a stop codon when a specific sequence is present, UGA codon and SECIS element for selenocysteine, UAG PYLIS downstream sequence for pyrrolysine. Formylmethionine is an amino acid encoded by the start codon AUG in bacteria, mitochondria and chloroplasts, but is often removed post-translationaly.

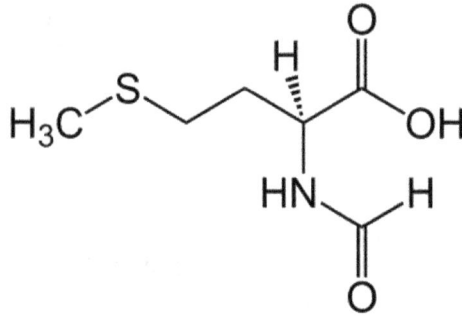

Formylmethionine

This amino acid is a methionine whose amino group has been protected by a formyl group.

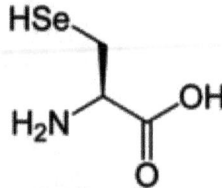

Selenocysteine

This amino acid contains a selenol group on its β-carbon

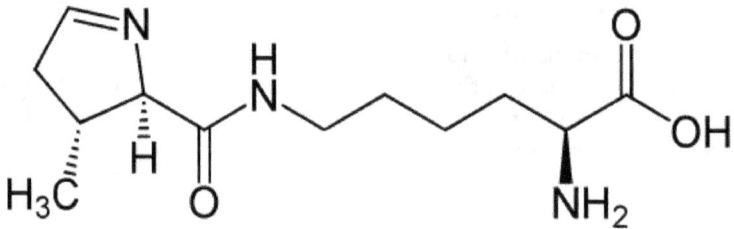

Pyrrolysine

This amino acid is formed by joining to the α-amino group of lysine a carboxylated pyrroline ring.

There are various groups of amino acids:

⭐ 20 standard amino acids

⭐ 23 proteinogenic amino acids

⭐ over 80 amino acids created abiotically in high concentrations

⭐ about 900 are produced by natural pathways

⭐ Over 118 engineered amino acids have been placed into protein

These groups overlap, but are not identical. All 23 proteinogenic amino acids are biosynthesised by organisms, but not all of them are abiotic (found in prebiotic experiments and meteorites), such as histidine. Many amino acids, such as ornithine, are metabolic intermediates produced biotically, but not coded. Others are only metabolic intermediates, such as citrulline. Others are solely found in abiotic mixes, such as α-methylnorvaline. Over 30 unnatural amino acids have been translationally inserted into protein in engineered systems, yet are not biosynthetic.

Nomenclature

In addition to the IUPAC numbering system to differentiate the various carbons in an organic molecule, by sequentially assigning a number to each carbon, including those forming a carboxylic group, the carbons along the side-chain of amino acids can also labelled with Greek letters, where the α-carbon is the central chiral carbon possessing a carboxyl group, a side chain and, in α-amino acid, an amine group the carbon in carboxylic groups is not counted. (Consequently, the IUPAC names of many non-proteinogenic α-amino acids start with 2-amino and end in -ic acid.)

Natural, but non L-α-Amino Acids

Most natural amino acids are α-amino acids in the L conformation, but some exceptions exist.

Non Alpha

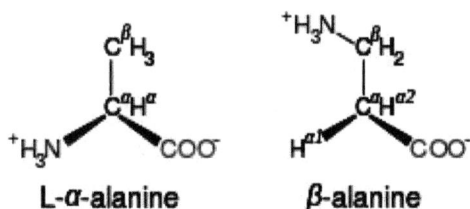

L-α-alanine   β-alanine

Some non-α amino acids exist in organisms, such as β-alanine, GABA and δ-Aminolevulinic acid.

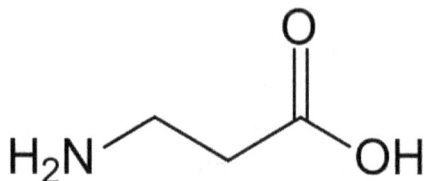

β-alanine: an amino acid produced by aspartate decarboxylase (*pan*Dencoded) and joined to pantoate via an amide bond, forming pantothenic acid, a precursor for cofactor A.

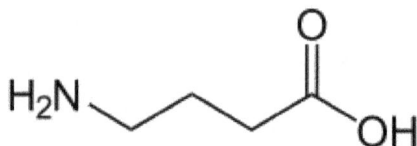

γ-Aminobutyric acid (GABA): a neurotransmitter in animals.

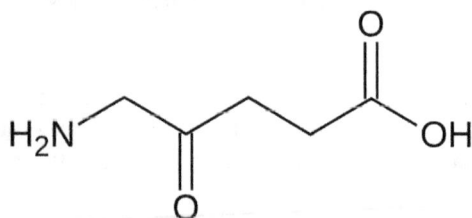

δ-Aminolevulinic acid: an intermediate in tetrapyrrole biosynthesis (haem, chlorophyll,cobalamin *etc.*).

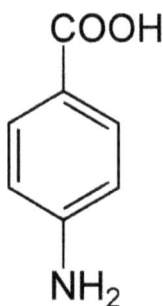

4-Aminobenzoic acid (PABA): an intermedia in folate biosynthesis.

The reason why α amino acids are using in protein has been attributed to their frequency in meteorites and prebiotic experiments. An initial speculation on the deleterious properties of β amino acids in terms of secondary structure, turned out to be incorrect. Additionally, several man-made inhibitors exist that are not α amino acids, such as isoserine.

D Amino Acids

Most bacterial cells walls are formed by peptidoglycan, a polymer composed of amino sugar crosslinked with short oligopeptides bridged between each other. The oligopeptide is non-ribosomally synthesised and contains several peculiarities, including D-amino acids, generally D-alanine and D-glutamate. A further peculiarity is that the former is racemised by a PLP-binding enzymes (encoded by *alr* or the homologue *dadX*), whereas the latter is racemised by a cofactor independent enzyme (*murI*). Some variants are present, in*Thermotoga* spp. D-lysine is present and in certain vancomycin-resistant strains D-serine is present (*vanT* gene). In animals, some D-amino acids are neurotransmitters.

Without a Hydrogen on the α-carbon

All proteinogenic amino acids have at least one hydrogen on the α-carbon: this is due to the different specificity the rybosomal transferase activity would require for a α-hydrogen versus a α-methyl and the biosynthetic problems faced with the quaternary carbon, which would block PLP-dependent catalysis (both SN2 and E2/attack). Nevertheless, some exceptions are present. In some fungi α-Amino isobutyric acid is produced as a monomer to synthesise some antibiotics. This compound is similar to alanine, but possess a methyl group instead of a hydrogen, given that it possess two methyl group on the α-carbon, the latter is therefore not a stereocentre. Another compound similar to alanine without an α-hydrogen is dehydroalanine, which possess a methene sidechain.

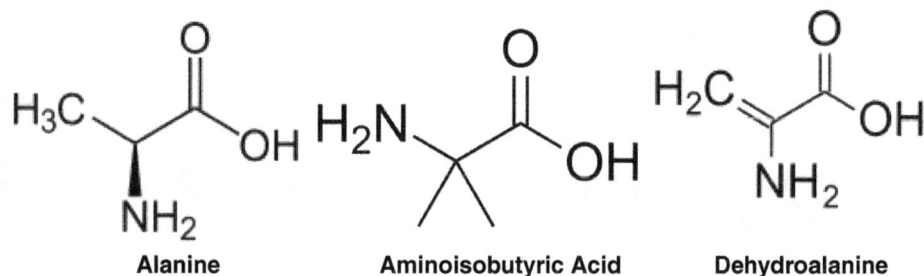

| Alanine | Aminoisobutyric Acid | Dehydroalanine |

Twin Amino Acid Stereocentres

A subset of L α amino acids possess two ends that could be considered α amino acids (obviously only one end is the α). In protein cysteine residues forms a disulfide bond with other cysteine residues therefore crosslinking the protein, two crosslinked cysteines form a cystine molecule. Cysteine and methionine are generally produced by direct sulfurylation, but in some species the can be produced by transfurylation, where the activated homoserine or serine is fused to a cysteine or homocysteine forming cystathionine, a molecule composed of a thioether bridged serine/cysteine moiety with a homoserine/homocysteine. A similar compound is lanthionine which can be seen as two cysteines joined via a thioether and is found in various organisms. Similarly, djenkolic acid, a plant toxin from jengkol beans, is composed of two cysteines joined via two thioethers separated by a methylene

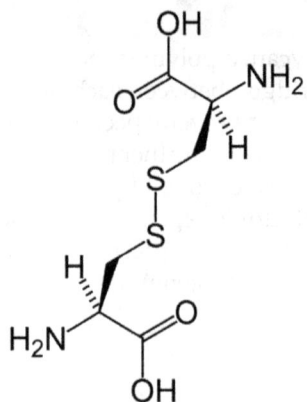

Cysteine

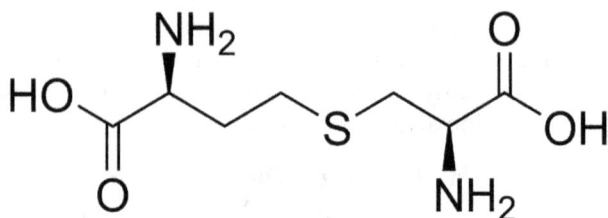

Cystathionine

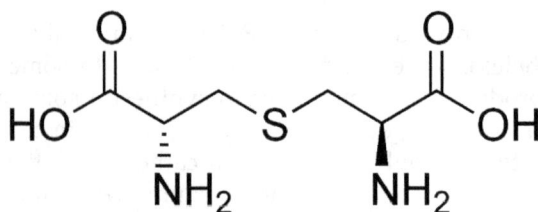

Lanthionine

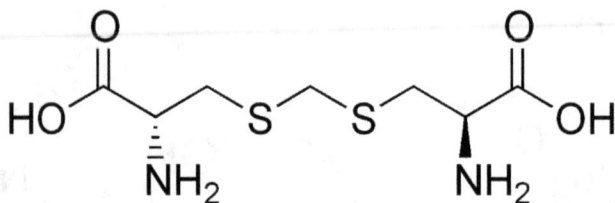

Djenkolic Acid

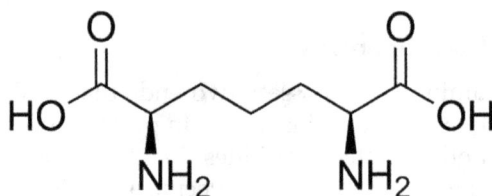

Diaminopimelic Acid

group. Diaminopimelic acid is both used as a bridge in petidoglycan and is used a precursor to lysine (via its decarboxylation).

Prebiotic Amino Acids and Alternative Biochemistries

In meteorites and in prebiotic experiments (*e.g.* Miller–Urey experiment) many more amino acids than the twenty standard amino acids are found, several of which at higher concentrations that the standard ones: it has been conjectured that if amino acid based life were to arise in parallel elsewhere in the universe, no more than 75 per cent of the amino acids would be in common. The most notable anomaly is the lack of aminobutyric acid.

Proportion of Amino Acids Relative to Glycine (per cent)

Molecule	Electric Discharge	Murchinson Meteorite
Glycine	100	100
Alanine	180	36
α-Amino-n-butyric acid	61	19
Norvaline	14	14
Valine	4.4	
Norleucine	1.4	
Leucine	2.6	
Isoleucine	1.1	
Alloisoleucine	1.2	
t-leucine	< 0.005	
α-Amino-n-heptanoic acid	0.3	
Proline	0.3	22
Pipecolic acid	0.01	11
α,β-diaminopropionic acid	1.5	
α,γ-diaminobutyric acid	7.6	
Ornithine	< 0.01	
lysine	< 0.01	
Aspartic acid	7.7	13
Glutamic acid	1.7	20
Serine	1.1	
Threonine	0.2	
Allothreonine	0.2	
Methionine	0.1	
Homocysteine	0.5	
Homoserine	0.5	
β-Alanine	4.3	10
β-Amino-n-butyric acid	0.1	5
β-Aminoisobutyric acid	0.5	7
γ-Aminobutyric acid	0.5	7
α-Aminoisobutyric acid	7	33

Contd...

Contd...

Molecule	Electric Discharge	Murchinson Meteorite
isovaline	1	11
Sarcosine	12.5	7
N-ethyl glycine	6.8	6
N-propyl glycine	0.5	
N-isopropyl glycine	0.5	
N-methyl alanine	3.4	3
N-ethyl alanine	< 0.05	
N-methyl β-alanine	1.0	
N-ethyl β-alanine	< 0.05	
isoserine	1.2	
α-hydroxy-γ-aminobutyric acid	17	

Straight Side Chain

The genetic code has been described as a frozen accident and the reasons why there is only one standard amino acid with a straight chain (alanine) could simply be redundancy with valine, leucine and isoleucine. However, straight chained amino acids are reported to form much more stable alpha helices.

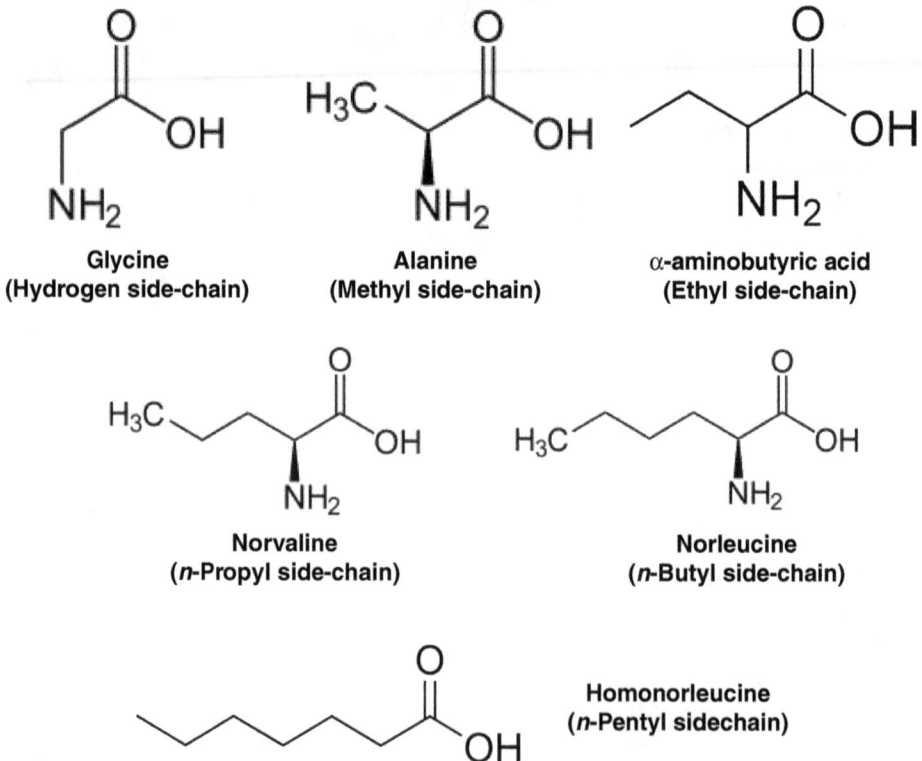

Glycine
(Hydrogen side-chain)

Alanine
(Methyl side-chain)

α-aminobutyric acid
(Ethyl side-chain)

Norvaline
(*n*-Propyl side-chain)

Norleucine
(*n*-Butyl side-chain)

Homonorleucine
(*n*-Pentyl sidechain)

Chalcogen

Serine, homoserine, O-methyl-homoserine and O-ethyl-homoserine possess a hydroxymethyl, hydroxyethyl, O-methyl-hydroxymethyl and O-methyl-hydroxyethyl side chain. Whereas cysteine, homocysteine, methionine and ethionine possess the thiol equivalents. The selenol equivalents are selenocysteine, selenohomocysteine, selenomethionine and selenoethionine. Amino acids with the next chalcogen down are also found in nature: several species such as Aspergillus fumigatus, Aspergillus terreus and Penicillium chrysogenum in the absence of sulfur are able to produce and incorporate into protein tellurocysteine and telluromethionine. Hydroxyglycine, an amino acid with a hydroxyl side-chain, is highly unstable.

Expanded Genetic Code

Roles

In cells, especially autotrophs, several non-proteinogenic amino acids are found as metabolic intermediates. However, despite the catalytic flexibility of PLP-binding enzymes, many amino acids are synthesised as keto-acids (*e.g.* 4-methyl-2-oxopentanoate to leucine) and aminated in the last step, thus keeping the number of non-proteinogenic amino acid intermediates fairly low. Ornithine and citrulline occur in the urea cycle, part of amino acid catabolism. In addition to primary metabolism, several non-proteinogenic amino acids are precursors or the final production in secondary metabolism to make small compounds or non-ribosomal peptides (such as some toxins).

Post-Translationally Incorporated into Protein

Despite not being encoded by the genetic code as proteinogenic amino acids, some non-standard amino acids are nevertheless found in proteins. These are formed by post-translational modification of the side chains of standard amino acids present in the target protein. These modifications are often essential for the function or regulation of a protein; for example, in Gamma-carboxyglutamate the carboxylation of glutamate allows for better binding of calcium cations and in hydroxyproline the hydroxylation of prolineis critical for maintaining connective tissues. Another example is the formation of hypusine in the translation initiation factor EIF5A, through modification of a lysine residue. Such modifications can also determine the localization of the protein, *e.g.*, the addition of long hydrophobic groups can cause a protein to bind to a phospholipid membrane.

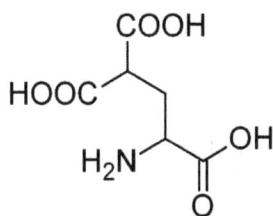

Carboxyglutamic Acid

Whereas glutamic acid possesses one γ-carboxyl group, Carboxyglutamic acid possess two.

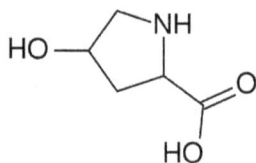

Hydroxyproline

This imino acid differs from proline due to a hydroxyl group on carbon 4.

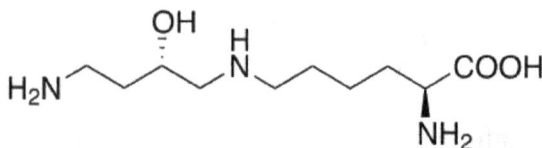

Hypusine

This amino acid is obtained by adding to the α-amino group of a lysine a 4-aminobutyl moiety (obtained fromspermidine)

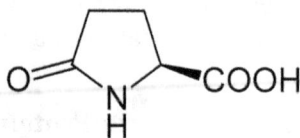

Pyroglutamic Acid

There is some preliminary evidence that aminomalonic acid may be present, possibly by misincorporation, in protein.

Toxic Analogues

Several non-proteinogenic amino acids are toxic due to their ability to mimic certain properties of proteinogenic amino acids, such as thialysine. Some non-proteinogenic amino acids are neurotoxic by mimicking amino acids used

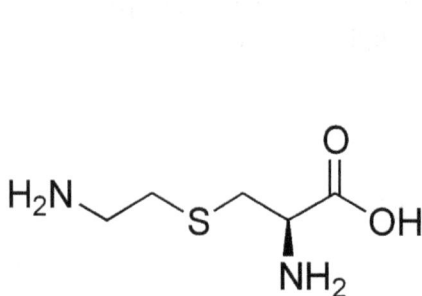

Thialysine

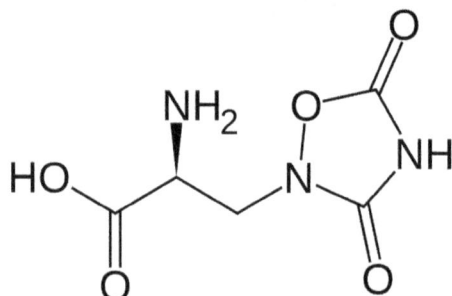

Quisqualic Acid

as neurotransmitters (*i.e.* not for protein biosynthesis), *e.g.* Quisqualic acid, canavanine or azetidine-2-carboxylic acid. Cephalosporin C has a α-aminoadipic acid (homoglutamate) backbone that is amidated with a cephalosporin moiety. Penicillamine is therapeutic amino acid, whose mode of action is unknown.

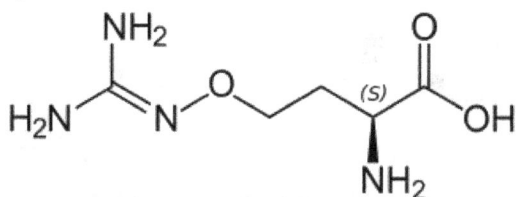

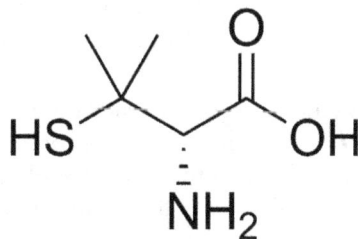

Canavanine

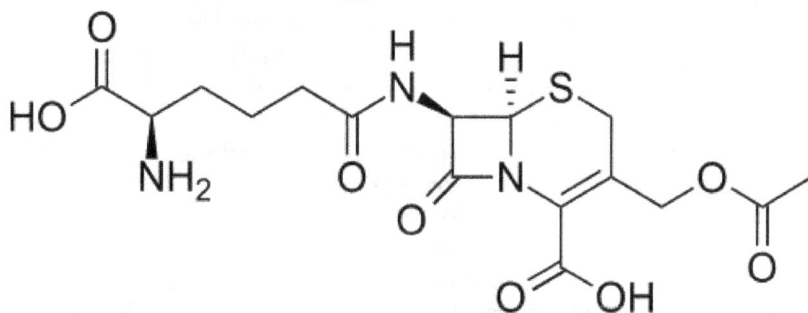

Cephalosporin C

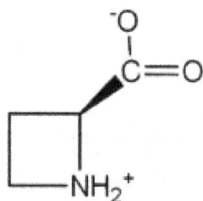

Azetidine-2-carboxylic Acid   **Penicillamine**

Naturally-occurring cyanotoxins can also include non-proteinogenic amino acids. Microcystin and nodularin, for example, are both derived from ADDA, a β-amino acid.

Not Amino Acids

Taurine is an amino sulfonic acid and not an amino acid, however it is occasionally considered as such as the amounts required to suppress the auxotroph in certain organisms (*e.g.* cats) are closer to those of "essential amino acids" (amino acid auxotrophy) than of vitamins (cofactor auxotrophy). The osmolytes, sarcosine and glycine betaine are derived from amino acids, but have an secondary and quaternary amine respectively.

Non-proteinogenic Amino Acids

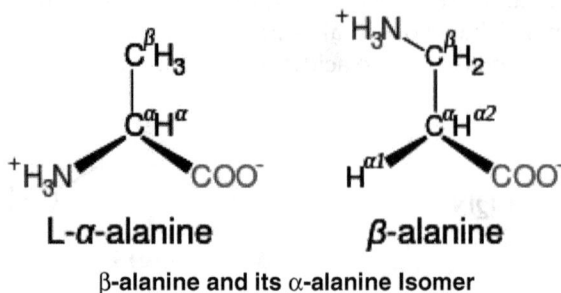

L-α-alanine　　　　　β-alanine

β-alanine and its α-alanine Isomer

Aside from the 23 proteinogenic amino acids, there are many other amino acids that are called *non-proteinogenic*. Those either are not found in proteins (for example carnitine, GABA) or are not produced directly and in isolation by standard cellular machinery (for example, hydroxyproline and selenomethionine). Non-proteinogenic amino acids that are found in proteins are formed by post-translational modification, which is modification after translation during protein synthesis. These modifications are often essential for the function or regulation of a protein; for example, the carboxylation of glutamate allows for better binding of calcium cations and the hydroxylation of proline is critical for maintainingconnective tissues. Another example is the formation of hypusine in the translation initiation factor EIF5A, through modification of a lysine residue. Such modifications can also determine the localization of the protein, *e.g.*, the addition of long hydrophobic groups can cause a protein to bind to a phospholipid membrane. Some non-proteinogenic amino acids are not found in proteins. Examples include lanthionine, 2-aminoisobutyric acid, dehydroalanine and the neurotransmitter gamma-aminobutyric acid. Non-proteinogenic amino acids often occur as intermediates in the metabolic pathways for standard amino acids – for example, ornithine and citrulline occur in the urea cycle, part of amino acid catabolism. A rare exception to the dominance of α-amino acids in biology is the β-amino acid beta alanine (3-aminopropanoic acid), which is used in plants and microorganisms in the synthesis of pantothenic acid (vitamin B_5), a component of coenzyme A.

Non-Standard Amino Acids

The 20 amino acids that are encoded directly by the codons of the universal genetic code are called *standard* or *canonical* amino acids. The others are called *non-standard* or *non-canonical*. Most of the non-standard amino acids are also non-proteinogenic (*i.e.* they cannot be used to build proteins), but three of them are proteinogenic, as they can be used to build proteins by exploiting information not encoded in the universal genetic code. The three non-standard proteinogenic amino acids are selenocysteine (present in many noneukaryotes as well as most eukaryotes, but not coded directly by DNA), pyrrolysine (found only in some archea and one bacterium) and N-formylmethionine (which is often the initial amino acid of proteins in bacteria, mitochondria and chloroplasts). For example, 25 human proteins include selenocysteine (Sec) in their primary structure and the structurally characterized enzymes (selenoenzymes) employ Sec as the catalytic moiety in their active sites. Pyrrolysine and selenocysteine are encoded via variant codons. For example, selenocysteine is encoded by stop codon and SECIS element.

In Human Nutrition

When taken up into the human body from the diet, the 22 standard amino acids either are used to synthesize proteins and other biomolecules or are oxidized to urea and carbon dioxide as a source of energy. The oxidation pathway starts with the removal of the amino group by a transaminase; the amino group is then fed into the urea cycle. The other product of transamidation is a keto acid that enters the citric acid cycle. Glucogenic amino acids can also be converted into glucose, through gluconeogenesis. Pyrrolysine trait is restricted to several microbes and only one organism has both Pyl and Sec. Of the 22 standard amino acids, 9 are called essential amino acids because thehuman body cannot synthesize them from other compounds at the level needed for normal growth, so they must be obtained from food. In addition, cysteine, taurine, tyrosine and arginine are considered semiessential amino-acids in children (though taurine is not technically an amino acid), because the metabolic pathways that synthesize these amino acids are not fully developed. The amounts required also depend on the age and health of the individual, so it is hard to make general statements about the dietary requirement for some amino acids.

Non-protein Functions

An amino acid neurotransmitter is an amino acid which is able to transmit a nerve message across a synapse.Neurotransmitters (chemicals) are packaged into vesicles that cluster beneath the axon terminal membrane on the presynaptic side of a synapse in a process called endocytosis. Amino acid neurotransmitter release (exocytosis) is dependent upon calcium Ca^{2+} and is a presynaptic response. There are inhibitory amino acids (IAA) or excitatory amino acids (EAA). Some EAA are L-Glutamate, L-Aspartate, L-Cysteine and L-Homocysteine. These neurotransmitter systems will activate post-synaptic cells. Some IAA includes GABA, Glycine, β-Alanine and Taurine. The IAA depress the activity of post-synaptic cells.

In humans, non-protein amino acids also have important roles as metabolic intermediates, such as in the biosynthesis of the neurotransmitter gamma-amino-butyric acid (GABA). Many amino acids are used to synthesize other molecules, for example:

☆ Tryptophan is a precursor of the neurotransmitter serotonin.

☆ Tyrosine (and its precursor phenylalanine) are precursors of the catecholamine neurotransmitters dopamine, epinephrine and norepinephrine.

☆ Glycine is a precursor of porphyrins such as heme.

☆ Arginine is a precursor of nitric oxide.

☆ Ornithine and S-adenosylmethionine are precursors of polyamines.

☆ Aspartate, glycine and glutamine are precursors of nucleotides.

☆ Phenylalanine is a precursor of various phenylpropanoids, which are important in plant metabolism.

However, not all of the functions of other abundant non-standard amino acids are known. Some non-standard amino acids are used as defenses against herbivores in plants. For example, canavanine is an analogue of arginine that is found in many legumes and in particularly large amounts in *Canavalia gladiata* (sword bean). This amino acid protects the plants from predators such as insects and can cause illness in people if some types of legumes are eaten without processing. The non-protein amino acid mimosine is found in other species of legume, in particular *Leucaena leucocephala*. This compound is an analogue of tyrosine and can poison animals that graze on these plants.

Monoamine neurotransmitters are neurotransmitters and neuromodulators that contain one amino group that is connected to an aromatic ring by a two-carbon chain ($-CH_2-CH_2-$). All monoamines are derived from aromatic amino acids like phenylalanine, tyrosine, tryptophan and the thyroid hormones by the action of aromatic amino acid decarboxylase enzymes. Monoaminergic systems, *i.e.*, the networks of neurons that utilize monoamine neurotransmitters, are involved in the regulation of cognitive processes such as emotion, arousal and certain types of memory. It has been found that monoamine neurotransmitters play an important role in the secretion and production of neurotrophin-3 by astrocytes, a chemical which maintains neuron integrity and provides neurons with trophic support. Drugs used to increase the effect of monoamine may be used to treat patients with psychiatric disorders, including depression, anxiety and schizophrenia.

Biodegradable Plastics

Biodegradable plastics are plastics that are capable of being decomposed by bacteria or other living organisms. Two basic classes of biodegradable plastics exist: Bioplastics, whose components are derived from renewable raw materials and plastics made from petrochemicals with biodegradable additives which enhance biodegradation.

Amino acids are under development as components of a range of biodegradable polymers. These materials have applications as environmentally friendly packaging and in medicine in drug delivery and the construction of prosthetic implants. These polymers include polypeptides, polyamides, polyesters, polysulfides and polyurethanes with amino acids either forming part of their main chains or bonded as side-chains. These modifications alter the physical properties and reactivities of the polymers. An interesting example of such materials is polyaspartate, a water-soluble biodegradable polymer that may have applications in disposable diapers and agriculture. Due to its solubility and ability to chelate metal ions, polyaspartate is also being used as a biodegradeable anti-scaling agent and a corrosion inhibitor. In addition, the aromatic amino acid tyrosine is being developed as a possible replacement for toxic phenols such as bisphenol A in the manufacture of polycarbonates.

Examples of Biodegradable Plastics

☆ While aromatic polyesters are almost totally resistant to microbial attack, most aliphatic polyesters are biodegradable due to their potentially hydrolysable ester bonds:

☆ Naturally Produced: Polyhydroxyalkanoates (PHAs) like the poly-3-hydroxybutyrate (PHB), polyhydroxyvalerate (PHV) and polyhydroxyhexanoate (PHH);

☆ Renewable Resource: Polylactic acid (PLA);

☆ Synthetic: Polybutylene succinate (PBS), polycaprolactone (PCL).

☆ Polyanhydrides

☆ Polyvinyl alcohol

☆ Most of the starch derivatives

☆ Cellulose esters like cellulose acetate and nitrocellulose and their derivatives (celluloid).

☆ Enhanced biodegradable plastic with additives.

References

Akiyama, M., Tsuge, T., Doi, Y. (2003). Polymer Degradation and Stability, 80: 183-194.

Ambrogelly, A., Palioura, S., Söll, D. (2007). "Natural expansion of the genetic code". *Nature Chemical Biology* 3 (1): 29–35. doi: 10.1038/nchembio847.

Arias, C. A., Martín-Martinez, M., Blundell, T. L., Arthur, M., Courvalin, P., Reynolds, P. E. (1999). "Characterization and modelling of VanT: A novel, membrane-bound, serine racemase from vancomycin-resistant Enterococcus gallinarum BM4174". *Molecular microbiology* 31 (6): 1653–1664. doi: 10.1046/j.1365-2958.1999.01294.x.

ASTM D5526 - 94 (2011) e1 Standard Test Method for Determining Anaerobic Biodegradation of Plastic Materials Under Accelerated Landfill Conditions". Astm.org. Retrieved 2011-06-30.

ASTM Subcommittee D20.96: Published standards under D20.96 jurisdiction. Astm. org. Retrieved 2011-06-30.

Axon Terminal : on Medical Dictionary Online. Archived from the original on 14 January 2009. Retrieved 2008-12-25.

Bhattacharjee, A., Bansal, M. (2005). "Collagen Structure: The Madras Triple Helix and the Current Scenario". *IUBMB Life (International Union of Biochemistry and Molecular Biology: Life)*57 (3): 161–172. doi: 10.1080/15216540500090710.

Biodegradable plastic and additives. Biosphere Biodegradable Plastic. Retrieved 2011-06-30.

Blenis, J., Resh, M. D. (1993). "Subcellular localization specified by protein acylation and phosphorylation". *Current opinion in cell biology* 5 (6): 984–989. doi: 10.1016/0955-0674 (93)90081-Z.

Böck, A., Forchhammer, K., Heider, J., Baron, C. (1991). "Selenoprotein synthesis: An expansion of the genetic code". *Trends in biochemical sciences* 16 (12): 463–467. doi: 10.1016/0968-0004 (91)90180-4.

Bohlmann, G. Biodegradable polymer life cycle assessment, Process Economics Program, 2001.

Bonhomme, S. Environmental biodegradation of polyethylene. *Polym. Deg. Stab* 81, 441-452 (2003).

Boniface, A., Parquet, C., Arthur, M., Mengin-Lecreulx, D., Blanot, D. (2009). "The Elucidation of the Structure of Thermotoga maritima Peptidoglycan Reveals Two Novel Types of Cross-link". *Journal of Biological Chemistry* 284 (33): 21856–21862. doi: 10.1074/jbc.M109.034363.

Chakauya, E., Coxon, K. M., Ottenhof, H. H., Whitney, H. M., Blundell, T. L., Abell, C., Smith, A. G. (2005). "Pantothenate biosynthesis in higher plants". *Biochemical Society Transactions*33 (4): 743–746. doi: 10.1042/BST0330743.

Copley, S. D., Frank, E., Kirsch, W. M., Koch, T. H. (1992). "Detection and possible origins of aminomalonic acid in protein hydrolysates". *Analytical biochemistry* 201 (1): 152–157.doi: 10.1016/0003-2697(92)90188-D.

Curis, E., Nicolis, I., Moinard, C., Osowska, S., Zerrouk, N., Bénazeth, S., Cynober, L. (2005). "Almost all about citrulline in mammals." *Amino Acids* 29 (3): 177–205. doi: 10.1007/s00726-005-0235-4.

Dasuri, K., Ebenezer, P. J., Uranga, R. M., Gavilán, E., Zhang, L., Fernandez-Kim, S. O. K., Bruce-Keller, A. J., Keller, J. N. (2011). "Amino acid analog toxicity in primary rat neuronal and astrocyte cultures: Implications for protein misfolding and TDP-43 regulation". *Journal of Neuroscience Research* 89 (9): 1471–1477. doi: 10.1002/jnr.22677.

D'haenen, Hugo, den Boer, Johan A. (2002). *Biological Psychiatry* (digitised online by Google books). Paul Willner. John Wiley and Sons. p. 415. ISBN 978-0-471-49198-9.

Foye, William O., Lemke, Thomas L. (2007). *Foye's Principles of Medicinal Chemistry*. David A. Williams. Lippincott Williams and Wilkins. p. 446. ISBN 978-0-7817-6879-5.

Frischknecht, R., Suter, P. Oko-inventare von Energiesystemen, third ed., 1997.

Gao, X., Chooi, Y. H., Ames, B. D., Wang, P., Walsh, C. T., Tang, Y. (2011). "Fungal Indole Alkaloid Biosynthesis: Genetic and Biochemical Investigation of the Tryptoquialanine Pathway in *Penicillium aethiopicum*". *Journal of the American Chemical Society* 133 (8): 2729–2741. doi: 10.1021/ja1101085.

Gerngross, T. U., Slater, S. C. Scientific American 2000, 283, 37-41.

Gerngross, Tillman U. (1999). "Can biotechnology move us toward a sustainable society?" *Nature Biotechnology* 17 (6): 541–544. doi: 10.1038/9843.

Koyack, M. J., Cheng, R. P. (2006). "Design and Synthesis of β-Peptides With Biological Activity". *Protein Design* 340. pp. 95–109. doi: 10.1385/1-59745-116-9: 95. ISBN 1-59745-116-9.

Lu, Y., Freeland, S. (2006). "On the evolution of the standard amino-acid alphabet". *Genome Biology* 7 (1): 102. doi: 10.1186/gb-2006-7-1-102. PMC 1431706.

Municipal Waste Factsheet. *PDF*. EPA. Retrieved 7 May 2013.

Padmanabhan, S., Baldwin, R. L. (1991). "Straight-chain non-polar amino acids are good helix-formers in water". *Journal of molecular biology* 219 (2): 135–137. doi: 10.1016/0022-2836(91)90553-I.

Park, M. H. (2006). "The Post-Translational Synthesis of a Polyamine-Derived Amino Acid, Hypusine, in the Eukaryotic Translation Initiation Factor 5A (eIF5A)". *Journal of Biochemistry* 139(2): 161–169. doi: 10.1093/jb/mvj034.

Pearce F. (2009). Oxo-degradable plastic bags carry more ecological harm than good. *The Guardian*.

Petkewich, R. (2003). "Technology Solutions: Microbes manufacture plastic from food waste". *Environmental Science and Technology* 37: 175A–. doi: 10.1021/es032456x.

Ramadan, S. E., Razak, A. A., Ragab, A. M., El-Meleigy, M. (1989). "Incorporation of tellurium into amino acids and proteins in a tellurium-tolerant fungi". *Biological trace element research*20 (3): 225–232. doi: 10.1007/BF02917437.

Sherman, F., Stewart, J. W., Tsunasawa, S. (1985). "Methionine or not methionine at the beginning of a protein". *BioEssays* 3 (1): 27–31. doi: 10.1002/bies.950030108.

Slater, S. C., Gerngross, T. U. (2000). "How Green are Green Plastics?". *Scientific American*.

Théobald-Dietrich, A., Giegé, R., Rudinger-Thirion, J. L. (2005). "Evidence for the existence in mRNAs of a hairpin element responsible for ribosome dependent pyrrolysine insertion into proteins". *Biochimie* 87 (9–10): 813–817. doi: 10.1016/j.biochi.2005.03.006

Trown, P. W., Smith, B., Abraham, E. P. (1963). "Biosynthesis of cephalosporin C from amino acids". *The Biochemical journal* 86 (2): 284–291.

Van Buskirk, J. J., Kirsch, W. M., Kleyer, D. L., Barkley, R. M., Koch, T. H. (1984). "Aminomalonic acid: Identification in *Escherichia coli* and atherosclerotic plaque". *Proceedings of the National Academy of Sciences of the United States of America* 81 (3): 722–725. doi: 10.1073/pnas.81.3.722.

Vermeer, C. (1990). "Gamma-carboxyglutamate-containing proteins and the vitamin K-dependent carboxylase". *The Biochemical journal* 266 (3): 625–636.

Vink, E. T. H., Glassner, D. A., Kolstad, J. J., Wooley, R. J., O'Connor, R. P. (2007). *Industrial Biotechnology*. 3, 58-81.

Vink, E. T. H., Rabago, K. R., Glassner, D. A., Gruber, P. R. (2003). Polymer Degradation and Stability. 80, 403-419.

Voet, D., Voet, J. G. (2004). *Biochemistry* (3rd ed.). John Wiley and Sons. ISBN 047119350X.

WCI student isolates microbe that lunches on plastic bags. News.therecord.com. 2010-04-21. *Retrieved* 2011-06-30.

Weber, A. L., Miller, S. L. (1981). "Reasons for the occurrence of the twenty coded protein amino acids". *Journal of molecular evolution* 17 (5): 273–284.doi: 10.1007/ BF01795749.

William Harris. "How long does it take for plastics to biodegrade?". How Stuff Works. *Retrieved* 2013-05-09.

Yabannavar, A. V. and Bartha, R. (1994). Methods for assessment of biodegradability of plastic films in soil. Appl. Environ. Microbiol. 60: 3608-3614.

Chapter 11

Oxalic Acid

Oxalic acid is an organic compound with the formula $H_2C_2O_4$. It is a colourless crystalline solid that forms a colourless solution in water. It is classified as a dicarboxylic acid. In terms of acid strength, it is much stronger than acetic acid. Oxalic acid is a reducing agent [3]and its conjugate base, known as oxalate ($C_2O_4^{2-}$), is a chelating agent for metal cations. Typically, oxalic acid occurs as the dihydrate with the formula $H_2C_2O_4.2H_2O$. Excessive ingestion of oxalic acid or prolonged skin contact can be dangerous.

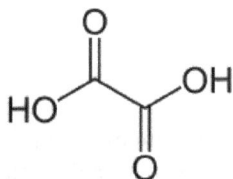

Preparation

Oxalic acid is mainly manufactured by the oxidation of carbohydrates or glucose using nitric acid or air in the presence of vanadium pentoxide. A variety of precursors can be used including glycolic acid and ethylene glycol. A newer method entails oxidativecarbonylation of alcohols to give the diesters of oxalic acid:

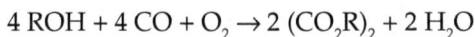

$$4 \ ROH + 4 \ CO + O_2 \rightarrow 2 \ (CO_2R)_2 + 2 \ H_2O$$

These diesters are subsequently hydrolyzed to oxalic acid. Approximately 120,000 tonnes are produced annually. Historically oxalic acid was obtained exclusively by using caustics, such as sodium or potassium hydroxide, on sawdust.

Laboratory Methods

Although it can be readily purchased, oxalic acid can be prepared in the laboratory by oxidizing sucrose using nitric acid in the presence of a small amount of vanadium pentoxide as a catalyst. The hydrated solid can be dehydrated with heat or by azeotropic distillation. Developed in the Netherlands, an electrocatalysis by a copper complex helps reduce carbon dioxide to oxalic acid.; This conversionuses carbon dioxide as a feedstock to generate oxalic acid. Of historical interest, Wöhler prepared oxalic acid by hydrolysis of cyanogen in 1824. This experiment may represent the first synthesis of a natural product.

Structure

Anhydrous oxalic acid exists as two polymorphs; in one the hydrogen-bonding results in a chain-like structure whereas the hydrogen bonding pattern in the other form defines a sheet-like structure. Because the anhydrous material is both acidic and hydrophilic(water seeking), it is used in esterifications.

Reactions

Oxalic acid is a relatively strong acid, despite being a carboxylic acid:

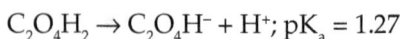

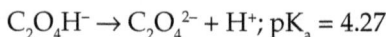

$$C_2O_4H_2 \rightarrow C_2O_4H^- + H^+; pK_a = 1.27$$
$$C_2O_4H^- \rightarrow C_2O_4^{2-} + H^+; pK_a = 4.27$$

Oxalic acid undergoes many of the reactions characteristic of other carboxylic acids. It forms esters such as dimethyl oxalate (m.p.52.5 to 53.5 °C (126.5 to 128.3 °F)). It forms an acid chloride called oxalyl chloride. Oxalate, the conjugate base of oxalic acid, is an excellent ligand for metal ions, *e.g.* the drug oxaliplatin. Oxalic acid and oxalates can be oxidized by permanganate in an autocatalytic reaction.

Occurrence

Biosynthesis

At least two pathways exist for the enzyme-mediated formation of oxalate. In one pathway, oxaloacetate, a component of the Krebs citric acid cycle, is hydrolyzed to oxalate and acetic acid by the enzyme oxaloacetase:

$$[O_2CC(O)CH_2CO_2]^{2-} + H_2O \rightarrow C_2O_4^{2-} + CH_3CO_2^-$$

It also arises from the dehydrogenation of glycolic acid, which is produced by the metabolism of ethylene glycol.

Occurrence in Foods and Plants

Calcium oxalate is the most common component of kidney stones. Early investigators isolated oxalic acid from wood-sorrel (*Oxalis*). Members of the spinach family and the brassicas (cabbage, broccoli, brussels sprouts) are high in oxalates, as are umbellifers likeparsley and sorrel.[14] Rhubarb leaves contain about 0.5 per cent oxalic acid and jack-in-the-pulpit (*Arisaema triphyllum*) contains calcium oxalate crystals. Bacteria produce oxalates from oxidation of carbohydrates. Plants of the fenestraria genus produce optical fibers made from crystalline oxalic acid to transmit light to subterranean photosynthetic sites.

Other

Oxidized bitumen or bitumen exposed to gamma rays also contains oxalic acid among its degradation products. Oxalic acid may increase the leaching of radionuclides conditioned in bitumen for radioactive waste disposal.

Biochemistry

The conjugate base of oxalic acid (oxalate) is a competitive inhibitor of the lactate dehydrogenase (LDH) enzyme. LDH catalyses the conversion of pyruvate to lactic acid (end product of the fermentation (anaerobic) process) oxidising the coenzyme NADH to NAD+ and H+ concurrently. Restoring NAD+ levels is essential to the continuation of anaerobic energy metabolism through glycolysis. As cancer cells preferentially use anaerobic metabolism (see Warburg effect) inhibition of LDH has been shown to inhibit tumor formation and growth, thus is an interesting potential course of cancer treatment.

Applications

About 25 per cent of produced oxalic acid is used as a mordant in dyeing processes. It is used in bleaches, especially for pulpwood. It is also used in baking powder.

Cleaning

Oxalic acid's main applications include cleaning or bleaching, especially for the removal of rust (iron complexing agent). Bar Keepers Friend is an example of a household cleaner containing oxalic acid. Its utility in rust removal agents is due to its forming a stable, water soluble salt with ferric iron, ferrioxalate ion.

Extractive Metallurgy

Oxalic acid is an important reagent in lanthanide chemistry. Hydrated lanthanide oxalates form readily in strongly acidic solutions in a densely crystalline, easily filtered form, largely free of contamination by nonlanthanide elements. Thermal decomposition of these oxalate gives the oxides, which is the most commonly marketed form of these elements.

Niche Uses

Vaporized oxalic acid, or a 3.2 per cent solution of oxalic acid in sugar syrup, is used by some beekeepers as a miticide against the parasitic varroa mite. Oxalic acid is rubbed onto completed marble sculptures to seal the surface and introduce a shine. Oxalic acid is also used to clean iron and manganese deposits from quartz crystals. Oxalic acid is used as bleach for wood, removing black stains caused by water penetration.

Content in Food Items

This table was originally published in *Agriculture Handbook No. 8-11, Vegetables and Vegetable Products*, 1984.

Vegetable	Oxalic Acid (g/100 g)	Vegetable	Oxalic Acid (g/100 g)
Amaranth	1.09	Asparagus	0.13
Beans, snap	0.36	Beet leaves	0.61
Broccoli	0.19	Brussels sprouts	0.36
Cabbage	0.10	Carrot	0.50
Cassava	1.26	Cauliflower	0.15
Celery	0.19	Chicory	0.2
Chives	1.48	Collards	0.45
Coriander	0.01	Corn, sweet	0.01
Cucumber	0.02	Eggplant	0.19
Endive	0.11	Garlic	0.36
Kale	0.02	Lettuce	0.33
Okra	0.05	Onion	0.05
Parsley	1.70	Parsnip	0.04
Pea	0.05	Bell pepper	0.04
Potato	0.05	Purslane	1.31
Radish	0.48	Rutabaga	0.03
Spinach	0.97	Squash	0.02
Sweet potato	0.24	Tomato	0.05
Turnip	0.21	Turnip greens	0.05
Watercress	0.31		

Toxicity and Safety

Oxalic acid has toxic effects through contact and if ingested; manufacturers provide details in Material Safety Data Sheets (MSDS). It is not identified as mutagenic or carcinogenic; there is a possible risk of congenital malformation in the fetus; may be harmful if inhaled and is extremely destructive to tissue of mucous membranes and upper respiratory tract; harmful if swallowed; harmful to and destructive of tissue and causes burns if absorbed through the skin or is in contact with the eyes. Symptoms and effects include a burning sensation, cough, wheezing, laryngitis, shortness of breath, spasm, inflammation and edema of the larynx, inflammation and edema of the bronchi, pneumonitis, pulmonary edema.

In humans, ingested oxalic acid has an oral LD_{Lo} (lowest published lethal dose) of 600 mg/kg. It has been reported that the lethal oral dose is 15 to 30 grams. The toxicity of oxalic acid is due to kidney failure caused by precipitation of solid calcium oxalate, the main component of kidney stones. Oxalic acid can also cause joint paindue to the formation of similar precipitates in the joints. Ingestion of ethylene glycol results in oxalic acid as a metabolite which can also cause acute kidney failure.

References

Bjerrum, J., et al. (1958) Stability Constants, Chemical Society, London.

Bouwman, Elisabeth, Angamuthu, Raja, Byers, Philip, Lutz, Martin, Spek, Anthony L. (July 15, 2010). "Electrocatalytic CO_2 Conversion to Oxalate by a Copper Complex" *Science* 327 (5393): 313–315. doi: 10.1126/science.1177981.

Bowden, E. (1943). "Methyl oxalate". *Org. Synth.*: 414., *Coll. Vol. 2*

CDC – Immediately Dangerous to Life or Health Concentrations (IDLH): Oxalic acid – NIOSH Publications and Products". cdc.gov

Clarke H. T.,. Davis, A. W. (1941). "Oxalic acid (anhydrous)". *Org. Synth.*: 421., *Coll. Vol. 1*

Dutton, M. V., Evans, C. S. (1996). "Oxalate production by fungi: Its role in pathogenicity and ecology in the soil environment". *Canadian Journal of Microbiology* 42(9): 881. doi: 10.1139/m96-114

Eiichi, Yonemitsu, Tomiya, Isshiki, Tsuyoshi, Suzuki and Yukio, Yashima "Process for the production of oxalic acid", U.S. Patent 3,678,107, priority date March 15, 1969.

EMEA Committee for veterinary medicinal products, oxalic acid summary report, December 2003.

Kovacs K.A., Grof P., Burai L., Riedel M. (2004). "Revising the mechanism of the permanganate/oxalate reaction". *J. Phys. Chem. A* 108 (50): 11026–11031. doi: 10.1021/jp047061u.

Le, Anne, Charles Cooper, Arvin Gouw, Ramani Dinavahi, Anirban Maitra, Lorraine Deck, Robert Royer, David Vander Jagt, Gregg Semenza, Chi Dang (2009). "Inhibition of lactate dehydrogenase A induces oxidative stress and inhibits tumor progression". *Proceedings of the National Academy of Sciences*. doi: 10.1073/pnas.0914433107.

Novoa, William, Alfred Winer andrew Glaid, George Schwert (1958). "Lactic Dehydrogenase V. inhibition by Oxamate and Oxalate". *Journal of Biological Chemistry* 234 (5): 1143–8.

Nutrient Data : Oxalic Acid Content of Selected Vegetables". ars.usda.gov

Oxalic acid dihydrate. MSDS. sigmaaldrich.com

Oxalic Acid Material Safety Data Sheet. Radiant Indus Chem. Retrieved 2014-05-20.

Practical Organic Chemistry by Julius B. Cohen, 1930 ed. preparation No.42

Pucher, GW, Wakeman, AJ, Vickery, HB (1938). "The organic acids of rhubarb (*Rheum hybridium*). III. The behavior of the organic acids during culture of excised leaves". *Journal of Biological Chemistry* 126 (1): 43.

Radiant Agro Chem. "Oxalic Acid MSDS".

Rock Currier – Cleaning Quartz. mindat.org

Rombauer, Rombauer Becker and Becker (1931/1997). *Joy of Cooking*, p.415. ISBN 0-684-81870-1.

Ullmann's Encyclopedia of Industrial Chemistry. Wiley. 2005. pp. 17624/28029. ISBN 9783527306732.

Von Wagner, Rudolf (1897). *Manual of chemical technology*. New York: D. Appleton and Co. p. 499.

Wells, A.F. (1984) *Structural Inorganic Chemistry*, Oxford: Clarendon Press. ISBN 0-19-855370-6.

Wilhelm Riemenschneider, Minoru Tanifuji "Oxalic acid" in *Ullmann's Encyclopedia of Industrial Chemistry*, 2002, Wiley-VCH, Weinheim. doi: 10.1002/14356007. a18_247.

Chapter 12

Phytic Acid

Phytic acid (known as inositol hexakisphosphate (IP6), inositol polyphosphate, or phytate when in salt form), discovered in 1903, a saturated cyclic acid, is the principal storage form of phosphorus in many plant tissues, especially bran and seeds. It can be found in cereals and grains. Catabolites of phytic acid are called lower inositol polyphosphates. Examples are inositol penta- (IP5), tetra- (IP4) and triphosphate (IP3).

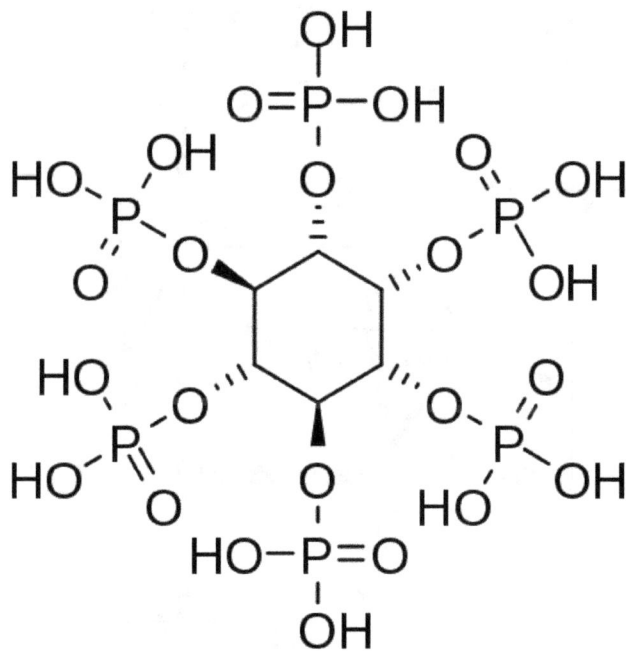

(1R,2R,3S,4S,5R,6S)-cyclohexane-1,2,3,4,5,6-hexayl hexakis [dihydrogen (phosphate)]

Phytic acid (myo-inositol-1,2,3,4,5,6-hexakis dihydrogen phosphate) is one of the typical antinutrient in legumes. It comprises approximately 5 per cent by weight of edible legumes, cereals, oilseeds, pollens and nuts. Phytic acid is the major storage form of phosphorus; nearly 60 to 90 per cent of the total phosphorus in seeds and is produced as a secondary product of carbohydrate metabolism. The proportion of phytic acid reaches up to 60 to 80 per cent of the dry weight of globoids of dicotyledons. Phytic acid exists as salts of calcium, magnesium or potassium. Phytic acid, which is a strong acid, forms a variety of salts with several heavy metals, such as zirconium, thorium, titanium and uranium in 6 N HCl. Phytic acid has 12 replaceable protons (Figure A) and is negatively charged at pH conditions generally encountered in food and feedstuff. Therefore, it is highly reactive towards positively charged groups such as metal ions and proteins. In general, one or two phosphate groups of phytic acid may bind with cations (Figure Ba). The mixed salt of phytic acid is formed when several cations complex within the same phytic acid molecule. The binding of phytic acid with minerals is pH dependent and complexes of varying solubilities are formed. Most polyvalent metal ions, especially calcium, magnesium, zinc and iron bind to phytic acid and form an insoluble complex which makes them unavailable for metabolism. The ability of phytic acid to complex with proteins depends on pH of the medium. At pH below isoelectric point of proteins, phytic acid binds directly with the positively charged proteins as a result of electrostatic attraction (Figure Bb). At intermediate pH above the isoelectric point of the protein, both phytic acid and protein molecules are negatively charged and phytic acid binds primarily with proteins mediated by polyvalent cations such as calcium or

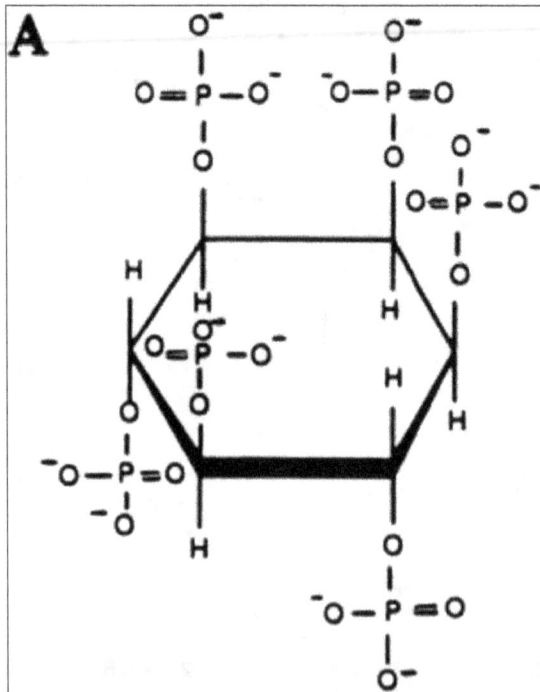

Charged Structure of Phytic Acid

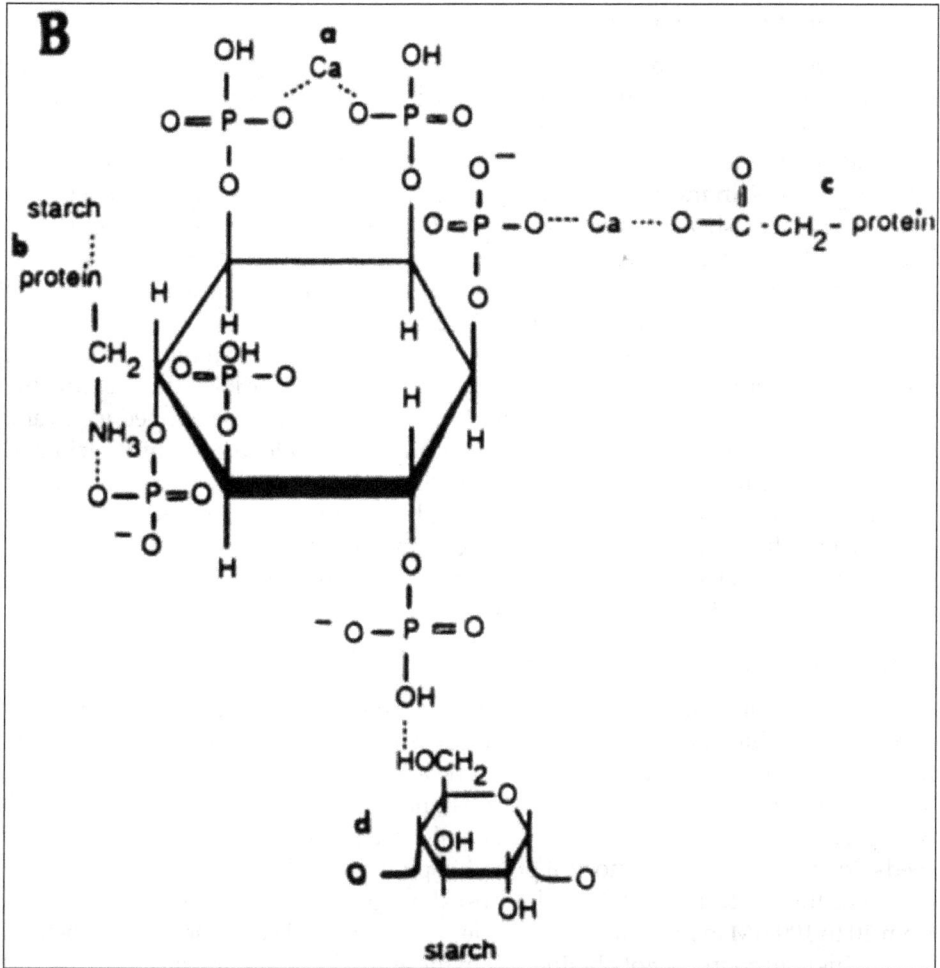

Interaction of Phytic Acid with Food Nutrients

magnesium (Figure Bc). However, direct binding of proteins with phytic acid does not take place to any considerable extent. Complexing of proteins with phytic acid, directly or through mediation by mineral ions, may alter the structure of proteins which may decrease their solubility, functionality and digestibility.

Phytic acid appears to be structurally capable of binding with starch through phosphate linkages or indirectly through its association with proteins (Figure Bd). The nutrient digestibility may also be affected by binding of phytic acid with digestive enzymes. These, in turn, may be responsible for both the adverse and beneficial health effects of phytic acid in foods passing through the digestive system. Phytates are not easily removed by traditional processing of pea seeds. The intact phytic acid remains embedded with proteins. The location and the strong association of phytic acid with proteins tend to concentrate it together with proteins during preparation of protein concentrates and isolates from pea seed meals.

Significance in Agriculture

Phosphorus and inositol in phytate form are not, in general, bioavailable to nonruminant animals because these animals lack the digestive enzyme phytase required to remove phosphate from the inositol in the phytate molecule. Ruminants are readily able to digest phytate because of the phytase produced by rumen microorganisms. In most commercial agriculture, nonruminant livestock, such as swine, fowl and fish, are fed mainly grains, such as maize, legumes and soybeans. Because phytate from these grains and beans is unavailable for absorption, the unabsorbed phytate passes through the gastrointestinal tract, elevating the amount of phosphorus in the manure. Excess phosphorus excretion can lead to environmental problems, such as eutrophication. The bioavailability of phytate phosphorus can be increased by supplementation of the diet with the enzyme phytase. Also, viable low-phytic acid mutant lines have been developed in several crop species in which the seeds have drastically reduced levels of phytic acid and concomitant increases in inorganic phosphorus. However, reported germination problems have hindered the use of these cultivars thus far. The use of sprouted grains will reduce the quantity of phytic acids in feed, with no significant reduction of nutritional value. Phytates also have the potential to be used in soil remediation, to immobilize uranium, nickel and other inorganic contaminants.

Biological and Physiological Roles

Although undigestable for many animals (as explained above), phytic acid and its metabolites as they occur in seeds and grains have several important roles for the seedling plant. Most notably, phytic acid functions as a phosphorus store, as an energy store, as a source of cations and as a source of myoinositol (a cell wall precursor). Phytic acid is the principal storage form of phosphorus in plant seeds. In animal cells, myoinositol polyphosphates are ubiquitous and phytic acid (myoinositol hexakisphosphate) is the most abundant, with its concentration ranging from 10 to 100 uM in mammalian cells, depending on cell type and developmental stage. This compound is not obtained from the animal diet, but must be synthesized inside the cell from phosphate and inositol (which in turn is produced from glucose, usually in the kidneys). The interaction of intracellular phytic acid with specific intracellular proteins has been investigated *in vitro* and these interactions have been found to result in the inhibition or potentiation of the physiological activities of those proteins. The best evidence from these studies suggests an intracellular role for phytic acid as a cofactor in DNA repair by nonhomologous end-joining. Other studies using yeast mutants have also suggested intracellular phytic acid may be involved in mRNA export from the nucleus to the cytosol. There are still major gaps in the understanding of this molecule and the exact pathways of phytic acid and lower inositol phosphate metabolism are still unknown. As such, the exact physiological roles of intracellular phytic acid are still a matter of debate.

Food Science

Phytic acid is found within the hulls of nuts, seeds and grains. In-home food preparation techniques can break down the phytic acid in all of these foods. Simply cooking the food will reduce the phytic acid to some degree. More effective methods are soaking in an acid medium, lactic acid fermentation and sprouting. Phytic acid has a strong binding affinity to important minerals, such as calcium, iron and zinc, although the binding of calcium with phytic acid is pH-dependent. The binding of phytic acid with iron is more complex, although there certainly is a strong binding affinity, molecules like phenols and tannins also influence the binding. When iron and zinc bind to phytic acid they form insoluble precipitate and are far less absorbable in the intestines. This process can therefore contribute to iron and zinc deficiencies in people whose diets rely on these foods for their mineral intake, such as those in developing countries and vegetarians. Contrary to that, one study correlated decreased osteoporosis risk with phytic acid consumption. It also acts as an acid, chelating the vitamin niacin, the deficiency of which is known as pellagra. In this regard, it is an antinutrient, despite its possible therapeutic effects (see below).

Food Sources of Phytic Acid

Food	(Per cent minimum dry)	(Per cent maximum dry)
Linseed	2.15	2.78
Sesame seeds flour	5.36	5.36
Almonds	1.35	3.22
Brazilnuts	1.97	6.34
Coconut	0.36	0.36
Hazelnut	0.65	0.65
Peanut	0.95	1.76
Walnut	0.98	0.98
Maize (Corn)	0.75	2.22
Oat	0.42	1.16
Oat Meal	0.89	2.40
Brown rice	0.84	0.99
Polished rice	0.14	0.60
Wheat	0.39	1.35
Wheat flour	0.25	1.37
Wheat germ	0.08	1.14
Whole wheat bread	0.43	1.05
Beans, pinto	2.38	2.38
Chickpeas	0.56	0.56
Lentils	0.44	0.50
Soybeans	1.00	2.22
Tofu	1.46	2.90
Soy beverage	1.24	1.24
Soy protein concentrate	1.24	2.17
New potato	0.18	0.34
Spinach	0.22	NR

For people with a particularly low intake of essential minerals, especially those in developing countries, this effect can be undesirable. It has been hypothesized, but not tested, that probiotic lactobacilli and other species of endogenous digestive microflora may be a sufficient source of the enzyme phytase to improve mineral absorption. Phytase catalyzes the release of phosphate from phytate and hydrolyses the complexes formed by phytate and metal ions or other cations, rendering them more soluble, which improves and facilitates the absorption of zinc and magnesium.

Food Sources of Phytic Acid (Fresh weight)

Food	(Per cent minimum fresh weight)	(Per cent maximum fresh weight)
Taro	0.143	0.195
Cassava	0.114	0.152

Other Commercial Uses

Phytic acid has been marketed for its alleged anti-cancer properties, based on research by Professor Abulkalam Shamsuddin of the university of Maryland. The US Food and Drug Administration has listed it as a fake cancer "cure" and there is no good evidence that phytic acid has any beneficial role to play in cancer medicine. As a food additive, phytic acid is used as the preservative E391.

Antinutrients

Antinutrients are natural or synthetic compounds that interfere with the absorption of nutrients. Nutrition studies focus on those antinutrients commonly found in food sources and beverages.

Examples

Protease inhibitors are substances that inhibit the actions of trypsin, pepsin and other proteases in the gut, preventing the digestion and subsequent absorption of protein. For example, Bowman-Birk trypsin inhibitor is found in soybeans. Lipase inhibitors interfere with enzymes, such as human pancreatic lipase, that catalyze the hydrolysis of some lipids, including fats. For example, the anti-obesity drug orlistatcauses a percentage of fat to pass through the digestive tract undigested. Amylase inhibitors prevent the action of enzymes that break the glycosidic bonds of starches and other complex carbohydrates, preventing the release of simple sugars and absorption by the body. Amylase inhibitors, like lipase inhibitors, have been used as a diet aide and obesity treatment. Amylase inhibitors are present in many types of beans; commercially available amylase inhibitors are extracted from white kidney beans. Phytic acid has a strong binding affinity to minerals such as calcium, magnesium, iron, copper and zinc. This results in precipitation, making the minerals unavailable for absorption in the intestines. Phytic acids are common in the hulls of nuts, seeds and grains. Oxalic acid and oxalates are present in many plants, particularly in members of the spinach family. Oxalates bind to calcium and prevent its absorption in the human body. Glucosinolates prevent the uptake of iodine, affecting the function of the thyroid and thus are considered goitrogens.

They are found in broccoli, brussel sprouts, cabbage and cauliflower. Excessive intake of required nutrients can also result in them having an anti-nutrient action. Excessive intake of fiber can reduce the transit time through the intestines to such a degree that other nutrients cannot be absorbed. Because calcium, iron, zinc and magnesium share the same transporter within the intestine, excessive consumption of one of these minerals can lead to saturation of the transport system and reduced absorption of the other minerals. Some proteins can also be antinutrients, such as the trypsin inhibitors and lectins found in legumes. These enzyme inhibitors interfere with digestion. Another particularly widespread form of antinutrients is the flavonoids, which are a group of polyphenolic compounds that include tannins. These compounds chelate metals such as iron and zinc and reduce the absorption of these nutrients, but they also inhibit digestive enzymes and may also precipitate proteins. Saponins in plants may serve as anti-feedants.

Occurrence

Antinutrients are found at some level in almost all foods for a variety of reasons. However, their levels are reduced in modern crops, probably as an outcome of the process ofdomestication. The possibility now exists to eliminate antinutrients entirely using genetic engineering; but, since these compounds may also have beneficial effects, such genetic modifications could make the foods more nutritious but not improve people's health. Many traditional methods of food preparation such as fermentation, cooking and malting increase the nutritive quality of plant foods through reducing certain antinutrients such as phytic acid, polyphenols and oxalic acid. Such processing methods are widely used in societies where cereals and legumes form a major part of the diet. An important example of such processing is the fermentation of cassava to produce cassava flour: this fermentation reduces the levels of both toxins and antinutrients in the tuber.

Antimetabolites

An antimetabolite is a chemical that inhibits the use of a metabolite, which is another chemical that is part of normalmetabolism. Such substances are often similar in structure to the metabolite that they interfere with, such as the antifolatesthat interfere with the use of folic acid. The presence of antimetabolites can have toxic effects on cells, such as halting cell growth and cell division, so these compounds are used as chemotherapy for cancer.

Function

Cancer Treatment

Antimetabolites can be used in cancer treatment, as they interfere with DNA production and therefore cell division and the growth of tumors. Because cancer cells spend more time dividing than other cells, inhibiting cell division harms tumor cells more than other cells. Anti-metabolites masquerade as a purine (azathioprine, mercaptopurine) or a pyrimidine, chemicals that become the building-blocks of DNA. They prevent these substances becoming incorporated in to DNA during the S phase (of the cell cycle), stopping normal development and division. They also affect RNA synthesis. However, because thymidine is used in DNA but not in RNA (where uracil is used instead), inhibition of thymidine synthesis via thymidylate synthase selectively inhibits DNA synthesis over RNA synthesis. Due to their efficiency, these drugs are the most widely used cytostatics.

In the ATC system, they are classified under L01B.

Antibiotics

Antimetabolites may also be antibiotics, such as sulfanilamide drugs, which inhibit dihydrofolate synthesis in bacteria by competing with para-aminobenzoic acid.

Types

Main categories of these drugs include:

☆ Base analogs (altered nucleobases):

☆ Purine analogues

☆ Pyrimidine analogues

☆ Nucleoside analogues:

☆ Nucleosides with altered nucleobases

☆ Nucleosides with altered sugar component (e.g. Cytarabine)

☆ Nucleotide analogues

☆ Antifolates

Phytic Acid in Food

There is phytic acid in food and it matters. Researchers have found that if you can reduce the **phytic acid** in your food, you can improve your iron absorption markedly. A 2003 study examined the change in iron absorption when phytic acid was removed from various grains. Check out the graph at right that displays the results.

You are seeing correctly. The study found that participants absorbed 1160 per cent more iron when phytic acid was removed from wheat. Iron absorption was improved about **twelve times**.

Phytic Acid in Grains

Do grains have phytic acid (phytates) and should we care? Generally speaking, grains have high levels of *phytic acid*, a substance that reduces our absorption of minerals such as calcium, iron, zinc and magnesium. As an example, compare the milligrams of phytic acid in grains to a random collection of other foods. (This is a small sample of phytic acid levels as listed in a review article by Harland and Oberleas in a 1987 article.)

Phytic Acid in Grains Remedies

To reduce phytic acid in grains, you may soak or sprout them. You may also bake them using a long rise time and good pH content. Many people make "soaked

Increase In Iron Absorption

when phytic acid is removed from the grain

percent increase

0	500	1,000	1,500

Wheat — 1160

Oat — 836

Corn — 496

Rice — 309

Source: Hurrell et al., Am J Clin Nutr 77:1215

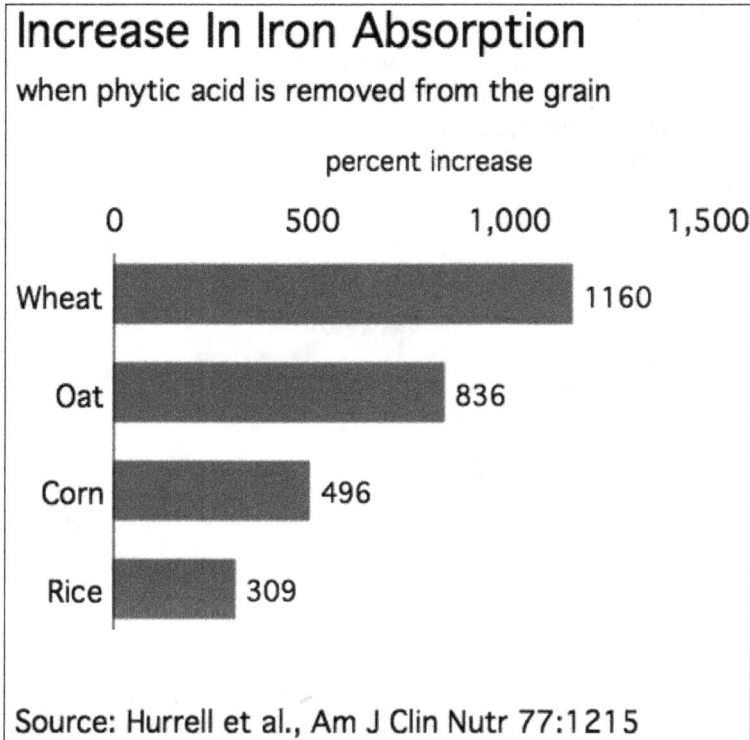

flour" breads to reduce phytic acid but this is not the optimal approach for flavor, texture, or phytic acid reduction. In the phytic acid paper there are a number of recipes that are far better and will open up a whole new world of baking. In the paper, I describe the research so that you know what techniques are best for you in your own kitchen.

Phytic Acid in Beans

Like grains, beans do have a high phytic acid content. There is also a wealth of scientific literature on reducing phytic acid in your beans. Some years ago I posted an article on the Internet about soaking beans, displaying the food science findings about the effect of temperature and time on reducing phytic acid.

Phytic Acid and Sprouting

Does sprouting grains reduce the phytic acid content of the grain? Absolutely. To sprout your grains, you soak them overnight in whole form (the whole spelt berry for instance), drain them well and place them in a container (usually a tray or a jar) with a a cheesecloth cover. In a day or two, they will begin to send out little tails, the sprouts. You grain seed is transforming itself from a seed to a seedling. The nutrition content of the seed changes, including a reduction in the phytic acid content.

Phytic Acid in Soy

When you turn over a tub of tofu or a carton of soy milk and marvel at the amount of magnesium in that bean curd, beware that precious little of it will make it into your brain cells. Likewise, little of the native calcium, zinc, or iron will nourish your body. Soy is high in phytic acid which binds to these minerals and keeps you from digesting them.

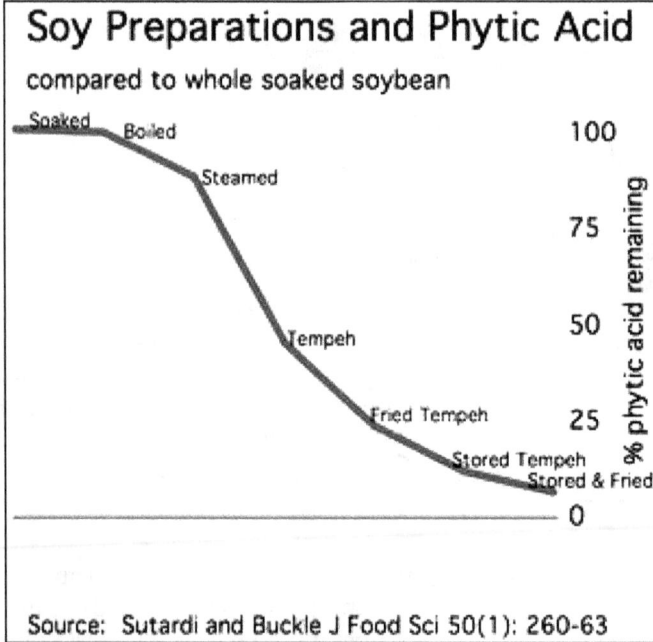

Soy Preparations and Phytic Acid
compared to whole soaked soybean

Soaked Boiled   100

Steamed   75   % phytic acid remaining

Tempeh   50

Fried Tempeh   25

Stored Tempeh
Stored & Fried   0

Source: Sutardi and Buckle J Food Sci 50(1): 260-63

Phytic Acid in Edamame

We know that soy has high levels of phytic acid. What about the immature form of soy, edamame?. Edamame is often enjoyed as a snack. You can purchase the pods and pop out the beans to eat as a crispy treat. Should we be concerned about the **phytic acid?**

Coconut Flour and Phytic Acid

With the increase in popularity of coconut flour in gluten-free diets, consumers have wondered if they should be concerned about phytic acid in coconut flour. There is not really any food science research on this topic to speak of but that in itself is telling.

Corn and Phytic Acid

Corn does contain phytic acid and it is actually an interesting grain because it is difficult to reduce the phytic acid in corn with typical techniques.

Phytic Acid White Paper

If food science and reducing phytic acid in your food interest you, check out the white paper on phytic acid. It displays the food science literature on phytic acid in grains and legumes (and the bit there is on nuts and seeds) and puts it together for you providing a kitchen process for you that is not only easy, but helps your food taste better. There are no "soaked flour" recipes in this paper (a popular approach to reducing phytic acid in flour). The method I provide is more efficient and has a better end-product.

References

Ali, M, Shuja, MN, Zahoor, M, Qadri, I (2010). "Phytic acid: how far have we come". *African Journal of Biotechnology* 9 (11): 1551–1554.

American Dietetic, A., Dietitians Of, C. (2003). "Position of the American Dietetic Association and Dietitians of Canada: Vegetarian diets". *Journal of the American Dietetic Association* 103 (6): 748–765. doi: 10.1053/jada.2003.50142.

Anderson, Eugene N. (2005). *Everyone eats: understanding food and culture.* New York: New York University Press. pp. 47–8. ISBN 0-8147-0496-4.

Antineoplastic Antimetabolites at the US National Library of Medicine Medical Subject Headings (MeSH)

Beecher GR (2003). "Overview of dietary flavonoids: nomenclature, occurrence and intake". *J. Nutr.* 133 (10): 3248S–54S.

Chavan JK, Kadam SS (1989). "Nutritional improvement of cereals by fermentation". *Crit Rev Food Sci Nutr* 28 (5): 349–400. doi: 10.1080/10408398909527507.PMID 2692608.

Cheryan, Munir, Rackis, Joseph (1980). "Phytic acid interactions in food systems". *Crit Rev Food Sci Nutr* 13 (4): 297–335. doi: 10.1080/10408398009527293.PMID 7002470.

Committee on Food Protection, Food and Nutrition Board, National Research Council (1973). "Phytates". *Toxicants Occurring Naturally in Foods.* National Academy of Sciences. pp. 363–371. ISBN 978-0-309-02117-3.

Dendougui, Ferial, Schwedt, Georg (2004). "*In vitro* analysis of binding capacities of calcium to phytic acid in different food samples". *European Food Research and Technology* 219 (4). doi: 10.1007/s00217-004-0912-7.

Ekholm, Päivi, Päivi Ekholm, Liisa Virkki, Maija Ylinen, Liisa Johansson (2003). "The effect of phytic acid and some natural chelating agents on the solubility of mineral elements in oat bran". *Food Chemistry* 80 (2): 165–70. doi: 10.1016/S0308-8146(02)00249-2.

Famularo G, De Simone C, Pandey V, Sahu AR, Minisola G (2005). "Probiotic lactobacilli: an innovative tool to correct the malabsorption syndrome of vegetarians?".*Med. Hypotheses* 65 (6): 1132–5. doi: 10.1016/j.mehy.2004.09.030.

Food and Drug Administration.: Listed as IP-6 Inositol Hexaphosphate

Gilani GS, Cockell KA, Sepehr E (2005). "Effects of antinutritional factors on protein digestibility and amino acid availability in foods". *J AOAC Int* 88 (3): 967–87.

Gordon, D. T., Chao, L. S. (1984). "Relationship of components in wheat bran and spinach to iron bioavailability in the anemic rat". *The Journal of nutrition* 114 (3): 526–35.

Guttieri, M. J., Peterson, K. M., Souza, E. J. (2006). "Milling and Baking Quality of Low Phytic Acid Wheat". *Crop Science* 46 (6): 2403–8. doi: 10.2135/cropsci2006. 03.0137.

Hanakahi LA, Bartlet-Jones M, Chappell C, Pappin D, West SC (2000). "Binding of inositol phosphate to DNA-PK and stimulation of double-strand break repair". *Cell* 102(6): 721–729. doi: 10.1016/S0092-8674(00)00061-1.

Heck, AM, Amy M. Heck, Jack A. Yanovski, Karim Anton Calis (2000). "Orlistat, a New Lipase Inhibitor for the Management of Obesity". *Pharmacotherapy* 20 (3): 270–9.doi: 10.1592/phco.20.4.270.34882.

Hotz C, Gibson RS (2007). "Traditional food-processing and preparation practices to enhance the bioavailability of micronutrients in plant-based diets". *J. Nutr.* 137 (4): 1097–100.

Hurrell RF (2003). "Influence of vegetable protein sources on trace element and mineral bioavailability". *The Journal of Nutrition* 133 (9): 2973S–7S.

Inositol Hexaphosphate. *Memorial Sloan Kettering Cancer Center.*

Ip-6: Uses, Side Effects, Interactions and Warnings – WebMD". webmd.com.

Klopfenstein, Terry J., Angel, Rosalina, Cromwell, Gary, Erickson, Galen E., Fox, Danny G., Parsons, Carl, Satter, Larry D., Sutton, Alan L., Baker, David H. (2002)."Animal Diet Modification to Decrease the Potential for Nitrogen and Phosphorus Pollution". *Council for Agricultural Science and Technology* 21.

López-González AA, Grases F, Roca P, Mari B, Vicente-Herrero MT, Costa-Bauzá A (2008). "Phytate (myo-inositol hexaphosphate) and risk factors for osteoporosis". *Journal of Medicinal Food* 11 (4): 747–52. doi: 10.1089/jmf 2008. 0087.

MacFarlane, B. J., Bezwoda, W. R., Bothwell, T. H., Baynes, R. D., Bothwell, J. E., MacPhail, A. P., Lamparelli, R. D., Mayet, F (1988). "Inhibitory effect of nuts on iron absorption". *The American journal of clinical nutrition* 47 (2): 270–4.

Malleshi, N. G., Desikachar, H. S. R. (1986). "Nutritive value of malted millet flours". *Plant Foods for Human Nutrition* 36 (3): 191–6. doi: 10.1007/BF01092036.

Mallin, M. A. (2003). "Industrialized Animal Production–A Major Source of Nutrient and Microbial Pollution to Aquatic Ecosystems". *Population and Environment* 24 (5): 369–385. doi: 10.1023/A: 1023690824045. JSTOR 27503850. edit

Mullaney, Edward J., Ullah, Abul H.J. "Phytases: attributes, catalytic mechanisms and applications". United States Department of Agriculture–Agricultural Research Service. Retrieved May 18, 2012.

Norris FA, Ungewickell E, Majerus PW (1995). "Inositol hexakisphosphate binds to clathrin assembly protein 3 (AP-3/AP180) and inhibits clathrin cage assembly *in vitro*". *J Biol Chem* 270 (1): 214–217. doi: 10.1074/jbc.270.1.214.

Nutrition: A Functional Approach, Canadian Edition. Pearson Benjamin Cummings. 2007.

Oboh G, Oladunmoye MK (2007). "Biochemical changes in micro-fungi fermented cassava flour produced from low- and medium-cyanide variety of cassava tubers". *Nutr Health* 18 (4): 355–67. doi: 10.1177/026010600701800405.

Oxford Dictionary of Biochemistry and Molecular Biology. Oxford University Press, 2006. ISBN 0-19-852917-1.

Oxford Dictionary of Biochemistry and Molecular Biology. Oxford University Press, 2006. ISBN 0-19-852917-1.

Peters GJ, van der Wilt CL, van Moorsel CJ, Kroep JR, Bergman AM, Ackland SP (2000). "Basis for effective combination cancer chemotherapy with antimetabolites". *Pharmacol. Ther.*87 (2–3): 227–53. doi: 10.1016/S0163-7258(00)00086-3.

Phillippy, B. Q., Bland, J. M., Evens, T. J. (2003). "Ion Chromatography of Phytate in Roots and Tubers". *Journal of Agricultural and Food Chemistry* 51 (2): 350.doi: 10.1021/jf025827m.

Phillips RD (1993). "Starchy legumes in human nutrition, health and culture". *Plant Foods Hum Nutr* 44 (3): 195–211. doi: 10.1007/BF01088314.

Phytates in cereals and legumes. fao.org.

Phytic acid. phytochemicals.info

Preuss, HG, Preuss HG (2009). "Bean amylase inhibitor and other carbohydrate absorption blockers: effects on diabesity and general health". *J Am Coll Nutr.* 28 (3): 266–76. doi: 10.1080/07315724.2009.10719781.

Prom-U-Thai, Chanakan, Huang, Longbin, Glahn, Raymond P, Welch, Ross M, Fukai, Shu, Rerkasem, Benjavan (2006). "Iron (Fe) bioavailability and the distribution of anti-Fe nutrition biochemicals in the unpolished, polished grain and bran fraction of five rice genotypes". *Journal of the Science of Food and Agriculture* 86 (8): 1209–15.doi: 10.1002/jsfa.2471.

Reddy NR, Sathe SK, Salunkhe DK (1982). "Phytates in legumes and cereals". *Adv Food Res* 28: 1–92. doi: 10.1016/s0065-2628(08)60110-x.

Reddy, N. R., Sathe, Shridhar K. (2001). *Food Phytates*. Boca Raton: CRC. ISBN 1-56676-867-5.

Romarheim, O.H., Zang, C., Penn, M., Liu, Y.-J., Tian, L.-H., Skrede, A., Krogdahl, Å., Storebakken, T. (2008). "Growth and intestinal morphology in cobia (*Rachycentron canadum*) fed extruded diets with two types of soybean meal partly replacing fish meal".*Aquaculture Nutrition* 14 (2): 174–180. doi: 10.1111/j.1365-2095.2007.00517.x.

Sasakawa N, Sharif M, Hanley MR (1995). "Metabolism and biological-activities of inositol pentakisphosphate and inositol hexakisphosphate". *Biochem Pharmacol* 50 (2): 137–146. doi: 10.1016/0006-2952(95)00059-9.

Seaman JC, Hutchison JM, Jackson BP, Vulava VM (2003). "In situ treatment of metals in contaminated soils with phytate". *Journal of Environmental Quality* 32 (1): 153–61. doi: 10.2134/jeq2003.0153.

Shears SB (2001). "Assessing the omnipotence of inositol hexakisphosphate". *Cell Signalling* 13 (3): 151–158. doi: 10.1016/S0898-6568(01)00129-2.

Smith, A. L. (1997). *Oxford dictionary of biochemistry and molecular biology*. Oxford [Oxfordshire]: Oxford University Press. p. 43. ISBN 0-19-854768-4.

SOM Faculty Profile : Abulkalam Shamsuddin. *umaryland.edu*.

Szwergold BS, Graham RA, Brown TR (1987). "Observation of inositol pentakis- and hexakis-phosphates in mammalian tissues by 31P NMR". *Biochem Biophys Res Commun* 149 (3): 874–881. doi: 10.1016/0006-291X(87)90489-X.

Takimoto CH, Calvo E. "Principles of Oncologic Pharmacotherapy" in Pazdur R, Wagman LD, Camphausen KA, Hoskins WJ (Eds) Cancer Management: A Multidisciplinary Approach. 11 ed. 2008.

Tan-Wilson, Anna L., Anna L. Tan-Wilson, Jean C. Chen, Michele C. Duggan, Cathy Chapman, R. Scott Obach, Karl A. Wilson (1987). "Soybean Bowman-Birk trypsin isoinhibitors: classification and report of a glycine-rich trypsin inhibitor class". *J. Agric. Food Chem* 35 (6): 974. doi: 10.1021/jf00078a028.

The Organic Chemistry of Drug Design and Drug Action" (2nd edition), R. B. Silverman, 2004.

Welch RM, Graham RD (2004). "Breeding for micronutrients in staple food crops from a human nutrition perspective". *J. Exp. Bot.* 55 (396): 353–64. doi: 10.1093/jxb/erh064.

York JD, Odom AR, Murphy R, Ives EB, Wente SR (1999). "A phospholipase C-dependent inositol polyphosphate kinase pathway{meaning?} required for efficient messenger RNA export". *Science (Washington, D.C.)* 285 (5424): 96–100. doi: 10.1126/science.285.5424.96.

Chapter 13

Phytoestrogens

Phytoestrogens are plant-derived xenoestrogens (see estrogen) not generated within the endocrine system but consumed by eating phytoestrogenic plants. Also called "dietary estrogens", they are a diverse group of naturally occurring nonsteroidal plant compounds that, because of their structural similarity with

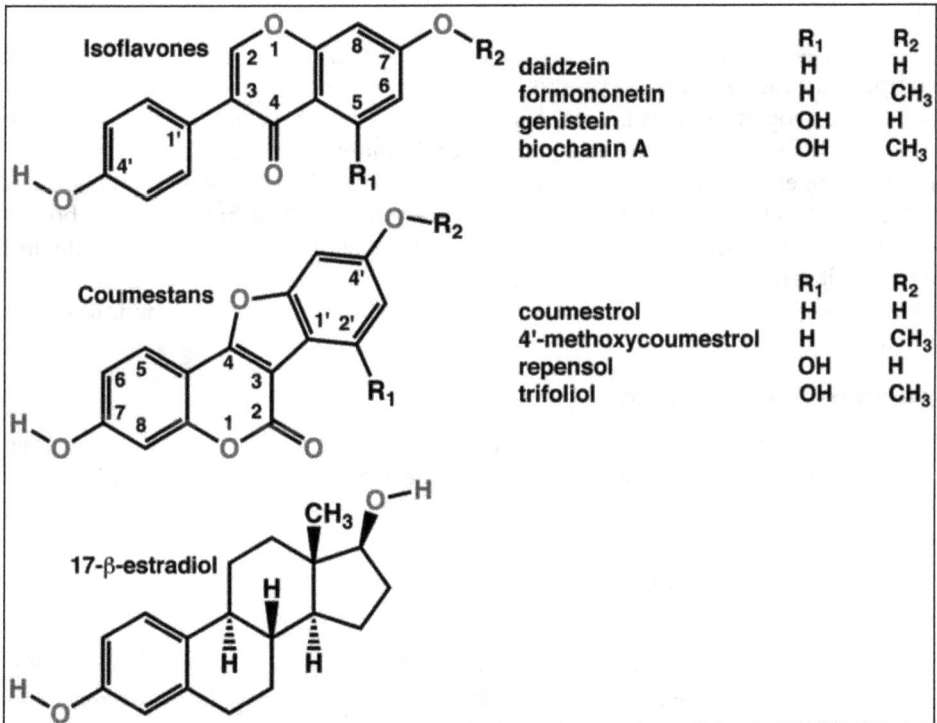

Isoflavones	R₁	R₂
	R$_1$	R$_2$
daidzein	H	H
formononetin	H	CH$_3$
genistein	OH	H
biochanin A	OH	CH$_3$

Coumestans	R$_1$	R$_2$
coumestrol	H	H
4'-methoxycoumestrol	H	CH$_3$
repensol	OH	H
trifoliol	OH	CH$_3$

Chemical Structures of the most Common Phytoestrogens found in Plants (Top and middle) Compared with Estrogen (Bottom) Found in Animals.

estradiol (17-β-estradiol), have the ability to cause estrogenic or/and antiestrogenic effects, by sitting in and blocking receptor sites against estrogen.

Their name comes from the Greek *phyto* ("plant") and *estrogen*, the hormone which gives fertility to female mammals. The word "estrus" - Greek ïßóôñïò - means "sexual desire" and "gene" - Greek γùñ - is "to generate". It has been proposed that plants use phytoestrogens as part of their natural defence against the overpopulation of herbivore animals by controlling male fertility. The similarities, at molecular level, of estrogens and phytoestrogens allow them to mildly mimic and sometimes act as antagonists of estrogen. Phytoestrogens were first observed in 1926, but it was unknown if they could have any effect in human or animal metabolism. In the 1940s, it was noticed for the first time that red clover (a phytoestrogens-rich plant) pastures had effects on the fecundity of grazing sheep. Researchers are exploring the nutritional role of these substances in the regulation of cholesterol and the maintenance of proper bone density post-menopause. Evidence is accruing that phytoestrogens may have protective action against diverse health disorders, such as prostate, breast, bowel and other cancers, cardiovascular disease,brain function disorders and osteoporosis, Phytoestrogens cannot be considered nutrients, given that the lack of these in the diet does not produce any characteristic deficiency syndrome nor do they participate in any essential biological function. Analytical methods are available to determine phytoestrogen content in plants and food.

Structure

Phytoestrogens mainly belong to a large group of substituted natural phenolic compounds: the coumestans, prenylflavonoids and isoflavones are three of the most active in estrogenic effects in this class. The best-researched are isoflavones, which are commonly found in soy and red clover. Lignans have also been identified as phytoestrogens, although they are not flavonoids. Mycoestrogens have similar structures and effects, but are not components of plants; these are mold metabolites of *Fusarium*, a fungus that is frequently found in pastures as well as in alfalfa and clover. Although mycoestrogens are rarely taken into account in discussions about phytoestrogens, these are the compounds that initially generated the interest on the topic.

Mechanism of Action

Phytoestrogens exert their effects primarily through binding to estrogen receptors (ER). There are two variants of the estrogen receptor, alpha (ER-α) and beta (ER-β) and many phytoestrogens display somewhat higher affinity for ER-β compared to ER-α.

The key structural elements that enable phytoestrogens to bind with high affinity to estrogen receptors and display estradiol-like effects are:

☆ The phenolic ring that is indispensable for binding to estrogen receptor

☆ The ring of isoflavones mimicking a ring of estrogens at the receptors binding site

☆ Low molecular weight similar to estrogens (MW=272)

☆ Distance between two hydroxyl groups at the isoflavones nucleus similar to that occurring in estradiol

☆ Optimal hydroxylation pattern

In addition to interaction with ERs, phytoestrogens may also modulate the concentration of endogenous estrogens by binding or inactivating some enzymes and may affect the bioavailability of sex hormones by depressing or stimulating the synthesis of sex hormone-binding globulin (SHBG). Emerging evidence shows that some phytoestrogens bind to and transactivate peroxisome proliferator-activated receptors (PPARs). *In vitro* studies show an activation of PPARs at concentrations above 1 µM, which is higher than the activation level of ERs. At the concentration below 1 µM, activation of ERs may play a dominant role. At higher concentrations (>1 µM), both ERs and PPARs are activated. Studies have shown that both ERs and PPARs influence each other and therefore induce differential effects in a dose-dependent way. The final biological effects of genistein are determined by the balance among these pleiotrophic actions.

Ecology

These compounds in plants are an important part of their defense system, mainly against fungi. Phytoestrogens are ancient naturally occurring substances and as dietary phytochemicals they are considered as co-evolutive with mammals. In the human diet, phytoestrogens are not the only source of exogenous estrogens. Xenoestrogens (novel, man-made), are found as food additives and ingredients and also in cosmetics, plastics and insecticides. Environmentally, they have similar effects as phytoestrogens, making it difficult to clearly separate the action of these two kind of agents in studies done on populations.

Avian Studies

The consumption of plants with unusual content of phytoestrogens under drought conditions, has been shown to decrease fertility in quail. Parrot food as available in nature has shown only weak estrogenic activity. Studies have been conducted on screening methods for environmental estrogens present in manufactured supplementary food, with the purpose to enable reproduction of endangered species.

Food Sources

According to a study by Canadian researchers about the content of nine common phytoestrogens in a Western diet, foods with the highest relative phytoestrogen content were nuts and oilseeds, followed by soy products, cereals and breads, legumes, meat products and other processed foods that may contain soy, vegetables, fruits, alcoholic and nonalcoholic beverages. Flax seed and other oilseeds contained the highest total phytoestrogen content, followed by soybeans and tofu. The highest concentrations of isoflavones are found in soybeans and soybean products followed by legumes, whereas lignans are the primary source of phytoestrogens found in nuts and oilseeds (*e.g.* flax) and also found in cereals, legumes, fruits and vegetables. Phytoestrogen content varies in different foods and may vary significantly within the same group of foods (*e.g.* soy beverages, tofu) depending on processing mechanisms and type of soybean used. Legumes (in particular soybeans), whole grain cereals and some seeds are high in phytoestrogens. A more comprehensive list of foods known to contain phytoestrogens includes:

☆ Soybeans and soy products

☆ Tempeh

☆ Linseed (flax)

☆ Sesame seeds

☆ Wheatberries

☆ Fenugreek (contains diosgenin, but also used to make Testofen®, a compound taken by men to increase testosterone).

☆ Oats

☆ Barley

☆ Beans

☆ Lentils

☆ Yams

☆ Rice

☆ Alfalfa

☆ Mung beans

☆ Apples

☆ Carrots

☆ Pomegranates

☆ Wheat germ

☆ Rice bran

☆ Lupin

☆ Kudzu

☆ Coffee

☆ Licorice root

☆ Mint

☆ Ginseng

☆ Hops,

☆ bourbon

☆ Beer,

☆ Fennel and

☆ Anise.

☆ Red clover (sometimes a constituent of green manure).

An epidemiological study of women in the United States found that the dietary intake of phytoestrogens in healthy post-menopausal Caucasian women is less than one milligram daily.

Effects on Humans

In human beings, phytoestrogens are readily absorbed, circulate in plasma and are excreted in the urine. Metabolic influence is different from that of grazing animals due to the differences between ruminant versus monogastric digestive systems.

Males

A 2010 meta-analysis of fifteen placebo-controlled studies said that "neither soy foods nor isoflavone supplements alter measures of bioavailable testosterone concentrations in men." Furthermore, isoflavone supplementation has no effect on sperm concentration, count or motility and it leads to no observable changes in testicular or ejaculate volume.

Females

There are conflicting studies and it is unclear if phytoestrogens have any effect on the cause or prevention of cancer in females. Epidemiological studies showed a protective effect against breast cancer. *In vitro'* studies concluded that females with current or past breast cancer should be aware of the risks of potential tumor growth when taking soy products since they can stimulate the growth of estrogen receptor-positive cells *in vitro*. The potential for tumor growth was found related only with small concentration of genistein and protective effects were found with larger concentrations of the same phytoestrogen. A 2006 review article stated the opinion that not enough information is available and that even if isoflavones have mechanisms to inhibit tumor growth, *in vitro* results justify the need to evaluate, at cellular level, the impact of isoflavones on breast tissue in females at high risk for breast cancer. Recent epidemiologic studies suggest that consumption of soy estrogens is safe for patients with breast cancer and that it may in fact decrease mortality and recurrence rates. A Cochrane Review of the use of phytoestrogens to relieve the vasomotor symptoms of menopause (hot flashes) demonstrated that there was no evidence to suggest any benefit to their use. It has been reported that phytoestrogens such as genistein may help prevent photoaging in human skin and promote formation of hyaluronic acid.

Infant Formula

Some studies have found that some concentrations of isoflavones may have effects on intestinal cells. At low doses, genistein acted as a weak estrogen and stimulated cell growth; at high doses, it inhibited proliferation and altered cell cycle dynamics. This biphasic response correlates with how genistein is thought to exert its effects. Some reviews express the opinion that more research is needed to answer the question of what effect phytoestrogens may have on infants, but their authors did not find any adverse effects. Multiple studies conclude there are no adverse effects in human growth, development, or reproduction as a result of the consumption of soy-based infant formula compared to conventional cow-milk formula. Soy formula presents no more risk than cow-milk formula. One of these studies, published at the Journal of Nutrition, concludes that: Comprehensive literature reviews and clinical studies of infants fed SBIFs [soy-based infant formulas] have resolved questions or raise no clinical concerns with respect to nutritional adequacy, sexual development, neurobehavioral development, immune development, or thyroid disease. SBIFs provide complete nutrition that adequately supports normal infant growth and development. FDA has accepted SBIFs as safe for use as the sole source of nutrition.

Clinical guidelines from the American Academy of Pediatrics state: "although isolated soy protein-based formulas may be used to provide nutrition for normal growth and development, there are few indications for their use in place of cow milk-based formula. These indications include (a) for infants with galactosemia and hereditary lactase deficiency (rare) and (b) in situations in which a vegetarian diet is preferred."

Ethnopharmacology

In some countries, phytoestrogenic plants have been used for centuries in the treatment of menstrual and menopausal problems, as well as for fertility problems. Plants used that have been shown to contain phytoestrogens include *Pueraria mirifica* and its close relative, kudzu, Angelica, fennel and anise. In a rigorous study, the use of one such source of phytoestrogen, red clover, has been shown to be safe, but ineffective in relieving menopausal symptoms (black cohosh is also used for menopausal symptoms, but does not contain phytoestrogens. Panax Ginseng contains phytoestrogens and has been used for menopausal symptoms.

References

Adlercreutz H (2002). "Phyto-oestrogens and cancer.". *Lancet Oncol.* 3 (6): 364–73. doi: 10.1016/S1470-2045(02)00777-5.

Albert-Puleo M (1980). "Fennel and anise as estrogenic agents". *J Ethnopharmacol* 2 (4): 337–344. doi: 10.1016/S0378-8741(80)81015-4.

Amadasi A, Mozzarelli A, Meda C, Maggi A, Cozzini P. Identification of Xenoestrogens in Food Additives by an Integrated in Silico and *in vitro* Approach. Chemical Research in Toxicology 2008,22: 52–63

Bhatia J, Greer F (2008). "Use of soy protein-based formulas in infant feeding". *Pediatrics* 121 (5): 1062–1068. doi: 10.1542/peds.2008-0564.

Brown, D. E., N.J. Walton, (1999). *Chemicals from Plants: Perspectives on Plant Secondary Products.* World Scientific Publishing. pp. 21, 141. ISBN 978-981-02-2773-9.

Chadwick LR, Nikolic D, Burdette JE, Overk CR, Bolton JL, van Breemen RB, Fröhlich R, Fong HH, Farnsworth NR, Pauli GF (2004). "Estrogens and congeners from spent hops (*Humulus lupulus*)". *Journal of Natural Products* 67 (12): 2024–2032. doi: 10.1021/np049783i.

Chen A, Rogan WJ (2004). "Isoflavones in soy infant formula: a review of evidence for endocrine and other activity in infants". *Annu. Rev. Nutr.* 24 (1): 33–54. doi: 10.1146/annurev.nutr.24.101603.064950.

Chen AC, Donovan SM (2004). "Genistein at a concentration present in soy infant formula inhibits Caco-2BBe cell proliferation by causing G2/M cell cycle arrest". *J. Nutr.*134 (6): 1303–1308.

Committee on Toxicity Group on Phytoestrogens. "Chemistry and Analysis of Phytoestrogens". *Draft Report.* United Kingdom Food Standards Agency. Retrieved 2011-11-11.

Dabrowski, Waldemar M. (2004). *Toxins in Food.* CRC Press Inc. p. 95. ISBN 978-0-8493-1904-4.

Dang Z, Löwik CW (2004). "The balance between concurrent activation of ERs and PPARs determines daidzein-induced osteogenesis and adipogenesis". *J. Bone Miner. Res.* 19 (5): 853–61. doi: 10.1359/jbmr.040120.

Dang ZC (2009). "Dose-dependent effects of soy phyto-oestrogen genistein on adipocytes: mechanisms of action". *Obes Rev* 10 (3): 342–9. doi: 10.1111/j.1467-789X.2008.00554.x.

Dang ZC, Audinot V, Papapoulos SE, Boutin JA, Löwik CW (2003). "Peroxisome proliferator-activated receptor gamma (PPARgamma) as a molecular target for the soy phytoestrogen genistein". *J. Biol. Chem.* 278 (2): 962–7. doi: 10.1074/jbc.M209483200.

Dang ZC, Lowik C (2005). "Dose-dependent effects of phytoestrogens on bone". *Trends Endocrinol. Metab.* 16 (5): 207–13. doi: 10.1016/j.tem.2005.05.001.

de Kleijn MJ, van der Schouw YT, Wilson PW, Adlercreutz H, Mazur W, Grobbee DE, Jacques PF (2001). "Intake of dietary phytoestrogens is low in postmenopausal women in the United States: the Framingham study(1–4)". *J. Nutr.* 131 (6): 1826–1832.

de Lemos ML (2001). "Effects of soy phytoestrogens genistein and daidzein on breast cancer growth". *Ann Pharmacother* 35 (9): 1118–1121. doi: 10.1345/aph.10257. PMID 11573864. Retrieved 2008-12-20.

Delmonte P, Rader JI (2006). "Analysis of isoflavones in foods and dietary supplements". *J AOAC Int* 89 (4): 1138–1146.

Fidler AE, Zwart S, Pharis RP, Weston RJ, Lawrence SB, Jansen P, Elliott G, Merton DV (2000). "Screening the foods of an endangered parrot, the kakapo (Strigops habroptilus), for oestrogenic activity using a recombinant yeast bioassay". *Reprod. Fertil. Dev.* 12 (3-4): 191–199. doi: 10.1071/RD00041.

Geller SE, Shulman LP, van Breemen RB, *et al.* (2009). "Safety and efficacy of black cohosh and red clover for the management of vasomotor symptoms: a randomized controlled trial". *Menopause* 16 (6): 1156–1166.doi: 10.1097/gme.0b013e3181ace49b.

Giampietro PG, Bruno G, Furcolo G, Casati A, Brunetti E, Spadoni GL, Galli E (2004). "Soy protein formulas in children: no hormonal effects in long-term feeding". *J. Pediatr. Endocrinol. Metab.* 17 (2): 191–196. doi: 10.1515/JPEM.2004.17.2.191.

Haffejee IE (1990). "Cow's milk-based formula, human milk and soya feeds in acute infantile diarrhea: a therapeutic trial". *J. Pediatr. Gastroenterol. Nutr.* 10 (2): 193–198. doi: 10.1097/00005176-199002000-00009.

Hamilton-Reeves JM, Vazquez G, Duval SJ, Phipps WR, Kurzer MS, Messina MJ (2010). "Clinical studies show no effects of soy protein or isoflavones on reproductive hormones in men: results of a meta-analysis". *Fertil Steril.* 94 (3): 997–1007.doi: 10.1016/j.fertnstert.2009.04.038.

Hughes CL (1988). "Phytochemical mimicry of reproductive hormones and modulation of herbivore fertility by phytoestrogens". *Environ. Health Perspect.* 78: 171–4. doi: 10.1289/ehp. 8878171.

Ingram D, Sanders K, Kolybaba M, Lopez D (1997). "Case-control study of phyto-oestrogens and breast cancer". *Lancet* 350 (9083): 990–994. doi: 10.1016/S0140-6736(97)01339-1.

Johnson, I (2003). *Phytochem Functional Foods.* CRC Press Inc. pp. 66–68.ISBN 978-0-8493-1754-5.

Kennelly EJ, Baggett S, Nuntanakorn P, Ososki AL, Mori SA, Duke J, Coleton M, Kronenberg F (2002). "Analysis of thirteen populations of black cohosh for formononetin". *Phytomedicine* 9 (5): 461–467. doi: 10.1078/09447110260571733.

Korach, Kenneth S. (1998). *Reproductive and Developmental Toxicology.* Marcel Dekker Ltd. pp. 278–279. ISBN 978-0-8247-9857-4.

Lee YS, Park JS, Cho SD, Son JK, Cherdshewasart W, Kang KS (2002)."Requirement of metabolic activation for estrogenic activity of Pueraria mirifica". *J. Vet. Sci.* 3 (4): 273–277.

Leopold AS, Erwin M, Oh J, Browning B (1976). "Phytoestrogens: adverse effects on reproduction in California quail". *Science* 191 (4222): 98–100. doi: 10.1126/science.1246602.

Lethaby AE, Brown J, Marjoribanks J, Kronenberg F, Roberts H, Eden J (2007). "Phytoestrogens for vasomotor menopausal symptoms". *Cochrane Database Syst Rev*(4): CD001395. doi: 10.1002/4651858.CD001395.pub3.

Mascie-Taylor, C. G. N., Bentley, Gillian R. (2000). *Infertility in the modern world: present and future prospects.* Cambridge, UK: Cambridge University Press. pp. 99–100.ISBN 0-521-64387-2.

Merritt RJ, Jenks BH (2004). "Safety of soy-based infant formulas containing isoflavones: the clinical evidence". *J. Nutr.* 134 (5): 1220S–1224S.

Messina M, McCaskill-Stevens W, Lampe JW (2006). "Addressing the soy and breast cancer relationship: review, commentary and workshop proceedings". *J. Natl. Cancer Inst.* 98 (18): 1275–1284. doi: 10.1093/jnci/djj356.

Miniello VL, Moro GE, Tarantino M, Natile M, Granieri L, Armenio L (2003). "Soy-based formulas and phyto-oestrogens: a safety profile". *Acta Paediatr Suppl* 91(441): 93–100. doi: 10.1111/j.1651-2227.2003.tb00655.x.

Mitchell JH, Cawood E, Kinniburgh D, Provan A, Collins AR, Irvine DS (2001)."Effect of a phytoestrogen food supplement on reproductive health in normal males". *Clin. Sci.* 100 (6): 613–618. doi: 10.1042/CS20000212.

Muller-Schwarze, Dietland (2006). *Chemical Ecology of Vertebrates.* Cambridge University Press. p. 287. ISBN 978-0-521-36377-8.

Naz, Rajesh K. (1999). *Endocrine Disruptors: Effects on Male and Female Reproductive Systems.* CRC Press Inc. p. 90. ISBN 978-0-8493-3164-0.

Richard C. Leegood, Per Lea (1998). *Plant Biochemistry and Molecular Biology.* John Wiley and Sons. p. 211. ISBN 978-0-471-97683-7.

Rosenblum ER, Stauber RE, Van Thiel DH, Campbell IM, Gavaler JS (1993)."Assessment of the estrogenic activity of phytoestrogens isolated from bourbon and beer". *Alcohol. Clin. Exp. Res.* 17 (6): 1207–1209. doi: 10.1111/j.1530-0277. 1993.tb05230.x.

Shu XO, Zheng Y, Cai H, Gu K, Chen Z, Zheng W, Lu W (2009). "Soy food intake and breast cancer survival". *JAMA* 302 (22): 2437–2443.doi: 10.1001/jama.2009.1783.

Strom BL, Schinnar R, Ziegler EE, Barnhart KT, Sammel MD, Macones GA, Stallings VA, Drulis JM, Nelson SE, Hanson SA (2001). "Exposure to soy-based formula in infancy and endocrinological and reproductive outcomes in young adulthood". *JAMA* 286 (7): 807–814. doi: 10.1001/jama.286.7.807.

Thompson LU, Boucher BA, Liu Z, Cotterchio M, Kreiger N (2006). "Phytoestrogen content of foods consumed in Canada, including isoflavones, lignans and coumestan".*Nutrition and Cancer* 54 (2): 184–201. doi: 10.1207/s15327914nc5402_5.

Turner JV, Agatonovic-Kustrin S, Glass BD (2007). "Molecular aspects of phytoestrogen selective binding at estrogen receptors". *J Pharm Sci* 96 (8): 1879–1885.doi: 10.1002/jps.20987.

van Elswijk DA, Schobel UP, Lansky EP, Irth H, van der Greef J (2004). "Rapid dereplication of estrogenic compounds in pomegranate (Punica granatum) using on-line biochemical detection coupled to mass spectrometry". *Phytochemistry* 65 (2): 233–241. doi: 10.1016/j.phytochem.2003.07.001.

Varner, J E, Bonner, J (1966). *Plant Biochemistry.* Academic Press. ISBN 978-0-12-114856-0.

Yildiz, Fatih (2005). *Phytoestrogens in Functional Foods.* Taylor and Francis Ltd. pp. 3–5, 210–211. ISBN 978-1-57444-508-4.

Zhao E, Mu Q (2011). "Phytoestrogen biological actions on Mammalian reproductive system and cancer growth". *Sci Pharm* 79 (1): 1–20.doi: 10.3797/scipharm.1007-15.

Chapter 14

Polyphenols

Polyphenols in Foods

Many common foods contain rich sources of polyphenols which have antioxidant properties only in test tube studies. As interpreted by the Linus Pauling Institute, dietary polyphenols have little or no direct antioxidant food value following digestion. Not like controlled test tube conditions, the fate of flavones or polyphenols *in vivo* shows they are poorly conserved (less than 5 per cent), with most of what is absorbed existing as metabolites modified during digestion and destined for rapid excretion. Spices, herbs and essential oils are rich in polyphenols in the plant itself and shown with antioxidant potential *in vitro*. Typical spices high in polyphenols (confirmed *in vitro*) areclove, cinnamon, oregano, turmeric, cumin, parsley, basil, curry powder, mustard seed, ginger, pepper, chili powder, paprika, garlic, coriander, onion and cardamom. Typical herbs are sage, thyme, marjoram, tarragon, peppermint, oregano, savory, basil and dill weed. Dried fruits are a good source of polyphenols by weight/serving size as the water has been removed making the ratio of polyphenols higher. Typical dried fruits are pears, apples, plums, peaches, raisins, figs and dates. Dried raisins are high in polyphenol count. Red wine is high in total polyphenol count which supplies antioxidant quality which is unlikely to be conserved following digestion (see section below). Deeply pigmented fruits like cranberries, blueberries, plums, blackberries, raspberries, strawberries, blackcurrants, figs, cherries, guava, oranges, mango, grape juice and pomegranate juice also have significant polyphenol content. Typical cooked vegetables rich in antioxidants are artichokes, cabbage, broccoli, asparagus, avocados, beetroot and spinach. Nuts are a moderate source of polyphenol antioxidants. Typical nuts are pecans, walnuts, hazelnuts, pistachio, almonds, cashew nuts, macadamia nuts and peanut butter. Sorghum bran, cocoa powder and cinnamon are rich sources of procyanidins, which are large molecular weight compounds found in many fruits and some vegetables. Partly due to the large molecular weight (size) of these compounds, their amount actually absorbed in the body is low, an effect also resulting from the action of stomach acids, enzymes and bacteria in the gastrointestinal tract where smaller derivatives are metabolized and prepared for rapid excretion.

Physiological Context

Despite the above discussion implying that ORAC-rich foods with polyphenols may provide antioxidant benefits when in the diet, there remains no physiological evidence that any polyphenols have such actions or that ORAC has any relevance in the human body. On the contrary, research indicates that although polyphenols are good antioxidants *in vitro*, antioxidant effects *in vivo* are probably negligible or absent. By non-antioxidant mechanisms still undefined, polyphenols may affect mechanisms of cardiovascular disease or cancer. The increase in antioxidant capacity of blood seen after the consumption of polyphenol-rich (ORAC-rich) foods is not caused directly by the polyphenols, but most likely results from increased uric acid levels derived from metabolism of flavonoids. According to Frei, "we can now follow the activity of flavonoids in the body and one thing that is clear is that the body sees them as foreign compounds and is trying to get rid of them." Another mechanism may be the increase in activities of paraoxonases by dietary antioxidants which can reduce oxidative stress.

Natural Phenols

Natural phenols are a class of molecules found in abundance in plants.

Flavonoids

Flavonoids, a subset of polyphenol antioxidants, are present in many berries, as well as in coffee and tea.

- ☆ Flavones:
  - ☆ Apigenin
  - ☆ Luteolin
  - ☆ Tangeritin
- ☆ Flavonols:
  - ☆ Isorhamnetin
  - ☆ Kaempferol
  - ☆ Myricetin - walnuts are a rich source
  - ☆ Proanthocyanidins, or condensed tannins
  - ☆ Quercetin and related, such as rutin
- ☆ Flavanones:
  - ☆ Eriodictyol
  - ☆ Hesperetin (metabolizes to hesperidin)
  - ☆ Naringenin (metabolized from naringin)
- ☆ Flavanols and their polymers:
  - ☆ Catechin, gallocatechin and their corresponding gallate esters

☆ Epicatechin, epigallocatechin and their corresponding gallate esters

☆ Theaflavin its gallate esters

☆ Thearubigins

☆ Isoflavone phytoestrogens - found primarily in soy, peanuts and other members of the Fabaceae family

☆ Daidzein

☆ Genistein

☆ Glycitein

☆ Stilbenoids:

☆ Resveratrol - found in the skins of dark-coloured grapes and concentrated in red wine.

☆ Pterostilbene - methoxylated analogue of resveratrol, abundant in Vaccinium berries

☆ Anthocyanins

☆ Cyanidin

☆ Delphinidin

☆ Malvidin

☆ Pelargonidin

☆ Peonidin

☆ Petunidin

Phenolic Acids and their Esters

☆ Chicoric acid - another caffeic acid derivative, is found in chicory and Echinacea.

☆ Chlorogenic acid - found in high concentration in coffee (more concentrated in robusta than arabica beans), blueberries and tomatoes. Produced from esterification of caffeic acid.

☆ Cinnamic acid and its derivatives, such as ferulic acid - found in seeds of plants such as in brown rice, whole wheat and oats, as well as in coffee, apple, artichoke, peanut, orange and pineapple.

☆ Ellagic acid - found in high concentration in raspberry and strawberry and in ester form in red wine tannins.

☆ Ellagitannins - hydrolyzable tannin polymer formed when ellagic acid, a polyphenol monomer, esterifies and binds with the hydroxyl group of a polyol carbohydrate such as glucose.

☆ Gallic acid - found in gallnuts, sumac, witch hazel, tea leaves, oak bark and many other plants.

☆ Gallotannins - hydrolyzable tannin polymer formed when gallic acid, a polyphenol monomer, esterifies and binds with the hydroxyl group of a polyol carbohydrate such as glucose.

☆ Rosmarinic acid - found in high concentration in rosemary, oregano, lemon balm, sage and marjoram.

☆ Salicylic acid - found in most vegetables, fruits and herbs; but most abundantly in the bark of willow trees, from where it was extracted for use in the early manufacture ofaspirin.

Other Nonflavonoid Phenolics

☆ Curcumin - Curcumin has low bioavailability, because, much of it is excreted through glucuronidation. However, bioavailability is substantially enhanced by solubilization in a lipid (oil or lecithin), heat, addition of piperine, or through nanoparticularization.

☆ Flavonolignans - e.g. silymarin - a mixture of flavonolignans extracted from milk thistle.

Rank	Food	Serving Size	Total Antioxidant Capacity per serving size
1	Small Red Bean	1/2 cup dried beans	13727
2	Wild blueberry	1 cup	13427
3	Red kidney bean	1/2 cup dried beans	13259
4	Pinto bean	1/2 cup	11864
5	Blueberry	1 cup cult-ivated berries	9019
6	Cranberry	1 cup whole berries	8983
7	Artichoke hearts	1 cup cooked	7904
8	Blackberry	1 cup	7701
9	Prune	1/2 cup	7291
10	Raspberry	1 cup	6058
11	Strawberry	1 cup	5938
12	Red Delicious apple	1	5900
13	Granny Smith	1	5381
14	Pecan	1 ounce	5095
15	Sweet cherry	1 cup	4873
16	Black plum	1	4844
17	Russet potato	1 cooked	4649
18	Black bean	1/2 cup dried beans	4181
19	Plum	1	4118
20	Gala apple	1	3903

☆ Xanthones - mangosteen is purported to contain a large variety of xanthones, but some of the xanthones like mangostin might be present only in the inedible shell.

☆ Eugenol

Polyphenols and Biochar

Polyphenols include several classes of compounds, such as phenols, phenolic acids, flavonoids, anthocyanins and others, with more complex structures, tannins and lignins. Polyphenols are secondary metabolites produced by plants in response to stress conditions (Bennet and Wallsgrove, 1994), such as infections, large amounts of UV rays or other factors (Popa *et al.*, 2007).

Oxidized polyphenols also inhibit growth and development of certain microbial strains. The toxicity mechanism of polyphenols may be explained by the inhibition of hydrolytic enzymes, or by other interactions, such as blocking protein transport, non-specific interactions with carbohydrates, etc. (Popa *et al.*, 2007). Stimulus or inhibition capacity on plant growth and development is closely correlated with the concentration of Polyphenolic compounds used (Anghel, 2001). Polyphenols, a large class of chemicals which are found in plants, have attracted much attention in

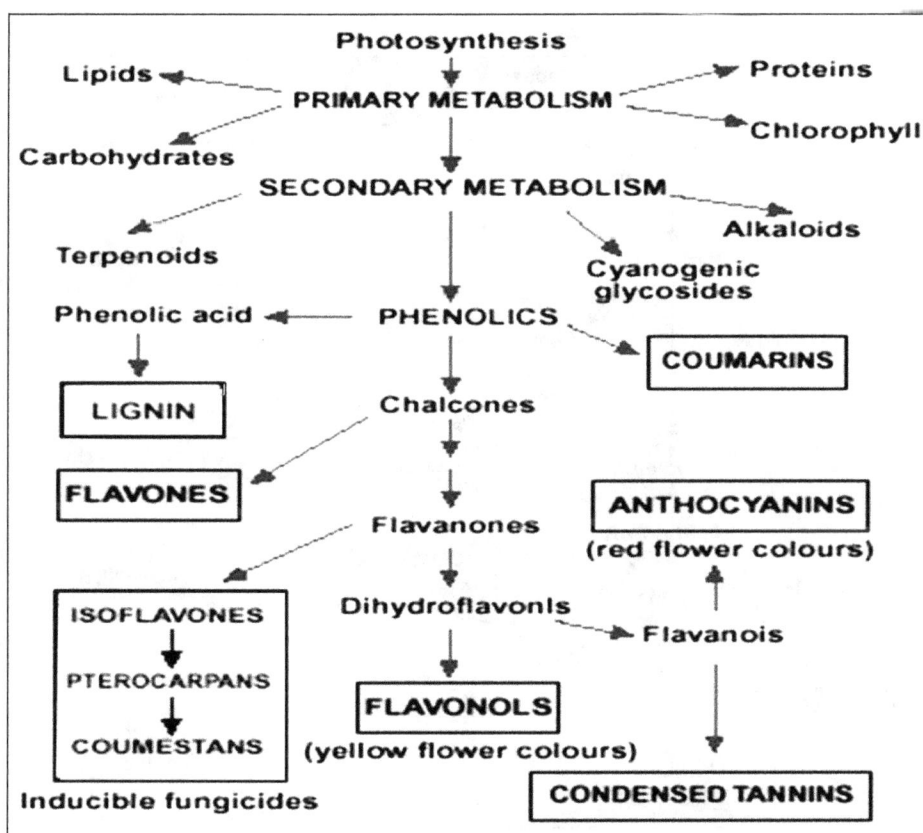

S assimilation

sulphate$_{out}$

sulphate
transporter

sulpho-
transferase

X-OH ⟶

S-containing metabolites
eg. glucosinolates
sulphoflavonoids
phytoalexins

sulphate$_{in}$

ATP sulphurylase

**APS
Kinase**

APS ⟶ **PAPS**

*APS
reductase*

serine

sulphite

*serine
acetyl-
transferase*

*sulphite
reductase*

O-acetylserine

sulphide

*O-acetylserine
(thiol)lyase*

cysteine

Plant Defence:
- Abiotic stress
- Wounding
- Pathogen attack
- Signalling
- ROS detox

Nutrition:
- Taste and flavour
- Medicinally important
  bioactive compounds
  -eg cancer prevention,
  anti-coagulants

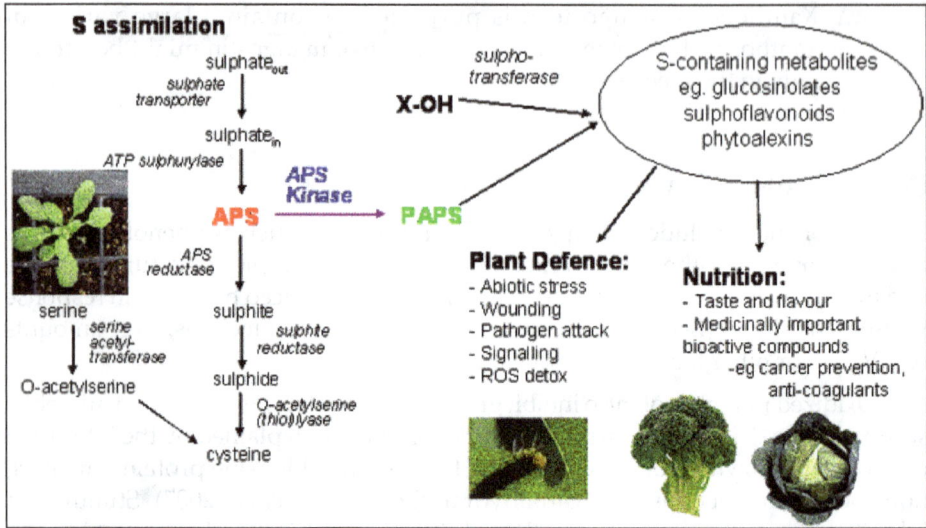

**APS Kinase Partitions S away from Primary Assimilation to be Used for
Incorporation into many Important Secondary Metabolites.**

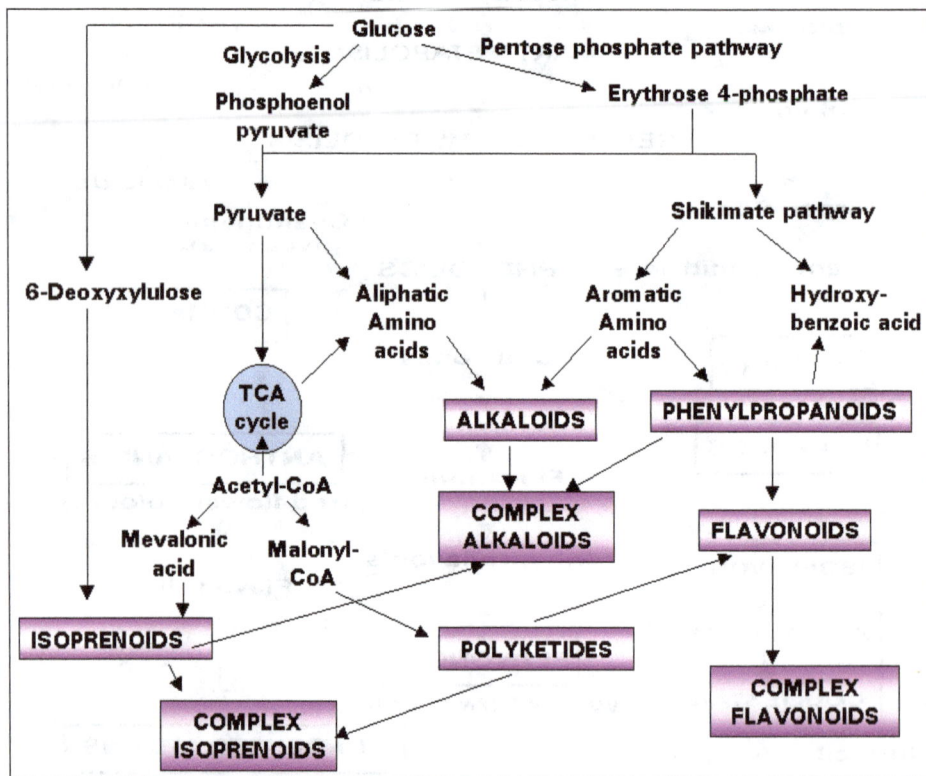

Glucose

Glycolysis

Pentose phosphate pathway

**Phosphoenol
pyruvate**

Erythrose 4-phosphate

Pyruvate

Shikimate pathway

6-Deoxyxylulose

**Aliphatic
Amino
acids**

**Aromatic
Amino
acids**

**Hydroxy-
benzoic acid**

**TCA
cycle**

ALKALOIDS

PHENYLPROPANOIDS

Acetyl-CoA

**COMPLEX
ALKALOIDS**

FLAVONOIDS

**Mevalonic
acid**

**Malonyl-
CoA**

ISOPRENOIDS

POLYKETIDES

**COMPLEX
FLAVONOIDS**

**COMPLEX
ISOPRENOIDS**

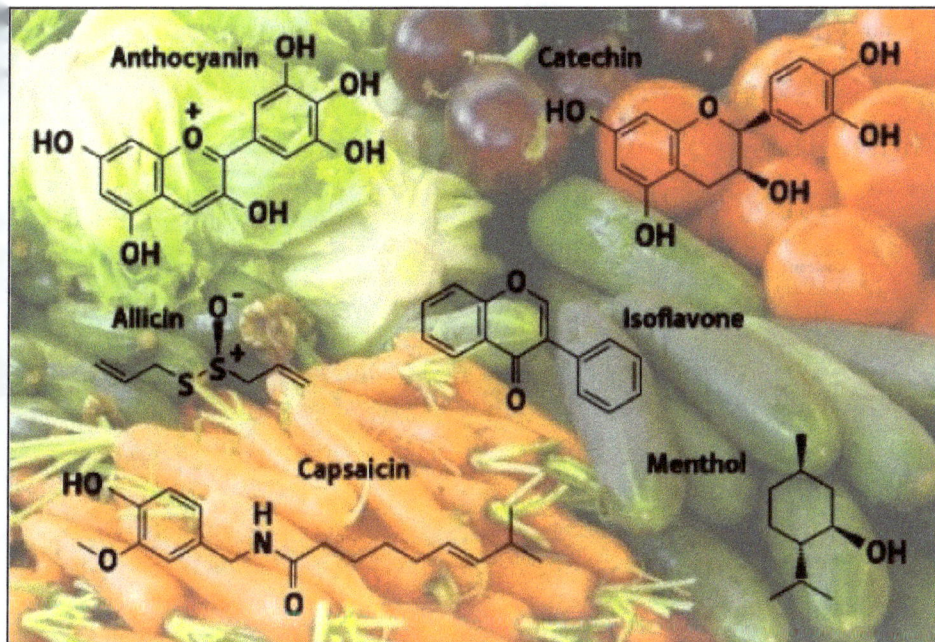

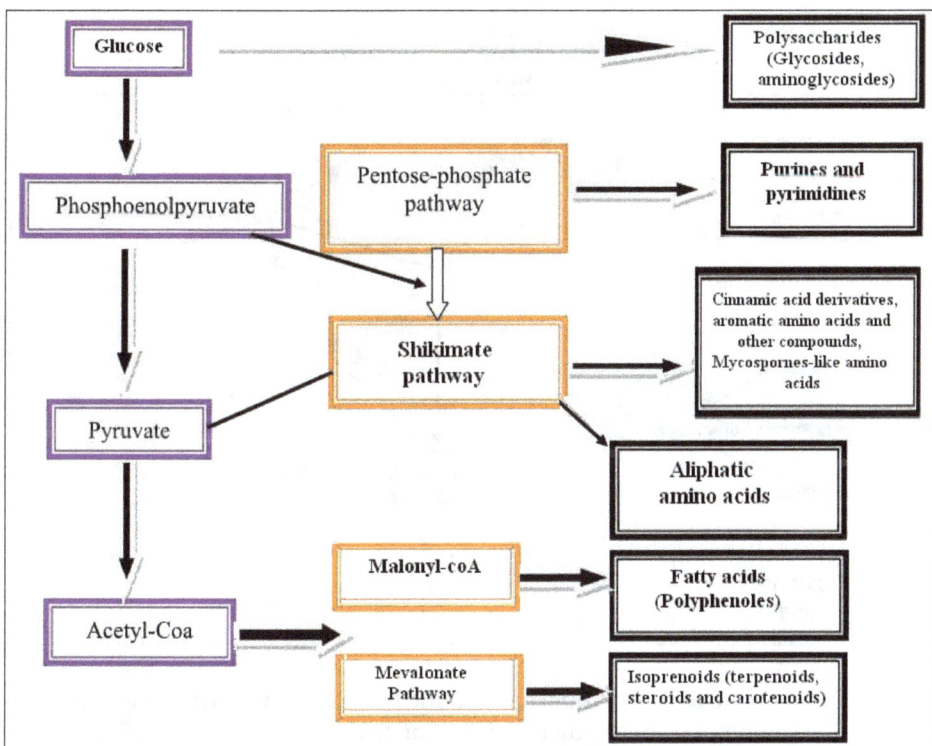

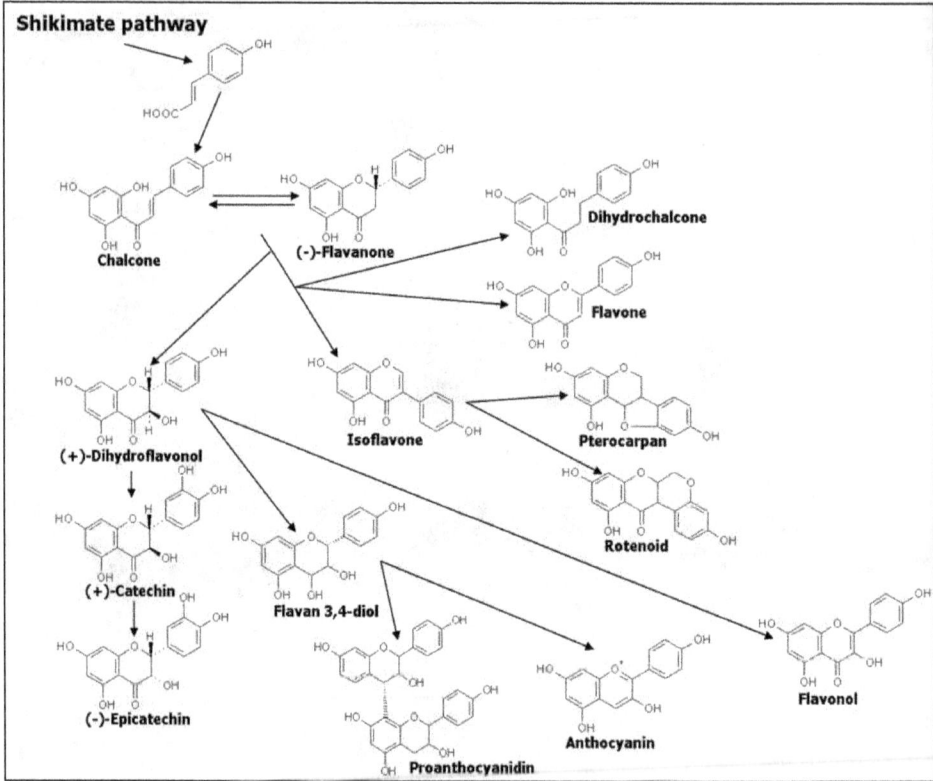

Shikimate pathway

Chalcone · (-)-Flavanone · Dihydrochalcone · Flavone · (+)-Dihydroflavonol · Isoflavone · Pterocarpan · Rotenoid · (+)-Catechin · Flavan 3,4-diol · (-)-Epicatechin · Proanthocyanidin · Anthocyanin · Flavonol

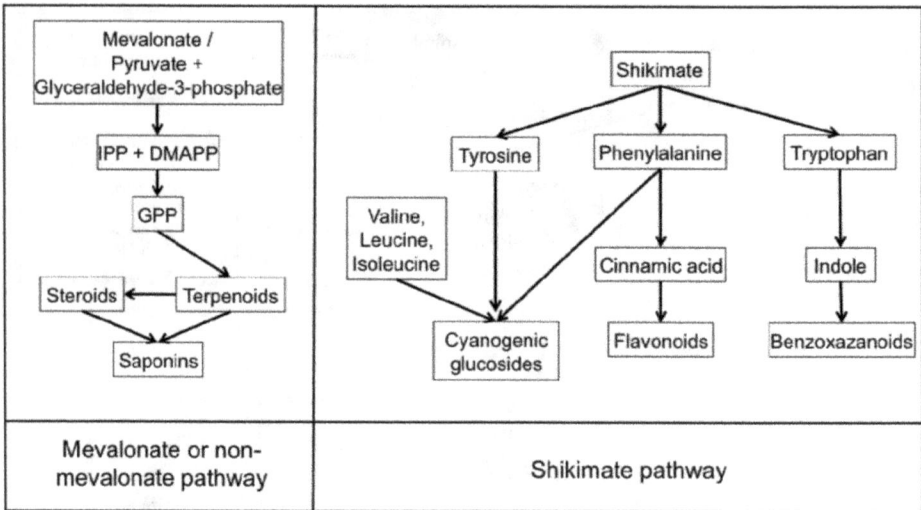

Mevalonate / Pyruvate + Glyceraldehyde-3-phosphate → IPP + DMAPP → GPP → Steroids ← Terpenoids → Saponins

Mevalonate or non-mevalonate pathway

Shikimate → Tyrosine · Phenylalanine · Tryptophan; Valine, Leucine, Isoleucine → Cyanogenic glucosides; Phenylalanine → Cinnamic acid → Flavonoids; Tryptophan → Indole → Benzoxazanoids

Shikimate pathway

the last decades due to their properties and the hope that they will show beneficial health effects, when taken as a dietary input or as complement (Hu, 2007).

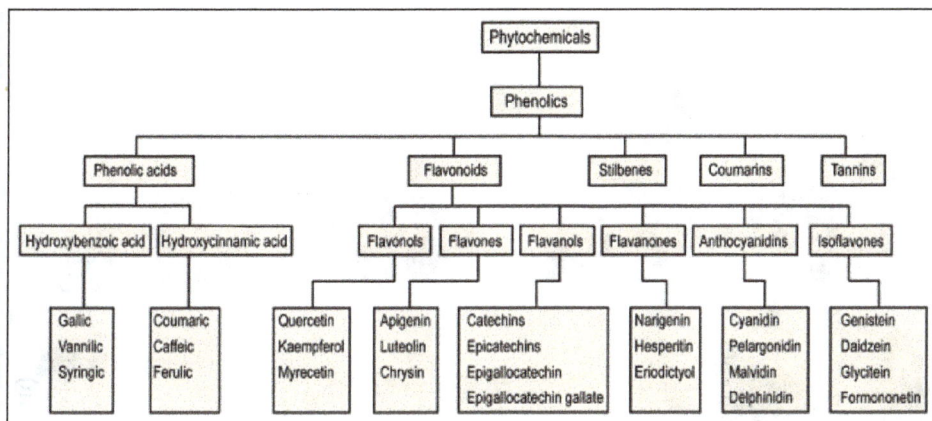

Phytochemicals

Phenolics

| Phenolic acids | | Flavonoids | | | | | | Stilbenes | Coumarins | Tannins |

| Hydroxybenzoic acid | Hydroxycinnamic acid | | Flavonols | Flavones | Flavanols | Flavanones | Anthocyanidins | Isoflavones |

Gallic	Coumaric		Quercetin	Apigenin	Catechins		Narigenin	Cyanidin	Genistein
Vannilic	Caffeic		Kaempferol	Luteolin	Epicatechins		Hesperitin	Pelargonidin	Daidzein
Syringic	Ferulic		Myrecetin	Chrysin	Epigallocatechin		Eriodictyol	Malvidin	Glycitein
					Epigallocatechin gallate			Delphinidin	Formononetin

Phenolic compounds constitute one of the most extensive groups of chemicals in the plant kingdom. It is estimated that more than 8000 compounds have been isolated and described (Ramos, 2007).

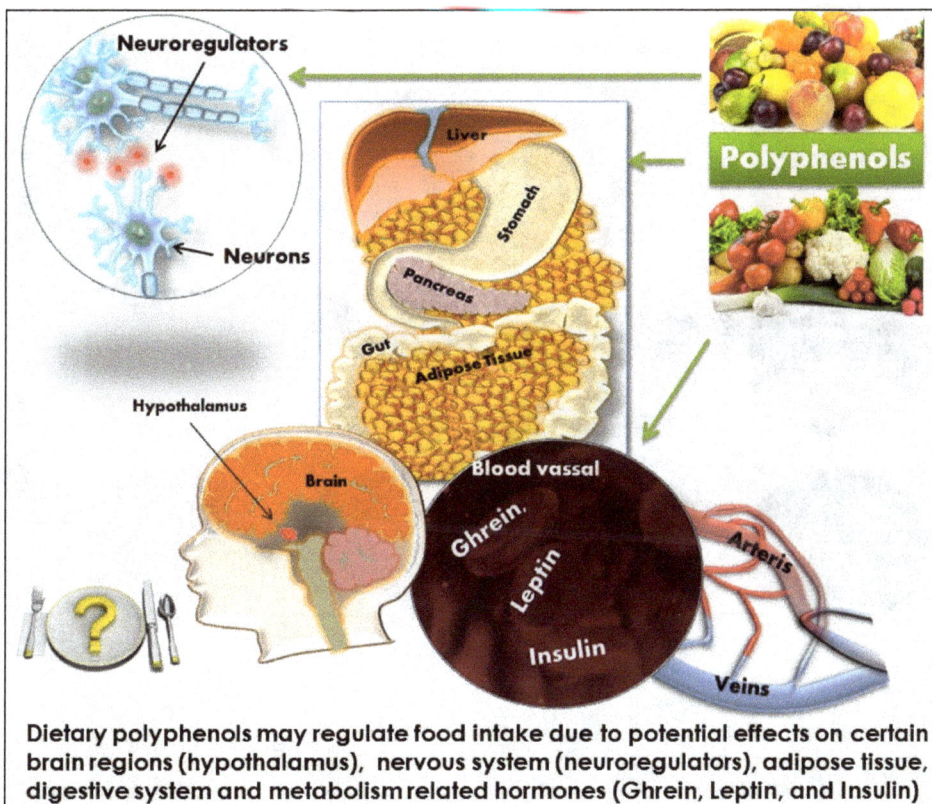

Dietary polyphenols may regulate food intake due to potential effects on certain brain regions (hypothalamus), nervous system (neuroregulators), adipose tissue, digestive system and metabolism related hormones (Ghrein, Leptin, and Insulin)

Cranberry-Whey   Cranberry-Oat Bran   Cranberry Wheat Bran   Cranberry Wheat Germ

Lemon-SPI   Apple-SPI   Pear-SPI   Kiwi-SPI

Pomegranate-SPI   Green Tea-Pea   Guava-SPI   Passion Fruit-SPI

Chokerry-SPI   Strawberry-SPI   Rhubarh-SPI   Cinnamon-SPI

Maqui Berry–Pea Grape-Pea   Grape-SPI   Cranberry-Pea   Cranberry-SPI

Soy Protein Isolate SPI   Blueberry-Pea   Blueberry-SPI   Black Currant-SPI

Defatted Soy Flour   Pea Protein   Hemp Protein

Ingredients are Standardized for Total Polyphenols (1-10 per cent)

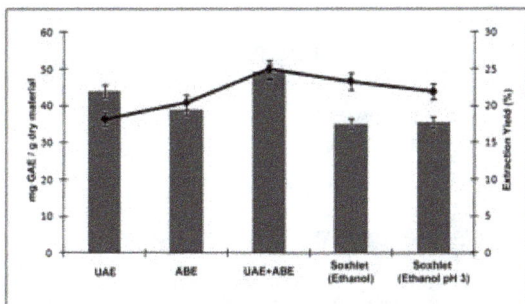

Total phenolic content and extraction yield from extracts of jambul peels (Syzygiumcumini) obtained through various methods of extraction.

In order to test the content of polyphenols in biochar derived from agriculture wastes research has been done, some investigations had found that the combination of charcoal and polyphenols addition have potential to improve the growth and yield of radish. The mixed combination of biochar and polyphenols applied at 1.5 per cent w/w to compost led to highest root yields (Jordan *et al.*, 2011), in a similar research Niggli and Schmidt (2010) tested biochar applications of biochar in Vineyards and found that grapes from biochar-treated plots had a 10 per cent higher polyphenol content. Together with the much higher amino acid content, this was an indication of a greater aromatic quality of the grapes, which is then passed into the wine.

Between species, the variation in polyphenol production by plants has been understood as a defense against herbivores (Haslam, 1981; Bernays *et al.*, 1989). Recent evidence proposes that the pools and variations of inorganic and organic soil nutrients can be influenced by polyphenols (Northup *et al.*, 1998; Schimel *et al.*, 1998). In terms of nutrient competition between plants and microbes, these effects

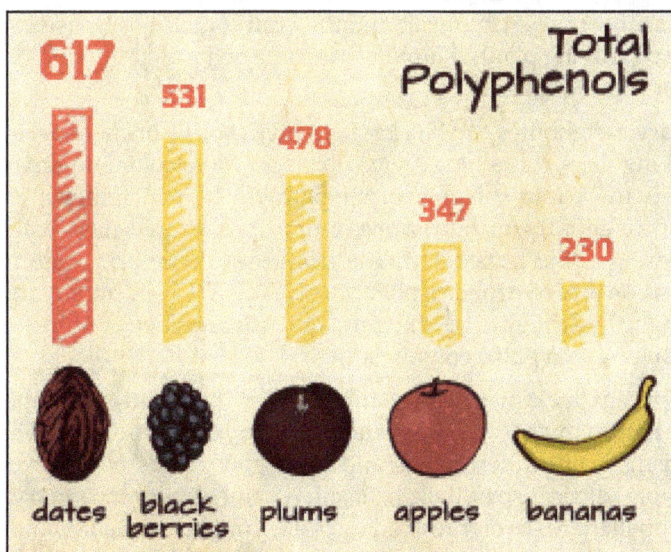

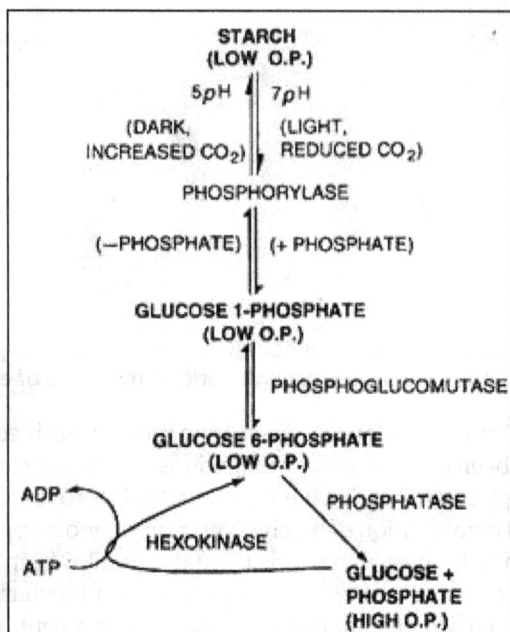

**Development of High and Low Osmotic Pressures in
Guard Cells throgh Starch Hydrolysis Theory.**

may possibly have wide-ranging consequences, also for retention and nutrient cycling in ecosystems. (Hättenschwiler and Vitousek, 2000).

Secondary metabolites are chemicals produced by plants for which no role has yet been found in growth, photosynthesis, reproduction, or other "primary" functions. These chemicals are extremely diverse; many thousands have been identified in several major classes. Each plant family, genus and species produces a characteristic mix of these chemicals and they can sometimes be used as taxonomic characters in classifying plants. Humans use some of these compounds as medicines, flavorings, or recreational drugs.

Secondary metabolites can be classified on the basis of chemical structure (for example, having rings, containing a sugar), composition (containing nitrogen or not), their solubility in various solvents, or the pathway by which they are synthesized (*e.g.*, phenylpropanoid, which produces tannins). A simple classification includes three main groups: the terpenes (made from mevalonic acid, composed almost entirely of carbon and hydrogen), phenolics (made from simple sugars, containing benzene rings, hydrogen and oxygen) and nitrogen-containing compounds (extremely diverse, may also contain sulfur).

The apparent lack of primary function in the plant, combined with the observation that many secondary metabolites have specific negative impacts on other organisms such as herbivores and pathogens, leads to the hypothesis that they have evolved because of their protective value. Many secondary metabolites are toxic or repellant to herbivores and microbes and help defend plants producing

Table 14.1: Different Types of Polyphenols.

Class	Example Compounds	Example Sources	Some Effects and Uses
NITROGEN-CONTAINING			
Alkaloids	Nicotine cocaine theobromine	Tobacco coca plant chocolate (cocao)	Interfere with neurotransmission, block enzyme action
NITROGEN-AND SULFUR-CONTAINING			
Glucosinolates	Sinigrin	Cabbage, relatives	
TERPENOIDS			
Monoterpenes	Menthol linalool	Mint and relatives, many plants	Interfere with neurotransmission, block ion transport, anesthetic
Sesquiterpenes	Parthenolid	Parthenium and relatives (*Asteraceae*)	Contact dermatitis
Diterpenes	Gossypol	Cotton	Block phosphorylation; toxic
Triterpenes, cardiac glycosides	Digitogenin	Digitalis (foxglove)	Stimulate heart muscle, alter ion transport
Tetraterpenoids	Carotene	Many plants	Antioxidant; orange colouring
Terpene polymers	Rubber	Hevea (rubber) trees, dandelion	Gum up insects; airplane tires
Sterols	Spinasterol	Spinach	Interfere with animal hormone action
PHENOLICS			
Phenolic acids	Caffeic, chlorogenic	All plants	Cause oxidative damage, browning in fruits and wine
Coumarins	Umbelliferone	Carrots, parsnip	Cross-link DNA, block cell division
Lignans	Podophyllin urushiol	Mayapple poison ivy	Cathartic, vomiting, allergic dermatitis
Flavonoids oxidants, estrogenic	Anthocyanin, catechin	Almost all plants	Flower, leaf colour; inhibit enzymes, anti- and pro-
Tannins	Gallotannin, condensed tannin	Oak, hemlock trees, birdsfoot trefoil, legumes	Bind to proteins, enzymes, block digestion, antioxidants
Lignin	Lignin	All land plants	Structure, toughness, fiber

them. Production increases when a plant is attacked by herbivores or pathogens. Some compounds are released into the air when plants are attacked by insects; these compounds attract parasites and predators that kill the herbivores. Recent research is identifying more and more primary roles for these chemicals in plants as signals, antioxidants and other functions, so "secondary" may not be an accurate description in the future.

Consuming some secondary metabolites can have severe consequences. Alkaloids can block ion channels, inhibit enzymes, or interfere with neurotransmission, producing hallucinations, loss of coordination, convulsions, vomiting and death. Some phenolics interfere with digestion, slow growth, block enzyme activity and cell division, or just taste awful.

Most herbivores and plant pathogens possess mechanisms that ameliorate the impacts of plant metabolites, leading to evolutionary associations between particular groups of pests and plants. Some herbivores (for example, the monarch butterfly) can store (sequester) plant toxins and gain protection against their enemies. Secondary metabolites may also inhibit the growth of competitor plants (allelopathy). Pigments (such as terpenoid carotenes, phenolics and flavonoids) colour flowers and, together with terpene and phenolic odors, attract pollinators.

Secondary chemicals are important in plant use by humans. Most pharmaceuticals are based on plant chemical structures and secondary metabolites are widely used for recreation and stimulation (the alkaloids nicotine and cocaine; the terpene cannabinol). The study of such plant use is called ethnopharmacology. Psychoactive plant chemicals are central to some religions and flavors of secondary compounds shape our food preferences. The characteristic flavors and aroma of cabbage and relatives are caused by nitrogen-and sulfur-containing chemicals, glucosinolates, which protect these plants from many enemies. The astringency of wine and chocolate derives from tannins. The use of spices and other seasonings developed from their combined uses as preservatives (since they are antibiotic) and flavorings.

References

Agosta, William (1996). Bombardier Beetles and Fever Trees: A Close-up Look at Chemical Warfare and Signals in Animals and Plants. Reading, MA: Addison-Wesley

Bidlack, Wayne R. (2000). Phytochemicals as Bioactive Agents. Lancaster, PA: Technomic Publishers, 2000.

Karban, Richard and Ian T. Baldwin (1997). Induced Responses to Herbivory. Chicago: University of Chicago Press.

Rosenthal, Gerald A. and May R. Berenbaum (1991). Herbivores, Their Interactions with Secondary Plant Metabolites. San Diego, CA: Academic Press.

Chapter 15

Saponin

Saponins are a class of chemical compounds found in particular abundance in various plant species. More specifically, they are amphipathic glycosides grouped phenomenologically by the soap-like foaming they produce when shaken in aqueous solutions and structurally by having one or more hydrophilic glycoside moieties combined with a lipophilic triterpene derivative.

What is Saponins?

Saponins are glucosides with foaming characteristics. Saponins consist of a polycyclic aglycones attached to one or more sugar side chains. The aglycone part, which is also called sapogenin, is either steroid (C27) or a triterpene (C30). The foaming ability of saponins is caused by the combination of a hydrophobic (fat-soluble) sapogenin and a hydrophilic (water-soluble) sugar part. Saponins have a bitter taste. Some saponins are toxic and are known as sapotoxin.

Distribution

Saponins are phytochemicals which can be found in most vegetables, beans and herbs. The best known sources of saponins are peas, soybeans and some herbs with names indicating foaming properties such as soapwort, saoproot, soapbark and soapberry. Commercial saponins are extracted mainly from *Yucca schidigera* and *Quillaja saponaria*.

Structural Variety and Biosynthesis

The aglycone (glycoside-free) portions of the saponins are termed sapogenins. The number of saccharide chains attached to the sapogenin/aglycone core can vary giving rise to another dimension of nomenclature (monodesmosidic, bidesmosidic, etc.) as can the length of each chain. A somewhat dated compilation has the range of saccharide chain lengths being 1–11, with the numbers 2-5 being the most frequent and with both linear and branched chain saccharides being represented. Dietary

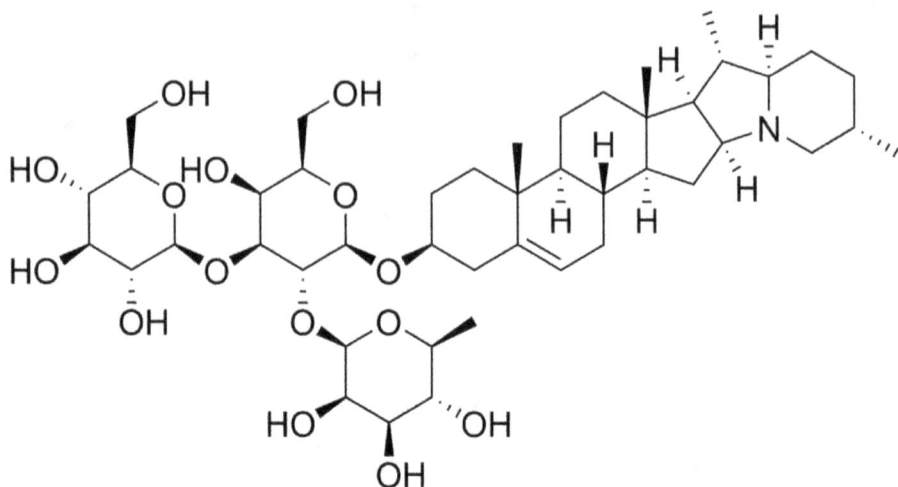

Figure 15.1: Chemical Structure of the Saponin Solanine.

monosaccharides such as D-glucose and D-galactose are among the most common components of the attached chains.

The lipophilic aglycone can be any one of a wide variety of polycyclic organic structures originating from the serial addition of 10-carbon (C10) terpene units to compose a C30 triterpene skeleton, often with subsequent alteration to produce a C27 steroidal skeleton. The subsets of saponins that are steroidal have been termed saraponins; Aglycone derivatives can also incorporate nitrogen, so some saponins also present chemical and pharmacologic characteristics of alkaloid natural products. The figure at right above presents the structure of the alkaloid phytotoxin solanine, a monodesmosidic, branched-saccharide steroidal saponin. (The lipophilic steroidal structure is the series of connected six- and five-membered rings at the right of the structure, while the three oxygen-rich sugar rings are at left and below. Note the nitrogen atom inserted into the steroid skeleton at right.)

Sources

Saponins occur naturally in soybeans, peas, ginseng, herbs, vegetables and yucca. They are phytochemicals, or plant chemicals, possessing detergent qualities that foam when mixed with water. Commercially, saponins appear in beverages and cosmetics as emulsifiers or sweeteners. They're also fed to livestock to cut down on odor because they bind to ammonia, which contributes to foul smells.

Saponins have historically been understood to be plant-derived, but they have also been isolated from marine organisms. Saponins are indeed found in many plants and derive their name from the soapwort plant (genus *Saponaria*, family Caryophyllaceae), the root of which was used historically as a soap.[2] Saponins are also found in the botanical family Sapindaceae, with its defining genus *Sapindus* (soapberry or soapnut) and in the closely related families Aceraceae (maples) and Hippocastanaceae (horse chestnuts; ref. needed). It is also found heavily in

Gynostemma pentaphyllum (*Gynostemma,* Cucurbitaceae) in a form called gypenosides and ginseng or red ginseng (*Panax,* Araliaceae) in a form called ginsenosides. Within these families, this class of chemical compounds is found in various parts of the plant: leaves, stems, roots, bulbs, blossom and fruit.commercial formulations of plant-derived saponins, *e.g.*, from the soap bark (or soapbark) tree, *Quillaja saponaria* and those from other sources are available via controlled manufacturing processes, which make them of use as chemical and biomedical reagents.

Test

Froth Test

Uses plant Gogo (bark) *Entada phaseoloides* as control. The positive result shows a honeycomb froth that is higher than 2 cm that persists for 10 minutes or longer.

Blood Agar Media (BAM): Is an agar cup semi-quantitative method that shows positive result of hemolytical halos.

Role in Plant Ecology and Impact on Animal Foraging

In plants, saponins may serve as anti-feedants and to protect the plant against microbes and fungi. Some plant saponins (*e.g.* from oat and spinach) may enhance nutrient absorption and aid in animal digestion. However, saponins are often bitter to taste and so can reduce plant palatability (*e.g.*, in livestock feeds), or even imbue them with life-threatening animal toxicity. Data makes clear that some saponins are toxic to cold-blooded organisms and insects at particular concentrations. Further research is needed to define the roles of these natural products in their host organisms, which have been described as "poorly understood" to date.

Ethnobotany

Most saponins, which readily dissolve in water, are poisonous to fish.[9] Therefore, in ethnobotany, they are primarily known for their use by indigenous people in obtaining aquatic food sources. Since prehistoric times, cultures throughout the world have used piscicidal plants, mostly those containing saponins, for fishing. Although prohibited by law, fish poison plants are still widely used by indigenous tribes in Guyana. On the Indian Subcontinent, the Gond tribes are known for their use of plant extracts in poison fishing.

Many of California's Native American tribes traditionally used soaproot, (genus *Chlorogalum*) and/or the root of various yucca species, which contain saponin, as a fish poison. They would pulverize the roots, mixing in water to create foam and then add the suds to a stream. This would kill or incapacitate the fish, which could be gathered easily from the surface of the water. Among the tribes using this technique were the Lassik, the Luiseño, the Yuki, the Yokut, the Chilula, the Wailaki, the Miwok, the Kato, the Mattole, the Nomlakiand the Nishinam.

Established Research Bioactivities and therapeutic claims

Bioactivities

One research use of the saponin class of natural products involves their complexation with cholesterol to form pores in cell membrane bilayers, *e.g.*, in red cell (erythrocyte) membranes, where complexation leads to red cell lysis (hemolysis) on intravenous injection. In addition, the amphipathic nature of the class gives them activity as surfactantsthat can be used to enhance penetration of macromolecules such as proteins through cell membranes. Saponins have also been used as adjuvants in vaccines.

Saponins from the *Gypsophila paniculata* (baby's breath) plant have been shown to significantly augment the cytotoxicity of immunotoxins and other targeted toxins directed against human cancer cells. The research groups of Professor Hendrik Fuchs (Charité University, Berlin, Germany) and Dr David Flavell (Southampton General Hospital, United Kingdom) are working together toward the development of *Gypsophila* saponins for use in combination with immunotoxins or other targeted toxins for patients with leukaemia, lymphoma and other cancers.

Medical Uses

Saponins are being promoted commercially as dietary supplements and nutraceuticals. There is evidence of the presence of saponins in traditional medicine preparations, where oral administrations might be expected to lead to hydrolysis of glycoside from terpenoid (and obviation of any toxicity associated with the intact molecule). But as is often the case with wide-ranging commercial therapeutic claims for natural products:

- ☆ The claims for organismal/human benefit are often based on very preliminary biochemical or cell biological studies; and
- ☆ Mention is generally omitted of the possibilities of individual chemical sensitivity, or to the general toxicity of specific agents and high toxicity of selected cases.

While such statements require constant review (and despite the myriad web claims to the contrary), it appears that there are very limited US, EU, etc. agency-approved roles for saponins in human therapy. In their use as adjuvants in the production of vaccines, toxicity associated with sterol complexation remains a major issue for attention. Therapeutic benefit is a result of careful administration of an appropriate dose. Very great care needs to be exercised in evaluating or acting on specific claims of therapeutic benefit from ingesting saponin-type and other natural products.

Use in Animal Feeding

Saponins are used widely for their effects on ammonia emissions in animal feeding. The mode of action seems to be an inhibition of the urease enzyme, which splits up excretedurea in feces into ammonia and carbon dioxide. Animal trials have shown that a reduced ammonia level in farming operations causes less damage to the respiratory tract of animals and may help to make them less vulnerable to diseases.

Health Benefits of Saponins

Saponins have many health benefits. Studies have illustrated the beneficial effects on blood cholesterol levels, cancer, bone health and stimulation of the immune system. Most scientific studies investigate the effect of saponins from specific plant sources and the results cannot be applied to other saponins.

Cholesterol Lowering

If you're working to lower your cholesterol level, eating foods rich in saponins may help. Your body uses cholesterol to make bile acids needed for proper digestion. When you eat, bile acids are released into your intestines. The detergent qualities of saponins allow them to bind to bile and prevent its reabsorption. Once bound to saponins, cholesterol leaves your body in waste. Peter R. Cheeke, Ph.D., from the Linus Pauling Institute, notes that many cholesterol-lowering medications perform the same role and over time excretion of bile may help lower your cholesterol. A lower cholesterol level means less risk of heart attack or stroke.

Improved Immune Function

Eating more saponins may boost your immune function and fight off fungal infections, according to an article published in "ACS Chemical Biology" in March 2010. The study noted that saponins cause death of fungal cells, such as Candida albicans, which is responsible for yeast infections, thrush and many hospital-acquired infections. Saponins appear to enhance your immune system's ability to fight off viruses and parasites as well. Pharmaceutical manufacturers often include saponins in vaccines to increase their effectiveness.

Preventing Cancer

According to an article published in the "Journal of Nutrition" in 1995, saponins found in soybeans slow the growth of human cancer cells. These plant compounds may also cause the death of tumor cells, according to an article published in the journal "Phytochemistry Reviews" in June 2010. The exact mechanism of these cell deaths varies depending on the source and dose of saponins. Few studies on saponins used human subjects. Animals and isolated cells in test tubes are the most common subjects. More research would provide a better picture of the potential role saponins play in cancer treatment and prevention.

Cholesterol Reduction

Saponins in beans, legumes and yucca may lower the cholesterol levels in your blood by blocking your body's absorption of cholesterol. Bile acid in your body binds with cholesterol to help your body absorb it. Saponins from the diet bind with bile acids and cholesterol so that they cannot enter your system. This reduces the amount of cholesterol your body absorbs and increases the amount it excretes. Saponins bind with bile salt and cholesterol in the intestinal tract. Bile salts form small micelles with cholesterol facilitating its absorption. Saponins cause a reduction of blood cholesterol by preventing its re-absorption.

Reduced Colon Cancer Risk

The same mechanism by which saponins may lower your cholesterol binding to bile acids may actually reduce your risk of colon cancer. According to the Linus Pauling Institute, some secondary bile acids promote colon cancer. Bacteria in your colon produce secondary bile acids from primary bile acids. By binding to primary bile acids, saponins reduce the amount of secondary bile acids your gut bacteria can produce, thereby reducing your risk of colon cancer. The Linus Pauling Institute states that feeding saponins to lab mice reduced the amount of precancer lesions in their colons. Researchers of a study published in 1995 in "Nutrition and Cancer" incubated human colon cancer tumor cells for one hour and 48 hours in various concentrations of saponins from soybeans and soapwort, an herb. They found the saponins inhibited tumor cell growth and reduced tumor cell activity in a dose-dependent manner the higher the concentration of saponins, the lower the tumor cell growth and activity.

Studies have shown that saponins have antitumor and anti-mutagenic activities and can lower the risk of human cancers, by preventing cancer cells from growing. Saponins seem to react with the cholesterol rich membranes of cancer cells, thereby limiting their growth and viability. Roa and colleagues (1995) found that saponins may help to prevent colon cancer and as shown in their article "Saponins as anti-carcinogens" published in The Journal of Nutrition. Some studies have shown that saponins can cause apoptosis of leukemia cells by inducing mitotic arrest.

Tumor Prevention

According to researchers of a review published in 2010 in the journal "Fitoterapia," around 150 natural saponins have anti-cancer properties. They explain that the various chemical structures of saponins help prevent tumor formation. Ginsenosides, a saponin found in the ginseng plant, has been found to suppress tumor growth and the spreading of tumor cells to other organs, according to researchers of the 2010 review. They also found that dioscin and diosgenin from wild yam may actually cause tumor cell cycle arrest and death.

Benefits of Saponins

Saponins have plenty of health benefits. Many studies have shown its beneficial effects on bone health, cholesterol levels and immune system. Most scientific studies look into the effects of saponins from particular plant sources and results can't be applied on other saponins.

Cholesterol Reduction

Saponins bind with cholesterol and bile salt in the gut. Bile salts are responsible for facilitating absorption of cholesterol. However, saponins prevent its re-absorption, causing reduction of cholesterol.

Antioxidant and Immunity Booster

The non-sugar part of saponins have also a direct antioxidant activity, which may results in other benefits such as reduced risk of cancer and heart diseases. Plants naturally produce saponins to combat parasitic infections. Such ability also applies when ingested by humans, as saponins seem to support the immune system and protect the body from bacteria and viruses. Furthermore, the non-sugar part of saponins serve as an antioxidant, helping the body fight free radicals to cause heart diseases and cancer. Plants produce saponins to fight infections by parasites. When ingested by humans, saponins also seem to help our immune system and to protect against viruses and bacteria.

Support Bone Health

Studies have shown that saponins have a protective role on reducing bone loss. Studies with ovariectomized induced rats have shown that some saponins, such as the steroidal saponins from Anemarrhena asphodeloides, a Chinese herb, have a protective role on bone loss.

Lower Cancer Risk

Studies have illustrated the anti-mutagenic and antitumor activities of saponins. Saponins appear to react with the cancer cells cholesterol-rich membranes, hence thwarting their viability and growth. Experts also found that saponins have the potential to prevent colon cancer and cause cell death on leukemia cells by stimulating mitotic arrest.

Saponins and Adaptogens

According to current studies on the biocomponents of adaptogens, plant compounds having adaptogenic abilities appear to fall into two groups the polyphenols and terpenes. Polyphenols include compounds called flavonoids and a lot of these substances have antioxidant properties. The biggest class of secondary metabolites, the terpenes enables plants to survive in their specific niche. The triterpenes include saponins and part of the saponins is a specific group called triterpenoids.

Chemical Study and Medical Spplication of Saponins as Anti-cancer Agents

Saponins are common in a variety of higher plants and usually found in roots, tubers, leaves, blooms or seeds. Based on the carbon skeletons, saponins were classified into triterpenes and steroids. Their glycone parts were mostly oligosaccharides, arranged either in a linear or branched fashion, attached to hydroxyl groups through an acetal linkage. Modern research found that saponins have antitumor effect on many cancer cells. Several saponins inhibit tumor cell growth by cell cycle arrest and apoptosis with IC50 values up to 0.2 mM. Meanwhile, saponins in combination with conventional tumor treatment strategies result in improved therapeutic success. Furthermore, a much clearer understanding of how the various saponin structures are related to each other is obtained with the use of the classification presented.

Mechanisms of the Antitumor Effect of Saponins

Cycloartanes

Cycloartane saponins displayed slight anti-cancer effect but they could be used as chemotherapeutic agent in the treatment of tumors. For example, total Astragalus saponins (AS) (Figure 15.2 and Table 15.1) possess antitumor properties in human colon cancer cells and tumor xenografts. They down-regulated expression of the HCC tumor marker α-fetoprotein and suppressed HepG2 cell growth by inducing apoptosis and modulating an ERK-independent NF-chachuuB signaling pathway. In addition, AS could be used as an adjuvant in combination with other orthodox chemotherapeutic drugs to reduce the side effects of the latter compounds. It would target at NSAID-activated gene (NAG-1) to reduce the additive effects when used along with PI3K-Akt inhibitors. The information obtained could facilitate future development of a novel target-specific chemotherapeutic agent with known molecular pathways.

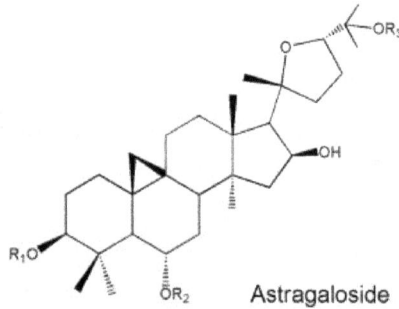

Figure 15.2: The Structure of Astragalosides.

Table 15.1: Astragalosides

	R1	R2	R3
Astragaloside I	Xyl (2,3-diAc)	Glc	H
Astragaloside IV	Xyl	H	H
Astragaloside VII	Xyl	Glc	Glc

Dammaranes

Most dammarane saponins showed anti-cancer effect. The naturally occurring compound OSW-1 (Figure 15.3) is found in the bulbs of *Ornithogalum saudersiae* and is highly cytotoxic against tumor cell lines. Nonmalignant cells were statistically significantly less sensitive to OSW-1 than cancer cells, with concentrations that cause a 50 per cent loss of cell viability 40–150-fold greater than those observed in malignant cells. Electron microscopy and biochemical analyses revealed that OSW-1 damaged the mitochondrial membrane and cristae in both human leukemia and pancreatic cancer cells, leading to the loss of transmembrane potential, increase of cytosolic calcium and activation of calcium-dependent apoptosis.

Oleananes

Oleananes own most kinds of saponins in the nature. Their antitumor effect worked through various pathways, such as anti-cancer, anti-metastasis, immunostimulation, chemoprevention and so forth. Avicins, tubeimoside, saikosaponins, platycodigenins, soybean saponin and *Pulsatilla koreana* saponins

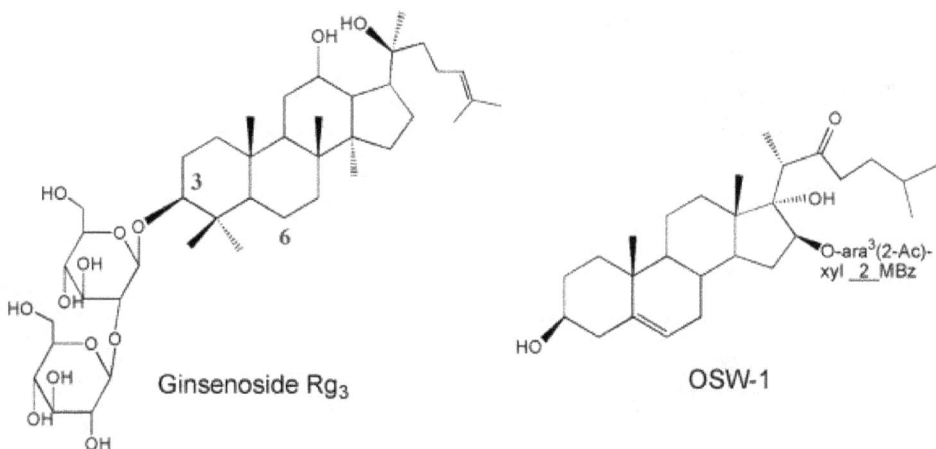

Figure 15.3: Structures of Ginsenosides and OSW-1.

showed anti-cancer effect through different signaling transductions. In addition, all of them except tubeimoside and *Pulsatilla koreana* saponins displayed immunostimulation. Saikosaponins, platycodigenins and soybean saponin also have anti-metastatic activity. The detailed mechanisms of saponins listed as following. Avicins (Figure 15.4), derived from the Cactus plant *Acacia victoriae* found in Australia's deserts, can dephosphorylate Stat3 in a variety of human tumor cell lines and lead to a decrease in the transcriptional activity of Stat3, which regulated proteins such as c-myc, cyclin D1, Bcl2, survivin and VEGF. Avicins D and G, as the main components of avicins, induced growth inhibition of human T lymphocytes, promoted apoptosis and triggered autophagic cell death. Meanwhile, they decreased

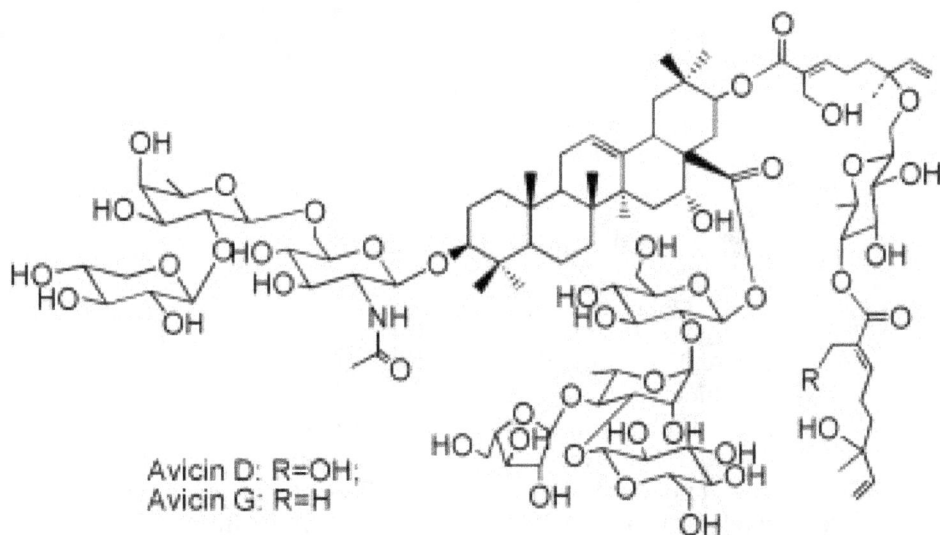

Figure 15.4: Structure of Avicins.

respiratory activity and induced ATP efflux after inhibition of the voltage dependent anion channel in the outer mitochondrial membrane.

Tubeimoside I (Figure 15.5), one of the triterpenoid saponins from the bulb of *Bolbostemma paniculatum* (Maxim)*Franquet*, appears to be a promising agent for cancer chemoprevention. It exerts cytotoxicity in HeLa cells through both mitochondrial dysfunction and ER stress cell death pathways. As an anti-microtubule agent, it can bind to the colchicine binding site of tubulin.

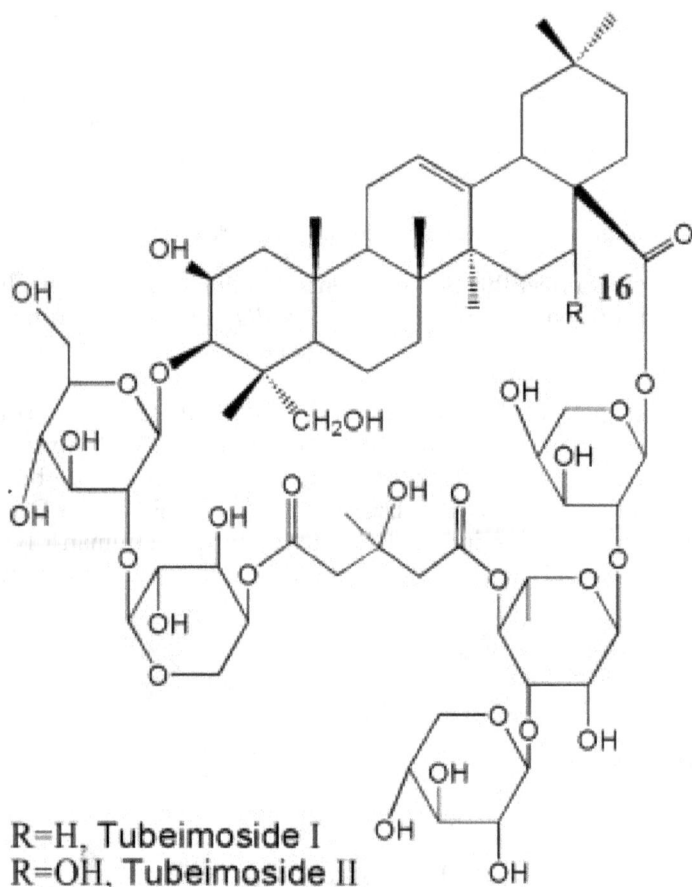

R=H, Tubeimoside I
R=OH, Tubeimoside II

Figure 15.5: Structure of Tubeimosides.

Saikosaponin A (Figure 15.6) activates ERK together with its downstream transcriptional machinery mediated p15 (INK4b) and p16(INK4a) expression that led to HepG2 growth inhibition. It inhibited the proliferation or viability of the MDA-MB-231 and MCF-7 cells in a dose-dependent manner and caused an obvious increase in the sub-G1 population of cell cycles. Treatment with saikosaponin D (Fig. 6) decreased the cell proliferation of Hep G2 and Hep 3B cells in a dose-dependent manner. It therefore decreased the cell proliferation and inducted apoptosis both in p53-positive Hep G2 and p53-negative Hep 3B cells. In addition, it inhibited the proliferation of A549 by inducing apoptosis and blocking cell cycle progression in the G1 phase.

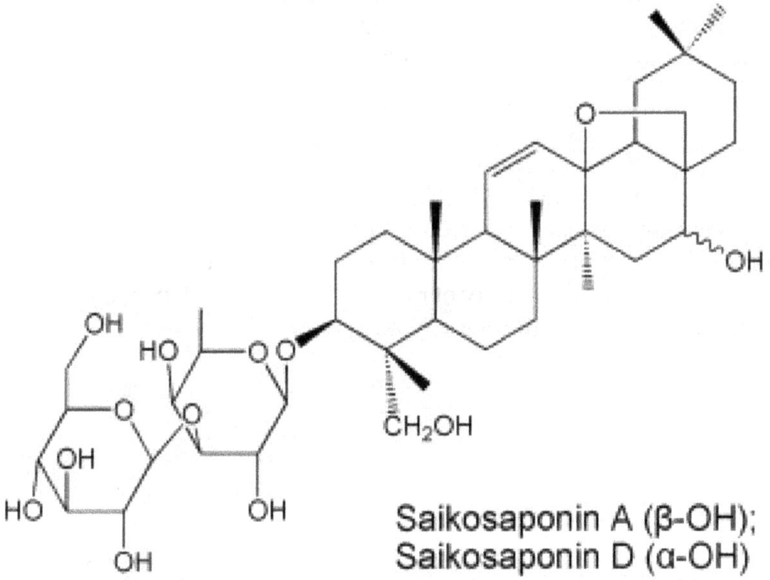

Figure 15.6: Structure of Saikosaponins.

Saponins derived from *Platycodon grandiflorum* may suppress tumor invasion and migration by inhibiting MMP-2 and MMP-9 activation. Platycodon D (Figure 15.7) as one of the platycodigenins, is a potentially interesting candidate for use in cancer chemotherapy. Its exposure induced apoptosis through caspase-3 dependent PARP, lamin A cleavage and ROS induced through Egr-1 activation.

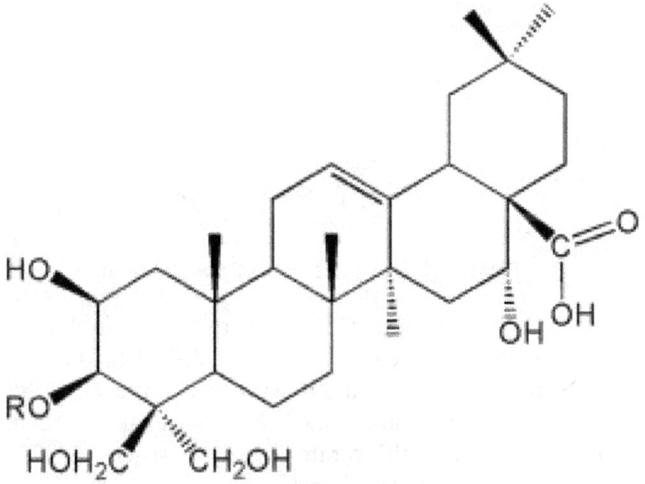

Figure 15.7: Structure of Platycodigenin.

The primary antileukemia activity is induction of endoreduplication and mitotic arrest, as a consequence of suppressing spindle MT dynamics and promoting apoptosis in human leukemia cells. Furthermore, it has direct cytotoxic effect on human leukemia cells and suppresses telomerase activity through transcriptional and post-translational suppression of hTERT.

Soybean saponin (Figure 15.8) inhibits tumor cell metastasis by suppressing MMP-2 and MMP-9 productions and stimulating TIMP-2 secretion. At physiologically relevant doses, it can suppress HCT-15 colon cancer cell proliferation through S-phase cell cycle delay and induce macroautophagy, the hallmark of Type II PCD. B-group soyasaponins may be another colon cancer suppressive component of soy that warrants further examination as a potential chemopreventive phytochemical. It significantly increased activity of raf-1 by a maximal 200 per cent, suggesting that this enzyme in part modulates the enhanced ERK1/2 activity.

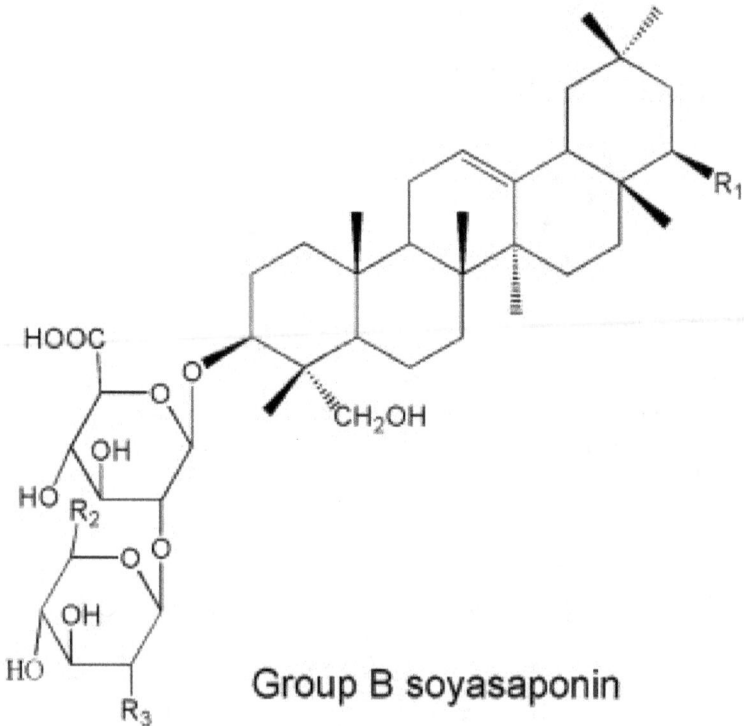

Group B soyasaponin

Figure 15.8: Structure of Soyasaponin.

Pulsatilla koreana saponins (Figure 15.9) were examined for their *in vitro* cytotoxic activity against the human solid cancer cell lines, A-549, SK-OV-3, SK-MEL-2 and HCT15, using the SRB assay method and there *in vivo* antitumor activity using BDF1 mice bearing Lewis lung carcinoma (LLC). Pulsatilla saponin D as an antitumor component showed potent inhibition rate of tumor growth (IR, 82 per cent) at the dose of 6.4 mg/kg on the BDF1 mice bearing LLC cells.

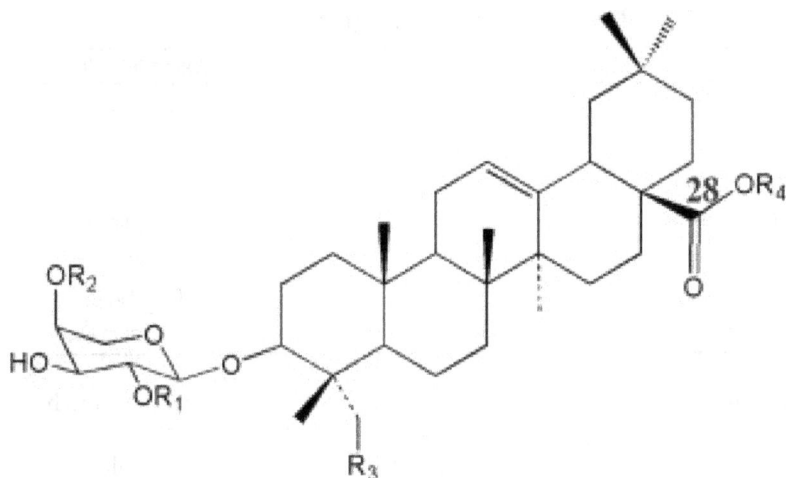

Figure 15.9: Structure of *Pulsatilla koreana* Saponin.

Spirostanes

Polyphyllin D (PD), formosanin C and dioscin belonging to the diosgenyl saponins, showed strong anti-cancer and immunostimulative activity. With the ascertained chemical structure and the improved synthesis of polyphyllin D, both *in vitro* and *in vivo* studies were performed on its effect. Recent research showed that PD is a potent apoptosis inducer through mitochondrial dysfunction and ER stress. Meanwhile, dioscin is a preclinical drug showing potent antiproliferative activities against most cell lines from leukemia and solid tumors. Proteomic analysis revealed that it induced apoptosis via the mitochondrial and some other pathway (Figure 15.10 and Table 15.2).

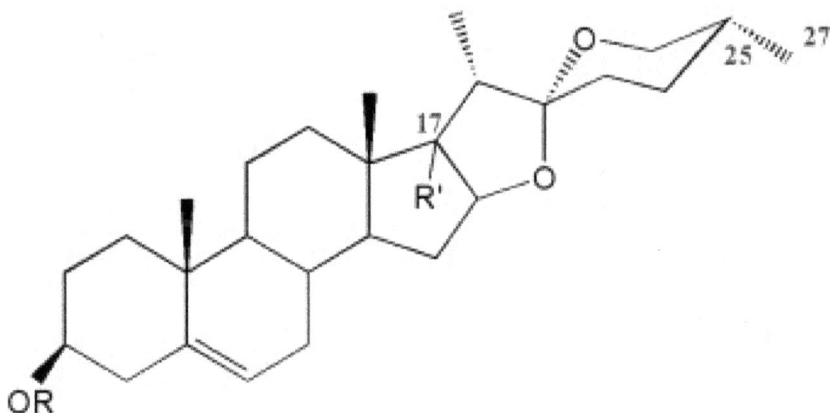

Figure 15.10: Structures of Diosgenyl and Pennogenyl Saponins.

Table 15.2: Diosgenyl Saponins (R'=H)

R	Name (R'=H)
-H	Diosgenin
-3-O-Glc	Trillin
-3-O-Rha (1→2)-Glc	ParisV
-3-O-Ara (1→4)-[Rha(1→2)]-Glc	Polyphyllin D
-3-O-Rha(1→4)-[Rha(1→2)]-Glc	Dioscin
-3-O-Rha(1→2)-[Glc (1→3)]-Glc	Gracillin
-3-O-Rha(1→4)-Rha(1→4)-[Rha(1→2)]-Glc	Formosanin C

Formosanin C, mainly a constituent in *Rhizoma Paris* saponins either, had some effect on the immune responses. Intraperitoneal treatment with 1–2.5 mg/kg of formosanin C would retard the growth of subcutaneously transplanted MH134 mouse hepatoma. The mechanism of its antitumor effect might be associated with its modification of the immune system. It can also enhance the antitumor effect of 5-fluorouracil. Activation of caspase-2 and the dysfunction of mitochondria maybe also contributed to its antitumor effect in human colourectal cancer HT-29 cells.

Furostanes

Most of furostanes only showed some anti-cancer activity. Protoneodioscin, protodioscin, protoneogracillin and protogracillin, along with their corresponding artifacts: methyl protoneodioscin, methyl protodioscin, methyl protoneogracillin and methyl protogracillin showed cytotoxic activities against K562 cancer cell as antineoplastic agents (Figure 15.11).

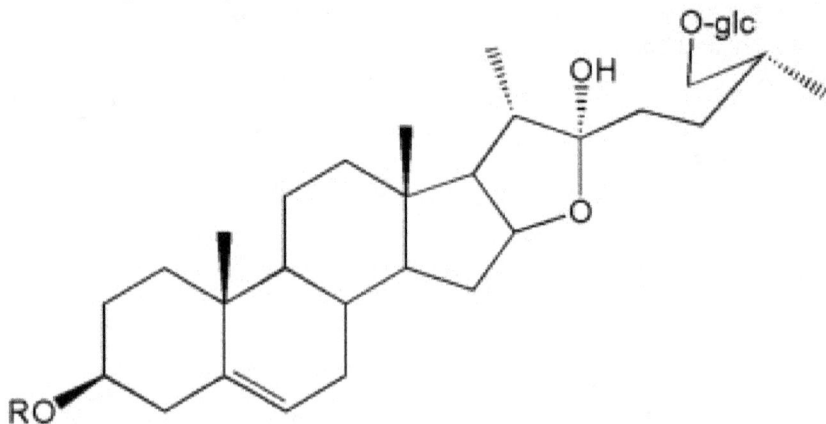

Figure 15.11: Structure of Proto-type Saponins.

Methyl protogracillin was cytotoxic against all the tested cell lines from leukemia and solid tumors in the NCI's human cancer panel; it showed particular selectivity against one colon cancer line (KM12), one central nervous system (CNS) cancer line (U251), two melanoma lines (MALME-3M and M14), two renal cancer lines (786-0 and UO-31) and one breast cancer line (MDA-MB-231).

Structure Function Relationship of Saponins with the Antitumor Properties

Differences in saponin structure which include the type, position and number of sugar moieties attached by a glycosidic bond at different positions of the rings can characteristically influence biological responses, especially for the antitumor activity. We could draw the following structure activity relationships in the succeeding sections.

Influence of the Aglycone on the Antitumor Activities of Saponins

Comparing different kinds of saponins, it shows that small changes such as different positions or the number of the hydroxyl groups, R-S configuration on the aglycone led to slight changes in activity and more sizable changes diminished the activity.

The Site of the Hydroxyl Group

Changes on the agycone could change the antitumor activity of saponins. C-16 hydroxyl group of tubeimoside II plays an important role in enhancing the biological activity of tubeimoside II and in decreasing its toxicity (Figure 15.5). C-17α-hydroxyl group to the aglycone of the active saponins slightly reduced their cytotoxicities, such as pennogenyl saponins and diosgenyl saponins (Figure 15.10). C-27 of the aglycone of the furostanol saponins, which also bore an additional monosaccharide at C-27 (compared to the spirostan saponins mentioned above), showed less antitumor effect (Figure 15.10).

The Number of the Hydroxyl Group in Aglycone

Ginsenosides with a dammarane structure have two main classes: panaxadiols (PPD) and panaxatriols (PPT). The activities of PPD compounds are greater than those of the PPT compounds. And the aglycones are more effective than the ginsenosides Rh1 (PPT type) and Rh2 (PPD type), which possess sugar moieties at C-6 and C-3, respectively. All the ginsenosides have similar chemical structures, but their effects on B16 melanoma cells were remarkably different (Figure 15.3).

Else

Based on structure–activity relationship, C-25 R/S configuration was critical for leukemia selectivity between methyl protoneogracillin and methyl protogracillin. Meanwhile, F-ring was critical to selectivity between furostanol (methyl protoneogracillin and methyl protogracillin) and spirostanol (gracillin) saponins. Methyl protoneogracillin has been selected as a potential anti-cancer candidate for hollow fiber assay to nude mice, which is slightly better than methyl protogracillin, but gracillin would not be pursued due to the lack of selectivity against human cancer diseases (Figure 15.10).

Influence of the Sugar Side Chain on the Antitumor Activities of Saponins

In the comparison with the saponins bearing saccharide groups, different characteristics (sugar linkage, the number, lipophilicity, or different kinds) of sugar side chain play important roles in their antitumor effect.

The Sugar Linkage

With the same aglycone and length of sugar chain, the sugar linkage determines the antitumor potency. This point is clearly demonstrated by the four disaccharide congeners. 1→3 linkage had much lower activity than 1→2 and 1→4 linkages, respectively.

The Lipophilicity of the Sugar

Some saponins showed no activity with the exceptions of those possessing some acyl groups at the glycosyl moiety. Meanwhile, two deoxypyranoses, including D-fucose and L-rhamnose, were also cytotoxic. These data led us to assume that the presence of a certain degree of lipophilicity in the sugar moiety is essential for exhibiting the cytotoxic activity.

The Number of the Sugar

The number of the sugar also influences the antitumor effect of saponins. The activity of the various ginsenosides has been demonstrated to be in the order: monosaccharide glycoside > disaccharide glycoside > trisaccharide glycoside > tetrasaccharide glycoside, indicating that increasing the number of sugar moieties reduces the potency of the compound. In the contrast, diosgenyl saponins showed the contrary rule. Diosgenin β-D-glucoside showed no cytotoxic activity against HL-60 cells (IC50 20 mg/ml) and the attachment of an α-L-rhamnosyl group at C-2 of the glucosyl moiety led to the appearance of considerable activity. Further addition of a α-L-rhamnosyl, a α-L-arabinofuranosyl or a β-D-glucosyl, with the exception of a β-D-galactosyl, to C-3 or C-4 of the inner glucosyl moiety either gave no influence on the activity or slightly increased the activity; the attachment of a β-D-galactosyl at C-3 of the glucosyl residue led to a decrease in the activity.

The Kinds of Sugar Sequence

C-3 of oleanolic acid and hederagenin linked with a sugar sequence O-α-L-rhamnopyranosyl-(1→2)-α-L-arabinopyranoside showed a good effect, suggesting that the two elements are essential factors for the antitumor activity. Meanwhile, in some disaccharide derivatives, it was used as a nontoxic carrier moiety to enhance the activity of anti-cancer drugs.

Conclusion and Perspective

The identification and development of saponins have greatly contributed to medical treatment of cancer and many of these compounds are now being used in clinical practice. Almost all saponins induce apoptosis in tumor cells; they are preferable drugs for the treatment of cancer, because eliminating tumor cells by apoptosis is helpful to lower side effects in patients by avoiding necrosis. A good understanding of the antitumor mechanisms of saponins is necessary for a directed improvement of saponin based tumor therapies in the future. Meanwhile, special attention should be given to combinations of saponins and other anticarcinogenic drugs, since these offer very efficient treatment regimens against cancer. The most important is the saponin-mediated potentiation of tumor growth inhibition and the possibility to circumvent drug resistance. Furthermore, the elucidation of structure–activity relationships between different saponins in combination with conventional drugs is much more complicated than for saponins alone. Thus, it is not surprising that no mechanistic processes for these effects are known, however, detailed information on this basis is necessary for a directed improvement of saponin-based tumor therapies in the future.

References

Ahn, M. J., C.Y. Kim and K.D. Yoon. (2006). Steroidal saponins from the rhizomes of *Polygonatum sibiricum*. *J Nat Prod*, 69, pp. 360–364.

Akinjogunla OJ, Yah CS, Eghafona NO and Ogbemudia FO, (2010). Antibacterial activity of leave extracts of *Nymphaea lotus* (Nymphaeaceae) on Methicillin resistant *Staphylococcus aureus* (MRSA) and Vancomycin resistant *Staphylococcus aureus* (VRSA) isolated from clinical samples. *Annals of Biological Research*, 1 (2), pp. 174-184.

Ali, B. H. and G. Blunden. (2003). Pharmacological and toxicological properties of *Nigella sativa*. *Phytother Res*, 17, pp. 299–305.

Asl, Marjan Nassiri, Hossein Hosseinzadeh (2008). "Review of pharmacological effects of *Glycyrrhiza* sp. and its bioactive compounds". *Phytotherapy Research* 22 (6): 709–24.

Auyeung, K. K., C.H. Cho and J.K. Ko. (2009). A novel anticancer effect of *Astragalus* saponins: transcriptional activation of NSAID-activated gene. *Int J Cancer*, 125, pp. 1082–1091.

Auyeung, K. K., P.C. Law and J.K. Ko. (2009). Astragalus saponins induce apoptosis via an ERK-independent NF-kappaB signaling pathway in the human hepatocellular HepG2 cell line. *Int J Mol Med*, 23, pp. 189–196.

Bader, G., B. Plohmann, K. Hiller and G. Franz. (1996). Cytotoxicity of triterpenoid saponins. Part 1: activities against tumor cells *in vitro* and hemolytical index. *Pharmazie*, 51, pp. 414–417.

Bai, H., Y. Zhong and Y.Y. Xie. (2007). A major triterpenoid saponin from *Gypsophila oldhamiana*. *Chem Biodivers*, 4, pp. 955–960.

Bang, S. C., H.H. Seo and H.R. Shin. (2008). A convenient preparation of a disaccharide motif and its role in the cytotoxicity of the triterpenoid saponin, alpha-hederin. *Arch Pharm Res*, 31, pp. 555–561.

Bang, S. C., H.H. Seo, H.Y. Yun and S.H. Jung. (2007). Facile synthesis of trisaccharide moiety corresponding to antitumor activity in triterpenoid saponins isolated from *Pullsatilla* roots. *Chem Pharm Bull*, 55, pp. 1734–1739.

Bang, S. C., J.H. Lee and G.Y. Song. (2005). Antitumor activity of *Pulsatilla koreana* saponins and their structure activity relationship. *Chem Pharm Bull*, 53, pp. 1451–1454.

Barthomeuf, C., E. Debiton, V. Mshvildadze, E. Kemertelidze and G. Balansard. (2002). *In vitro* activity of hederacolchisid A1 compared with other saponins from *Hedera colchica* against proliferation of human carcinoma and melanoma cells. *Planta Med*, 68, pp. 672–675.

Bedir, E., I.A. Khan and L.A. Walker. (2002). Biologically active steroidal glycosides from *Tribulus terrestris*. *Pharmazie*, 57, pp. 491–493.

Beutler, J. A., Y. Kashman and L.K. Pannell. (1997). Isolation and characterization of novel cytotoxic saponins from *Archidendron ellipticum*. *Bioorg Med Chem*, 5, pp. 1509–1517.

Cai, J., M. Liu, Z. Wang and Y. Ju. (2002). Apoptosis induced by dioscin in Hela cells. *Biol Pharm Bull*, 25, pp. 193–196.

Calabria, L. M., S. Piacente and I. Kapusta. (2008). Triterpene saponins from *Silphium radula*. *Phytochemistry*, 69, pp. 961–972.

Campbell, Paul (1999). *Survival skills of native California*. Gibbs Smith. p. 433.

Candra, E., K. Matsunaga and H. Fujiwara. (2001). Two steroidal saponins from *Camassia cusickii* induce L1210 cell death through the apoptotic mechanism. *Can J Physiol Pharmacol*, 79, pp. 953–958.

Cao, S., P. Brodie and M. Callmander. (2009). Antiproliferative triterpenoid saponins of *Dodonaea viscosa* from the Madagascar dry forest. *J Nat Prod*, 72, pp. 1705–1707.

Cao, S., P.J. Brodie and M. Callmander. (2009). Saponins and a lignan derivative of *Terminalia tropophylla* from the Madagascar Dry Forest. *Phytochemistry*.

Chan, P. K. (2007). Acylation with diangeloyl groups at C21–22 positions in triterpenoid saponins is essential for cytotoxicity towards tumor cells. *Biochem Pharmacol*, 73, pp. 341–350.

Chan, P. K., M. Zhao, C.T. Che and E. Mak. (2008). Cytotoxic acylated triterpene saponins from the husks of *Xanthoceras sorbifolia*. *J Nat Prod*, 71, pp. 1247–1250.

Chen, H., G. Wang and N. Wang. (2007). New furostanol saponins from the bulbs of *Allium macrostemon* Bunge and their cytotoxic activity. *Pharmazie*, 62, pp. 544–548.

Chen, J. C., N.W. Chang, J.G. Chung and K.C. Chen. (2003). Saikosaponin-A induces apoptotic mechanism in human breast MDA-MB-231 and MCF-7 cancer cells. *Am J Chin Med*, 31, pp. 363–377.

Cheng, G., Y. Zhang and X. Zhang. (2006). Tubeimoside V (1), a new cyclic bisdesmoside from tubers of *Bolbostemma paniculatum*, functions by inducing apoptosis in human glioblastoma U87MG cells. *Bioorg Med Chem Lett*, 16, pp. 4575–4580.

Cheng, Z. X., B.R. Liu and X.P. Qian. (2008). Proteomic analysis of anti-tumor effects by *Rhizoma Paridis* total saponin treatment in HepG2 cells. *J Ethnopharmacol*, 120, pp. 129–137.

Cheung, J. Y., R.C. Ong and Y.K. Suen. (2005). Polyphyllin D is a potent apoptosis inducer in drug-resistant HepG2 cells. *Cancer Lett*, 217, pp. 203–211.

Chiang, H. C., T.H. Tseng, C.J. Wang, C.F. Chen and W.S. Kan. (1991). Experimental antitumor agents from *Solanum indicum* L. *Anticancer Res*, 11, pp. 1911–1917.

Chwalek, M., N. Lalun, H. Bobichon, K. Ple and L. Voutquenne-Nazabadioko. (2006). Structure–activity relationships of some hederagenin diglycosides: haemolysis, cytotoxicity and apoptosis induction. *Biochim Biophys Acta*, 1760, pp. 1418–1427.

Cioffi, G., F. Dal Piaz and A. Vassallo. (2008). Antiproliferative oleanane saponins from *Meryta denhamii*. Journal of natural products, 71, pp. 1000–1004.

Danloy, S., J. Quetin-Leclercq and P. Coucke. (1994). Effects of alpha-hederin, a saponin extracted from *Hedera helix*, on cells cultured *in vitro*. *Planta Med*, 60, pp. 45–49.

Dastager, S. G., J.C. Lee, Y.J. Ju, D.J. Park and C.J. Kim. (2008). *Microbacterium kribbense* sp. nov., isolated from soil. *Int J Syst Evol Microbiol*, 58, pp. 2536–2540.

Dong, M., X.Z. Feng, B.X. Wang, T. Ikejima and L.J. Wu. (2004). Steroidal saponins from *Dioscorea panthaica* and their cytotoxic activity. *Pharmazie*, 59, pp. 294–296.

Dong, M., X.Z. Feng, L.J. Wu, B.X. Wang and T. Ikejima. (2001). Two new steroidal saponins from the rhizomes of *Dioscorea panthaica* and their cytotoxic activity. *Planta Med*, 67, pp. 853–857.

Elbandy, M., T. Miyamoto, B. Chauffert, C. Delaude and M.A. Lacaille-Dubois. (2002). Novel acylated triterpene glycosides from *Muraltia heisteria*. *J Nat Prod*, 65, pp. 193–197.

Ellington, A. A., M. Berhow and K.W. Singletary. (2005). Induction of macroautophagy in human colon cancer cells by soybean B-group triterpenoid saponins. *Carcinogenesis*, 26, pp. 159–167.

Ellington, A. A., M.A. Berhow and K.W. Singletary. (2006). Inhibition of Akt signaling and enhanced ERK1/2 activity are involved in induction of macroautophagy by triterpenoid B-group soyasaponins in colon cancer cells. *Carcinogenesis*, 27, pp. 298–306.

ERK signaling pathway is involved in p15INK4b/p16INK4a expression and HepG2 growth inhibition triggered by TPA and Saikosaponin a Oncogene, 22, pp. 955–963.

Fan, C. Q., H.F. Sun and S.N. Chen. (2002). Triterpene saponins from *Craniotome furcata*. *Nat Prod Lett*, 16, pp. 161–166.

Fang, J. B., Z. Yao and J.C. Chen. (2009). Cytotoxic triterpene saponins from *Alternanthera philoxeroides*. *J Asian Nat Prod Res*, 11, pp. 261–266.

Feng, X. Z., M. Dong, Z.J. Gao and S.X. Xu. (2003). Three new triterpenoid saponins from *Ixeris sonchifolia* and their cytotoxic activity. *Planta Med*, 69, pp. 1036–1040.

Fish-poison plants, *Bulletin of Miscellaneous Information (Royal Gardens, Kew)*1930 (4), 1930: 129–153.

Foerster, Hartmut (2006). "MetaCyc Pathway: saponin biosynthesis I". Retrieved 23 February 2009.

Francis, George, Zohar Kerem, Harinder P. S. Makkar and Klaus Becker (2002). "The biological action of saponins in animal systems: a review". *British Journal of Nutrition* 88 (6): 587–605.

Fujioka, T., K. Yoshida and H. Fujii. (2003). Antiproliferative constituents from Umbelliferae plants VI. New ursane-type saikosaponin analogs from the fruits of *Bupleurum rotundifolium*. *Chemical and Pharmaceutical Bulletin*, 51, pp. 365–372.

Fukumura, M., H. Ando and Y. Hirai. (2009). Achyranthoside H methyl ester, a novel oleanolic acid saponin derivative from *Achyranthes fauriei* roots, induces apoptosis in human breast cancer MCF-7 and MDA-MB-453 cells via a caspase activation pathway. *J Nat Med*, 63, pp. 181–188.

Gaidi, G., A. Marouf and B. Hanquet. (2000). A new major triterpene saponin from the roots of *Cucurbita foetidissima*. *J Nat Prod*, 63, pp. 122–124.

Gaidi, G., M. Correia and B. Chauffert. (2002). Saponins-mediated potentiation of cisplatin accumulation and cytotoxicity in human colon cancer cells. *Planta Med*, 68, pp. 70–72.

Galanty, A., M. Michalik, L. Sedek and I. Podolak. (2008). The influence of LTS-4, a saponoside from *Lysimachia thyrsiflora* L., on human skin fibroblasts and human melanoma cells. *Cell Mol Biol Lett*, 13, pp. 585–598.

Gao, X. K., W.C. Ye and A.C. Yu. (2003). Pulsatilloside A and anemoside A3 protect PC12 cells from apoptosis induced by sodium cyanide and glucose deprivation. *Planta Med*, 69, pp. 171–174.

Gnoula, C., V. Megalizzi and N. De Neve. (2008). Balanitin-6 and -7: diosgenyl saponins isolated from *Balanites aegyptiaca* Del. display significant anti-tumor activity *in vitro* and *in vivo*. *Int J Oncol*, 32, pp. 5–15.

Gonzalez, A. G., J.C. Hernandez and F. Leon. (2003). Steroidal saponins from the bark of *Dracaena draco* and their cytotoxic activities. *J Nat Prod*, 66, pp. 793–798.

Guo, H., K. Koike and W. Li. (2004). Saponins from the flower buds of *Buddleja officinalis*. *J Nat Prod*, 67, pp. 10–13.

Hamed, A. I., S. Piacente and G. Autore. (2005). Antiproliferative hopane and oleanane glycosides from the roots of *Glinus lotoides*. *Planta Med*, 71, pp. 554–560.

Haridas, V., G. Nishimura and Z.X. Xu. (2009). Avicin D: a protein reactive plant isoprenoid dephosphorylates Stat 3 by regulating both kinase and phosphatase activities. *PLoS ONE*, 4, p. e5578.

Haridas, V., M. Higuchi and G.S. Jayatilake. (2001). Avicins: triterpenoid saponins from *Acacia victoriae* (Bentham) induce apoptosis by mitochondrial perturbation. *Proc Natl Acad Sci USA*, 98, pp. 5821–5826.

Haridas, V., X. Li and T. Mizumachi. (2007). Avicins, a novel plant-derived metabolite lowers energy metabolism in tumor cells by targeting the outer mitochondrial membrane. *Mitochondrion*, 7, pp. 234–240.

Hernandez, J. C., F. Leon, F. Estevez, J. Quintana and J. Bermejo. (2006). A homo-isoflavonoid and a cytotoxic saponin from *Dracaena draco*. *Chem Biodivers*, 3, pp. 62–68.

Hibasami, H., H. Moteki and K. Ishikawa. (2003). Protodioscin isolated from fenugreek (*Trigonella foenumgraecum* L.) induces cell death and morphological change indicative of apoptosis in leukemic cell line H-60, but not in gastric cancer cell line KATO III. *Int J Mol Med*, 11, pp. 23–26.

Hostettmann, K., A. Marston (1995). *Saponins*. Cambridge: Cambridge University Press. p. 3ff.

Hsu, Y. L., P.L. Kuo and C.C. Lin. (2004). The proliferative inhibition and apoptotic mechanism of saikosaponin D in human non-small cell lung cancer A549 cells. *Life Sci*, 75, pp. 1231–1242.

Hsu, Y. L., P.L. Kuo, L.C. Chiang and C.C. Lin. (2004). Involvement of p53, nuclear factor kappaB and Fas/Fas ligand in induction of apoptosis and cell cycle arrest by saikosaponin d in human hepatoma cell lines. *Cancer Lett*, 213, pp. 213–221.

Hu, K. and X. Yao. (2001). Methyl protogracillin (NSC-698792): the spectrum of cytotoxicity against 60 human cancer cell lines in the National Cancer Institute's anticancer drug screen panel. *Anticancer Drugs*, 12, pp. 541–547.

Hu, K. and X. Yao. (2003). The cytotoxicity of methyl protodioscin against human cancer cell lines *in vitro*. *Cancer Investig*, 21, pp. 389–393.

Hu, K. and X. Yao. (2003). The cytotoxicity of methyl protoneogracillin (NSC-698793) and gracillin (NSC-698787), two steroidal saponins from the rhizomes of *Dioscorea collettii* var. hypoglauca, against human cancer cells *in vitro*. *Phytother Res*, 17: 620–626.

Hu, K., A. Dong, X. Yao, H. Kobayashi and S. Iwasaki. (1996). Antineoplastic agents, I. three spirostanol glycosides from rhizomes of *Dioscorea collettii* var. *hypoglauca*. *Planta Medica*, 62, pp. 573–575.

Hu, K., A. Dong, X. Yao, H. Kobayashi and S. Iwasaki. (1997). Antineoplastic agents. II. Four furostanol glycosides from rhizomes of *Dioscorea collettii* var. *hypoglauca*. *Planta Medica*, 63, pp. 161–165.

Huang, H. C., M.D. Wu and W.J. Tsai. (2008). Triterpenoid saponins from the fruits and galls of *Sapindus mukorossi*. *Phytochemistry*, 69, pp. 1609–1616.

Huang, T. H., V.H. Tran, B.D. Roufogalis and Y. Li. (2007). Gypenoside XLIX, a naturally occurring gynosaponin, PPAR-alpha dependently inhibits LPS-induced tissue factor expression and activity in human THP-1 monocytic cells. *Toxicol Appl Pharmacol*, 218, pp. 30–36.

Ivanova, A., J. Serly and D. Dinchev. (2009). Screening of some saponins and phenolic components of *Tribulus terrestris* and *Smilax excelsa* as MDR modulators. *In vivo* (Athens, Greece), 23, pp. 545–550.

Jayatilake, G. S., D.R. Freeberg and Z. Liu. (2003). Isolation and structures of avicins D and G: *in vitro* tumor-inhibitory saponins derived from *Acacia victoriae. J Nat Prod*, 66, pp. 779–783.

Jonathan G. Cannon, Robert A. Burton, Steven G. Wood and Noel L. Owen (2004),"Naturally Occurring Fish Poisons from Plants", *J. Chem. Educ.* 81 (10): 1457.

Kang, J. H., I.H. Han and M.K. Sung. (2008). Soybean saponin inhibits tumor cell metastasis by modulating expressions of MMP-2, MMP-9 and TIMP- 2. *Cancer Lett*, 261, pp. 84–92.

Kim, G. S., H.T. Kim and J.D. Seong. (2005). Cytotoxic steroidal saponins from the rhizomes of *Asparagus oligoclonos. J Nat Prod*, 68, pp. 766–768.

Kim, M. O., D.O. Moon and Y.H. Choi (2006). Platycodin D induces mitotic arrest *in vitro*, leading to endoreduplication, inhibition of proliferation and apoptosis in leukemia cells. *Int J Cancer*, 122, pp. 2674–2681.

Kim, Y. J., P. Wang and M. Navarro-Villalobos. (2006). Synthetic studies of complex immunostimulants from *Quillaja saponaria*: synthesis of the potent clinical immunoadjuvant QS-21Aapi. *J Am Chem Soc*, 128, pp. 11906–11915.

Kuljanabhagavad, T., P. Thongphasuk, W. Chamulitrat and M. Wink. (2008). Triterpene saponins from *Chenopodium quinoa* Willd. *Phytochemistry*, 69, pp. 1919–1926.

Kuo, Y. H., H.C. Huang and L.M. Yang Kuo. (2005). New dammarane-type saponins from the galls of *Sapindus mukorossi. J Agric Food Chem*, 53, pp. 4722–4727.

Lee, J. C., C.L. Su, L.L. Chen and S.J. Won. (2009). Formosanin C-induced apoptosis requires activation of caspase-2 and change of mitochondrial membrane potential. *Cancer Sci*, 100, pp. 503–513.

Lee, K. J., S.J. Hwang, J.H. Choi and H.G. Jeong. (2008). Saponins derived from the roots of *Platycodon grandiflorum* inhibit HT-1080 cell invasion and MMPs activities: regulation of NF-kappaB activation via ROS signal pathway. *Cancer Lett*, 268, pp. 233–243.

Lee, M. S., J.C. Yuet-Wa and S.K. Kong. (2005). Effects of polyphyllin D, a steroidal saponin in *Paris polyphylla*, in growth inhibition of human breast cancer cells and in xenograft. *Cancer Biol Ther*, 4, pp. 1248–1254.

Lemeshko, V. V., V. Haridas, J.C. Quijano Perez and J.U. Gutterman. (2006). Avicins, natural anticancer saponins, permeabilize mitochondrial membranes. *Arch Biochem Biophys*, 454, pp. 114–122.

Li, Z. L., B.Z. Yang and X. Li. (2006). Triterpenoids from the husks of *Xanthoceras sorbifolia* Bunge. *J Asian Nat Prod Res*, 8, pp. 361–366.

Liang, B., J.K. Tian, L.Z. Xu and S.L. Yang. (2006). Triterpenoid saponins from *Lysimachia davurica*. *Chem Pharm Bull*, 54, pp. 1380–1383.

Liener, Irvin E (1980). *Toxic constituents of plant foodstuffs*. New York City: Academic Press. p. 161.

Liu, H. W., K. Hu and Q.C. Zhao. (2002). Bioactive saponins from *Dioscorea futschauensis*. *Pharmazie*, 57, pp. 570–572.

Liu, M. J., P.Y. Yue, Z. Wang and R.N. Wong (2005). Methyl protodioscin induces G2/M arrest and apoptosis in K562 cells with the hyperpolarization of mitochondria. *Cancer Lett*, 224, pp. 229–241.

Liu, W., X.F. Huang and Q. Qi. (2009). Asparanin A induces G(2)/M cell cycle arrest and apoptosis in human hepatocellular carcinoma HepG2 cells. *Biochem Biophys Res Commun*, 381, pp. 700–705.

Lu, J., B. Xu and S. Gao. (2009). Structure elucidation of two triterpenoid saponins from rhizome of *Anemone raddeana* Regel. *Fitoterapia*, 80, pp. 345–348.

Lu, Y., J. Luo, X. Huang and L. Kong (2009). Four new steroidal glycosides from *Solanum torvum* and their cytotoxic activities. *Steroids*, 74, pp. 95–101.

Luo, L., Z. Li, Y. Zhang and R. Huang. (1998). Triterpenes and steroidal compounds from *Momordica dioica*. *Acta Pharmaceutica Sinica*, 33, pp. 839–842.

Ma, L., Y.C. Gu and J.G. Luo. (2009). Triterpenoid saponins from *Dianthus versicolour*. *J Nat Prod*, 72, pp. 640–644.

Man, S., W. Gao and Y. Zhang. (2009). Antitumor and antimetastatic activities of *Rhizoma Paridis* saponins. *Steroids*, 74, pp. 1051–1056.

Marquina, S., N. Maldonado and M. L. Garduno-Ramirez. (2001). Bioactive oleanolic acid saponins and other constituents from the roots of *Viguiera decurrens*. *Phytochemistry*, 56, pp. 93–97.

MetaCyc Pathway: saponin biosynthesis IV". Retrieved 23 February 2009.

Mimaki, Y., A. Yokosuka and M. Hamanaka. (2004). Triterpene saponins from the roots of *Clematis chinensis*. *J Nat Prod*, 67, pp. 1511–1516.

Mimaki, Y., A. Yokosuka, M. Kuroda and Y. Sashida. (2001). Cytotoxic activities and structure–cytotoxic relationships of steroidal saponins. *Biol Pharm Bull*, 24, pp. 1286–1289.

Mimaki, Y., K. Watanabe and Y. Ando. (2001). Flavonol glycosides and steroidal saponins from the leaves of *Cestrum nocturnum* and their cytotoxicity. *J Nat Prod*, 64, pp. 17–22.

Miyagawa, T., T. Ohtsuki, T. Koyano, T. Kowithayakorn and M. Ishibashi. (2009). Cardenolide glycosides of *Thevetia peruviana* and triterpenoid saponins of *Sapindus emarginatus* as TRAIL resistance-overcoming compounds. *J Nat Prod*, 72, pp. 1507–1511.

Mskhiladze, L., J. Legault and S. Lavoie. (2008). Cytotoxic steroidal saponins from the flowers of *Allium leucanthum*. *Molecules* (Basel, Switzerland), 13, pp. 2925–2934.

Murthy E N, Pattanaik, Chiranjibi, Reddy, C Sudhakar, Raju, V S (2010), *Piscicidal plants used by Gond tribe of Kawal wildlife sanctuary andhra Pradesh, India*, pp. 97–101

Nam, D. H., H.M. Jeon and S. Kim. (2008). Activation of notch signaling in a xenograft model of brain metastasis. *Clin Cancer Res*, 14, pp. 4059–4066.

Nartowska, J., E. Sommer, K. Pastewka, S. Sommer and E. Skopinska-Rozewska. (2004). Anti-angiogenic activity of convallamaroside, the steroidal saponin isolated from the rhizomes and roots of *Convallaria majalis* L. *Acta Pol Pharm*, 61, pp. 279–282.

Nguyen, V. T., N. Darbour and C. Bayet. (2008). Selective modulation of P-glycoprotein activity by steroidal saponines from *Paris polyphylla. Fitoterapia*, 44, pp. 132-135.

Ohtsuki, T., T. Koyano and T. Kowithayakorn. (2004). New chlorogenin hexasaccharide isolated from Agave fourcroydes with cytotoxic and cell cycle inhibitory activities. *Bioorg Med Chem*, 12, pp. 3841–3845.

Patlolla, J. M., J. Raju, M.V. Swamy and C.V. Rao. (2006). Beta-escin inhibits colonic aberrant crypt foci formation in rats and regulates the cell cycle growth by inducing p21(waf1/cip1) in colon cancer cells. *Mol Cancer Ther*, 5, pp. 1459–1466.

Perrone, A., A. Plaza and E. Bloise. (2005). Cytotoxic furostanol saponins and a megastigmane glucoside from *tribulus parvispinus. J Nat Prod*, 68, pp. 1549–1553.

Pettit, G. R., Q. Zhang and V. Pinilla. (2005). Antineoplastic agents. 534. Isolation and structure of sansevistatins 1 and 2 from the African *Sansevieria ehrenbergii. J Nat Prod*, 68, pp. 729–733.

Project Summary: (2009). Functional Genomics of Triterpene Saponin Biosynthesis in *Medicago truncatula*. Retrieved 23 February.

Raju, J., J.M. Patlolla, M.V. Swamy and C.V. Rao. (2004). Diosgenin, a steroid saponin of *Trigonella foenumgraecum* (Fenugreek), inhibits azoxymethane-induced aberrant crypt foci formation in F344 rats and induces apoptosis in HT-29 human colon cancer cells. *Cancer Epidemiol Biomark Prev*, 13, pp. 1392–1398.

Riguera, Ricardo (1997). "Isolating bioactive compounds from marine organisms". *Journal of Marine Biotechnology* 5 (4): 187–193.

Rooney, S. and M.F. Ryan. (2005). Effects of alpha-hederin and thymoquinone, constituents of *Nigella sativa*, on human cancer cell lines. *Anticancer Res*, 25, pp. 2199–2204.

Saponin from quillaja bark. Sigma-Aldrich. Retrieved 23 February 2009.

Saponin". J.T. Baker. Retrieved 23 February 2009.

Saponins. (2009). Cornell University. 14 August 2008. Retrieved 23 February.

Seo, Y., J. Hoch and M. Abdel-Kader. (2002). Bioactive saponins from *Acacia tenuifolia* from the suriname rainforest. *J Nat Prod*, 65, pp. 170–174.

Shin, D. Y., G.Y. Kim and W. Li. (2009). Implication of intracellular ROS formation, caspase-3 activation and Egr-1 induction in platycodon D-induced apoptosis of U937 human leukemia cells. *Biomedicine and Pharmacotherapy*, 63, pp. 86–94.

Siu, F. M., D.L. Ma and Y.W. Cheung. (2008). Proteomic and transcriptomic study on the action of a cytotoxic saponin (Polyphyllin D): induction of endoplasmic reticulum stress and mitochondria-mediated apoptotic pathways. *Proteomics*, 8, pp. 3105–3117.

Skene, Caroline D., Philip Sutton (2006). "Saponin-adjuvanted particulate vaccines for clinical use". *Methods* 40 (1): 53–9.

Son, I. H., Y.H. Park, S.I. Lee, H.D. Yang and H.I. Moon. (2007). Neuroprotective activity of triterpenoid saponins from *Platycodi radix* against glutamate-induced toxicity in primary cultured rat cortical cells. *Molecules* (Basel, Switzerland), 12, pp. 1147–1152.

Sparg, S. G., M.E. Light and J. van Staden. (2004). Biological activities and distribution of plant saponins. *J Ethnopharmacol*, 94, pp. 219–243.

Sun, B., W. Qu and Z. Bai. (2003). The inhibitory effect of saponins from *Tribulus terrestris* on Bcap-37 breast cancer cell line *in vitro*. *Journal of Chinese medicinal materials*, 26, pp. 104–106.

Sun, B., W.J. Qu and X.L. Zhang. (2004). Investigation on inhibitory and apoptosis-inducing effects of saponins from *Tribulus terrestris*on hepatoma cell line BEL-7402. *China Journal of Chinese Materia Medica*, 29, pp. 681–684.

Sy, L. K., S.C. Yan, C.N. Lok, R.Y. Man and C.M. Che. (2008). Timosaponin A-III induces autophagy preceding mitochondria-mediated apoptosis in HeLa cancer cells. *Cancer Res*, 68, pp. 10229–10237.

Tang, H. F., S.Y. Zhang and Y.H. Yi. (2006). Isolation and structural elucidation of a bioactive saponin from tubers of *Bolbostemma paniculatum*. *China Journal of Chinese Materia Medica*, 31, pp. 213–217.

Tian, Y., H.F. Tang and F. Qiu. (2009). Triterpenoid saponins from *Ardisia pusilla* and their cytotoxic activity. *Planta Med*, 75, pp. 70–75.

Tian, Z., Y.M. Liu and S.B. Chen. (2006). Cytotoxicity of two triterpenoids from *Nigella glandulifera*. *Molecules* (Basel, Switzerland), 11, pp. 693–699.

Tin, M. M., C.H. Cho, K. Chan, A.E. James and J.K. Ko. (2007). Astragalus saponins induce growth inhibition and apoptosis in human colon cancer cells and tumor xenograft. *Carcinogenesis*, 28, pp. 1347–1355.

Tran, Q. L., Y. Tezuka and A.H. Banskota. (2001). New spirostanol steroids and steroidal saponins from roots and rhizomes of *Dracaena angustifolia* and their antiproliferative activity. *J Nat Prod*, 64, pp. 1127–1132.

Tundis, R., M. Bonesi and B. Deguin. (2009). Cytotoxic activity and inhibitory effect on nitric oxide production of triterpene saponins from the roots of *Physospermum verticillatum* (Waldst and Kit) (Apiaceae). *Bioorg Med Chem*, 17, pp. 4542–4547.

Vincken, J. P., L. Heng, A. de Groot and H. Gruppen. (2007). Saponins, classification and occurrence in the plant kingdom. *Phytochemistry*, 68, pp. 275–297.

Wang, G., H. Chen and M. Huang. (2006). Methyl protodioscin induces G2/M cell cycle arrest and apoptosis in HepG2 liver cancer cells. *Cancer Lett*, 241, pp. 102–109.

Wang, S. L., B. Cai and C.B. Cui. (2003). Apoptosis of human chronic myeloid leukemia k562 cell induced by prosapogenin B of dioscin (P.B) *in vitro*. *Chinese Journal of Cancer*, 22 (2003), pp. 795–800.

Wang, S., J. Li and H. Huang. (2009). Anti-hepatitis B virus activities of astragaloside IV isolated from *radix Astragali*. *Biol Pharm Bull*, 32, pp. 132–135.

Wang, T., Z. Liu and J. Li. (2007). Determination of protodioscin in rat plasma by liquid chromatography-tandem mass spectrometry. *J Chromatogr*, 848, pp. 363–368.

Wang, W., Y. Zhao and E.R. Rayburn. (2007). *In vitro* anti-cancer activity and structure–activity relationships of natural products isolated from fruits of *Panax ginseng*. *Cancer Chemother Pharmacol*, 59, pp. 589–601.

Wang, Y., D. Zhang and W. Ye. (2008). Triterpenoid saponins from *Androsace umbellata* and their anti-proliferative activities in human hepatoma cells. *Planta Med*, 74, pp. 1280–1284.

Wang, Y., Y. Zhang and Z. Zhu. (2007). Exploration of the correlation between the structure, hemolytic activity and cytotoxicity of steroid saponins. *Bioorg Med Chem*, 15, pp. 2528–2532.

Wang, Y., Y.H. Cheung and Z. Yang. (2006). Proteomic approach to study the cytotoxicity of dioscin (saponin). *Proteomics*, 6, pp. 2422–2432.

Wen, P., X.M. Zhang, Z. Yang, N.L. Wang and X.S. Yao. (2008). Four new triterpenoid saponins from *Ardisia gigantifolia* Stapf. and their cytotoxic activity. *Journal of Asian natural products research*, 10, pp. 873–880.

Weng, X. Y., R.D. Ma and L.J. Yu. (2003). Apoptosis of human nasopharyngeal carcinoma CNE-2Z cells induced by tubeimoside I. *Chinese Journal of Cancer*, 22, pp. 806–811.

Wu, R. T., H.C. Chiang and W.C. Fu. (1990). Formosanin-C, an immunomodulator with antitumor activity. *Int J Immunopharmacol*, 12, pp. 777–786.

Xu R, Zhao W, Xu J, Shao B, Qin G (1996). "Studies on bioactive saponins from Chinese medicinal plants". *Advances in Experimental Medicine and Biology* 404: 371–82.

Xu, Y., J.F. Chiu, Q.Y. He and F. Chen. (2009). Tubeimoside-1 exerts cytotoxicity in HeLa cells through mitochondrial dysfunction and endoplasmic reticulum stress pathways. *J Proteome Res*, 8, pp. 1585–1593.

Xu, Z. X., J. Liang and V. Haridas. (2007). A plant triterpenoid, avicin D, induces autophagy by activation of AMP-activated protein kinase. *Cell Death Differ*, 14, pp. 1948–1957.

Yan, L. H., L.Z. Xu, J. Lin, S.L. Yang and Y.L. Feng. (2009). Triterpenoid saponins from the stems of *Clematis parviloba*. *J Asian Nat Prod Res*, 11, pp. 332–338.

Yan, M. C., Y. Liu and H. Chen. (2006). Synthesis and antitumor activity of two natural N-acetylglucosamine-bearing triterpenoid saponins: lotoidoside D and E. *Bioorg Med Chem Lett*, 16, pp. 4200–4204.

Yang, C. X., S.S. Huang, X.P. Yang and Z.J. Jia. (2004). Nor-lignans and steroidal saponins from *Asparagus gobicus*. *Planta Med*, 70, pp. 446–451.

Yang, H. J., W.J. Qu and B. Sun. (2005). Experimental study of saponins from *Tribulus terrestris* on renal carcinoma cell line. *China Journal of Chinese Materia Medica*, 30, pp. 1271–1274.

Yang, M. F., Y.Y. Li, X.P. Gao, B.G. Li and G. L. Zhang. (2004). Steroidal saponins from *Myriopteron extensum* and their cytotoxic activity. *Planta Med*, 70, pp. 556–560.

Yang, S. L., X.K. Liu, H. Wu, H.B. Wang and C. Qing. (2009). Steroidal saponins and cytoxicity of the wild edible vegetable *Smilacina atropurpurea*. *Steroids*, 74, pp. 7–12.

Yao, N., X. Gu and Y. Li. (2009). Effects of three C21 steroidal saponins from *Cynanchum auriculatum* on cell growth and cell cycle of human lung cancer A549 cells. *China Journal of Chinese Materia Medica*, 34, pp. 1418–1421.

Yin, F., Y.N. Zhang, Z.Y. Yang and L.H. Hu. (2006). Nine new dammarane saponins from *Gynostemma pentaphyllum*. *Chem Biodivers*, 3, pp. 771–782.

Yokosuka, A., M. Jitsuno, S. Yui, M. Yamazaki and Y. Mimaki. (2009). Steroidal glycosides from *Agave utahensis* and their cytotoxic activity. *J Nat Prod*, 72, pp. 1399–1404.

Yoo, H. H., S.W. Kwon and J.H. Park. (2006). The cytotoxic saponin from heat-processed *Achyranthes fauriei* roots. *Biol Pharm Bull*, 29, pp. 1053–1055.

Yu, L. J., R.D. Ma and Y.Q. Wang. (1992). Potent anti-tumorigenic effect of tubeimoside 1 isolated from the bulb of *Bolbostemma paniculatum* (Maxim) Franquet. *Int J Cancer*, 50, pp. 635–638.

Yu, T. X., R.D. Ma and L.J. Yu. (2001). Structure–activity relationship of tubeimosides in anti-inflammatory, antitumor and antitumor-promoting effects. *Acta Pharmacol Sin*, 22, pp. 463–468.

Yui, S., K. Ubukata and K. Hodono. (2001). Macrophage-oriented cytotoxic activity of novel triterpene saponins extracted from roots of *Securidaca inappendiculata*. *Int Immunopharmacol*, 1, pp. 1989–2000.

Yun, H., J.C. Li and H.Z. Wen. (2007). Separation and identification of steroidal compounds with cytotoxic activity against human gastric cancer cell lines *in vitro* from the rhizomes of *Paris polyphylla* var. *chinensis*. *Chemistry of Natural Compounds*, 43, pp. 672–677.

Zentner, Eduard (2011). "Effects of phytogenic feed additives containing quillaja saponaria on ammonia in fattening pigs". Retrieved 27 November 2012.

Zhang, C., B. Li and A.S. Gaikwad. (2008). Avicin D selectively induces apoptosis and downregulates p-STAT-3, bcl-2 and survivin in cutaneous T-cell lymphoma cells. *J Investig Dermatol*, 128, pp. 2728–2735.

Zhang, W. and D.G. Popovich. (2008). Effect of soyasapogenol A and soyasapogenol B concentrated extracts on HEP-G2 cell proliferation and apoptosis. *J Agric Food Chem*, 56, pp. 2603–2608.

Zhang, Y., H.Z. Li and Y.J. Zhang. (2006). Atropurosides A–G, new steroidal saponins from *Smilacina atropurpurea*. *Steroids*, 71, pp. 712–719.

Zhang, Z. and S. Li. (2007). Cytotoxic triterpenoid saponins from the fruits of *Aesculus pavia* L. *Phytochemistry*, 68, pp. 2075–2086.

Zheng, Z. F., J.F. Xu, Z.M. Feng and P.C. Zhang. (2008). Cytotoxic triterpenoid saponins from the roots of *Ardisia crenata*. *J Asian Nat Prod Res*, 10, pp. 833–839.

Zheng, Z. X., L.J. Zhang and C.X. Huang. (2007). Antitumour effect of total saponins of *Rubus parvifolius* on malignant melanoma. *China Journal of Chinese Materia Medica*, 32, pp. 2055–2058.

Zhong, L., G. Qu, P. Li, J. Han and D. Guo. (2003). Induction of apoptosis and G2/M cell cycle arrest by Gleditsioside E from *Gleditsia sinensis* in HL-60 cells. *Planta Med*, 69, pp. 561–563.

Zhou, L. B., T.H. Chen and K.F. Bastow. (2007). Filiasparosides A–D, cytotoxic steroidal saponins from the roots of *Asparagus filicinus*. *J Nat Prod*, 70, pp. 1263–1267.

Zhou, X., X. He and G. Wang. (2006). Steroidal saponins from *Solanum nigrum*. *J Nat Prod*, 69, pp. 1158–1163.

Zhou, Y., C. Garcia-Prieto and D.A. Carney. (2005). OSW-1: a natural compound with potent anticancer activity and a novel mechanism of action. *J Natl Cancer Inst*, 97, pp. 1781–1785.

Chapter 16

Steroids

Steroids comprise a group of cyclic organic compounds whose basis is a characteristic arrangement of seventeen carbon atoms in a four-ring structure linked together from three 6-carbon rings followed by a 5-carbon ring and an eight-carbon side chain on carbon 17. These rings are synthesized by biochemical processes from cyclization of a thirty-carbon chain, squalene, into lanosterol or cycloartenol. Hundreds of distinct steroids are found in animals, fungi, plants and elsewhere and many steroids are necessary to life at all levels. They include cholesterol, the sex hormones estradiol andtestosterone, bile acids and drugs such as the anti-inflammatory agent dexamethasone. The three cyclohexanerings are designated as rings A, B and C in the figure to the right and the one cyclopentane ring as ring D. Individual steroids vary, first and primarily, by the oxidation state of the carbon atoms in the rings and by the chains and functional groups attached to this four-ring system; second, steroids can vary more markedly via changes to the ring structure (*e.g.*, via ring scissions that produce secosteroids such as vitamin D$_3$, see below). Sterols are a particularly important form of steroids, with sterols having a cholestane-derived framework and a hydroxyl group at the C-3 ring position being the most prominent (*e.g.*, as in cholesterol, shown at right).

This structural lipid and key steroid biosynthetic precursor is shown in line-angle representation with its IUPAC-approved ring lettering and atom numbering conventions; the rings of the molecule are projected onto the plane of the page in a horizontal orientation, also IUPAC-specified. Hydrogens in this representation are implied at every vertex (fulfilling the additional bonds required for carbon, up to its valenceof four). Bonds extending above the plane are indicated with bold wedges and those extending below the plane with cross-hatched wedges. Alternative representations of the extending bonds are as bold and cross-hatched *lines*, or simple straight lines (where the stereochemical relationships are inferred).

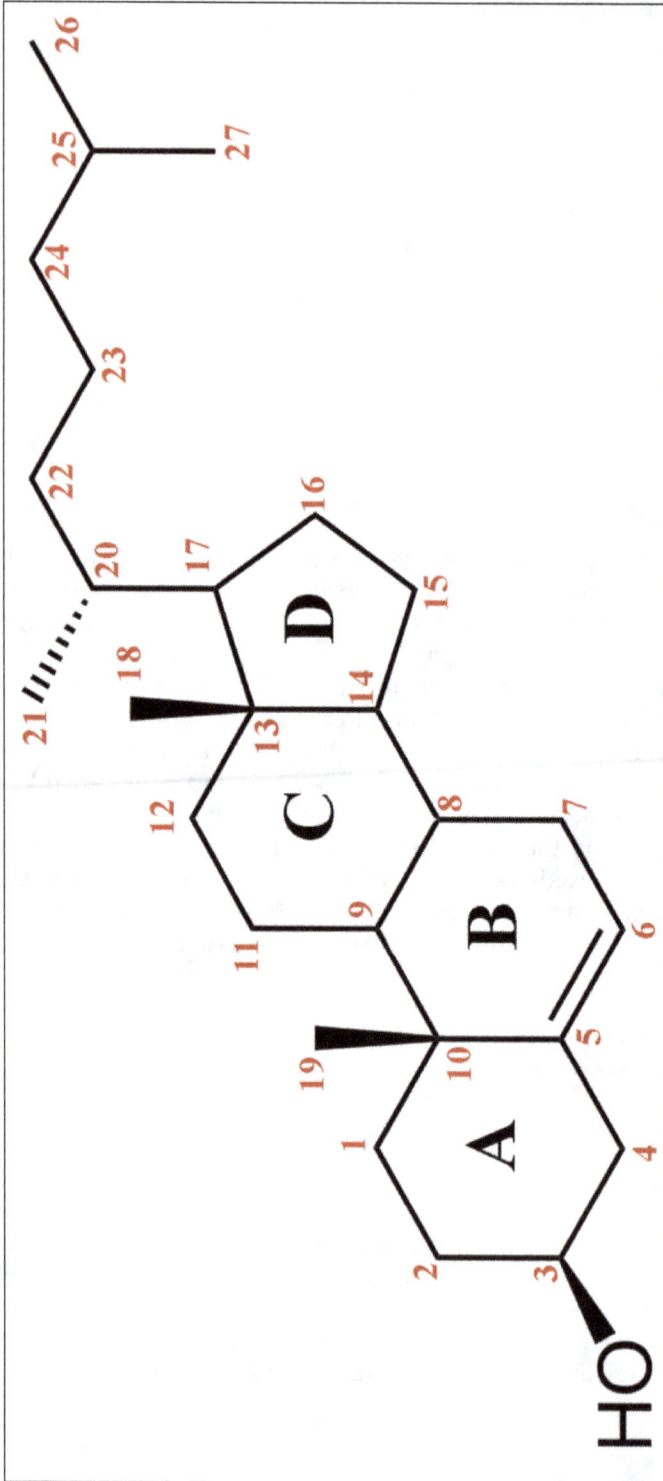

Cholesterol, a Prototypical Animal Sterol

Nomenclature and Examples

As IUPAC guidance notes (and is explained more fully following the quote), Steroids are compounds possessing the skeleton of cyclopenta phenanthrene or a skeleton derived therefrom by one or more bond scissions or ring expansions or contractions. Methyl groups are normally present at C-10 and C-13. An alkyl side chain may also be present at C-17. Sterols are steroids carrying a hydroxyl group at C-3 and most of the skeleton of cholestane. Additional carbon atoms may be present in the side chain.

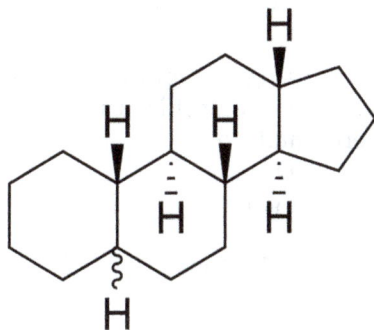

Gonane

Gonane is the simplest possible steroid and is composed of seventeen carbon atoms in carbon-carbon bonds that form four fused rings in a defined three-dimensional shape. The three cyclohexane rings (designated as rings A, B and C

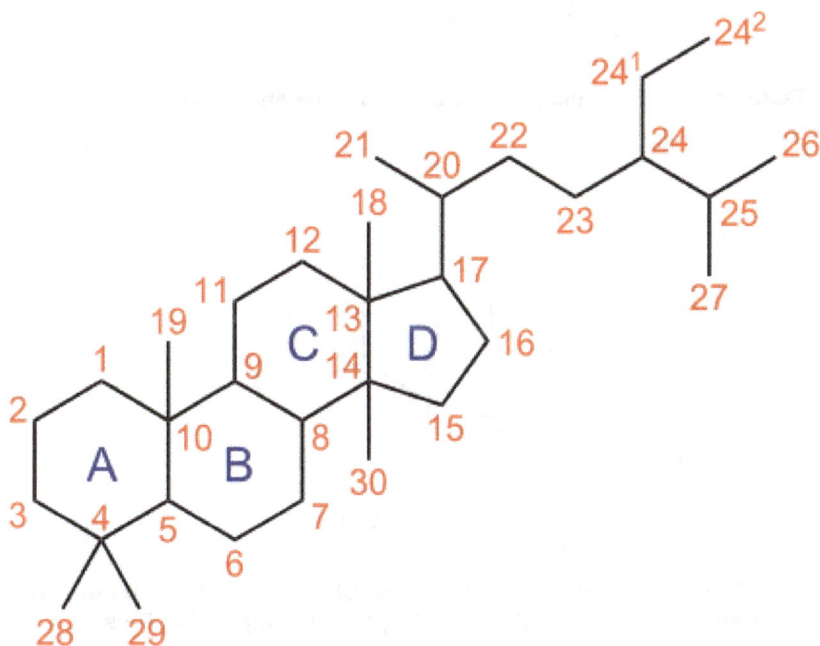

Steroid Ring System with Branchings

in the figures above form the skeleton of a perhydro- derivative of phenanthrene. The D-ring has a cyclopentane structure; hence, though it is uncommon, per IUPAC steroids can also be named as various hydro-derivatives of cyclopenta[a] phenanthrene. When the two methyl groups and 8 carbon side chain (at C-17, as shown for cholesterol) are present, the steroid is said to have a cholestane framework. The two common 5α and 5β stereoisomeric forms of steroids exist because of differences in the side of the largely planar ring system that the hydrogen (H) atom at carbon-5 is attached, which results in a change in steroid A-ring conformation.

The following are further important examples of steroid structures, in line-angle representation (see cholesterol image above, for explanation):

The parent ABCD steroid ring system (hydrocarbon framework) is shown again except with all normally seen branchings, in line-angle representation, with IUPAC-approved ring lettering and atom numbering. At its core is the hydrocarbon composed only of the four ABCD-rings, without branches or methyl substituents; this is the 17-carbon compound gonane, the simplest steroid and a substructure present in most steroids.

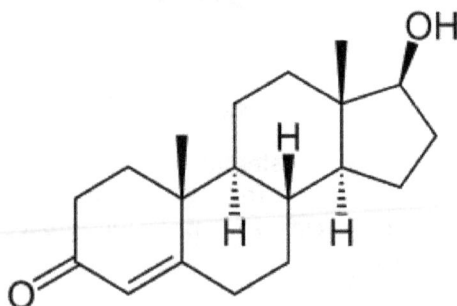

Testosterone, the Principal Male Sex Hormone and an Anabolic Steroid

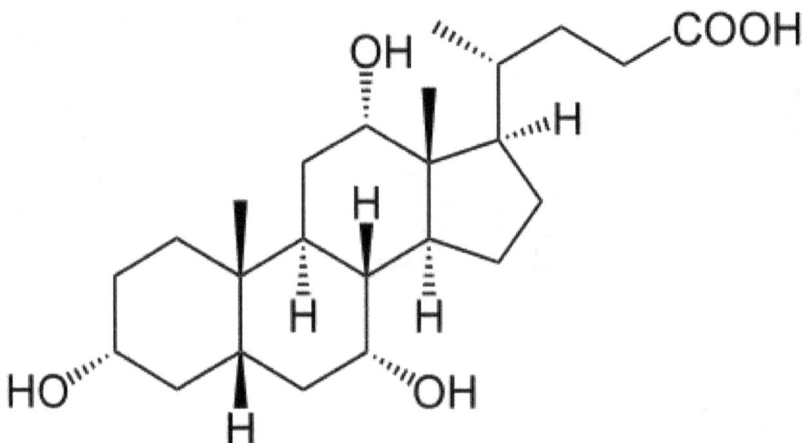

The Bile Acid with Common Name Cholic Acid, showing the Carboxylic Acid and Additional Hydroxyl groups often Present

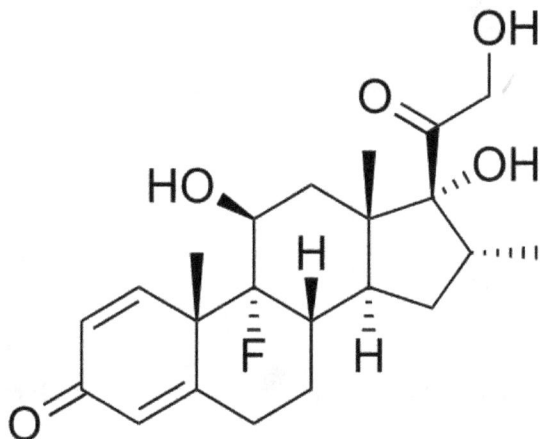

**Dexamethasone, a Synthetic Corticosteroid Drug,
the most Commonly Prescribed Steroid for Pain**

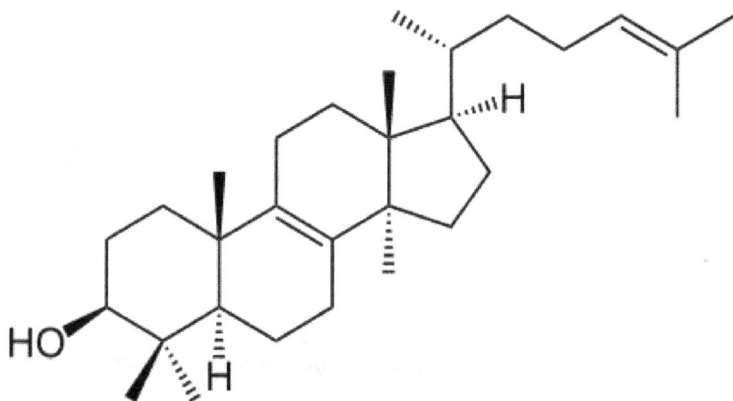

**Lanosterol, the biosynthetic precursor to animal steroids.
Note, the total number of carbons (30) makes clear its triterpenoid classification.**

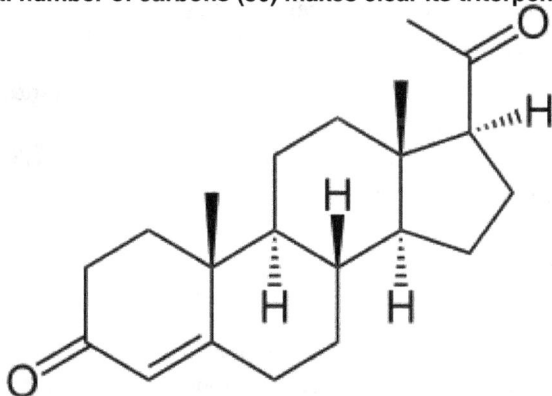

**Progesterone, a steroid hormone involved in the female menstrual cycle,
pregnancy and embryogenesis.**

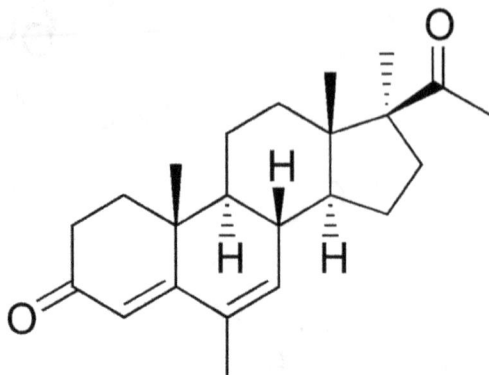

Medrogestone, a Synthetic Drug with Similar Effects as Progesterone

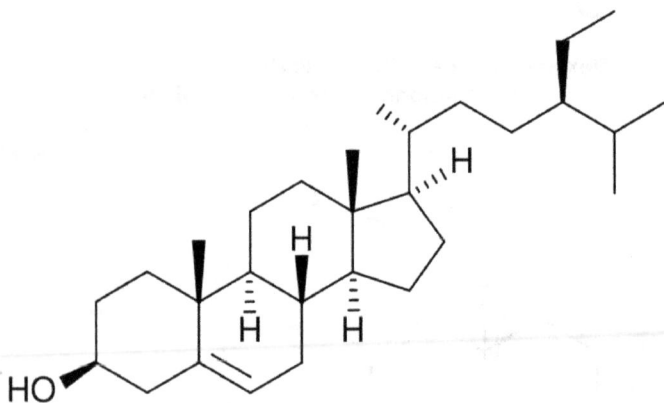

β-Sitosterol, a plant or phytosterol, with a fully branched hydrocarbon side chain at C-17 and a hydroxyl group at C-3.

In addition to the ring scissions (cleavages) and expansions and contractions (cleavage and reclosing to larger or smaller rings) noted in the IUPAC definition all variations in the carbon –carbon bond framework steroids can also vary:

☆ In the bond orders within the rings,

☆ In the number of methyl groups attached to the ring (and, when present, on the prominent side chain at C17),

☆ In the functional groups attached to the rings and side chain and

☆ In the configuration of groups attached to the rings and chain.

For instance, sterols such as cholesterol and lanosterol have an hydroxyl group attached at position C-3, while testosterone and progesterone have a carbonyl (oxo substituent) at C-3; of these examples, lanosterol alone has two methyl groups at C-4 and cholesterol with a C-5 to C-6 double bond differs from testosterone and progesterone, which have a C-4 to C-5 double bond.

Species Distribution and Function

The following are some of the common categories of steroids. In eukaryotes, steroids are found in the fungi, animals and plants. Fungal steroids include the ergosterols. The animal steroids include compounds of vertebrate and insect origin, in the latter case including ecdysteroids such as ecdysterone, which is involved in the control of molting in some species. Vertebrate examples include the steroid hormones and cholesterol, the latter of which is a structural component of cell membranes that is involved in determining the fluidity of cell membranes and is a principal constituent of plaques implicated in atherosclerosis. The steroid hormones include:

☆ The sex hormones that influence sex differences and support reproduction; these include androgens, estrogens and progestagens;

☆ The corticosteroids, including the preponderance of synthetic steroid drugs, with natural product classes being the glucocorticoids that regulate many aspects of metabolism and immune function and the mineralocorticoids that help maintain blood volume and control renal excretion of electrolytes; and

☆ The anabolic steroids, natural and synthetic, that interact with androgen receptors to increase muscle and bone synthesis, where in popular expressions, use of the term "steroids" may refer to anabolic steroids.

Plant steroids include steroidal alkaloids found in Solanaceae, the phytosterols and the brassinosteroids (which include several plant hormones).

In prokaryotes, biosynthetic pathways exist both for producing the tetracyclic steroid framework (*e.g.*, in mycobacteria) where its origin from eukaryotes is conjectured as well as the more common pentacyclic triterpinoid hopanoid framework.

Types

It is also possible to classify steroids based upon their chemical composition. One example of how MeSH performs this classification is available at the Wikipedia MeSH catalog. Examples from this classification include:

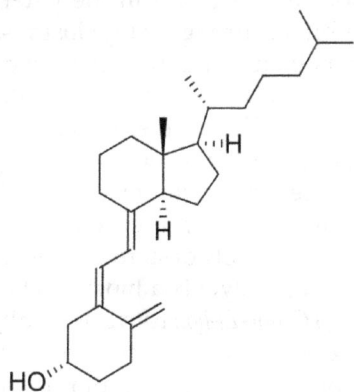

Cholecalciferol (vitamin D_3), an example of a 9,10-secosteroid. The triene substructure attached to the ring bearing the hydroxyl group is a result of the B-ring scission (cleavage) of the parent steroid framework, giving rise to this secosteroid. (The hydroxyl group is in position C3 of the parent steroid A-ring.)

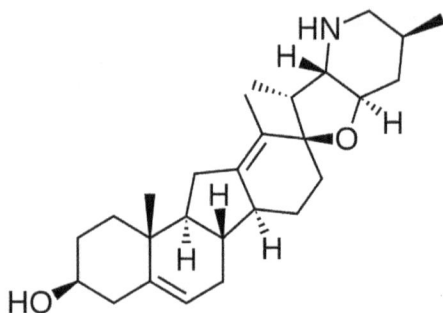

Cyclopamine, an example of a complex C-nor-D-homosteroid. In this steroid natural product responsible forcyclopia in lambs whose mothers ingest it in corn lily the C-12 atom is migrated from the C-ring into the D-ring during the course of biosynthesis; other significant changes take place on the C-17 side chain.

Class	Examples	Number of Carbon Atoms
Cholestanes	cholesterol	27
Cholanes	cholic acid	24
Pregnanes	progesterone	21
Androstanes	testosterone	19
Estranes	estradiol	18

The *gonane* (or steroid nucleus) is the parent (17-carbon tetracyclic) hydrocarbon molecule without any alkyl sidechains.

Cleaved, Contracted and Expanded Rings

Secosteroids (L. *seco*, "to cut") are a subclass of steroidal compounds resulting, biosynthetically or conceptually, via scission (cleavage) of parent steroid rings, generally one of the four. Major secosteroid subclasses are defined by the steroid carbon atoms where this scission has taken place. For instance, the prototypical secosteroid cholecalciferol, vitamin D$_3$ (shown), is in the important 9,10-secosteroid subclass, derived via cleavage between carbon atoms C-9 and C-10 of the steroid B-ring (similarly 5,6-secosteroids, 13,14-steroids, etc.). *Norsteroids* (nor-, L. *norma*, from "normal" in chemistry, indicating carbon removal) and *homosteroids* (homo-, Gk. *homos* for same, indicating carbon addition) are two structural subclasses of steroids formed via biosynthetic or bench chemistry steps, in the former case involving enzymic ring expansion/contraction reactions and in the latter accomplished similarly (biomimetically) or, more often, through ring closures of acyclic precursors with more or fewer ring atoms than in the parent steroid framework. These two classes represent further unique classes of steroids with important biological activities and societal impacts; the effect of these chemical operations on the ring structures is such that recognition of the parent tetracyclic ring system can be challenging. Combinations of these ring alterations are also possible and are known in nature. For instance, ewes that graze on corn lily ingestcyclopamine (shown) and veratramine, two of a sub-family of steroids where the C- and D-rings are contracted and expanded, respectively, via a biosynthetic migration of the original C-13 atom. Ingestion of these *C-nor-D-homosteroids* result in birth defects in progeny lambs: cyclopia in the case of cyclopamine and leg deformity with veratramine. A further C-nor-D-homosteroid, nakiterpiosin, is excreted by Okinawan cyanobacteriosponges, *Terpios hoshinota*, leading to coral mortality from black coral disease. Nakiterpiosin-type steroids are active against the *smoothened - hedgehog* pathway that is hyperactive in various cancers; Merck chemists established that C-13 atom migration could be achieved by "bench chemistry" and thisbiomimetic synthesis allows medicinal chemistry to proceed on this steroidal anticancer hypothesis.

Biological Significance

Steroid and their metabolites are frequently used signalling molecules. The most notable examples are the steroid hormones. Steroids along with phospholipids function as components of cell membranes. Steroids such as cholesterol decrease membrane fluidity. Similar to lipids, steroids represent highly concentrated energy

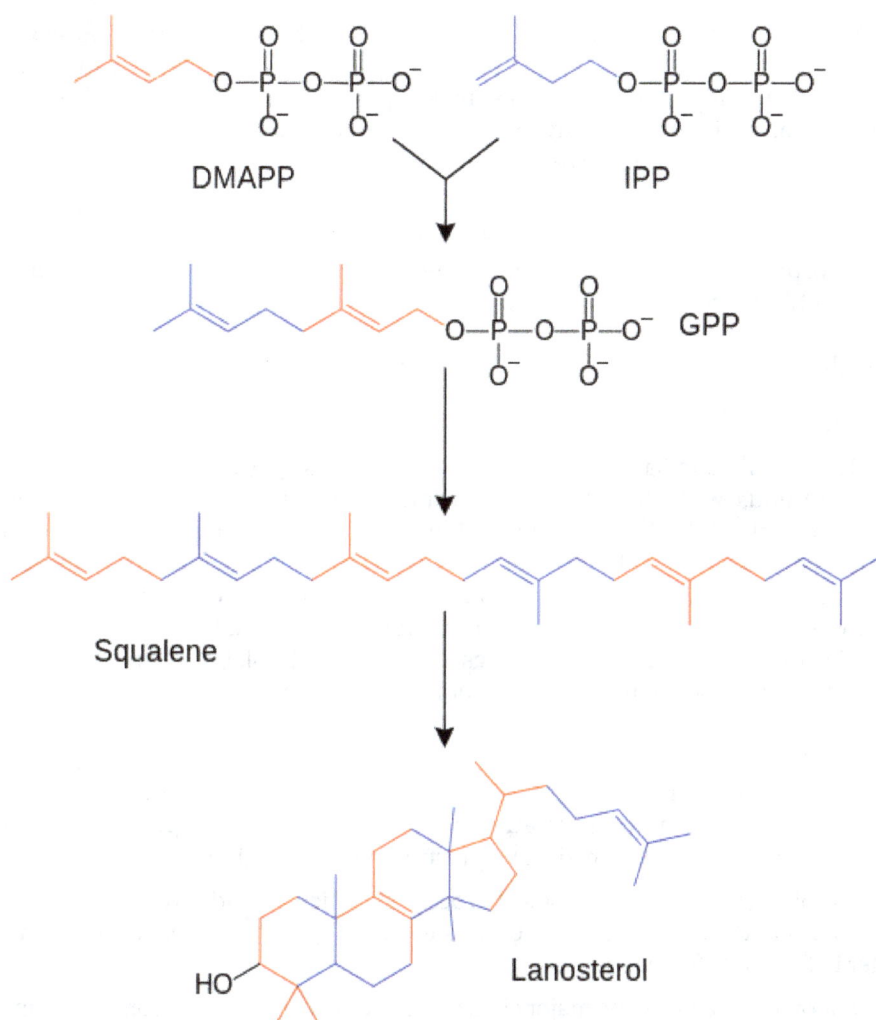

stores. However, steroids are not typically used as sources of energy. In mammals, they are normally metabolized and excreted.

Biosynthesis and Metabolism

The hundreds of distinct steroids found in animals, fungi and plants are made either from lanosterol (in animals and fungi, see examples above) or from cycloartenol (in plants). Both lanosterol and cycloartenol are derived via cyclization of the triterpenoid squalene.

Simplified version of latter part of steroid synthesis pathway, where the intermediates isopentenyl pyrophosphate (PP or IPP) anddimethylallyl pyrophosphate (DMAPP) form geranyl pyrophosphate (GPP),squalene and, finally, lanosterol, the first steroid in the pathways. Some intermediates are omitted for clarity.

Steroid biosynthesis is an anabolic metabolic pathway that produces steroids from simple precursors. A unique biosynthetic pathway is followed in animals compared to many other organisms, making the pathway a common target for antibiotics and other anti-infectivedrugs. In addition, steroid metabolism in humans is the target of cholesterol-lowering drugs such as statins.

In humans and other animals, the biosynthesis of steroids follows the mevalonate pathway that uses acetyl-CoA as building-blocks to formdimethylallyl pyrophosphate (DMAPP) and isopentenyl pyrophosphate (IPP). In subsequent steps, DMAPP and IPP are joined to form geranyl pyrophosphate (GPP), which in turn is used to synthesize the steroid lanosterol. Further modifications of lanosterol into other steroids are classified steroidogenesis transformations.

Mevalonate Pathway

The mevalonate pathway or HMG-CoA reductase pathway starts with acetyl-CoA and ends with dimethylallyl pyrophosphate (DMAPP) and isopentenyl pyrophosphate (IPP). DMAPP and IPP in turn donate isoprene units, which are assembled and modified to form terpenes and isoprenoids, which are a large class of lipids that include the carotenoids and form the largest class of plant natural products. Here, the isoprene units are joined together to make squalene and then folded up and formed into a set of rings to make lanosterol. Lanosterol can then be converted into other steroids such as cholesterol and ergosterol.

Steroidogenesis

The human steroidogenesis, with the major classes of steroid hormones, individual steroids and enzymaticpathways. Note that changes in molecular structure compared to the respective precursor are highlighted with white circles.

Steroidogenesis is the biological process by which steroids are generated from cholesterol and transformed into other steroids. Thepathways of steroidogenesis differ between different species.

Following is a list of the major classes of steroid hormones and some prominent members, with examples of major related functions:

Mevalonate pathway

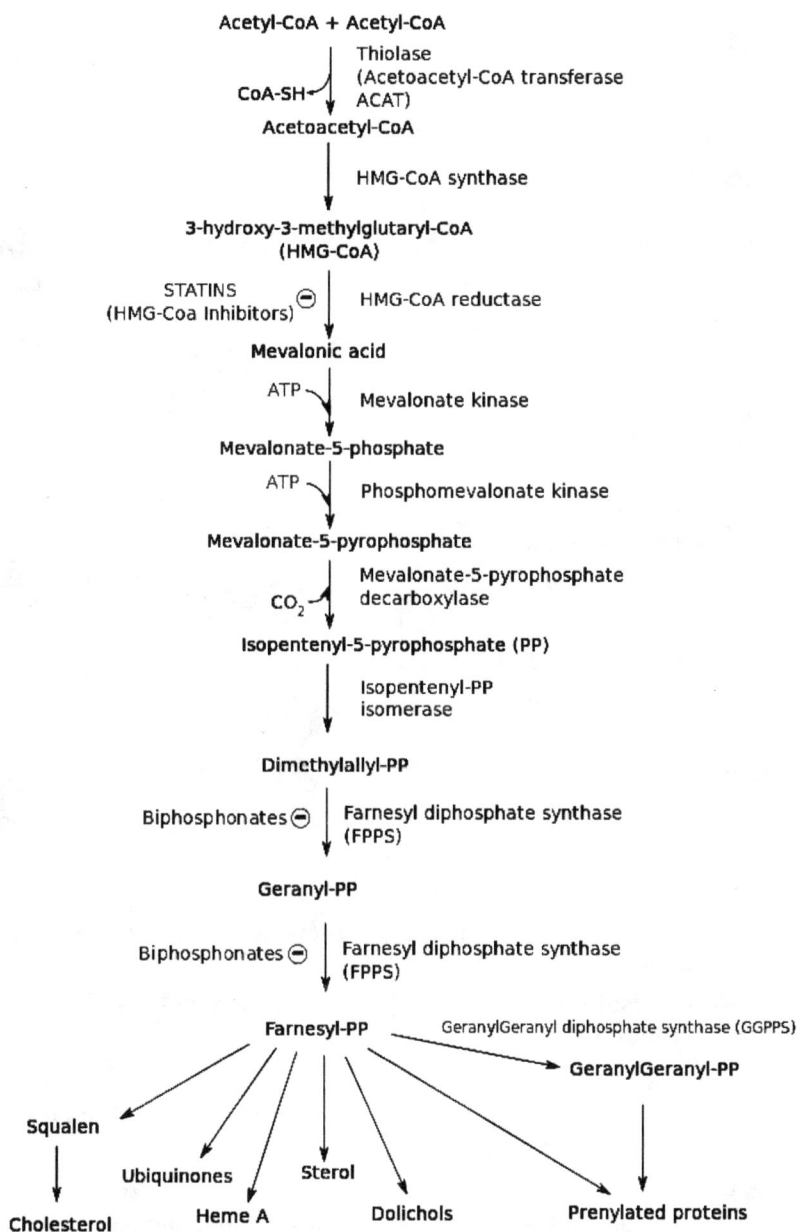

Acetyl-CoA + Acetyl-CoA

Thiolase
(Acetoacetyl-CoA transferase
CoA-SH — ACAT)

Acetoacetyl-CoA

HMG-CoA synthase

**3-hydroxy-3-methylglutaryl-CoA
(HMG-CoA)**

STATINS ⊖ HMG-CoA reductase
(HMG-Coa Inhibitors)

Mevalonic acid

ATP — Mevalonate kinase

Mevalonate-5-phosphate

ATP — Phosphomevalonate kinase

Mevalonate-5-pyrophosphate

Mevalonate-5-pyrophosphate
CO_2 — decarboxylase

Isopentenyl-5-pyrophosphate (PP)

Isopentenyl-PP
isomerase

Dimethylallyl-PP

Biphosphonates ⊖ Farnesyl diphosphate synthase
(FPPS)

Geranyl-PP

Biphosphonates ⊖ Farnesyl diphosphate synthase
(FPPS)

Farnesyl-PP — GeranylGeranyl diphosphate synthase (GGPPS)

GeranylGeranyl-PP

Squalen

Ubiquinones   **Sterol**

Cholesterol   **Heme A**   **Dolichols**   **Prenylated proteins**

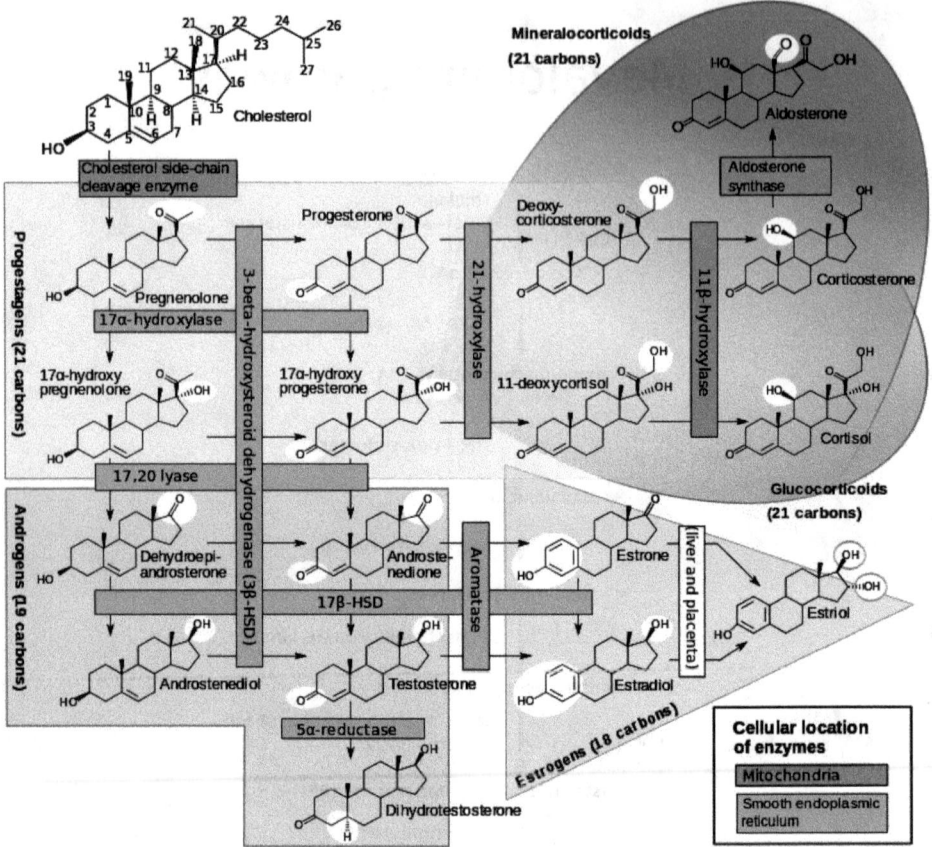

- ☆ Progestogens:
  - ☆ Progesterone, which regulates the cyclical changes of the endometrium of the uterus and the maintenance of pregnancy
- ☆ Corticosteroids (Corticoids):
  - ☆ Aldosterone (mineralocorticoids), which contributes to the regulation of blood pressure
  - ☆ Cortisol (glucocorticoids), whose functions include acting as an immunosuppressant
- ☆ Androgens:
  - ☆ Testosterone, which contributes to the development and maintenance of male secondary sex characteristics
- ☆ Estrogens:
  - ☆ Estrogen, which contributes to the development and maintenance of female secondary sex characteristics

Locations of Human Steroidogenesis

☆ Progestogens serve as precursors to all other human steroids thus all human tissues which produce steroids must first convertcholesterol to pregnenolone. This conversion is the rate-limiting step of steroid synthesis, which occurs inside the mitochondrion of the respective tissue.

☆ Corticosteroids are produced in the adrenal cortex.

☆ Estrogen and progesterone are made primarily in the ovary and in the placenta during pregnancy and testosterone in the testes.

☆ Testosterone is also converted into estrogen to regulate the supply of each, in the bodies of both females and males.

☆ In addition, certain neurons and glia in the central nervous system (CNS) express the enzymes that are required for the local synthesis of pregnane neurosteroids, either de novo or from peripherally derived sources.

Regulation

Several key enzymes can be activated through DNA transcriptional regulation on activation of SREBP (sterol regulatory element-binding protein-1 and -2). This intracellular sensor detects low cholesterol levels and stimulates endogenous production by the HMG-CoA reductase pathway, as well as increasing lipoprotein uptake by up-regulating the LDL receptor. Regulation of this pathway is also achieved by controlling the rate of translation of the mRNA, degradation of reductase and phosphorylation.

Alternative Pathways

In plants and bacteria, the non-mevalonate pathway uses pyruvate and glyceraldehyde 3-phosphate as substrates.

Metabolism

Steroids are oxidized mainly by cytochrome P450 oxidase enzymes, such as CYP3A4. These reactions introduce oxygen into the steroid ring and allow the structure to be broken up by other enzymes, to form bile acids as final products. These bile acids can then be eliminated through secretion from the liver in the bile. The expression of this oxidase gene can be upregulated by the steroid sensor PXR when there is a high blood concentration of steroids.

Isolation, Structure Determination and Methods of Analysis

The *isolation* of steroids refers, depending on context, either to the isolation of the considerable quantities of pure chemical matter required for chemical structure elucidation, derivitzation/degradation chemistry, biological testing and other research needs (generally milligrams to grams, but historically, often more), or to the isolation of "analytical quantities" of the substance of interest, where the focus is on identification and quantitation of the substance (*e.g.*, in biological tissue or fluid) and where the amount isolated depends on the analytical method applied (but is generally always sub-microgram in scale). The methods of isolation applied toward achieving these two distinct scales of product are likewise distinct, but generally involve extraction, precipitation, adsorptions, chromatography and sometimes crystallizations. In both cases, the isolated substance is purified to *chemical homogeneity, i.e.,* specific combined separation and analytical methods such as LC-MS methods are chosen to be "orthogonal" achieving their separations based on distinct modes of interaction between substance and isolating matrix with the goal being detection of only a single species present in the purportedly pure sample. The expression *structure determination* refers to methods that are applied to determine the chemical structure of an isolated, pure steroid, a process that involves an array of chemical and physical methods that have changed markedly over the history of steroid research, but that have included NMR and small molecule crystallography. *Methods of analysis* include samplings of both of these prior areas, but especially analytical methods aimed at determining if a steroid is present in an analytical mixture and determining its quantity in that medium.

References

American Chemical Society, International Historic Chemical Landmark (1999). "Russell Marker and the Mexican Steroid Hormone Industry," see [8], accessed 10 May 2104.

Bode HB, Zeggel B, Silakowski B, Wenzel SC, Reichenbach H, Müller R (2003). "Steroid biosynthesis in prokaryotes: identification of myxobacterial steroids and cloning of the first bacterial 2,3(S)-oxidosqualene cyclase from the myxobacterium *Stigmatella aurantiaca*". *Mol. Microbiol.* 47 (2): 471–81. doi: 10.1046/j.1365-2958. 2003.03309.x.

Conner AH, Nagaoka M, Rowe JW, Perlman D (1976). "Microbial conversion of tall oil sterols to C19 steroids". *Appl. Environ. Microbiol.* 32 (2): 310–1.

Desmond E, Gribaldo S (2009). "Phylogenomics of sterol synthesis: insights into the origin, evolution and diversity of a key eukaryotic feature". *Genome Biol Evol* 1: 364–81.doi: 10.1093/gbe/evp036.

Dubey V, Bhalla R, Luthra R (2003). "An overview of the non-mevalonate pathway for terpenoid biosynthesis in plants". *J Biosci* 28 (5): 637–46. doi: 10.1007/BF02703339.

Edgren RA, Stanczyk FZ (1999). "Nomenclature of the gonane progestins". *Contraception* 60 (6): 313. doi: 10.1016/S0010-7824 (99)00101 -8.

Grochowski L, Xu H, White R (2006). "*Methanocaldococcus jannaschii* uses a modified mevalonate pathway for biosynthesis of isopentenyl diphosphate". *J Bacteriol* 188 (9): 3192–8. doi: 10.1128/JB.188.9.3192-3198.2006.

Hanson JR (2010). "Steroids: partial synthesis in medicinal chemistry". *Nat Prod Rep* 27 (6): 887–99. doi: 10.1039/c001262a.

Hanukoglu I (1992). "Steroidogenic enzymes: structure, function and role in regulation of steroid hormone biosynthesis.". *J Steroid Biochem Mol Biol* 43 (8): 779–804.doi: 10.1016/0960-0760(92)90307-5.

Here and following, see Shuanhu Gao and Chio Chen (2012). "Nakiterpiosin", in *Total Synthesis of Natural Products: At the Frontiers of Organic Chemistry* (Jie Jack Li and E.J. Corey, eds.), Berlin: Springer, pp. 25-38, esp. 25-28, *e.g.*, [6], accessed 20 May 2014.

International Union of Pure and Applied Chemistry (IUPAC), 1999, "RF-4.1 Removal of Skeletal Atoms," in "RF-4. Skeletal Modifications" in Revised Section F: Natural Products and Related Compounds (IUPAC Recommendations 1999), see [3], accessed 20 May 2014. See also IUPAC, 1976, "Nomenclature of Organic Chemistry: Section F - Natural Products and Related Compounds, Recommendations 1976", IUPAC Information Bulletin Appendices on Tentative Nomenclature, Symbols, Units and Standards, No. 53, December, 1976, also in *Eur. J. Biochem.* 1978, 86, 1-8.

International Union of Pure and Applied Chemistry (IUPAC) (1999). "RF-4.2. Addition of Skeletal Atoms," in "RF-4. Skeletal Modifications" in Revised Section F: Natural Products and Related Compounds (IUPAC Recommendations 1999), see [4], accessed 20 May 2014. See also IUPAC, 1976, "Nomenclature of Organic Chemistry: Section F - Natural Products and Related Compounds, Recommendations 1976", IUPAC Information Bulletin Appendices on Tentative Nomenclature, Symbols, Units and Standards, No. 53, December, 1976, also in *Eur. J. Biochem.* 1978, 86, 1-8.

János Wölfling (2007). "Recent developments in the isolation and synthesis of D-homosteroids and related compounds" (Issue in Honor of Prof. Lutz F. Tietze), ARKIVOC (v) 210-230.

Kliewer S, Goodwin B, Willson T (2002). "The nuclear pregnane X receptor: a key regulator of xenobiotic metabolism". *Endocr. Rev.* 23 (5): 687–702. doi: 10.1210/er.2001-0038.

Kuzuyama T, Seto H (2003). "Diversity of the biosynthesis of the isoprene units". *Nat Prod Rep* 20 (2): 171–83. doi: 10.1039/b109860h. PMID 12735695.

Lanosterol biosynthesis. *Recommendations on Biochemical and Organic Nomenclature, Symbols and Terminology.* International Union of Biochemistry and Molecular Biology.

Lednicer D (2011). *Steroid Chemistry at a Glance.* Hoboken: Wiley. ISBN 978-0-470-66084-3.

Lees N, Skaggs B, Kirsch D, Bard M (1995). "Cloning of the late genes in the ergosterol biosynthetic pathway of *Saccharomyces cerevisiae*–a review". *Lipids* 30 (3): 221–6. doi: 10.1007/BF02537824.

Lichtenthaler H (1999). "The 1-Dideoxy-D-xylulose-5-phosphate pathway of isoprenoid biosynthesis in plants". *Annu Rev Plant Physiol Plant Mol Biol* 50: 47–65.doi: 10.1146/annurev.arplant.50.1.47. PMID 15012203.

Makin HLJ, Gower DB (2010). *Steroid analysis.* Dordrecht, New York: Springer.ISBN 978-1-4020-9774-4.

Moss GP (1989). "Nomenclature of Steroids (Recommendations 1989)". *Pure and Appl. Chem.* 61 (10): 1783–1822. doi: 10.1351/pac198961101783. PDF "IUPAC-IUB Joint Commission on Biochemical Nomenclature (JCBN). The nomenclature of steroids. Recommendations 1989". *Eur. J. Biochem.* 186 (3): 429–58. December 1989. doi: 10.1111/j.1432-1033.1989. tb15228.x.

Moss, G. P. and the Working Party of the IUPAC-IUB Joint Commission on Biochemical Nomenclature, "The Nomenclature of Steroids", hosted at Queen Mary University of London, Section 3S-1 (esp. 3S-1.4, incl. note 4) See [1] and [2], accessed 10 May 2014. Also available from same authors at Pure Appl. Chem. 1989, 61, 1783-1822 (esp. p. 1785f), or R.A. Hill, D.N. Kirk, H.L.J. Makin, H.L.J. and G.M. Murphy, 1991, "Dictionary of Steroids" London: Chapman and Hall, pp. xxx-lix. The Working Party of the IUPAC-IUB JCBN were P. Karlson (chairman), J.R. Bull, K. Engel, J. Fried[+], H.W. Kircher, K.L. Loening, G.P. Moss, G. Popják and M.R. Uskokovic.

Pikuleva IA (2006). "Cytochrome P450s and cholesterol homeostasis". *Pharmacol. Ther.*112 (3): 761–73. doi: 10.1016/j.pharmthera.2006.05.014. PMID 16872679.

Rhen T, Cidlowski JA (2005). "Antiinflammatory action of glucocorticoids – new mechanisms for old drugs". *N. Engl. J. Med.* 353 (16): 1711–23. doi: 10.1056/NEJMra 050541.

Rossier MF (2006). "T channels and steroid biosynthesis: in search of a link with mitochondria". *Cell Calcium.* 40 (2): 155–64. doi: 10.1016/j.ceca.2006.04.020.

Sadava D, Hillis DM, Heller HC, Berenbaum MR (2011). *Life: The Science of Biology 9th Edition.* San Francisco: Freeman. pp. 105–114. ISBN 1-4292-4646-4.

Schroepfer G (1981). "Sterol biosynthesis". *Annu Rev Biochem* 50: 585–621. doi: 10.1146/annurev.bi.50.070181.003101.

Siedenburg G, Jendrossek D (2011). "Squalene-hopene cyclases". *Appl. Environ. Microbiol.* 77 (12): 3905–15. doi: 10.1128/AEM.00300-11.

Uemura, D., M. Kita, H. Arimoto and M. Kitamura (2009). "Recent aspects of chemical ecology: Natural toxins, coral communities and symbiotic relationships," Pure Appl. Chem., 81(6), 1093–1111, esp. 1101. DOI 10.1351/PAC-CON-08-08-12. See [7], accessed 20 May 2014.

Wang F-Q, Yao K, Wei D-Z. (2010). "From Soybean Phytosterols to Steroid Hormones, Soybean and Health". In El-Shemy H. *Soybean and Health.* InTech. doi: 10.5772/18808. ISBN 978-953-307-535-8.

Zollner G, Marschall HU, Wagner M, Trauner M (2006). "Role of nuclear receptors in the adaptive response to bile acids and cholestasis: pathogenetic and therapeutic considerations". *Mol. Pharm.* 3 (3): 231–51. doi: 10.1021/mp060010s.

Chapter 17

Tannins

A tannin (also known as *vegetable tannin, natural organic tannins*, or sometimes *tannoid, i.e.* a type of biomolecule, as opposed to modern synthetic tannin) is an astringent, bitter plant polyphenolic compound that binds to and precipitates proteins and various other organic compounds including amino acids and alkaloids. The term tannin (from *tanna*, an Old High German word for oak or fir tree, as in Tannenbaum) refers to the use of wood tannins from oak in tanning animal hides into leather; hence the words "tan" and "tanning" for the treatment of leather. However, the term "tannin" by extension is widely applied to any large polyphenolic compound containing sufficient hydroxyls and other suitable groups (such as carboxyls) to form strong complexes with various macromolecules. The tannin compounds are widely distributed in many species of plants, where they play a role in protection from predation and perhaps also as pesticides and in

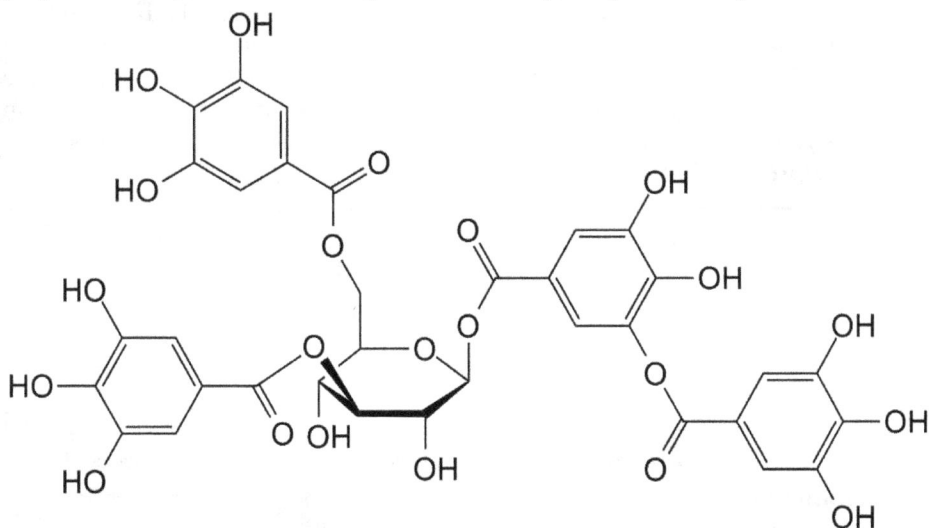

Tannic Acid, a Type of Tannin

plant growth regulation. The astringency from the tannins is what causes the dry and puckery feeling in the mouth following the consumption of unripened fruit or red wine. Likewise, the destruction or modification of tannins with time plays an important role in the ripening of fruit and the aging of wine. Tannins have molecular weights ranging from 500 to over 3,000 (gallic acid esters) and up to 20,000 (proanthocyanidins).

Tannin Powder (Mixture of compounds)

Structure and Classes of Tannins

There are three major classes of tannins: Shown below are the base units or monomer of the tannin. Particularly in the flavone-derived tannins, the base shown must be (additionally) heavily hydroxylated and polymerized in order to give the high molecular weight polyphenol motif that characterizes tannins. Typically, tannin molecules require at least 12 hydroxyl groups and at least five phenyl groups to function as protein binders.

Base Unit:

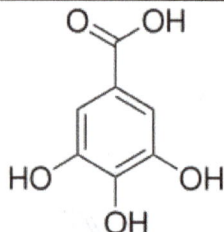
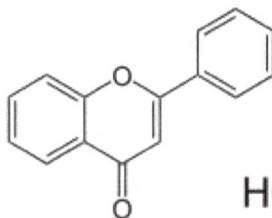
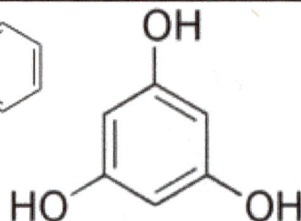

	Gallic Acid	Flavone	Phloroglucinol
Class/Polymer:	Hydrolyzable tannins	Non-Hydrolyzable or condensed tannins	Phlorotannins
Sources	Plants	Plants	Brown algae

Oligostilbenoids (oligo- or polystilbenes) are oligomeric forms of stilbenoids and constitute a class of tannins.

Pseudo Tannins

Pseudo tannins are low molecular weight compounds associated with other compounds. They do not change colour during the Goldbeater's skin test, unlike hydrolysable and condensed tannins and cannot be used as tanning compounds. [4] Some examples of pseudo tannins and their sources are:

Pseudo Tannin	Source(s)
Gallic acid	Rhubarb
Flavan-3-ols (*Catechins*)	Tea, acacia, catechu, cocoa, guarana
Chlorogenic acid	Nux-vomica, coffee, mate
Ipecacuanhic acid	*Carapichea ipecacuanha*

History

Ellagic acid, gallic acid and pyrogallic acid were first discovered by chemist Henri Braconnot in 1831. Julius Löwe was the first person to synthesize ellagic acid by heatinggallic acid with arsenic acid or silver oxide. Maximilian Nierenstein studied natural phenols and tannins found in different plant species. Working with Arthur George Perkin, he prepared ellagic acid from algarobilla and certain other fruits in 1905. He suggested its formation from galloyl-glycine by *Penicillium* in 1915. Tannase is an enzyme that Nierenstein used to produce m-acid from gallotannins. He proved the presence of catechin in cocoa beans in 1931. He showed in 1945 that luteic acid, a molecule present in the myrobalanitannin, a tannin found in the fruit of *Terminalia chebula*, is an intermediary compound in the synthesis of ellagic acid. At these times, molecule formulas were determined through combustion analysis. The discovery in 1943 by Martin and Synge of paper chromatography provided for the first time the means of surveying the phenolic constituents of plants and for their separation and identification. There was an explosion of activity in this field after 1945, none more so than that of Edgar Charles Bate-Smith and Tony Swain at Cambridge University. In 1966, Edwin Haslam proposed a first comprehensive definition of plant polyphenols based on the earlier proposals of Bate-Smith, Swain and White, which includes specific structural characteristics common to all phenolics having a tanning property. It is referred to as the White–Bate-Smith–Swain–Haslam (WBSSH) definition.

Occurrence

Tannins are distributed in species throughout the plant kingdom. They are commonly found in both gymnosperms as well as angiosperms. Mole (1993) studied the distribution of tannin in 180 families of dicotyledons and 44 families of monocotyledons (Cronquist). Most families of dicot contain tannin-free species (tested by their ability to precipitate proteins).

The best known families of which all species tested contain tannin are: Aceraceae, Actinidiaceae, Anacardiaceae, Bixaceae, Burseraceae, Combretaceae, Dipterocarpaceae,Ericaceae, Grossulariaceae, Myricaceae for dicot and Najadaceae and Typhaceae in Monocot. To the family of the oak, Fagaceae, 73 per cent of the species tested (N = 22) contain tannin. For those of acacias, Mimosaceae, only 39 per cent of the species tested (N = 28) contain tannin, among Solanaceae rate drops to 6 per cent and 4 per cent for the Asteraceae. Some families like the Boraginaceae, Cucurbitaceae and Papaveraceae contain no tannin-rich species. The most abundant polyphenols are the condensed tannins, found in virtually all families of plants and comprising up to 50 per cent of the dry weight of leaves. Tannins of tropical woods tend to be of a cathetic nature rather than of the gallic type present in temperate woods. There may be a loss in the bio-availability of still other tannins in plants due to birds, pests and other pathogens.

Plants Secondary Metabolites

For million years, humankind is completely dependent on plants as source of carbohydrates, proteins and fats for food and shelter. In addition, plants are a valuable source of a wide range of secondary metabolites, which are used as pharmaceuticals, agrochemicals, flavours, fragrances, colours, biopesticides and food additives. The number of known chemical structures is estimated to be nearly fourfold greater than that in the microbial kingdom. The United State market sales of plant medicinal have risen up about US$ 3 billion per year (Ramachandra Rao andRavishankar, 2002).During growth and maturation period in plants some substances can be found in structure of them which they have essential role in plant fortune. These substances called plants secondary metabolites (Hagerman and Buther, 1981; Hassanpour *et al.*, 2011). It has been suggested that accumulation of secondary compounds in plants is dependent upon photosynthetic capacity, season, rain and temperature (Mooney *et al.*, 1975).Groups of natural plants secondary metabolites of higher plants are shown in Table 17.1. One of the most important secondary metabolites is polyphenols (*e.g.* tannins) (Hagerman and Buther, 1981; ChaichiSemsari *et al.*, 2011; Hassanpour *et al.*, 2011; Maheri-sis *et al.*, 2011). A great deal of research with tannins has followed an approach that looks at biological relationships: taxonomy, phylogeny, biosynthesis, etc

Polyphenols are the most widely distributed class of plant secondary metabolites and several thousand different compounds have been identified. Polyphenols play many different roles in plant biology and human life, including UV protective agents, defensive compounds against herbivores and pathogens, contributors to plant colours, contributors to the taste of food and drink and pharmaceuticals (Haslam,, 1989; Hassanpour *et al.*, 2011).

What is Tannin?

Tannins are defined as phenolic compounds of high molecular weight ranging from 500 Da to more than 3000 Da which they found in plants leaves, bark, fruit, wood and roots located basically in the tissues in thevacuoles. They have been closely associated with plant defense mechanisms against mammalian herbivores, birdsand insects (Hagerman and Buther, 1981; Hassanpour *et al.*, 2011). Except of some higher molecular weight structures tannins are soluble in water (20- 35°C). Oligomeric compounds with multiple structure units with free phenolic groups can complex with proteins, starch, cellulose and minerals. In the plant kingdom tannins are found in both flowering plants and non-flowering plants. They are found in many plant species such as Acacia spp, Sericealespedeza as well as pasture species such as

Table 17.1: Groups of Natural Secondary Metabolites of Higher Plants

Phenylpropanoids	Alkaloids	Terpenoids	Quinones	Steroids
Anthocyanins	Acridines	Carotenes	Anthroquinones	Cardiac glycosides
Coumarins	Betalaines	Monoterpenes	Benzoquinones	Pregnenolone
Flavonoids	Quinolizidines	Sesquiterpenes	Naphthoquinones	
Hydroxycinnamoyl derivatives	Furonoqui nones	Diterpenes		
Isoflavonoids	Harringtonines	Triterpenes		
Lignans	Isoquinolines			
Phenolenones	Indoles			
Proanthocyanidins	Purines			
Stilbenes	Pyridines			
Tannins	Tropane alkaloids			

Adapted from Ramachandra Rao and Ravishankar, 2002.

Lotus spp. (Hassanpour *et al.*, 2011).Tanniniferous plants are widespread in nature and although a lot of attention has been given to their study in recent years, the term "tannin" continues to be difficult to define accurately. Indeed, whereas related phenolic compounds such as simple phenolics, neolignans and flavonoids are characterised and classified according to their chemical structure, tannins are a diverse group of compounds that are related primarily in their ability to complex with proteins (Fahey and Jung, 1989; Hassanpour *et al.*, 2011). Thus, tannins are usually defined as water-soluble polyphenolic substances and have ability bound to proteins that form insoluble or soluble tannin-protein complexes. As a consequence, tannins able to make complex with polysaccharides (cellulose, hemicelluloses and pectin) and nucleic acids, steroids, alkaloids and saponins (Haslam, 1986; ChaichiSemsari *et al.*, 2011; Maheri-sis *et al.*, 2011).There are some observations with regard to the presence of tannins that deserve some attention. For example, within plant cells, tannins are found in the vacuole (Chafe and Durzan, 1973; Lees *et al.*, 1995) and this has been suggested to be a method to preventing inhibition of the cell metabolism by tannins (Haslam, 1974). Also, one must as found about the energetic costs and on the reasons for such a practice, especially when plants devote so much carbon to the production of tannins. Haslam (1986) was suggested secondary metabolism serves to maintain primary metabolism in circumstances not propitious for growth. In recent years many researchers demonstrated that tannins have positive effects on animals by anti microbial, anthelmintic, protein bypassed effects in ruminants (Athanasiadou *et al.*, 2001; ChaichiSemsari *et al.*, 2011; Hassanpour *et al.*, 2011; Maheri-sis *et al.*, 2011; Sadaghian *et al.*, 2011).

Types of Tannin

According to their chemical structure and properties, tannins are divided into two main groups: hydrolysable (HT) and condensed tannins (CT) (Athanasiadou *et al.*, 2001; ChaichiSemsari *et al.*, 2011; Hassanpour *et al.*, 2011; Maheri-Sis *et al.*, 2011). The characteristics of the two groups are different in molecular weight, structure and produce a different effect on the herbivorous animals especially on ruminant when ingested. According to the chemical structure (Figure 17.1) of HTs (gallotannins and ellagitannins) are molecules which contain a carbohydrate, generally D-glucose, as a central core (Min and Hart, 2003). The hydrolysable groups of these carbohydrates areesterified with phenolic groups, such as ellagic acid or gallic acid (Mangan, 1988; Haslem, 1989). Hydrolysable tannins are usually found in lower concentrations in plants than CTs. Hydrolysable tannins are subdivided into taragallo tannins (gallic and quinic acid) and caffetannins (caffeic and quinic acid) (Mangan, 1988). They are hydrolyzed by tanninase enzymes which engage in ester bond hydrolysis. HTs can form compounds such as pyrogallol which is toxic to ruminants. Toxic compounds from more than 20 per cent HT in the diet can cause liver necrosis, kidney damage with proximal tuberal necrosis, lesions associated with hemorrhagic gastroenteritis and high mortality, which were observed in sheep and cattle (Reed, 1995). Hydrolysable tannins can also affect monogastrics by reducing growth rates, protein utilization and causing damage to the mucosa of the digestive tract and increasing the excretion of protein and amino acids. Condensed tannins (CT or proanthocyanidins), are the most common type of tannins found in forage, legumes, trees and stems (Barry and McNabb, 1999). These types of tannins are widely distributed in legume pasture species such as:

Lotus corniculatus and in several kinds of acacia and other plant species (Degen *et al.*, 1995). Condensed tannins have a variety of chemical structures affecting

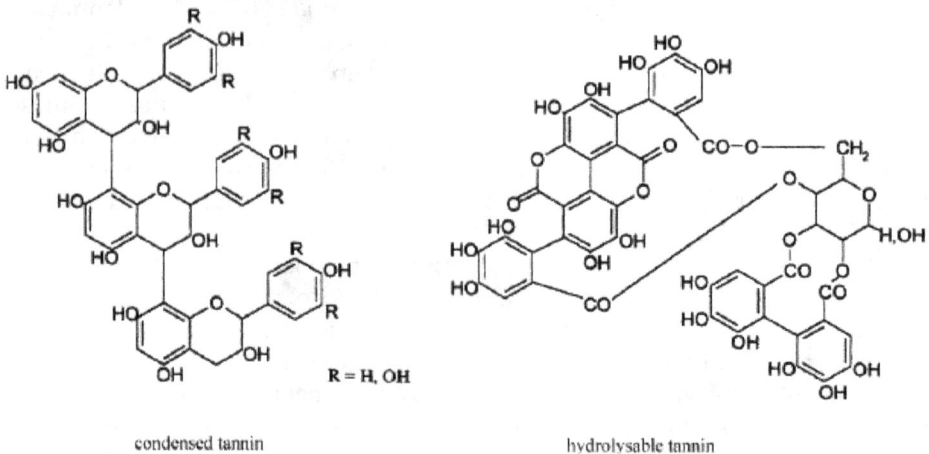

condensed tannin   hydrolysable tannin

Figure 17.1: Chemical Structure of CTs and HTs.
Adapted from McSweeney *et al.* (2001).

their physical and biological properties (Min *et al.*, 2003). They are consist of flavanoid units (flavan-3-ol) linked by carbon-carbon bonds. The complexity of CT depends on the flavanoid units which vary among constituents and within sites for interflavan bond formation.the term proanthocyanidins (PAs) is derived from the acid-catalyzed oxidation reaction producing red anthocyanidins upon heating PAs in acidic alcohol solutions. Anthocyanidin pigment is responsible for the colours observed in flowers, leaves, fruits juices and wines. The astringent taste of some leaves, fruits and wines is due to the presence of tannin.

Condensed Tannin Biosynthesis

Biosynthesis of tannins is shown in Figure 17.2. According to Gottlieb (1990) which provided a clear and concise summary of plant metabolism when he stated: "The basic metabolism of autotrophic plants combines photosynthesis with respiration, leading from CO_2 via the sugars of the Calvin cycle, Pyruvic acid and acetic acid either to the fatty acids of the Lynen spiral (a reversible process) or to the simple aliphatic acids of the Krebs cycle and thence back to CO_2. Connected by mostly reversible pathways are some essential intermediates such as the Krebs-cycle-derivedaliphatic amino acids, the purines and the pyrimidines, the acetic acid-derived Mevalonic acid, the sugar-derived glycerol and the sugar-plus pyruvic acid-derived Shikimic acid, the latter functioning as a precursor to the aromatic amino acids." Indeed, the shikimic acid and acetate-malonate pathways are the major metabolic routes of polyphenolic synthesis in plants. Also, two precursors are necessary for flavonoid synthesis, acetate and phenylalanine (Van Soest, 1982; Jung and Fahey, 1983).

Additionally, Stafford (1990) observed for the synthesis of CTs, a flavan-3, 4,-diol is one of the immediate precursors and the other is usually a flavan-3-ol acting as a nucleophile. In tracer studies with a range of fruit-bearing plants using a variety of labeled cinnamate precursors, this author showed that the C6-C3 carbon skeleton of the cinnamate precursor was incorporated intact into the flavan units. Many factors can affect CTs biosynthesis in plants. Foo *et al.* (1982) suggested that because polymers isolated from the leaves and roots of the same plant show structural variation, biosynthesis of tannins may be under different control in the two tissues. They reported two independent processes of tannin biosynthesis in mutants of *L. pedunculatus*: (a) light mediated and occurring in the apical meristem and (b) nutritional and occurring in the root system. Biosynthesis of CTs in leaves was controlled by light quality and in the roots by stressing the plants by applying conditions of nitrogen deficiency. Condensed tannin concentration in plant tissue has been shown to vary with many factors. These include plant species (Jackson *et al.*, 1996), plant part (Foo *et al.*, 1982, Barahona *et al.*, 1997), plant maturity (Lees *et al.*, 1995), growing season (Feeny, 1970) and soil fertility (Barry and Forss, 1983; Barry, 1989).

Biosynthesis of tannins, Adapted from Hättenschwiler and Vitousek (2000). The diagram shows a simplified over view of biosynthesis [(a) represents any living plant tissues], release into the environment and fate of polyphenols in the soil (b). The unbroken lines indicate the biosynthetic pathways of polyphenols and their fluxes

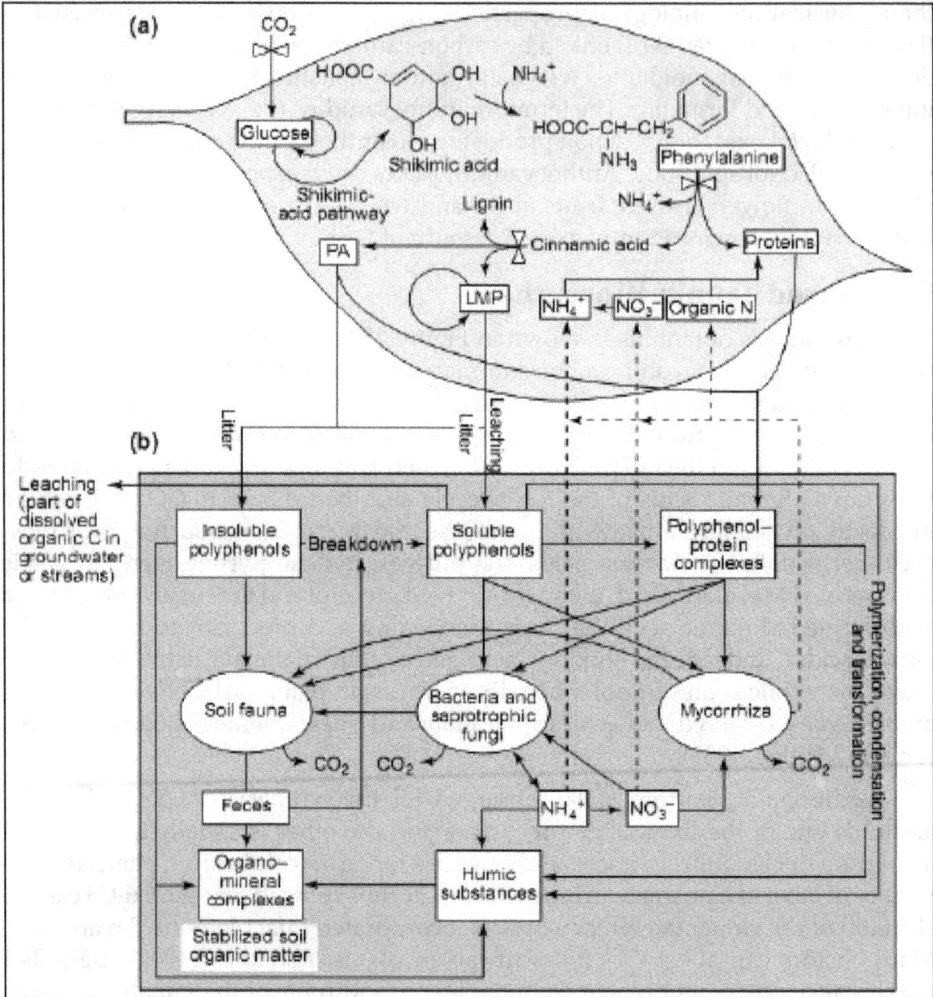

Figure 17.2: Biosynthesis of Tannins.

and transformations into and within the soil. The broken lines indicate nitrogen (N) uptake by the plant. The aromatic amino acid phenylalanine, synthesized in the shikimic-acid pathway, is the common precursor of proteins and phenolic compounds. Low molecular weight phenolics (LMP) might undergo a high turnover in living plant tissues, whereas high molecular weight PAs are considered to be metabolic end products with minimal turnover and a tendency to accumulate with the aging of plant tissues. Major control mechanisms (indicated by the regulation symbol ▶◀) occur at the level of the availability of glucose and phenylalanine (mainly quantity of polyphenols) and at the level of cinnamic acid (polyphenol quality). Soil organisms influence not only the uptake and metabolism of phenolic compounds, but also the fragmentation, mixing and translocation of polyphenol-containing litter material (soil fauna) and the production of extracellular enzymes (microorganisms) that drive either the breakdown of insoluble polyphenols or the formation of humic substances from low molecular weight polyphenols (according to the polyphenol theory of humus synthesis). The uptake, transformation and/or metabolism of polyphenol–protein complexes by soil organisms might be a major link between polyphenols and nutrient cycling (Hättenschwiler and Vitousek, 2000).

Benefits of Tannins

In the tropics, the efficiency of forage-based ruminant production systems is limited by forage quality and forage quantity during the dry season (NAS, 1979). In most tropical regions, native grasses from permanent pastures constitute the most important feed resource for ruminants. Unfertilized and unmanaged native grasses generally have poor nutritive value: 2.5-7.0 per cent crude protein (dry matter basis) and low dry matter digestibility 40-50 per cent (Patraa and Saxena, 2010). During the dry season, which ranges from two to six months, the dry matter availability from the sepastures diminishes dramatically. Furthermore, the low levels of crude protein and minerals in tropical grasses tend to decline rapidly during the dry season. As a result, cattle lose weight and milk production drops (NAS, 1979; Patraa and Saxena, 2010). During the wet season, the protein level in most grasses is lower than the level necessaryfor adequate animal growth (Patraa and Saxena, 2010). In addition, the low digestibility of tropical grasses can limit animal production, being on average 13 per cent less digestible than temperate grasses (Minson and McLeod, 1970).Although not yet fully exploited, the strategy of incorporating forage legumes into feeding schemes such as cut and carry systems or protein banks has enormous potential towards solving the severe nutritional limitations that ruminants face within tropical production systems (Devendra, 1990). Indeed, improved animal performance has been frequently reported in response to the use of high quality tanniniferous forages as supplements for ruminants fed low-quality roughage diets. Additionally (Gonzalo Hervás *et al.*, 2003) was demonstrate that moderate amounts of CTs have been reported to exert beneficial effects on protein metabolism in ruminants, decreasing rumen degradation of dietary protein and increasing absorption of amino acids in the small intestine. Hence, The CT may enable dietary protein bypass from the rumen for digestion in the lower digestive tract (Hassanpour *et al.*, 2011). An increase in flow of metabolizable protein or essential amino acids to the small intestine has been observed in animals grazing forages of high CT content compared to those grazing a low CT diet (Waghorn, 2008).

Tannin Sources

Indeed, tannins can be found in nearly all of the legumes, shrubs, vegetables and fruits in the world. Forexample sorghum (Reed, 1995), tea, wine (Corder *et al.*, 2006) and pomegranate (Scalbert *et al.*, 2003). Also, Previous researchers observed that high levels of CT have found in Acacia species (*e.g.* saligna, mearnsii, decurren, dealbata, pyonantha) (Athanasiadou *et al.*, 2001; Waghorn, 2008; Max, 2010; Hassanpour *et al.*, 2011; Sadaghian *et al.*, 2011), red wood (Quebracho tannins) (Athanasiadou *et al.*, 2001; Waghorn, 2008; ChaichiSemsari *et al.*, 2011; Maheri-sis *et al.*, 2011), birdsfoot trefoil (*Lotus corniculatus*), sainfoin, (*Onobrychis*), Sulla (*Hedysarum coronarium*) and lotus major (*L. pedunculatus*) (Waghorn, 2008).

Localization in Plant Organs

Tannins are found in leaf, bud, seed, root and stem tissues. An example of the location of the tannins in stem tissue is that they are often found in the growth areas of trees, such as the secondary phloem and xylem and the layer between the cortex and epidermis. Tannins may help regulate the growth of these tissues.

Cellular Localization

In all vascular plants studied so far, tannins are manufactured by a chloroplast-derived organelle, the tannosome. Tannins are mainly physically located in the vacuoles or surface wax of plants. These storage sites keep tannins active against plant predators, but also keep some tannin from affecting plant metabolism while the plant tissue is alive; it is only after cell breakdown and death that the tannins are active in metabolic effects. Tannins are classified as ergastic substances, *i.e.*, non-protoplasm materials found in cells. Tannins, by definition, precipitate proteins. In this condition, they must be stored in organelles able to withstand the protein precipitation process. Idioblasts are isolated plant cells which differ from neighboring tissues and contain non-living substances. They have various functions such as storage of reserves, excretory materials, pigments and minerals. They could contain oil, latex, gum, resin or pigments etc. They also can contain tannins. In Japanese persimmon (*Diospyros kaki*) fruits, tannin is accumulated in the vacuole of tannin cells, which are idioblasts of parenchyma cells in the flesh.

Presence in Soils

The convergent evolution of tannin-rich plant communities has occurred on nutrient-poor acidic soils throughout the world. Tannins were once believed to function as anti-herbivore defenses, but more and more ecologists now recognize them as important controllers of decomposition and nitrogen cycling processes. As concern grows about global warming, there is great interest to better understand the role of polyphenols as regulators of carbon cycling, in particular in northern boreal forests. Leaf litter and other decaying parts of a kauri (*Agathis australis*), a tree species found in New Zealand, decompose much more slowly than those of most other species. Besides its acidity, the plant also bears substances such as waxes and phenols, most notably tannins, that are harmful to microorganisms.

Presence in Water and Wood

The leaching of highly water soluble tannins from decaying vegetation and leaves along a stream may produce what is known as a black water river. Water flowing out of bogs has a characteristic brown colour from dissolved peat tannins. The presence of tannins (or humic acid) in well water can make it smell bad or taste bitter, but this does not make it unsafe to drink. Tannins leaching from an unprepared driftwood decoration in an aquarium can cause pH lowering and colouring of the water to a tea-like tinge. A way to avoid this is to boil the wood in water several times, discarding the water each time. Using peat as an aquarium substrate can have the same effect. Many hours of boiling the driftwood may need to be followed by many weeks or months of constant soaking and many water changes before the water will stay clear. Adding baking soda to the water to raise its pH level will accelerate the process of leaching, as the more alkaline solution can draw out tannic acid from the wood faster than the pH-neutral water. Softwoods, while in general much lower in tannins than hardwoods, are usually not recommended for use in an aquarium so using a hardwood with a very light colour, indicating a low tannin content, can be an easy way to avoid tannins. Tannic acid is brown in colour, so in general white woods have low tannin content. Woods with a lot of yellow, red, or brown colouration to them (like southern yellow pine, cedar, redwood, red oak, etc.) tend to contain a lot of tannin.

Extraction

There is no single protocol for extracting tannins from all plant material. The procedures used for tannins are widely variable. It may be that acetone in the extraction solvent increases the total yield by inhibiting interactions between tannins and proteins during extraction or even by breaking hydrogen bonds between tannin-protein complexes.

Tests for Tannins

There are three groups of methods for the analysis of tannins: precipitation of proteins or alkaloids, reaction with phenolic rings and depolymerization.

Biosynthesis and Genetic Regulation of Proanthocyanidins in Plants

Proanthocyanidins (PAs), also known as condensed tannins, are oligomers or polymers of flavan-3- ol units, which have been used as medicines or tanning agents by humans since ancient times and whose chemistry and synthesis have been studied for many decades (Dixon *et al.*, 200 and Bruyne *et al.*, 1999). As one of the most ubiquitous groups of all plant phenolics, PAs are synthesized via the phenylpropanoid and the flavonoid pathways and are widespread throughout the plant kingdom, presenting diverse biological and biochemical activities, including protection against predation (herbivorous animals) and pathogen attack (both bacterial and fungal), as well as restricting the growth of neighboring plants (Koes *et al.*, 2005; Feeny, 1970; Scalbert, 1991; Bais *et al.*, 2003). PAs are also widely distributed in foods of plant origin, particularly in fruits, legume seeds, cereal grains and different beverages such as juice, wine, cider, tea and cocoa, where they contribute to the bitter flavor and astringency and have a significant influence on the mouth feel (Santos-Buelga and Scalbert, 2000 and Peleg, *et al.*, 1999). In recent years, considerable attention has been drawn to PAs and their monomers because of their potential beneficial effects on human health such as immunomodulatory and anticancer activities, antioxidant and radical scavenging functions, anti-inflammatory activities, cardio-protective properties, vasodilating and antithrombotic effects, UV-protective functions, etc (Prior, *et al.*, 2005; Zhao, *et al.*, 2007; Rao, *et al.*, 2004; Subarnas and Wagner, 2000; Sato, *et al.*, 1999; Sano, *et al.*, 2005 and Sharma, *et al.*, 2007). Although PAs, extracted from various plants, especially grapes, have been widely used as nutritional supplements, their safety and potential long term toxicity still need to be further investigated to allow a systematic evaluation (Yamakoshi, *et al.*, 2002). Nowadays, relying upon modern analytical techniques such as high pressure liquid chromatography (HPLC), mass spectrometry (MS), circular dichroism (CD) and nuclear magnetic resonance (NMR), researchers have been able to detect the low molecular weight compositions of PAs and their derivatives in either natural products or in artificial solutions and provide a comprehensive understanding of their structures and chemical properties (Gu, *et al.*, 2003; Tarascou, *et al.*, 2006; Slade, *et al.*, 2005). On the other hand, with the application of acid-catalyzed PA cleavage in the presence of a nucleophile, such as phloroglucinol, benzyl mercaptan or other nucleophilic agent, researchers can use a post-HPLC analysis to estimate PA polymers from natural products (Matthews, *et al.*, 1997; Kennedy and Jones, 2001). Furthermore, researchers have identified a series of landmarks in PA biosynthesis by means of biochemical, genetic and biogenetic methods, including the recent isolation of the gene for leucoanthocyanidin reductase (LAR), the functional identification of BANYULS (BAN, the gene for anthocyanidin reductase in Arabidopsis thaliana) and the discovery of a host of new biochemical and regulatory elements and transport factors (Marles, *et al.*, 2003). However, some important questions in the field of PA biosynthesis remain unanswered. For example, how do flavan-3-ol units polymerize to produce PAs and whether this polymerization is enzyme-mediated or not, as well as what is the exact "enzyme" (Dixon, *et al.*, 2005 and Xie and Dixon, 2005). As early as the 1980s, Stafford commented on the similarities between PAs and lignins, primarily in their common origins as polyphenolic polymers and their potential functions in plant defense (Stafford, 2008). Dixon *et al.*, highlighted this comment and further speculated on the potential commonalities between them with respect to their mode of assembly and the importance of their stereochemistry (Dixon, *et al.*, 2005). Besides these, other previous reviews summarized the foregoing achievement in almost every aspect of the study of PAs mainly in the past 20 years (Schofield, *et al.*, 2001; Kelm, *et al.*, 2005; Tian, *et al.*, 2007; Buzzini, *et al.*, 2007; Rasmussen, *et al.*, 2005; Ferreira and Li, 2000; Ferreira and Slade, 2002). This paper will review the important concepts and updates in the research of PAs with regard to the elucidation of structure and stereochemistry, the general biosynthetic pathway, the potential polymerization mechanisms, their transport, regulation and genetic manipulation. The future trends in this field are also discussed.

Structures of Proanthocyanidins

PAs are oligomers and polymers composed of elementary flavan-3-ol units. As they occur widely in the plant kingdom and are considered the second most abundant group of natural phenolics after lignins, the structure complexity of PAs has been studied for many decades (Gu, *et al.*, 2003). However, their structures were still ill defined until the recent application of modern technologies such as NMR, CD and MS. Generally, the structure variability of PAs depends upon the nature (the stereochemistry at the chiral centers and the hydroxylation pattern) of the flavan-3-ol extension and end units, the location and stereochemistry of the interflavan linkage (IFL) between the monomeric units and the degree of polymerization (DP) (Dixon, *et al.*, 2005). Additionally, derivatizations such as O-methylation, O-acylation,

C- and Oglycosylation are involved in the rearrangement products of PAs (Xie and Dixon, 2005).

As the building blocks of PAs, the flavan-3-ol units have the typical C6-C3-C6 flavonoid skeletons. The heterocyclic benzopyran ring is referred to as the C ring, the fused aromatic ring as the A ring and the phenyl constituent as the B ring (Ribéreau-Gayon, 1972). They differ structurally according to the nature of the stereochemistry of the asymmetric carbons on the C rings and the number of hydroxyl groups on the B rings. 2, 3-trans-(+)-Catechin and 2,3-cis-(-)-epicatechin are stated to be the most usual monomeric units in PAs, which have the opposite stereochemistry of the chiral C3 carbon on the C rings (Figure 17.3, 1-2). As observed in (+)-catechin (2R, 3S) and (-)-epicatechin (2R, 3R), the C2 configuration is almost always R, but some exceptions are found in monocotyledons and in selected dicotyledonous families in succession, such as Rhus, Uncaria, Polygonum, Rapheolepsis and Schinopsis (Geiss, *et al.*, 1995; Santos-Buelga, *et al.*, 1995; Ferreira and Bekker, 1996) where flavan-3- ols with 2S type configuration are consequently distinguished by the prefix enantio (ent-), which may further enhance the structure complexity of PAs (Figure 17.3, 3-4). Acting in the early steps of the flavonoid pathway, the cytochrome P450 monooxygenases flavonoid 3'-hydroxylase (F3'H) and flavonoid 3', 5'-hydroxylase (F3'5'H) catalyze the formation of the 3'-hydroxyl and 3',5'-hydroxyl groups on the B rings, respectively (Holton, *et al.*, 1993; Ayabe and Akashi, 2006). Thus, the presence or absence of these two enzymes determines the B-ring hydroxylation pattern of the monomeric units in PAs (Figure 17.1). Generally, the term proanthocyanidins refers to the release of anthocyanidins from extension positions after being boiled with strong mineral acid. Correspondingly, procyanidins designate oligomers and polymers with 3',4'-dihydroxyl pattern ((+)-catechin and/or (-)-epicatechin units) extension units, while propelargonidins or prodelphinidins designate oligomers and polymers with extension units of 4'-hydroxyl pattern ((+)-afzelechin and/or (-)-epiafzelechin units) or 3',4',5'- trihydroxyl pattern ((+)-gallocatechin and/or (-)-epigallocatechin), respectively. Therefore, PAs can thus be classified, according to the differences in hydroxylation patterns, into several subgroups: propelargonidins (3,4',5,7-hydroxyl), procyanidins (3,3',4',5,7-hydroxyl), prodelphinidins (3,3',4',5,5',7-hydroxyl), proguibourtinidins (3,4',7-hydroxyl), profisetinidins (3,3',4',7-hydroxyl), prorobinetinidins (3,3',4',5',7-hydroxyl), proteracacidins (4',7,8-hydroxyl), promelacacidins (3',4',7,8-hydroxyl), proapigeninidins (4',5,7-hydroxyl) and proluteolinidins (3',4',5,7-hydroxyl), some of which have been synthesized by chemical methods and have not been found to date in Nature. In fact, when several types of units are present in a plant species or organ they are found together in mixed polymers. PAs are linked between the C4 position of the upper unit and the C8 or C6 position of the lower unit and the type of IFL can be either α or β type. Generally, the IFL between C4 and C8 position are stereochemically predominant in procyanidins and prodelphinidins, as the C4→C8 IFL and the C4→C6 IFL are usually present in a ratio of 3:1, but in 5-deoxy PAs, the C4→C6 IFL is predominant (Malan, *et al.*, 1990; Steynberg, *et al.*, 1994). Oligomeric and polymeric PAs which are composed of flavan-3-ol units linked mainly through C4→C8 and/or C4→C6 IFL are categorized as B-type PAs. Among the dimers, procyanidins B1, B2, B3 and B4 are linked by the C4→C8 IFL and are the most

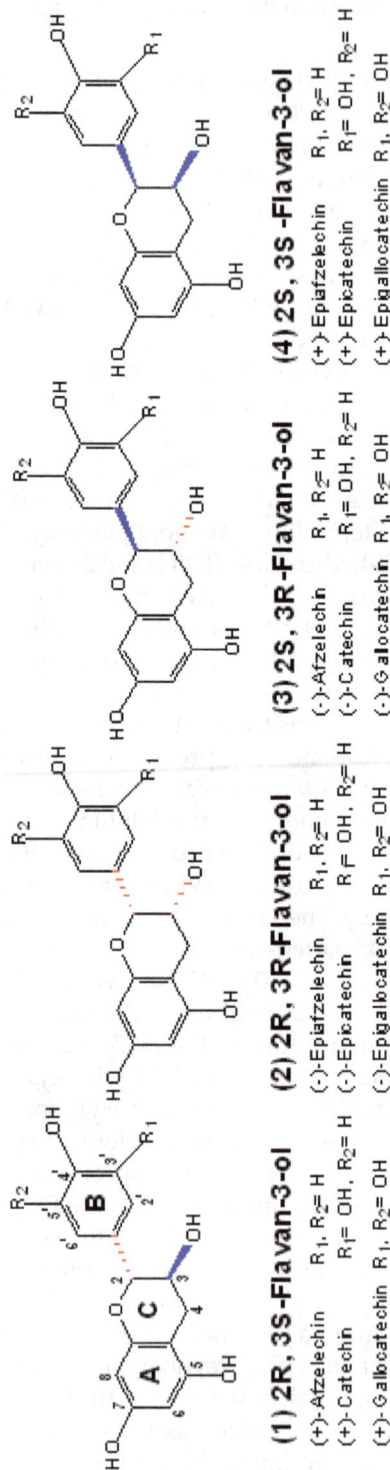

(1) 2R, 3S -Flavan-3-ol

(+)-Afzelechin	$R_1, R_2 = H$	
(+)-Catechin	$R_1 = OH, R_2 = H$	
(+)-Gallocatechin	$R_1, R_2 = OH$	

(2) 2R, 3R-Flavan-3-ol

(-)-Epiafzelechin	$R_1, R_2 = H$	
(-)-Epicatechin	$R_1 = OH, R_2 = H$	
(-)-Epigallocatechin	$R_1, R_2 = OH$	

(3) 2S, 3R-Flavan-3-ol

(-)-Afzelechin	$R_1, R_2 = H$	
(-)-Catechin	$R_1 = OH, R_2 = H$	
(-)-Gallocatechin	$R_1, R_2 = OH$	

(4) 2S, 3S -Flavan-3-ol

(+)-Epiafzelechin	$R_1, R_2 = H$	
(+)-Epicatechin	$R_1 = OH, R_2 = H$	
(+)-Epigallocatechin	$R_1, R_2 = OH$	

Figure 17.3: Structures of the 2R-type flavan-3-ols 1, 2 and the 2S-type ent-flavan-3-ols 3, 4.

frequently occurring in plants, whereas procyanidins B5, B6, B7 and B8 are linked by the C4→C6 IFL and are also widespread. On the other hand, the flavan-3-ol units can also be doubly linked by an additional ether bond between C2 position of the upper unit and the oxygen at C7 or C5 position of the lower unit. The oligomers or polymers which contain both C2β→O→7 ether-type IFL and C4→C8 or C4→C6 IFL, as well as those which contain both C2β→O→5 ether-type IFL and C4→C6 IFL are categorized as A-type PAs, which are found in various food of plant origin, such as cranberry, plum, avocado, peanut, etc. The structures of the B-type procyanidins (B1-B8 and C1) and A-type procyanidins (A1-A2) are shown in Figure 17.4. However, with more and more new PA oligomers being identified and characterized, the old nomenclature system of PAs with simple letters and numbers no longer meets the needs of researchers. As a result, it is difficult to use it to present a clear understanding of a PA's structure (Thompson, *et al.*, 1972). A new nomenclature, derived from polysaccharides, was later introduced to name the increasing number of new structures. In this nomenclature, the elementary units of the oligomers are designated with the name of the corresponding flavan-3-ol monomers (Hemingway, *et al.*, 2008). The IFL and its location and direction are indicated in parentheses with an arrow (→) and its configuration is described as α or β. In A-type PAs, both linkages are indicated within the parentheses, but it is unnecessary to indicate the oxygen in the additional ether bond since it is obvious from the substitution pattern of flavan-3-ol extension units. For instance, according to this nomenclature, procyanidin dimer B1 is named as epicatechin-(4β→8)- catechin, dimer A1 is named as epicatechin-(4β→8, 2β→7)-catechin and trimer C1 is named as [epicatechin-(4β→8)]2-epicatechin.

The degree of polymerization (DP) is another variable factor for the structure complexity of PAs. Since PAs were first elucidated in 1960s, more than 200 oligomers with DPs of no more than 5 have been well identified and characterized (Weinges and Freundenberg, 1965). However, plant PA polymers usually have higher DPs. For example, the DP of PAs in cider apple skin and pulp range from 7 to 190, in brown or black soybean coat it can be up to 30 and even more (Guyot, *et al.*, 2001 and Takahata, *et al.*, 2001). Researchers also usually use mean degree of polymerization (mDP) to evaluate the molecular weight of PAs. For example, the mDP of PAs in pear is from 13 to 44 and in grape seeds it is from 2.3 to 16.7 (Ferreira, *et al.*, 2002 and Prieur, *et al.*, 1994). Interestingly, if all constitutive units and linkages can be distributed at random within a polymer, the number of possible isomers increases exponentially with the chain length, which can be calculated: a given N-DP with x types of constitutive units and y types of linkages is able to generate Nx (N-1) y possible isomers (Cheynier, 2005). The presence or absence of modifications of the monomeric flavan-3-ol units further enhances the structure complexity of PAs. Various methyl, acyl or glycosyl substituents of the monomeric units of PAs all occur in natural products. For example, the 3-hydroxyl group of the flavan-3-ol units of B-type PAs in grape seeds is usually esterified with gallic acid. A recent report even shows evidence for the existence of galloylated A-type procyanidins in grape seeds (Passos, *et al.*, 2007). Besides, free (-)- epigallo-catechin gallate is abundant in tea (Forrest and Bendall, 1969) and (-)-epicatechin gallate is contained in grape seeds. Furthermore, gallic acid substituted (+)-catechin and (+)-gallocatechin also

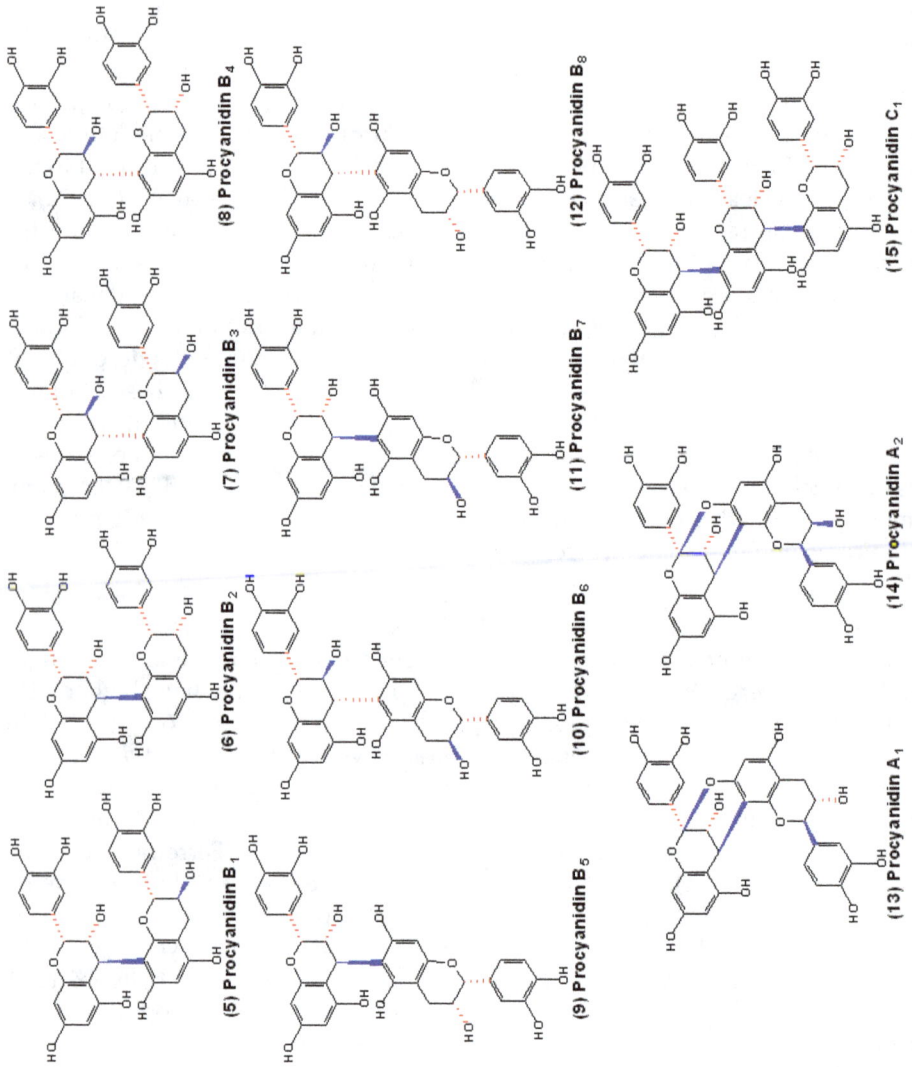

Figure 17.4: Structures of the B-type Procyanidins 5-12, 15 and A-type Procyanidins 13-14.

exist in grape (Lee and Jaworski, 1987 and Lee and Jaworski, 1990). In addition, glycosyl substituent is another common modification of the flavan-3-ol units, for the sugar is usually linked to the 3-hydroxyl group and also to the 5-hydroxyl group sometimes (Achmadi, *et al.*, 1994 and Iida, *et al.*, 2007). However, PA heterosides are less frequently detected than other flavonoid glycosides, such as anthocyanins. It is noteworthy to point that as early as the 1970s, modern NMR technology was used to investigate the presence of PA dimers and was used to demonstrate the phenomenon of their conformational isomerism and to propose two different forms of restricted rotation about IFL: one compact and the other extended (Fletcher, *et al.*, 1977). In a rather recent study, through NMR and molecular modeling, researchers revealed the three-dimensional (3D) structures of five B-type PA dimers and their conformational exchange in water and hydro-alcoholic media. Interestingly, they found that the compact form for these PA dimers dominated in most cases and further discussed this finding in relation with tannin-saliva protein interactions which rendered PAs the flavor of astringency in wine. It is well known that the structure determines the property for chemicals (Zucker, 1983). The structure complexity of PAs endows them with multiple biochemical properties, mainly including the interaction of proteins, the chelation of metals and the antioxidant bioactivity (Sarni-Manchado, *et al.*, 1999; Kennedy and Powell, 1985; Weber, *et al.*, 2007). Surely, these essential properties are not only the basis of their various protective functions for plants, but also their potential nutritional effects for human beings. General biosynthetic pathway of flavonoids PAs are synthesized as oligomeric or polymeric end products of one of several branches of the flavonoid pathway, which shares the same upstream pathway with anthocyanins (Figure 17.5). Because of their great contribution to red colour of plants, the biosynthesis of anthocyanins has been extensively investigated at the biochemical and genetic levels (Saito and Yamazaki, 2002; Springob, *et al.*, 2003). In recent years, the cloning and identification of a series of key structure genes relating to the formation of the direct precursors of PAs also lead us to a new era to understand the pathway. Generally, investigations with plant mutants have largely promoted the identification of key structure genes in this pathway, as shown in Table 17.2 (Abrahams, *et al.*, 2002; Cone, 2007 and Jende-Strid, 1993). As is usual for flavonoid formation, the first committed step of the pathway is the condensation and subsequent intramolecular cyclization of three malonyl-CoA molecules with one 4-coumaroyl-CoA molecule to produce a naringenin chalcone (Kreuzaler and Hahlbrock, 1972). This process is catalyzed by the ubiquitous plant enzyme chalcone synthase (CHS, EC 2.3.1.74, locus for AtCHS: At5g13930), which possesses extensive biological functions. CHS is found to mainly localize on the endoplasmic reticulum and tonoplast of the epidermal cells in Arabidopsis roots and also in the chloroplasts or chromoplast of grape berry cells (Saslowsky and Winkel-Shirley, 2001 and Tian, *et al.*, 2008). Additionally, the two kinds of precursors are derived from phenylalanine and acetyl-CoA, respectively.

The second step of the pathway is the isomerization of the naringenin chalcone to the naringenin, which can occur spontaneously, without enzymatic activity. However, chalcone isomerase (CHI, EC 5.5.1.6, locus for AtCHI: At3g55120) stereospecifically directs and greatly accelerates the intramolecular cyclization of chalcones to form the flavanones in the cytoplasm of plant cells (about 107 -fold

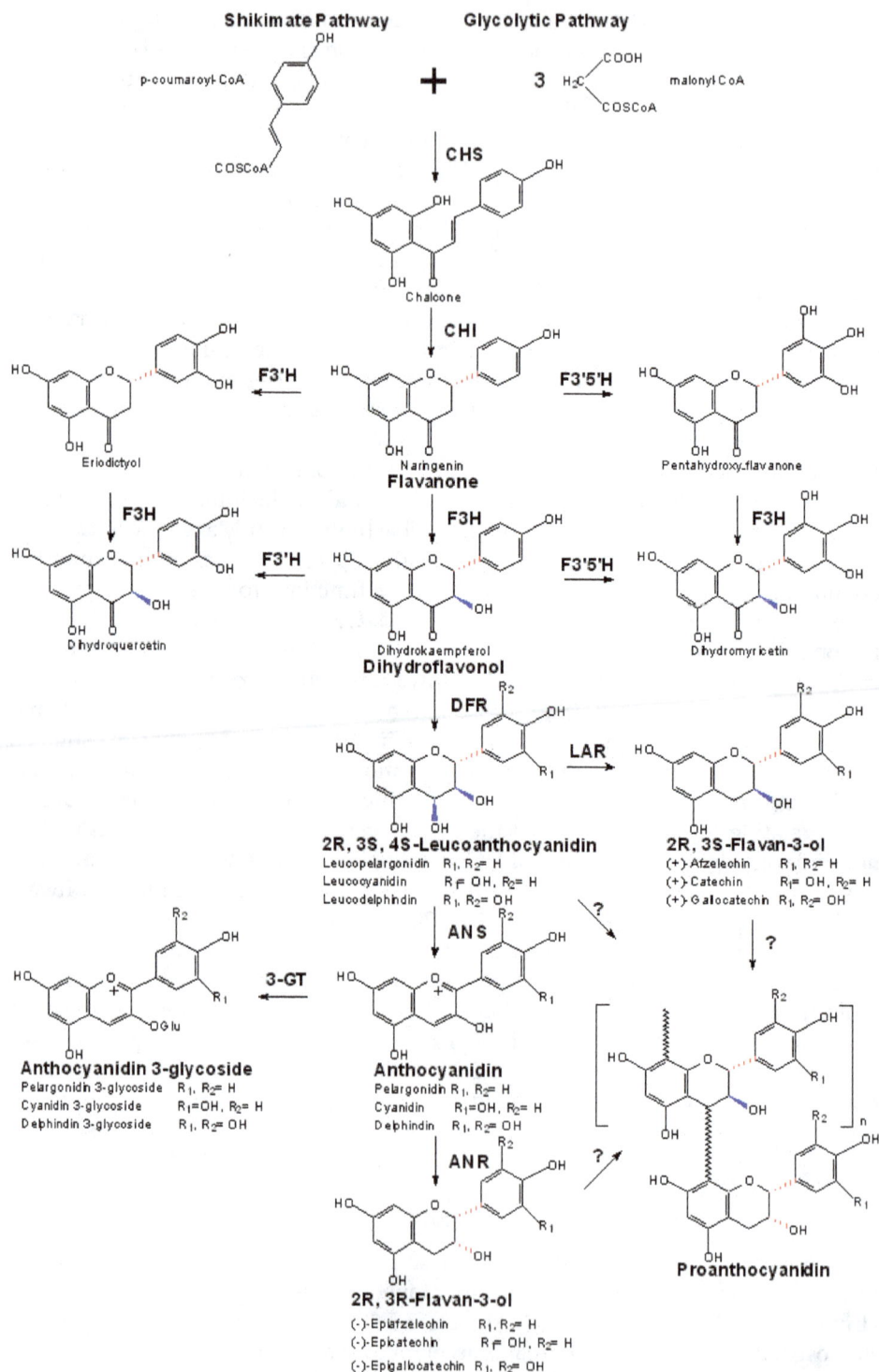

Figure 17.5: The General Flavonoid Pathway Leading to the Biosynthesis of PAs.

more efficiently), which serve as exclusive substrates for downstream reactions (Cain, *et al.*, 1997). In addition, the production of (2S)-flavanone by the catalysis of CHI is highly stereoselective, for the fact that pure CHI catalyzes the formation of (2S)-flavanone over 105 -fold faster than to the (2R)-flavanone and yields over 99.999 per cent of the (2S)-flavanone in the products (Bednar and Hadcock, 1988). Following studies of the reaction mechanism of CHI also support the statement of that CHI controls the preferred formation of the biologically active (2S)-flavanones. For example, only the (2S)-isomer bounds in the CHI active site and the steric clash prevents the formation (2R)-naringenin (Jez, *et al.*, 2000 and Jez and Noel, 2002). Furthermore, the localization of CHI is also mainly on the endoplasmic reticulum and tonoplast of the epidermal cells in Arabidopsis roots, just like CHS. Hither, the basic skeleton of all flavonoids which consist of three C6-C3-C6 aromatic rings has been generated through the catalysis of CHS and CHI.

Table 17.2: Partial Mutants Used for Identifying Structure Genes in the General Biosynthetic Pathway of Flavonoids in Arabidopsis Thaliana, Maize and Barley

Structure Proteins	Arabidopsis	Plant Species	
		Maize	Barley
CHS	tt4	c2	
CHI	tt5	chi	ant30
F3'H	tt7	pr1	ant1, ant2, ant 5
F3H	tt6	fht1	ant17, ant22
DFR	tt3	a1	ant18
ANS	ans/tt18	a2	
ANR	ban		
Transporter	tt12		

The following steps are related to the "Gird" section with the B-ring hydroxylation. Flavonoid 3'-hydroxylase (F3'H, EC 1.14.13.21, locus for AtF3'H: At5g07990) or flavonoid 3',5'-hydroxylase (F3'5'H, EC 1.14.13.88, no F3'5'H gene in Arabidopsis thaliana) can catalyze the conversion of naringenin into eriodictyol or pentahydroxyflavanone, respectively. These two enzymes both belong to the P450 monooxygenase superfamily of membrane-bound hemoproteins and work dependent upon NADPH and O2. All of the three (2S)-flavanones mentioned above can be oxidated to yield dihydroflavonols (dihydrokaempferol, dihydroquercetin and dihydromyricetin, respectively) stereospecifically by flavanone 3-β-hydroxylase (F3H, EC 1.14.11.9, locus for AtF3H: At3g51240), which is a non-heme iron enzyme and acts dependent upon Fe2+, O2, 2-oxoglutarate and ascorbate (Britsch, *et al.*, 1993 and Lukaèin, *et al.*, 2000). However, dihydrokaempferol can also be the potential substrate of F3'H or F3'5'H, which can convert it to dihydroquercetin or dihydromyricetin, respectively. Thus, both flavanones and dihydroflavonols can be hydroxylated by F3'H and F3'5'H and the presence or absence of the two enzymes determines the B-ring hydroxylation pattern of dihydroflavonol, as well as the subsequent monomers in PAs and anthocyanins.

Following these, dihydroflavonol 4-reductase (DFR, EC 1.1.1.219, locus for AtF3H: At5g42800), depending upon NADPH, reduces dihydroflavonols (dihydrokaempferol, dihydroquercetin and dihydromyricetin) to form leucoanthocyanidins (also named flavan-3,4-diols, leucopelargonidin, leucocyanidin and leucodelphinidin, respectively) (Martens, *et al.*, 2002). Different from the key enzymes in the early steps of the flavonoid pathway, DFR belongs to the plant reductase-epimerase-dehydrogenase (RED) superfamily. It is also worthwhile to point out that the conversion of (2R,3R)-dihydroflavonols to (2R,3S,4S)-leucoanthocyanidins by DFR is also stereospecific (Petit, *et al.*, 2007). Anthocyanidin synthase (ANS, EC 1.14.11.19, Locus for AtANS: At4g22880), also known as leucoanthocyanidin dioxygenase (LDOX), is also a non-haem Fe^{2+}, O_2, 2-oxoglutarate and ascorbate dependent oxygenase (Nakajima, *et al.*, 2001). ANS used to be recognized as the first pivotal enzyme of anthocyanin formation, but now it has been considered to play a significant role in PA biosynthesis as well (Pelletier, *et al.*, 1997). In one pathway branch, leucoanthocyanidin molecules are oxidized by the catalysis of ANS to form coloured anthocyanidins (pelargonidin, cyanidin and delphinidin, respectively) firstly (Saito, *et al.*, 1999). These unstable anthocyanidins could be further converted into colourless (2R, 3R)-flavan-3-ols [(-)- epiafzelechin, (-)-epicatechin and (-)-epigallocatechin, respectively] by the action of anthocyanidin reductase (ANR, EC 1.3.1.77, locus for BAN: At1g61720) (Xie, *et al.*, 2003 and Xie, *et al.*, 2004). Recently, researchers demonstrated that ANS from Arabidopsis thaliana is also able to convert its natural substrate leucoanthocyanidin to cis- and trans- dihydroquercetin *in vitro*, as well as quercetin, showing its flavonol synthase (FLS) activity with leucoanthocyanidin as substrate (Welford, *et al.*, 2001). A more recent report reveals that beyond the oxidation of leucoanthocyanidins, ANS from Gerbera hybrida can also catalyze the oxidation of (+)- catechin to form an novel 4,4-dimer (93 per cent) and the conversion of (+)-catechin to cyanidin (7 per cent) *in vitro*, suggesting its FLS activity with (+)-catechin as substrate (Wellmann, *et al.*, 2006). In another pathway branch, leucoanthocyanidins can be converted into (2R, 3S)-flavan-3-ols [(+)- afzelechin, (+)-catechin and (+)-gallocatechin, respectively] by leucoanthocyanidin reductase (LAR, EC.1.17.1.3, no LAR gene in *Arabidopsis thaliana*) (Tanner, *et al.*, 2003). Interestingly, some plants such as *Vitis vinifera*, *Gossypium arboretum* and *Gossypium raimondii*, contain two homologous LAR genes, whereas other plants such as *Hordeum vulgare*, *Phaseolus coccineus*, *Pinus taeda*, *Vitis shuttleworthii* only have a single LAR gene in the cells (Bogs, *et al.*, 2005). No LAR gene has ever been found in *Arabidopsis thaliana*, so PAs in this plant are composed of only (-)-epicatechin units. Both LAR and ANR are two members of the isoflavone reductase-like (IFR-like) group of the plant RED super family, which are both localized in the cytosol (Bogs, *et al.*, 2005 and Pang, *et al.*, 2007). Thus, the two familiar potential precursors, 2,3-cis-2R,3R-(-)- epicatechin and 2,3-trans-2R,3S-(+)-catechin are synthesized from two distinct pathway branches by two different substrates and two distinct enzymes, although the difference between these two potential precursors is only in the cis- or trans- stereochemistry at the C2 and the C3 positions of the C-ring. However, there are still no exact answers at present to the existence and the formation of the biosynthetic ent-flavan-3-ols, including 2,3-cis-2S,3S-(+)-epicatechin and 2,3-trans-2S,3R-(-)- catechin. Additionally, the unstable anthocyanidins can be also glycosylated at the 3-O-position to produce their corresponding anthocyanidin 3-glycosides by the catalysis of UDP-glucose: anthocyanidin/flavonoid 3-glucosyltransferase (UFGT or 3-GT, EC 2.4.1.115, locus for AtUFGT: At5g54060), leading to the absolute branch pathway of anthocyanin biosynthesis (Ford, *et al.*, 1998 and Kim, *et al.*, 2006).

Both (2R, 3R)-flavan-3-ols and (2R, 3S)-flavan-3-ols, as well as (2R, 3S, 4S)-flavan-3,4-diols are proposed as the potential precursors of PAs. However, the identity of the final "enzyme" which catalyzes the polymerization reaction of these precursors to form PAs and the exact mechanism whereby the elementary units of flavanols are assembled *in vivo* remain unknown. Thus, a unilateral genetic or molecular genetics research approach seems to be not enough and the debate concerning the operation of an enzymatic or nonenzymatic mechanism for PA condensation still continues. Recently, a plasma membrane H+ -ATPase (AHA10, locus for AtAHA10: At1g17260) is found to be required for the formation of PAs in the seed coat endothelium of *Arabidopsis thaliana*, the mutant of which can disrupt both the PA biosynthesis and the vacuole biogenesis. Further studies support the hypothesis that AHA10 or even other P-type proton pumps in plants could help acidify cytoplasmic or vacuolar compartments (Baxter, *et al.*, 2005). According to the proposed mechanism of acid catalysis of PA formation, flavan-3-ol (terminal unit) can be added to a quinone methide or its protonated carbocation (the precursor of extension unit) nonenzymatically under acidic environment. Thus, the founding of AHA10 is essential to promote such nonenzymatic polymerization mechanisms for PAs.

Transport Factors of Proanthocyanidin Precursors

Generally, the known enzymes in the biosynthetic pathway of flavonoid are observed to mostly locate on the endoplasmic reticulum membranes or in the cytoplasm. However, it is well known that PAs accumulate in the vacuole. Thus, the intracellular transport of PA precursors, flavan-3-ols and/or flavan-3, 4-diols from the site of synthesis to the site of storage is a crucial problem in PA biosynthesis.

Up to now, only two genes involved in transport processes required for PA biosynthesis have been identified from the seed coat of Arabidopsis thaliana by genetic analysis methods. Both of the genes are obtained from the transparent testa mutations (light-coloured seeds) which are caused by reductions of PA deposition in vacuoles of endothelial cells and are named as tt12 and tt19, respectively. However, the functional relation between TT12 and TT19 remains unclear until now (Debeaujon, *et al.*, 2001 and Kitamura, *et al.*, 2004). The TT12 gene (locus: At3g59030) encodes a protein with 12 transmembrane segments exhibiting similarity to prokaryotic and eukaryotic secondary MATE (multidrug and toxic compound extrusion) transporters. This gene is expressed specifically in the endothelial layer of the developing seed coats, suggesting that TT12 may be involved in the accumulation of PA precursors in vacuoles. Further studies reveal that flavan-3-ols and PAs are absent and quercetin-3-O-rhamnosides are reduced in the vacuoles of tt12 mutant seeds and this transporter is localized to the tonoplast. Finally, TT12 is demonstrated to act as a vacuolar flavonoid/H+ -antiporter on the vacuolar membrane of PA synthesizing cells of the seed coat. It mediates the MgATP-dependent transport of cyanidin-3-Oglucoside *in vitro* but not the aglycones cyanidin and epicatechin. Furthermore, TT12 transports neither glycosylated flavonols and procyanidin dimers nor nonglycosylated flavan-3-ol monomers and dimers (Marinova, *et al.,* 2007). As a consequence, a series of new questions about the biosynthesis of PAs are put forward. The TT19 (locus: At5g17220) gene is a member of the Arabidopsis thaliana glutathione Stransferase (GST) gene family, which is required for transferring anthocyanins into vacuoles. Because the deposition pattern of PA precursors in the tt19 mutant is also different from that in the wild type, TT19 may participate in the PA pathway with an unclear role in flavonoid accumulation. Similar GST genes are also found in other plants. In maize, the Bronze-2 gene encoding a GST is related to the deposition of red and purple pigments (anthocyanins) in the vacuoles of maize issues (Marrs, *et al.,* 1995). In petunia, the AN9 gene encoding a GST involved in anthocyanin transport is quite similar to the TT19 gene functionally, which can complement the Arabidopsis thaliana tt19 mutation with respect to allowing vacuolar uptake of anthocyanins, but cannot restore the deposition of PAs (Mueller, *et al.,* 2000). Recently, four specific genes are identified as systematically co-expressed with anthocyanin accumulation in grape berries. One of them is a type-I GST gene which is orthologous to the Bronze-2 gene and the AN9 gene and encodes a flavonoid binding GST protein required for vacuolar transport of anthocyanins (Ageorges, *et al.,* 2006). However, no such MATE transporter factors have been found in grapes until now. Interestingly, some very recent research shows the evidence for a putative flavonoid translocator similar to mammalian bilitranslocase (BTL, TC 2.A.65.1.1) in grape (*Vitis vinifera* L.) berries of cultivar Merlot (red grapes) during ripening, which is responsible for the transport of anthocyanins (Braidot, *et al.,* 2008).

Potential Polymerization Mechanisms for Proanthocyanidins

Although the biosynthesis of the precursors of PAs has been well known for several years, a clear polymerization mechanism is still unknown. Most of models state that the electrophilic C4 position of the extension unit (flavan-3, 4-diol) condenses with the nucleophilic C8 or C6 position of the start/terminal unit (flavan-3-ol) to produce PAs (Botha, *et al.,* 1981 and Delcour, *et al.,* 1983). However, these models neglect the fact that the enzymatic formation of flavan-3,4-diol is stereospecific 2,3-trans, whereas lots of the extension units are 2,3-cis in plants. To support this, Porter characterized 58 procyanidin oligomers from plants of more than two dozen different species and found that (-)-epicatechin accounted for 55 per cent and 81 per cent as the start and extension units, respectively (Porter, 1988 and Porter, 1993). This paradox has been standing for a long time and partially been solved by the functional identification of BANYULS in recent years. Up to now, some highly speculative solutions have been developed for the mechanism of PAs condensation. In one previously proposed route, 2R, 3S, 4S-leucoanthocyanidins are first converted into 2R, 3Squinone methides, which can be used as the 2R,3S-extension units directly. Meanwhile, they can also be converted into 2R, 3R-quinone methides via the flavan-3-en-3-ol intermediates, which can be condensed as the 2R,3R- extension units directly (Hemingway and Laks, 1985). Furthermore, these trans- or cis- quinone methides could be converted to their corresponding carbocations which could be attacked by (+)- catechins or (-)-epicatechins directly to produce the PAs (Figure 17.6). However, these hypotheses are lack of direct experimental proof *in vivo* until now and the only experimental basis comes from the reports of Creasey and Swain, who first chemically synthesized leucocyanidin from (+)- dihydroquercetin and make it possible to synthesize procyanidins through the condensation reactions of leucocyanidin with either (+)-catechin or (-)-epicatechin *in vitro* (Creasey and Swain, 1965).

In another model, flavan-3-ols [(+)-catechins or (-)-epicatechins] are converted to their corresponding quinone methides by the catalysis of polyphenol oxidase (PPO). These quinone methides can be further converted to carbocations via flavan-3-en-3-ol intermediates or be reduced to carbocations through coupled non-enzymatic oxidation and these carbocations can be accepted as the direct extension units. Interestingly, o-quinones are also supposed to be derived from (+)-catechins or (-)-epicatechins by enzymatic or nonenzymatic reduction. And also via flavan-3-en-3-ol intermediates, o-quinones could be converted to their corresponding carbocations which may participate in the condensation reaction (Figure 17.7). However, this hypothesis is highly speculative and does not have any direct experimental support.

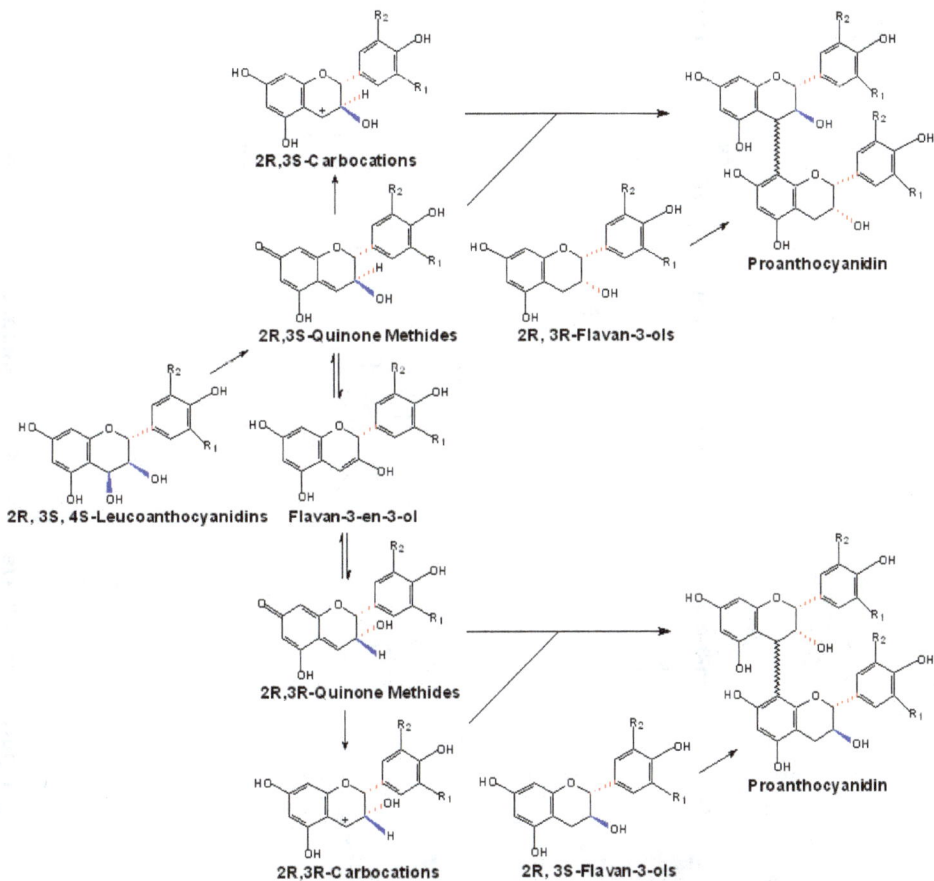

Figure 17.6: Putative Route of the Generation of Quinone Methides or Carbocations from Flavan-3,4-diols.

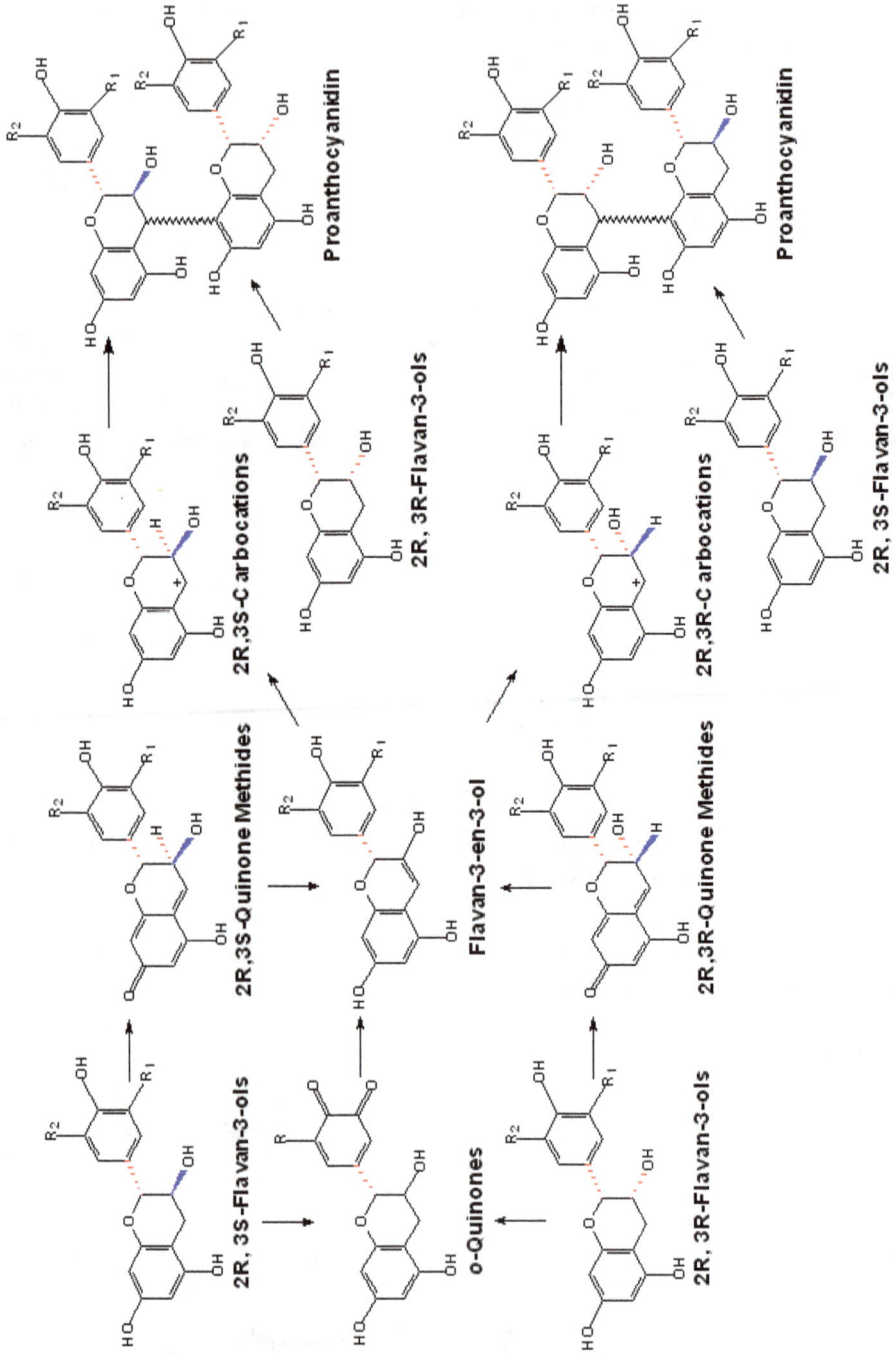

Figure 17.7: Putative Route of the Generation of Quinone Methides or Carbocations from Flavan-3-ols.

The general localization of PPO is on the plastids (or chloroplasts), whereas the formation and accumulation of PAs is in the vacuoles (Vaughn, *et al.*, 1988). Therefore, the model of PPO mediated PA polymerization requires the existence a novel form of PPO with alternative localization. Recently, aureusidin synthase, a flavonoid biosynthetic PPO that catalyzes the oxidative formation of aurones from chalcones in snapdragon (Antirrhinum majus) and is responsible for the yellow colouration of flowers, is found to localize within the vacuole lumen, showing us the variability of the PPO localization (Ono, *et al.*, 2006).

However, in the PPO mediated oxidative polymerization of flavan-3-ols, researchers do not obtain PAs with natural structures, but rather a series of oxidative flavan-3-ols oligomers with particular configurations. For example, in the oxidation of (+)-catechins in aqueous systems carried out using grape PPO as catalysts, some yellow pigments and colourless products of oligomeric (+)-catechins are formed (Guyot, *et al.*, 1996). In another report of Oszmianski and Lee, the enzymatic oxidation of (+)-catechin, chlorogenic acid and their mixture with PPO also produce unnatural dimers and other polymers of (+)- catechin, polymers of chlorogenic acid and copolymers of (+)-catechin and chlorogenic acid, respectively (Oszmianski and Lee, 1990). Here, such oxidative dimers of flavan-3-ols are identified as dehydrodicatechins, which are usually formed as a result of enzymatic oxidation, chemical oxidation or autoxidation of flavan-3-ols (Oszmianski, *et al.*, 1996 and Sun and Miller, 2003). Generally, the oxidative dimers of flavan-3-ols linked by C6'→C8 or C6'→C6 IFL are classified as B-type dehydrodicatechins, resulting from the repeated condensation reactions between the A-ring of the lower unit and the B-ring of the upper unit through a mechanism of so-called "head to tail" polymerization. Correspondingly, the oxidative dimers which contain additional C-O-C ether-type IFL are classified as A-type dehydrodicatechins, as shown in Figure 17.8. Furthermore, a new TT10 gene is obtained through the Arabidopsis thaliana tt10 mutant, which encodes a protein that may be involved in the oxidative polymerization of flavonoids and functions as a laccase-type flavonoid oxidase. Interestingly, the major products resulting from TT10 activity are also yellow quinone-methide epicatechin dimers and trimers, which are not characterized but possibly related to dehydrodicatechins (Pourcel, *et al.*, 2005). Therefore, the model of PPO mediated PA polymerization cannot obtain direct experimental support

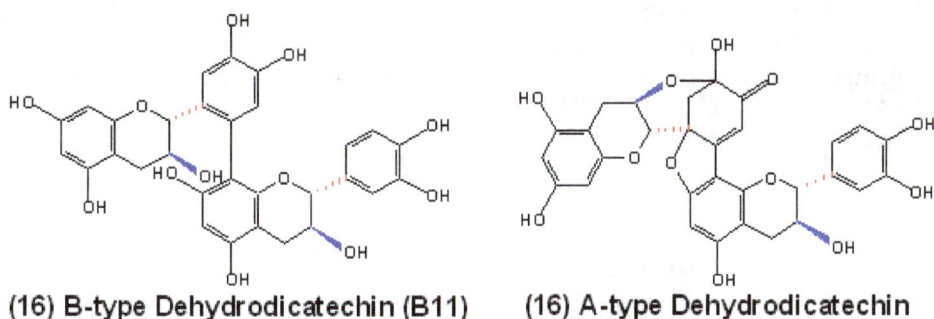

(16) B-type Dehydrodicatechin (B11)   (16) A-type Dehydrodicatechin

Figure 17.8: Structures of PPO Mediated Oligomeric Flavan-3-ols: B-type and A-type Dehydrodicatechins.

from the previous studies and PAs might not be simply formed through the PPO-mediated condensation mechanism.

Anthocyanins and anthocyanidins are other groups of potential substrates for PA polymerization. On one hand, anthocyanins are shown to condense with PA monomers (flavan-3-ols), oligomers or polymers in wine to change its tannin composition, as well as its colour and mouth-feel (Vidal, *et al.*, 2002 and Remy-Tanneau, *et al.*, 2003) and on the other hand, anthocyanidins are mainly present as their corresponding flavylium ions under acidic conditions. These ions may be converted to quinone methides and then to carbocations depending upon PPO (Yokotsuka and Singleton, 1997). Carbocations can condense with each other as the extension units and also condense with flavan-3-ols through non-enzymatic oxidation (Escribano-Bailon, *et al.*, 1996). Furthermore, the biogenesis of 2,3-cis-2R,3R-PAs from 2,3-trans-2R,3R-dihydroflavonols may also be accounted for by tautomerism between quinone methide and flav-3-en-3-ol intermediates (Hemingway and Laks, 1985). Although flavan-3-en-3-ols may be released from ANR theoretically, no dimeric or oligomeric PAs are detected to form with ANR reactions *in vitro*.

Regulation of Proanthocyanidins Biosynthesis

Similar to other secondary metabolism pathways in plants, the biosynthetic pathway of PA is under complex control by multiple regulatory genes at the transcriptional level. According to gene structure, these regulatory genes are categorized into six different families: Myc transcriptional factors (encoding basic helix–loop-helix proteins, bHLH), Myb transcriptional factors, WD40-like protein, WRKY transcription factors, MADS homeodomain genes and TFIIIA-like proteins "WIP", as summarized in Table 17.3. Generally, Myc genes, together with other regulatory genes such as Myb genes, make a widely control of the biosynthesis of PAs, as well as some other flavonoid end products like anthocyanins (Nesi, *et al.*, 2000; Ramsay, *et al.*, 2003; Nesi, *et al.*, 2001 and Borevitz, *et al.*, 2000). WD40-like protein is required for the accumulation of anthocyanins and PAs, as well as the formation of root hair and trichome (Walker, *et al.*, 1999). Recently, researchers found that in Arabidopsis thaliana, several Myb and Myc proteins form ternary MYB-BHLH-WDR (MBW) complexes with WD-repeat proteins that regulate the transcription of genes involved in biosynthesis of anthocyanins and PAs (Gonzalez, *et al.*, 2008). For example, TT2 (Myb family), TT8 (Myc family) and TTG1 (WD-like protein) form a transcriptional complex capable of directly activating the expression of BAN (Baudry, *et al.*, 2004). WRKY transcription factor is a zinc finger-like protein which acts downstream of WD40-like protein and also affects the accumulation of PAs (Johnson, *et al.*, 2002). Furthermore, a recent study also suggests that the WRKY transcription factor is directly regulated by Myb transcription factors, such as TT2, etc (Ishida, *et al.*, 2007). MADS homeodomain genes directly regulate BAN and may act upstream of other regulatory genes, but they affect the accumulation of PAs regionally (Nesi, *et al.*, 2002 and Lalusin, *et al.*, 2006). WIP proteins specifically control the assembly of PAs polymer, rather than the formation of monomeric flavan-3-ol units (Sagasser, *et al.*, 2001). Additionally, WIP proteins also interact with developmental genes, such as BAN, to control PA biosynthesis. However, besides these six major families, there are also regulatory genes of other types which also have direct or indirect influences on the genes of biosynthesis of anthocyanins and / or PAs, such as ANTHOCYANINLESS2 (ANL2) which is a homeobox gene of the homedomain-leucine zipper (HDZIP IV) family, AtDOF4; 2 which is a member of DNA-binding-with-one-finger (DOF) transcription factor family, etc (Kubo, *et al.*, 2008 and Skirycz, *et al.*, 2007).

Table 17.3: Transcription Factors Involved in Proanthocyanidin Biosynthesis in *Arabidopsis thaliana*

Families	Genes	Locus	Gene Regulated
MYC	TT8	At4g09820	BAN, DFR
	GL3/EGL3	At5g41315/At1g63650	DFR
MYB	TT2	At5g35550	BAN, DFR, ANS, TT12, AHA10, TT8, TTG2
	PAP1/PAP2	At1g56650/At1g66390	PAL, CHS, DFR, ANS, TT8
WD40	TTG1	At5g24520	BAN, DFR, TT8, TTG2
WRKY	TTG2	At2g37260	BAN, TT12
MADS	TT16	At5g23260	BAN, TT2
WIP	TT1	At1g34790	BAN

On the other hand, these regulatory genes can also be classified into three different categories according to their functions: (1) regulation of the general genes in the pathway of flavonoid biosynthesis down to anthocyanidins, (2) modulation of the branch pathway of PAs, including the biosynthesis, transport and polymerization of the subunits of PAs; and (3) control of the developmental processes for the generation of the cells accumulating PAs specially. So far, most of the regulatory genes and regulation mechanism have been characterized in Arabidopsis thaliana by using several mutants with low level of PAs or with low levels of both anthocyanins and PAs. In Arabidopsis thaliana, the first category of transcriptional factors which regulate the general flavonoid biosynthesis down to anthocyanidins includes PAP1 and PAP2 (Myb family), GL3 and EGL3 (Myc family) and TTG1, which are connected in a hierarchical network to regulate flavonoid synthesis (Zhang, *et al.*, 2003). The second category of regulatory genes in Arabidopsis thaliana includes TT2, TT8 and TTG1, modulating the expression of the late flavonoid biosynthetic genes, such as DFR, ANS, BAN, the transport factor TT12 and the proton pump AHA10 (Baudry, *et al.*, 2006). The third category of transcription factors includes TT1, TT16 and TTG2 in Arabidopsis thaliana and they regulate organ and cell development for PA deposition, as well as the transcription of PA specific genes. It is worth pointing out that a series of Myb and Myc genes are found to work as negative regulators of the biosynthetic pathway of anthocyanins and PAs in Arabidopsis thaliana. MYBL2 encodes a R3- MYB-related protein that interacts with MBW complexes and directly modulates the expression of flavonoid target genes. The loss of MYBL2 activity leads to a dramatic increase in the accumulation of anthocyanins, while overexpression of MYBL2 in seeds inhibits the biosynthesis of PAs, showing its negative regulation on the pathway (Dubos, *et al.*, 2008). MYB4 is a R2R3-MYB that can interact with BHLH proteins controlling the expression of flavonoid structural genes and repress the transcription of early genes in phenylpropanoid metabolism (Zimmermann, *et al.*, 2004 and Jin, *et al.*, 2000). Myc negative regulators affecting flavonoid TTG1- dependent pathways also exist in Arabidopsis thaliana. BHLH32 in Arabidopsis negatively regulates root hair formation, anthocyanin accumulation and phosphoenolpyruvate carboxylase kinase (PPCK) expression, all of which are induced by (inorganic phosphate) Pi starvation. For example, BHLH32 seems to act as a negative regulator of DFR in Pi-sufficient conditions (Chen, *et al.*, 2007).

The relationship between the transcription factors and the structure proteins in Arabidopsis thaliana is illustrated in Figure 17.9. As shown, a complex regulatory network controls PA accumulation in plants (Broun, 2005). Almost all of these transcription factors regulate the expression of BAN, suggesting its essential function among the structure genes in the biosynthetic pathway. Meanwhile, TT2, TT8 and TTG1 play a central role in the regulation of the PA biosynthesis and have complicated interactions. For example, TT8 controls its own expression in a feedback regulation involving TTG1 and homologous MYB (TT2, PAP1) and MYC (TT8, GL3 and EGL3) factors. However, the direct or indirect regulation of these regulators by internal or external signals leading to controlled responses is more perplexing (Endt, *et al.*, 2002). In plants, it is well known that fertilization is a key signal to start the PA biosynthesis, as it is required for the differentiation of PA-accumulating cells and the activation of the promoter of BAN, as well as most of other specific structure genes (Devic, *et al.*, 1999 and Debeaujon, *et al.*, 2003). Additionally, more and more environmental factors are found to have significant influences on the biosynthesis of PA and anthocyanins in plants, such as light stress and shading, atmospheric change (CO_2, N_2, O_2 and O_3), temperature (day and night), exogenous plant hormones (abscisic acid, naphthaleneacetic acid and ethylene), infection of pathogens (bacterial and fungal), UV (UV-A and UV-B) or solar radiation, nitrogen, water and Pi deficiency, etc. These environmental factors do not only manipulate the expression of the structure genes, but also widely affect the functions of the transcription factors (Cominelli, *et al.*, 2008; Peltonen, *et al.*, 2005; Mori, *et al.*, 2005; Ubi, *et al.*, 2006; Jeong, *et al.*, 2004; El-Kereamy, *et al.*, 2003; Punyasiri, *et al.*, 2004; Kortekamp, 2006; Hirner, *et al.*, 2001; Ryan, *et al.*, 2002; Lea, *et al.*, 2007; Castellarin, *et al.*, 2007). Similar regulatory genes are also identified in maize, which are coordinately active at the transcriptional level especially in the aleurone and belong to the first three types: Myc transcriptional factors (R, B-Peru, Sn, Lc), Myb transcriptional factors (Cl, P) and WD40-like protein (PAC1) (Taylor and Briggs, 1990; Hartmann, *et al.*, 2005; Carey, *et al.*, 2004). However, until now, only one major class of regulatory genes related to PA biosynthesis has been found in grape: the Myb transcriptional factors. VvMYBPA1, which was cloned and characterized from grapevine (*Vitis vinifera* L.) tissues of cultivar Shiraz (red grapes), controls the expression of the pathway genes of PAs including both VvLAR1 and VvANR, but do not activate the promoter of VvUFGT which is considered as an anthocyanin specific gene. That means VvMYBPA1 is specific to regulate the biosynthesis of PAs in grape (Bogs, *et al.*, 2007). Another group of Myb family transcriptional factors in grape named as VvMYBA, including VvMYBA1, VvMYBA2 and VvMYBA3, are only expressed after véraison (grape colour change) and demonstrated to regulate

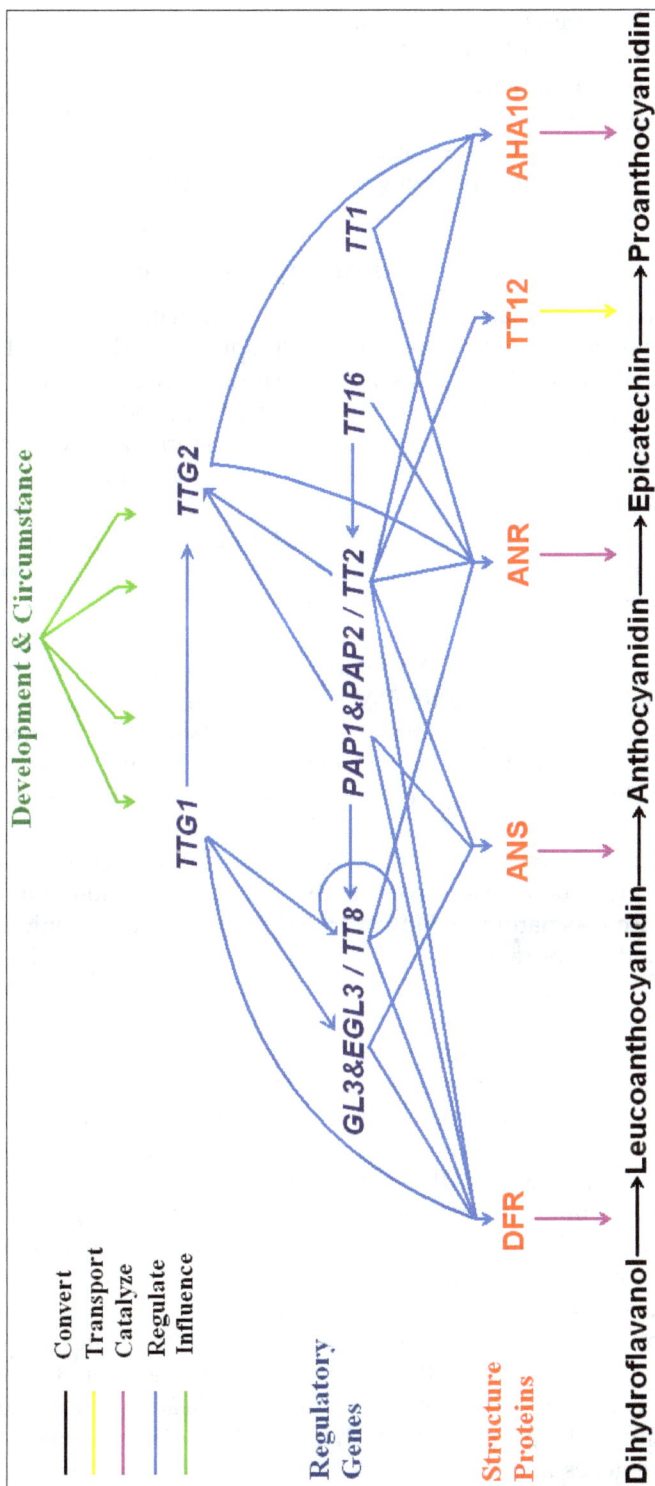

Figure 17.9: Interactions among the Transcription Factors and the Structure Proteins in *Arabidopsis thaliana*.

anthocyanin biosynthesis during ripening (Kobayashi, *et al.*, 2005; Walker, *et al.*, 2007; Kobayashi, *et al.*, 2002). VvMYB5a is the third grape transcriptional factor that is expressed mainly at the early steps of berry development (prior to véraison) and affects the metabolism of a series of polyphenols, including flavanols, anthocyanins, PAs and even lignins. It is thus suggested that VvMYB5a regulates a series of branches of phenylpropanoid pathway in grapevine, but not only that of flavonoids (Deluc, *et al.*, 2006).

Genetic Manipulation of Proanthocyanidin Biosynthesis

PAs have been recognized to play a significant role in the prevention of pasture bloat and the enhancement of bypass proteins (Goplen, *et al.*, 1980 and Lees, 1992). Thus, the induction of PA biosynthesis in some forages, such as alfalfa (Medicago sativa L.), can largely solve the problems of pasture blot and protein loss. On the other hand, PAs also have negative influences on non-ruminant livestock, because of their properties of non-specific interaction with protein and complexation of metal ions (Robbins, *et al.*, 1987 and Sarwar, *et al.*, 1981). As a result, the reduction of PA content in some crops, such as canola (*Brassica napus* L.), can also increase their values for non-ruminant livestock feed. However, it is difficult to manipulate the content of PAs in plants by using conventional breeding strategies, such as induced mutation, somaclonal variation and somatic hybridization. Genetic manipulation of PA biosynthesis has its unique advantage and a lot of the efforts have been drawn to manipulate the PA biosynthesis in the forage legumes, such as alfalfa and Lotus corniculatus (bird's foot trefoil). Generally, the genes of both structure enzymes and transcription factors are used in the attempts to genetically engineer PA biosynthesis (Ray, *et al.*, 2003; Wang, *et al.*, 2006; Carron and 1994; Bavage, *et al.*, 1997; Robbins, *et al.*, 2003; Paolocci, *et al.*, 2005; Paolocci, *et al.*, 2007). Three maize anthocyanin regulatory genes C1 (Myb family), Lc and B-peru (Myc family), are all introduced respectively to the genetic alternation of alfalfa which does not produce PAs naturally. However, only in the Lc- transgenic alfalfa both anthocyanins and PAs are accumulated in leaf tissue under stress conditions of high light intensity or low temperature. These results indicate that PA biosynthesis can be stimulated by some Myc transcription factors in alfalfa and offer us a feasible way to obtain bloat-safe alfalfa forage for ruminants. In contrast, much more efforts have been put into the genetic engineering of that in *Lotus corniculatus*, a common forage legume which naturally produces PAs. Antisense down-regulation of DFR expression successfully reduces its PA content of up to ≈80 per cent in hair root cultures, but overexpression of DFR in hairy roots results in variable influences on PA content and its monomer composition. In another research, over-expression of a maize anthocyanin regulatory gene Sn (Myc family) leads to unexpectedly increase in PA accumulation and to dramatically increase in the number of tannin-containing cells in its tissue. Further studies suggest that light and the transgene Sn synergistically affects the PA biosynthesis, as well as the structure genes in the pathway, such as DFR, ANS, ANR and LAR. Beside these, a lot of attempts have been performed in a series of model plants, such as *Arabidopsis thaliana*, tobacco and *Medicago truncatula*, the purpose of which is mainly to clarify the exact functions of the structure genes and the transcription factors, or their functional relations

(Sharma, *et al.*, 2005; Xie, *et al.*, 2006). In addition, it is worthwhile to mention that although grapevine is not suitable for being manipulated as the target plant, it is an ideal choice as the gene source, especially for the regulatory genes. Actually, some of Myb transcription factors in grapevine have already been introduced to other model plants. For example, the ectopic expression of VvMYB5a in tobacco has been used in the functional identification of this gene and the VvMYBPA1 gene can complement the Arabidopsis TT2 gene in its PA deficient mutant.

Outcome (conclusions)

In the past few years, our knowledge of the mechanism of PA biosynthesis has been changed by a series of advances. Nowadays, researchers have a deep understanding of the structure diversity of PAs, find and identify a host of structure genes and their regulatory factors and even genetically manipulate the accumulation of PAs or other end products of flavonoid pathway in some plants. However, a number of important questions remain unclear, some of which have been discussed emphatically by previous researcher, as illustrated below. What mechanism controls the flux at the interface between the anthocyanin and PA pathways, the known structure proteins and/or the regulatory factors? What are the functional relationships between them? Do the circumstance factors have influence on the flux control? Is LAR unnecessary for the biosynthesis of PAs, as its absence in Arabidopsis thaliana, as well as its product of (+)-catechin? Is this phenomenon a sign of evaluation for plants, or not? How are LAR and ANR regulated at the protein level (translation level)?

What are the exact extension units of PAs, flavan-3-ols, flavan-3, 4-diols or their derivates? Does the monomer availability mainly restrict the diversity of the PA composition? How are the ent-flavan- 3-ols formed, by nonenzymatic isomerization or enzymatic alternation? How exactly are the PA precursors transported into the vacuole? Is TT12 the only MATE transporter in plants? Does it cooperate with TT19? How are these glycosylated cyanidin monomers used in vacuole? Is there any homologous transporter which allow vacuolar uptake of nonglycosylated flavan-3-ol monomers as the direct precursors for PAs? How PA oligomers or polymers are actually assembled *in vivo*? Is it a nonenzymatic process or operated by one or more structure proteins? If it belongs to an enzymatic mechanism, what are the exact compositions of these enzymes? Is it a polymerase for the direct condensation, or a PPO for the formation of quinone methides? Furthermore, what is the location for the condensation within the vacuole? We speculate that the answers to the above questions will be obtained depending upon the applications of the biochemical, genetic and biogenetic technologies in no distant further, for example, the screening of the mutants specific to PA condensation in Arabidopsis thaliana or barley, the preparation and application of the antibodies of the key enzymes or transcription factors involved in the biosynthesis of PAs, the application of technologies of protein-protein interaction, etc (Abrahams, *et al.*, 2002: Tian, *et al.*, 2006; Hunt, *et al.*, 2008). On the other hand, further achievements in other areas of PA studies will also facilitate and accelerate our understanding of this group of polyphenolic secondary metabolites to extent our eyeshot of the natural products and to improve our daily life.

Goldbeater's Skin Test

When goldbeater's skin or ox skin is dipped in HCl, rinsed in water, soaked in the tannin solution for 5 minutes, washed in water and then treated with 1 per cent $FeSO_4$ solution, it gives a blue black colour if tannin was present.

Ferric Chloride ($FeCl_3$) Test

It is rather a test for phenolics in general. Powdered plant leaves of the test plant (1.0 g) are weighed into a beaker and 10 ml of distilled water are added. The mixture is boiled for five minutes. Two drops of 5 per cent $FeCl_3$ are then added. Production of a greenish precipitate was an indication of the presence of tannins. Alternatively, a portion of the water extract is diluted with distilled water in a ratio of 1:4 and few drops of 10 per cent ferric chloride solution is added. A blue or green colour indicates the presence of tannins (Evans, 1989).

Other Methods

The hide-powder method is used in tannin analysis for leather tannin and the Stiasny method for wood adhesives. Statistical analysis reveals that there is no significant relationship between the results from the hide-powder and the Stiasny methods.

Hide-powder Method

400 mg of sample tannins are dissolved in 100 ml of distilled water. 3 g of slightly chromated hide-powder previously dried in vacuum for 24h over $CaCl_2$ are added and the mixture stirred for 1 h at ambient temperature. The suspension is filtered without vacuum through a sintered glass filter. The weight gain of the hide-powder expressed as a percentage of the weight of the starting material is equated to the percentage of tannin in the sample.

Stiasny's Method

100 mg of sample tannins are dissolved in 10 ml distilled water. 1 ml of 10M HCl and 2 ml of 37 per cent formaldehyde are added and the mixture heated under reflux for 30 min. The reaction mixture is filtered while hot through a sintered glass filter. The precipitate is washed with hot water (5x 10 ml) and dried over $CaCl_2$. The yield of tannin is expressed as a percentage of the weight of the starting material.

Reaction with Phenolic Rings

The bark tannins of *Commiphora angolensis* have been revealed by the usual colour and precipitation reactions and by quantitative determination by the methods of Löwenthal-Procter and of Deijs (formalin-hydrochloric acid method). Colourimetric methods have existed such as the Neubauer-Löwenthal method which uses potassium permanganate as an oxidizing agent and indigo sulfate as an indicator, originally proposed by Löwenthal in 1877. The difficulty is that the establishing of a titer for tannin is not always convenient since it is extremely difficult to obtain the pure tannin. Neubauer proposed to remove this difficulty by establishing the titer not with regard to the tannin but with regard to crystallised oxalic acid, whereby he found that 83 g oxalic acid correspond to 41.20 g tannin. Löwenthal's method has been criticized. For instance, the amount of indigo used is not sufficient to retard noticeably the oxidation of the non-tannins substances. The results obtained by this method are therefore only comparative. A modified method, proposed in 1903 for the quantification of tannins in wine, Feldmann's method, is making use of calcium hypochlorite, instead of potassium permanganate and indigo sulfate.

Nutrition

Tannins have traditionally been considered antinutritional but it is now known that their beneficial or antinutritional properties depend upon their chemical structure and dosage. The new technologies used to analyze molecular and chemical structures have shown that a division into condensed and hydrolyzable tannins is far too simplistic. Recent studies have demonstrated that products containing chestnut tannins included at low dosages (0.15–0.2 per cent) in the diet of chickens may be beneficial. Some studies suggest that chestnut tannins have been shown to have positive effects on silage quality in the round bale silages, in particular reducing NPNs (non protein nitrogen) in the lowest wilting level. Improved fermentability of soya meal nitrogen in the rumen has also been reported by F. Mathieu and J. P. Jouany (1993). Studies by S. Gonzalez *et al.* (2002) on *in vitro* ammonia release and dry matter degradation of soybean meal comparing three different types of tannins (quebracho, acacia and chestnut) demonstrated that chestnut tannins are more efficient in protecting soybean meal from *in vitro* degradation by rumen bacteria.

Condensed tannins inhibit herbivore digestion by binding to consumed plant proteins and making them more difficult for animals to digest and by interfering with protein absorption and digestive enzymes (for more on that topic, see plant defense against her bivory). Many tannin-consuming animals secrete a tannin-binding protein (mucin) in their saliva. Tannin-binding capacity of salivary mucin is directly related to its proline content. Advantages in using salivary proline-rich proteins (PRPs) to inactivate tannins are:

☆ PRPs inactivate tannins to a greater extent than do dietary proteins; this results in reduced fecal nitrogen losses

☆ PRPs contain no specific nitrogen and nonessential amino acids; this makes them more convenient for an animal to exploit rather than using up valuable dietary protein

Drinks with Tannins

The best-known human dietary sources of tannins are tea and wine. Other sources include fruit juices, but not citrus. Wine is a particular case. Most wines aged in charred oak wooden barrels possess tannins absorbed from the naturally occurring tannins in the wood which seep out into the wine. This concentration gives wine its signature bitterness. The tannins draw residual proteins from the tongue and mouth which gives wine its aftertaste. Consumption of foods with higher fat content neutralizes this effect. Coffee pulp has been found to contain low to trace amounts of tannins.

Tannins in Tea and Tannins in Wine

The phenolic content in tea refers to the phenols and polyphenols, natural plant compounds which are found in tea. These chemical compounds affect the flavor and mouth feel and are speculated to provide potential health benefits. Polyphenols in tea include catechins, the aflavins, tannins and flavonoids.

Catechins

Catechins include epigallocatechin-3-gallate (EGCG), epicatechin (EC), epicatechin-3-gallate (ECg), epigallocatechin (EGC), catechin and gallocatechin (GC).The content of EGCG is higher in green tea. Catechins constitute about 25 per cent of the dry weight of fresh tea leaf, although total catechin content varies widely depending on species, clonal variation, growing location, season, light variation and altitude. They are present in nearly all teas made from *Camellia sinensis*, including white tea, green tea, black tea and oolong tea.

According to a report released by USDA, in a 200-ml cup of tea, the mean total content of flavonoids is 266.68 mg for green tea and 233.12 mg for black tea. Of all the catechins in tea, EGCG is the main subject of scientific study with regard to its potential health effects.

4-Hydroxybenzoic acid, 3,4-dihydroxybenzoic acid (protocatechuic acid), 3-methoxy-4-hydroxy-hippuric acid and 3-methoxy-4-hydroxybenzoic acid (vanillic acid) are the main catechins metabolites found in humans after consumption of green tea infusions.

Theaflavins

Catechin monomer structures are metabolized into dimers theaflavins and oligomers thearubigins with increasing degrees of oxidation of tea leaves. Theaflavins directly contribute to the bitterness and astringency of steeped black tea. The mean amount of theaflavins in a cup of black tea (200 ml) is 12.18 mg. Three main types of theaflavins are found in black tea, namely theaflavin (TF-1), theaflavin-3-gallate (TF-2) and theaflavin-3,3-digallate (TF-3). A number of laboratory studies on their possible health effects have been conducted. According to a report released by USDA, in a 200-ml cup of tea, the mean total content of flavonoids is 233.12 mg for black tea.

Tannins

Tannins are astringent, bitter polyphenolic compounds that bind to and precipitate organic compounds. Gallic acid conjugates of the catechins, such as EGCG (Epigallocatechin gallate), are tannins with astringent qualities.

Flavonoids

A group of natural phenols called the flavonoids are of popular interest because researchers have found them to have the potential to contribute to better health. Tea has one of the highest contents of flavonoids among common food and beverage products. Catechins are the largest type of flavonoids in growing tea leaves.

Phenolic Content in Wine

The phenolic content in wine refers to the phenolic compounds natural phenol and polyphenols in wine, which include a large group of several hundred chemical compounds that affect the taste, colour and mouthfeel of wine. These compounds include phenolic acids, stilbenoids, flavonols, dihydroflavonols, anthocyanins, flavanol monomers (catechins) and flavanol polymers (proanthocyanidins). This large group of natural phenols can be broadly separated into two categories, flavonoids and non-flavonoids. Flavonoids include the anthocyanins and tannins which contribute to the colour and mouthfeel of the wine. The non-flavonoids include the stilbenoids such as resveratrol and phenolic acids such as benzoic, caffeic and cinnamic acids.

Origin of the Phenolic Compounds

The natural phenols are not evenly distributed within the fruit. Phenolic acids are largely present in the pulp, anthocyanins and stilbenoids in the skin and other phenols (catechins, proanthocyanidins and flavonols) in the skin and the seeds. During the growth cycle of the grapevine, sunlight will increase the concentration of phenolics in the grape berries, their development being an important component of canopy management. The proportion of the different phenols in any one wine will therefore vary according to the type of vinification. Red wine will be richer in phenols abundant in the skin and seeds, such as anthocyanin, proanthocyanidins and flavonols, whereas the phenols in white wine will essentially originate from the pulp and these will be the phenolic acids together with lower amounts of catechins and stilbenes. Red wines will also have the phenols found in white wines.

Wine simple phenols are further transformed during wine aging into complex molecules formed notably by the condensation of proanthocyanidins and anthocyanins, which explains the modification in the colour. Anthocyanins react with catechins, proanthocyanidins and other wine components during wine aging to form new polymeric pigments resulting in a modification of the wine colour and a lower astringency. Average total polyphenol content measured by the Folin method is 216 mg/100 ml for red wine and 32 mg/100 ml for white wine. The content of phenols in rosé wine (82 mg/100 ml) is intermediate between that in red and white wines. In winemaking, the process of maceration or "skin contact" is used to increase the concentration of phenols in wine. Phenolic acids are found in the pulp or juice of the wine and can be commonly found in white wines which usually do not go through a maceration period. The process of oak aging can also introduce phenolic compounds into wine, most notably vanillin which adds vanilla aroma to wines. Most wine phenols are classified as secondary metabolites and were not thought to be active in the primary metabolism and function of the grapevine. However, there is evidence that in some plants flavonoids play a role as endogenous regulators of auxin transport. They are water soluble and are usually secreted into the vacuole of the grapevine asglycosides.

Grape Polyphenols

Vitis vinifera produces many phenolic compounds. There is a varietal effect on the relative composition. In red wine, up to 90 per cent of the wine's phenolic content falls under the classification of flavonoids. These phenols, mainly derived from the stems, seeds and skins are often leached out of the grape during the maceration period of winemaking. The amount of phenols leached is known as extraction. These compounds contribute to the astringency, colour and mouthfeel of the wine. In white wines the number of flavonoids is reduced due to the lesser contact with the skins that they receive during winemaking. There is on-going study into the health benefits of wine derived from the antioxidant and chemopreventive properties of flavonoids.

Flavonols

Within the flavonoid category is a subcategory known as flavonols, which includes the yellow pigment - quercetin. Like other flavonoids, the concentration of flavonols in the grape berries increases as they are exposed to sunlight. Some viticulturalists will use measurement of flavonols such as quercetin as an indication of a vineyard's sun exposure and the effectiveness of canopy management techniques.

Anthocyanins

Anthocyanins are phenolic compounds found throughout the plant kingdom, being frequently responsible for the blue to red colours found in flowers, fruits and leaves. In wine grapes, they develop during the stage of *veraison* when the skin of red wine grapes changes colour from green to red to black. As the sugars in the grape increase during ripeningso does the concentration of anthocyanins. In most grapes anthocyanins are found only in the outer cell layers of the skin, leaving the grape juice inside virtually colourless. Therefore to get colour pigmentation in the wine, the fermenting must needs to be in contact with the grape skins in order for the anthocyanins to be extracted. Hence, white wine can be made from red wine grapes in the same way that many white sparkling wines are made from the red wine grapes of Pinot noir and Pinot Meunier. The exception to this is the small class of grapes known as teinturiers, such as Alicante Bouschet, which have a small amount of anthocyanins in the pulp that produces pigmented juice.

There are several types of anthocyanins (as the glycoside) found in wine grapes which are responsible for the vast range of colouring from ruby red through to dark black found in wine grapes. Ampelographers can use this observation to assist in the identification of different grape varieties. The European vine family *Vitis vinifera* is characterized by anthocyanins that are composed of only one molecule of glucose while non-*vinifera* vines such as hybrids and the American *Vitis labrusca* will have anthocyanins with two molecules. This phenomenon is due to a double mutation in the anthocyanin 5-O-glucosyl transferase gene of *V. vinifera*. In the mid-20th century, French ampelographers used this knowledge to test the various vine varieties throughout France to identify which vineyards still contained non-*vinifera* plantings. Red-berried Pinot grape varieties are also known to not synthesize para-coumaroylated or acetylated anthocyanins as other varieties do.

The colour variation in the finished red wine is partly derived from the ionization of anthocyanin pigments caused by the acidity of the wine. In this case, the three types of anthocyanin pigments are red, blue and colourless with the concentration of those various pigments dictating the colour of the wine. A wine with low pH (and such greater acidity) will have a higher occurrence of ionized anthocyanins which will increase the amount of bright red pigments. Wines with a higher pH will have a higher concentration of blue and colourless pigments. As the wine ages, anthocyanins will react with other acids and compounds in wines such as tannins, pyruvic acid and acetaldehyde which will change the colour of the wine, causing it to develop more "brick red" hues. These molecules will link up to create polymers that eventually exceed their solubility and become sediment at the bottom of wine bottles. Pyranoanthocyanins are chemical compounds formed in red wines by yeast during fermentation processes or during controlled oxygenation processes during the aging of wine.

Tannins

Tannins refer to the diverse group of chemical compounds in wine that can affect the colour, aging ability and texture of the wine. While tannins cannot be smelled or tasted, they can be perceived during wine tasting by the tactile drying sensation and sense of bitterness that they can leave in the mouth. This is due to the tendency of tannins to react with proteins, such as the ones found in saliva. In food and wine pairing, foods that are high in proteins (such as red meat) are often paired with tannic wines to minimize the astringency of tannins. However, many wine drinkers find the perception of tannins to be a positive trait especially as it relates to mouthfeel. The management of tannins in the winemaking process is a key component in the resulting quality.

Tannins are found in the skin, stems and seeds of wine grapes but can also be introduced to the wine through the use of oak barrels and chips or with the addition of tannin powder. The natural tannins found in grapes are known as proanthocyanidins due to their ability to release red anthocyanin pigments when they are heated in an acidic solution. Grape extracts are mainly rich in monomers and small oligomers (mean degree of polymerization <8). Grape seed extracts contain three monomers (catechin, epicatechin and epicatechin gallate) and procyanidin oligomers. Grape skin extracts contain four monomers (catechin, epicatechin, gallocatechin and epigallocatechin), as well as procyanidins and prodelphinidins oligomers. The tannins are formed by enzymes during metabolic processes of the grapevine. The amount of tannins found naturally in grapes varies depending on the variety with Cabernet Sauvignon, Nebbiolo, Syrah and Tannat being 4 of the most tannic grape varieties. The reaction of tannins and anthocyanins with the phenolic compound catechins creates another class of tannins known as pigmented tannins which influence the colour of red wine. Commercial preparations of tannins, known as *enological tannins*, made from oak wood, grape seed and skin, plant gall, chestnut, quebracho, gambier and myrobalan fruits, can be added at different stages of the wine production to improve colour durability. The tannins derived from oak influence are known as "hydrolysable tannins" being created from the ellagic and gallic acid found in the wood.

In the vineyards, there is also a growing distinction being made between "ripe" and "unripe" tannins present in the grape. This "physiological ripeness", which is roughly determined by tasting the grapes off the vines, is being used along with sugar levels as a determination of when to harvest. The idea is that "riper" tannins will taste softer but still impart some of the texture components found favorable in wine. In winemaking, the amount of the time that they must spends in contact with the grape skins, stems and seeds will influence the amount of tannins that are present in the wine with wines subjected to longer maceration period having more tannin extract. Following harvest, stems are normally picked out and discarded prior to fermentation but some winemakers may intentionally leave in a few stems for varieties low in tannins (like Pinot noir) in order to increase the tannic extract in the wine. If there is an excess in the amount of tannins in the wine, winemakers can use various fining agents like albumin, casein and gelatin that can bind to tannins molecule and precipitate them out as sediments. As wine ages, tannins will form long polymerized chains which come across to a taster as "softer" and less tannic. This process can be accelerated by exposing the wine to oxygen, which oxidizes tannins to quinone-like compounds that are polymerization-prone. The winemaking technique of micro-oxygenation and decanting wine use oxygen to partially mimic the effect of aging on tannins. A study in wine production and consumption has shown that tannins, in the form of proanthocyanidins, have a beneficial effect on vascular health. The study showed that tannins suppressed production of the peptide responsible for hardening arteries. To support their findings, the study also points out that wines from the regions of southwest France and Sardinia are particularly rich in proanthocyanidins and that these regions also produce populations with longer life spans.

Reactions of tannins with the phenolic compound anthocyanidins creates another class of tannins known as *pigmented tannins* which influences the colour of red wine.

Addition of Enological Tannins

Commercial preparations of tannins, known as *enological tannins*, made from oak wood, grape seed and skin, plant gall, chestnut, quebracho, gambier and myrobalan fruits, can be added at different stages of the wine production to improve colour durability.

Effects of Tannins on the Drinkability and Aging Potential of Wine

Tannins are a natural preservative in wine. Un-aged wines with high tannin content can be less palatable than wines with a lower level of tannins. Tannins can be described as leaving a dry and puckered feeling with "furriness" in the mouth that can be compared to a stewed tea, which is also very tannic. This effect is particularly profound when drinking tannic wines without the benefit of food. Many wine lovers see natural tannins (found particularly in varietals such as Cabernet Sauvignon and often accentuated by heavy oak barrel aging) as a sign of potential longevity and age ability. Tannins impart a mouth-puckering astringency when the wine is young but "resolve" (through a chemical process called polymerization) into delicious and complex elements of "bottle bouquet" when the wine is cellared under appropriate temperature conditions, preferably in the range of a constant 55 to 60 °F (13 to 16 °C). Such wines mellow and improve with age with the tannic "backbone" helping the wine survive for as long as 40 years or more. In many regions (such as in Bordeaux), tannic grapes such as Cabernet Sauvignon are blended with lower-tannin grapes such as Merlot or Cabernet Franc, diluting the tannic characteristics. White wines and wines that are vinified to be drunk young (for examples, see nouveau wines) typically have lower tannin levels.

Other Flavonoids

Flavan-3-ols (catechins) are flavonoids that contribute to the construction of various tannins and contribute to the perception of bitterness in wine. They are found in highest concentrations in grape seeds but are also in the skin and stems. Catechins play a role in the microbial defense of the grape berry, being produced in higher concentrations by the grape vines when it is being attacked by grape diseases such as downy mildew. Because of that grape vines in cool, damp climates produce catechins at high levels than vines in dry, hot climates. Together with anthocyanins and tannins they increase the stability of a wines colour-meaning that a wine will be able to maintain its colouring for a longer period of time. The amount of catechins present varies among grape varieties with varietals like Pinot noir having high concentrations while Merlot and especially Syrah have very low levels. As an antioxidant, there are some studies into the health benefits of moderate consumption of wines high in catechins. In red grapes, the main flavonol is on average quercetin, followed by myricetin, kaempferol, laricitrin, isorhamnetin and syringetin. In white grapes, the main flavonol is quercetin, followed by kaempferol and isorhamnetin. The delphinidin-like flavonols myricetin, laricitrin and syringetin are missing in all white varieties, indicating that the enzyme flavonoid 3',5'-hydroxylase is not expressed in white grape varieties. Myricetin, laricitrin and syringetin, flavonols which are present in red grape varieties only, can be found in red wine.

Non-flavonoids

Hydroxycinnammic Acids

Hydroxycinnamic acids are the most important group of nonflavonoid phenols in wine. The four most abundant ones are the tartaric acidesters *trans*-caftaric, *cis*- and *trans*-coutaric and *trans*-fertaric acids. In wine they are present also in the free form (*trans*-caffeic, *trans*-p-coumaric and *trans*-ferulic acids).

Stilbenoids

Resveratrol is found in highest concentration in the skins of wine grapes. The accumulation in ripe berries of different concentrations of both bound and free resveratrols depends on the maturity level and is highly variable according to the genotype. Both red and white wine grape varieties contain resveratrol, but more frequent skin contact and maceration leads to red wines normally having ten times more resveratrol than white wines. Resveratrol produced by grape vines provides defense against microbes and production can be further artificially stimulated by ultraviolet radiation. Grapevines in cool, damp regions with higher risk of grape diseases, such as Bordeaux and Burgundy, tend to produce grapes with higher levels of resveratrol than warmer, drier wine regions such as California and Australia. Different grape varieties tend to have differing levels, with Muscadines and the Pinot family having high levels while the Cabernet family has lower levels of resveratrol. In the late 20th century interest in the possible health benefits of resveratrol in wine was spurred by discussion of the French paradoxinvolving the health of wine drinkers in France. Piceatannol is also present in grape from where it can be extracted and found in red wine.

Phenolic Acids

Vanillin is a phenolic aldehyde most commonly associated with the vanilla notes in wines that have been aged in oak. Trace amounts of vanillin are found naturally in grapes, but they are most prominent in the lignin structure of oak barrels. Newer barrels will impart more vanillin, with the concentration present decreasing with each subsequent usage.

Phenolic Compounds found in Wine

LC chromatograms at 280 nm of a pinot red wine (top), a Beaujolais rosé(middle) and a white wine (bottom). The picture shows peaks corresponding to the different phenolic compounds. The hump between 9 and 15 minutes corresponds to the presence of tannins, mostly present in the red wine (Figure 17.10).

Depending on the methods of production, wine type, grape varieties, ageing processes, the following phenolics can be found in wine. The list, sorted in alphabetical order of common names, is not exhaustive.

- ☆ Acutissimin A
- ☆ aesculetin
- ☆ Anthocyanidin-caftaric acid adducts
- ☆ Astilbin
- ☆ Astringin
- ☆ B type proanthocyanidin dimers
- ☆ B type proanthocyanidin trimers
- ☆ Caffeic acid
- ☆ Caftaric acid
- ☆ Castalagin
- ☆ Castavinol C1
- ☆ Castavinol C2
- ☆ Castavinol C3
- ☆ Castavinol C4
- ☆ Catechin
- ☆ Catechin-(4,8)-malvidin-3-O-glucoside
- ☆ Compound NJ2

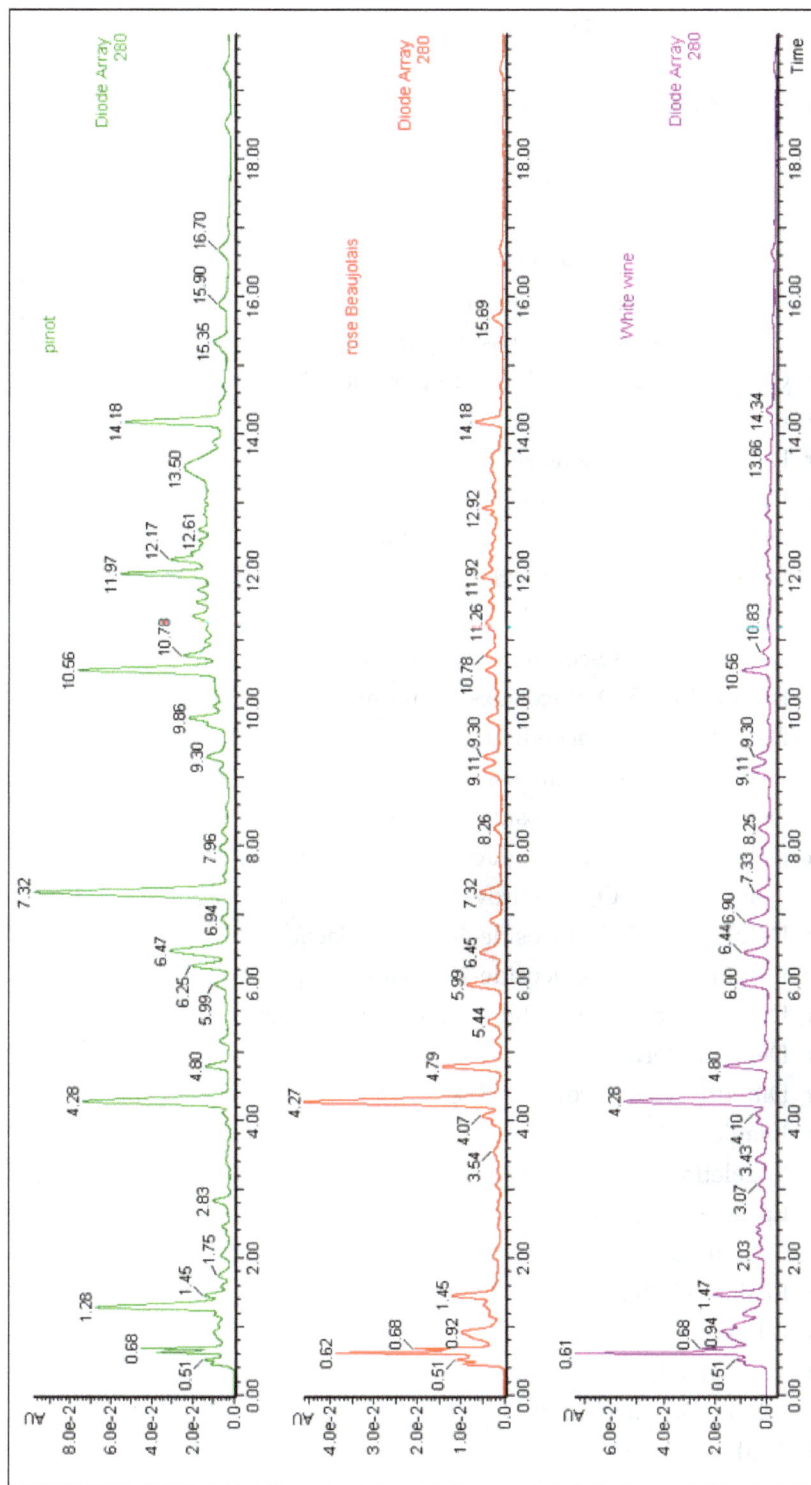

Figure 17.10: Chromatograms of Phenolic Compounds from in Red Wine, Rose Wine and White Wine.

- ☆ Coniferyl aldehyde
- ☆ Coumaric acid
- ☆ Coutaric acid
- ☆ Cyanidin
- ☆ Cyanidin 3-O-glucoside
- ☆ Cyanidin acetyl 3-O-glucoside
- ☆ Cyanidin coumaroyl 3-O-glucoside
- ☆ Cyanidin-3-O-glucoside-pyruvic acid
- ☆ Cyanidin-3-O-acetylglucoside-pyruvic acid
- ☆ Cyanidin-coumaroylglucoside-pyruvic acid
- ☆ Delphinidin
- ☆ Delphinidin 3-O-glucoside
- ☆ Delphinidin acetyl-3-O-glucoside
- ☆ Delphinidin coumaroyl 3-O-glucoside
- ☆ Delphinidin-3-O-glucoside-pyruvic acid
- ☆ Delphinidin-3-O-acetylglucoside-pyruvic acid
- ☆ Delphinidin-3-O-coumaroylglucoside-pyruvic acid
- ☆ Delphinidin-3-O-glucoside-4-vinylcatechol
- ☆ Delphinidin-3-O-acetylglucoside-4-vinylcatechol
- ☆ Delphinidin-3-O-coumaroylglucoside-4-vinylcatechol
- ☆ Delphinidin-3-O-glucoside-4-vinylphenol
- ☆ Delphinidin-3-O-acetylglucoside-4-vinylphenol
- ☆ Delphinidin-3-O-coumaroylglucoside-4-vinylphenol
- ☆ Delphinidin-3-O-glucoside-4-vinylguaiacol
- ☆ Delphinidin-3-O-glucoside-4-vinyl(epi)catechin
- ☆ Delphinidin-3-O-acetylglucoside-4-vinyl(epi)catechin
- ☆ Delta-viniferin
- ☆ Dihydro-resveratrol
- ☆ Ellagic acid
- ☆ Engeletin
- ☆ Epicatechin gallate
- ☆ Epigallocatechin
- ☆ Epsilon-viniferin
- ☆ Ethyl caffeate
- ☆ Ethyl gallate
- ☆ Ethyl protocatechuate
- ☆ 4-Ethylguaiacol

☆ 4-Ethylphenol

☆ Fertaric acid

☆ Ferulic acid

☆ Gallic acid

☆ Gentisic acid

☆ Grandinin

☆ Grape reaction product (GRP)

☆ Guaiacol

☆ Hopeaphenol

☆ p-Hydroxybenzoic acid

☆ Isorhamnetol 3-glucoside

☆ Kaempferol

☆ Kaempferol glucoside

☆ Kaempferol glucuronide

☆ Malvidin

☆ Malvidin 3-O-glucoside (oenin)

☆ Malvidin acetyl-3-O-glucoside

☆ Malvidin cafeoyl-3-O-glucoside

☆ Malvidin coumaroyl-3-Oglucoside

☆ Malvidin glucoside-ethyl-catechin

☆ Malvidin-3-O-glucoside-pyruvic acid

☆ Malvidin-3-O-acetylglucoside-pyruvic acid

☆ Malvidin-3-O-coumaroylglucoside-pyruvic acid

☆ Malvidin-3-O-glucoside-acetaldehyde

☆ Malvidin-3-O-acetylglucoside-acetaldehyde

☆ Malvidin-3-O-coumaroylglucoside-acetaldehyde

☆ Malvidin-3-O-glucoside-4-vinylcatechol

☆ Malvidin-3-O-acetylglucoside-4-vinylcatechol

☆ Malvidin-3-O-coumaroylglucoside-4-vinylcatechol

☆ Malvidin-3-O-glucoside-4-vinylphenol

☆ Malvidin-3-O-acetylglucoside-4-vinylphenol

☆ Malvidin-3-O-coumaroylglucoside-4-vinylphenol

☆ Malvidin-3-O-caffeoylglucoside-4-vinylphenol

☆ Malvidin-3-O-glucoside-4-vinylguaiacol

☆ Malvidin-3-O-acetylglucoside-4-vinylguaiacol

☆ Malvidin-3-O-coumaroylglucoside-vinylguaiacol

☆ Malvidin-3-O-glucoside-4-vinyl(epi)catechin

- ☆ Malvidin-3-O-acetylglucoside-4-vinyl(epi)catechin
- ☆ Malvidin-3-O-coumaroylglucoside-4-vinyl(epi)catechin
- ☆ Methyl gallate
- ☆ Myricetol
- ☆ Myricetol 3-glucoside
- ☆ Myricetol 3-glucuronide
- ☆ Oxovitisin A
- ☆ Pallidol
- ☆ Peonidin 3-O-glucoside
- ☆ Peonidin acetyl-3-O-glucoside
- ☆ Peonidin-3-(6-p-caffeoyl)-glucoside
- ☆ Peonidin coumaroyl 3-O-glucoside
- ☆ Peonidin-3-O-glucoside-pyruvic acid
- ☆ Peonidin-3-O-acetylglucoside-pyruvic acid
- ☆ Peonidin-3-O-coumaroylglucoside-pyruvic acid
- ☆ Peonidin-3-O-glucoside-4-vinylcatechol
- ☆ Peonidin-3-O-acetylglucoside-4-vinylcatechol
- ☆ Peonidin-3-O-coumaroylglucoside-4-vinylcatechol
- ☆ Peonidin-3-O-glucoside-4-vinylphenol
- ☆ Peonidin-3-O-acetylglucoside-4-vinylphenol
- ☆ Peonidin-3-O-coumaroylglucoside-4-vinylphenol
- ☆ Peonidin-3-O-glucoside-4-vinylguaiacol
- ☆ Peonidin-3-O-glucoside-4-vinyl(epi)catechin
- ☆ Peonidin-3-O-acetylglucoside-4-vinyl(epi)catechin
- ☆ Petunidin
- ☆ Petunidin 3-O-glucoside
- ☆ Petunidin acetyl-3-O-glucoside
- ☆ Petunidin coumaroyl-3-O-glucoside
- ☆ Petunidin-3-O-glucoside-pyruvic acid
- ☆ Petunidin-3-O-acetylglucoside-pyruvic acid
- ☆ Petunidin-3-O-coumaroylglucoside-pyruvic acid
- ☆ Petunidin-3-O-glucoside-4-vinylcatechol
- ☆ Petunidin-3-O-acetylglucoside-4-vinylcatechol
- ☆ Petunidin-3-O-coumaroylglucoside-4-vinylcatechol
- ☆ Petunidin-3-O-glucoside-4-vinylphenol
- ☆ Petunidin-3-O-acetylglucoside-4-vinylphenol
- ☆ Petunidin-3-O-coumaroylglucoside-4-vinylphenol

☆ Petunidin-3-O-glucoside-4-vinylguaiacol

☆ Petunidin-3-O-glucoside-4-vinyl(epi)catechin

☆ Petunidin-3-O-acetylglucoside-4-vinyl(epi)catechin

☆ Phloroglucinol carboxylic acid

☆ Piceatannol

☆ Piceids

☆ Pinotin A

☆ Oligomeric procyanidins :

   ◆ Procyanidin B1

   ◆ Procyanidin B2

   ◆ Procyanidin B3

   ◆ Procyanidin B4

   ◆ B1-3-O-gallate

   ◆ B2-3-O-gallate

   ◆ B2-3'-O-gallate

   ◆ procyanidin C1 (epicatechin-(4β→8)-epicatechin-(4β→8)-epicatechin)

   ◆ Procyanidin C2 (catechin-(4α→8)-catechin-(4α→8)-catechin)

   ◆ procyanidin T2 (trimer)

☆ Protocatechuic acid

☆ protocatechuic aldehyde

☆ Quercetin

☆ Quercetol glucoside

☆ Quercetol glucuronide

☆ Resveratrol

☆ Roburin A

☆ Roburin E

☆ Scopoletin

☆ Sinapic aldehyde

☆ Sinapinic acid

☆ Syringic acid

☆ Tyrosol

☆ Vanillic acid

☆ vanillin

☆ Vescalagin

☆ 4-Vinylphenol

☆ Vitisin A

☆ Vitisin B

☆ Vinylpyranomalvidin-3-O-glucoside-procyanidin dimer

☆ VinylpyranoMv-3-coumaroylglucoside-procyanidin dimer

☆ Vinylpyranomalvidin-3-O-glucoside-catechin

☆ Vinylpyranomalvidin-3-O-coumaroylglucoside-catechin

☆ Vinylpyranomalvidin-3-O-phenol

☆ Vinylpyranopetunidin-3-O-glucoside-catechin

☆ Vinylpyranopeonidin-3-O-glucoside-catechin

☆ Vinylpyranomalvidin-3-O-acetylglucoside-catechin

Fruit Juices

Although citrus fruits do not themselves contain tannins, orange-coloured juices often contain food dyes with tannins. Apple juice, grape juices and berry juices are all high in tannins. Sometimes tannins are even added to juices and ciders to create a more astringent feel to the taste.

Beer

In addition to the alpha acids extracted from hops to provide bitterness in beer, condensed tannins are also present. These originate both from the malt and hops. Especially in Germany, trained brew masters consider the presence of tannins as a flaw. In some styles, the presence of this astringency is acceptable or even desired, as, for example, in a Flanders red ale.

In lager type beers, the tannins can form a precipitate with specific haze-forming proteins in the beer resulting in turbidity at low temperature. This chill haze can be prevented by removing part of the tannins or part of the haze-forming proteins. Tannins are removed using PVPP, haze-forming proteins by using silica or tannic acid.

Health Effects of Tea

According to legend, the health effects of tea have been examined ever since the first infusions of *Camellia sinensis* about 4700 years ago in China. Emperor Shennong claimed in *The Divine Farmer's Herb-Root Classic* that *Camellia sinensis* infusions were useful for treating a variety of disease conditions. Historically before modern safe drinking water, it was thought to be beneficial to make tea with boiled water which may reduce microorganisms, but over-boiling may actually increase the amount of consumed contaminants.

Content

Tea leaves contains over 700 various chemicals including alkaloids such as caffeine, dietary minerals, linoleic acid, carbohydrates and polyphenols, (specifically catechins, a type of flavanol). In a freshly picked tea leaf, catechins can compose up to 30 per cent of the dry weight. Catechins are highest in concentration in white tea and green tea, while black tea has substantially fewer. The amounts of carbohydrates, fat and protein found in tea are negligible. Although tea contains various types of polyphenols, including a high content of epigallocatechins, tea does not contain tannic acid. Tannic acid is not an appropriate standard for any type of tannin analysis because of its poorly defined composition.

Theanine and Caffeine

Tea also contains theanine and the stimulant caffeine at about 3 per cent of its dry weight, translating to between 30 mg and 90 mg per 8 oz (250 ml) cup depending on type, brand and brewing method. Tea also contains small amounts of theobromine and theophylline. Due in part to modern-day environmental pollution fluoride and aluminium have also been found to occur in tea, with certain types of brick teamade from old leaves and stems having the highest levels. This occurs due to the tea plant's high sensitivity to and absorption of environmental pollutants. Dry tea has more caffeine by weight than dry coffee; nevertheless, more dry coffee than dry tea is used in typical drink preparations which results in a cup of brewed tea containing significantly less caffeine than a cup of coffee of the same size.

Potential Benefits

Several randomized controlled trials suggest green tea can reduce body fat by a small amount for a short time, though it is uncertain whether the reduction would be meaningful for most people.

Anti-cancer Properties

In 2011, the Food and Drug Administration (FDA) reported that there was very little evidence to support the claim that green tea consumption may reduce the risk of breast and prostate cancer. The US National Cancer Institute reports that in epidemiological studies and the few clinical trials of tea for the prevention of cancer, the results have been inconclusive. The institute does not recommend for or against the use of tea to reduce the risk of any type of cancer. Inconsistencies in study findings regarding tea and cancer risk may be due to variability in tea preparation, tea consumption, the bioavailability of tea compounds (the amounts that can be absorbed by the body), lifestyle differences and individual genetic differences. Though there is some positive evidence for risk reduction of breast, prostate, ovarian and endometrial cancers with green tea, it is weak and inconclusive.

Weight Loss

Green tea consumption has no meaningful effect in aiding weight loss.

Potential Drawbacks

Fluoride

All tea leaves contain fluoride; however, mature leaves contain as much as 10 to 20 times the fluoride levels of young leaves from the same plant. Although low concentrations of fluoride are maintained in many public water supplies for dental health, very high fluoride intake (over 2 mg per day for children, 4 mg adults) increases the risk of osteofluorosis and fractures. There is evidence that over-intake of teas produced using mature leaves (*e.g.* brick tea) can cause fluorosis in humans. The fluoride content of made tea depends on the picking method and fluoride content of the soil in which it is grown; tea plants absorb this element at a greater rate than other plants. Care in the choice of the location where the plant is grown may reduce the risk. It is speculated that hand-picked tea would contain less fluoride than machine-harvested tea, because there is a much lower chance of harvesting older leaves during the harvest process. A 2013 British study of 38 teas found that cheaper UK supermarket tea blends had the highest levels of fluoride with about 580 mg per kilogram, green teas averaged about 397 mg per kg and pure blends about 132 mg per kg. The researchers suggested that economy teas may use older leaves which contain more fluoride. They calculated a person drinking a liter of economy tea per day would consume about 4 mg of fluoride, the maximum recommended amount of fluoride per day but below the maximum tolerable amount of 10 mg fluoride per day.

Aluminum and Heavy Metals

Tea drinking accounts for a high proportion of aluminum in the human diet. The levels are safe, but there has been some concern that aluminum traces may be associated with Alzheimer's disease. A recent study additionally indicated that some teas contained possibly risky amounts of lead (mostly Chinese) and aluminum (Indian/Sri Lanka blends, China). There is still insufficient evidence to draw firm conclusions on this subject.

Caffeine

The caffeine in tea is a mild diuretic. However, the British Dietetic Association has suggested that tea can be used to supplement normal water consumption and that "the style of tea and coffee and the amounts we drink in the UK are unlikely to have a negative effect [on hydration]".

Oxalates

Tea contains oxalate, overconsumption of which can cause kidney stones, as well as binding with free calcium in the body. The bioavailability of oxalate from tea is low, thus a possible negative effect requires a large intake of tea.

Prostate Cancer

A study, published in 2012, suggested that men who drink large quantities of black tea more than seven cups per day – *increase* their risk of prostate cancer by 50 per cent. The story was relayed in the media, but according to the National Health Service, "this study had many limitations that call into question the reliability of its results".

Food Items with Tannins

Fruits

Pomegranates

The pomegranate ellagitannins, which include punicalagin isomers, are ellagitannins found in the fruit, rind (peel), bark or heartwood of pomegranates (*Punica granatum*).

Chemistry

As the chemistry of punicalagins became known it was found to be not unique to pomegranate. Punicalagins are present in numerous species of the genus *Terminalia*, species *chebula* Retz. ("Fructus Chebulae"), *myriocarpa*, *catappa* and citrina (tropical flowering trees historically used in African traditional medicine for antiobiotic and antifungal purposes). They have also been isolated from *Cistus salvifolius* (a Mediterranean shrub) and *Combretum molle* (an African shrub). Pomegranate fruits natural phenols can be extracted with ethyl acetate and fractionation can afford the ellagitannin punicalagins.

Dietary Supplementation

A few dietary supplements and nutritional ingredients are available that contain extracts of whole pomegranate and/or are standardized to punicalagins, the marker compound of pomegranate. Extracts of pomegranate are also 'Generally Recognized As Safe' (GRAS) by the United States.

List of Compounds

☆ Pedunculagin, a compound found in the pericarp of the pomegranate (*Punica granatum*).

☆ Punicacortein A, a compound found in the bark of *P. granatum* and in Osbeckia chinensis

☆ Punicacortein B, a compound found in the bark of *P. granatum*

☆ Punicacortein C, a compound found in the bark of *P. granatum*

☆ Punicacortein D, a compound found in the bark and heartwood of *P. granatum*

☆ Punicafolin, a compound found in the leaves of *Punica granatum*

☆ Punigluconin, a compound found in the bark of *P. granatum* and in *Emblica officinalis*

☆ Punicalagin, a compound found in the pericarps of *P. granatum*

☆ Punicalin, a compound found in pomegranates or in the leaves of *Terminalia catappa*

☆ Granatin A, a compound found in the pericarp of *P. granatum*

☆ Granatin B, a compound found in the fruit of *P. granatum*

☆ Diellagic acid rhamnosyl (1→4) glucopyranoside

☆ 5-O-galloylpunicacortein D

☆ 2-O-galloylpunicalin

☆ Casuarinin, a compound found in the pericarp of *P. granatum*

☆ Gallagyldilactone, a compound found in the pericarp of *P. granatum*

☆ Corilagin, a compound found in the leaves of *P. granatum* and other species

☆ Strictinin, a compound found in the leaves of *P. granatum*

Other Phenolics

☆ Ellagic acid, a constitutive compound of ellagitannins

☆ Gallagic acid, a compound found in many ellagitannins

☆ 1,2,4,6-tetra-O-galloyl-β-D-glucose, a compound found in the leaves of *P. granatum*

☆ 1,2,3,4,6-penta-O-galloyl-β-D-glucose, a compound found in the leaves of *P. granatum* and the common precursor of gallotannins and the related ellagitannins

☆ Brevifolin, a compound found in the leaves of *P. granatum*

☆ Brevifolin carboxylic acid, a compound found in the leaves of *P. granatum*

☆ 3,6-(R)-hexahydroxydiphenoyl-(α/β)-C-glucopyranose, a compound found in the leaves of *P. granatum*

☆ 1,2,6-tri-O-galloyl-β-C-glucopyranose, a compound found in the leaves of *P. granatum*

☆ 1,4,6-tri-O-galloyl-β-C-glucopyranose, a compound found in the leaves of *P. granatum*

☆ 3,4,8,9,10-pentahydroxydibenzo[b,d]pyran-6-one, a compound found in the leaves of *P. granatum*

Persimmons

Some persimmons are highly astringent and therefore inedible when they are not extremely ripe (to be specific, the Korean, American and Hachiya or Japanese).

Berries

Most berries, such as cranberries, strawberries and blueberries, contain both hydrolyzable and condensed tannins.

Nuts

Nuts that can be consumed raw, such as hazelnuts, walnuts and pecans, contain high amounts of tannins. Almonds have a lower content. Tannin concentration in the crude extract of these nuts did not directly translate to the same relationships for the condensed fraction. Acorns contain such high concentrations of tannins that they must be processed before they can be consumed safely. The areca nut also contains tannin.

Smoked Foods

Tannins from the wood of mesquite, cherry, oak and other woods used in smoking are present on the surface of smoked fish and meat.

Herbs and spices

Cloves, tarragon, cumin, thyme, vanilla and cinnamon all contain tannins.

Legumes

Most legumes contain tannins. Red-coloured beans contain the most tannin and white-coloured beans have the least. Peanuts without shells have very low tannin content. Chickpeas (garbanzo beans) have a smaller amount of tannins.

Chocolate

Chocolate liquor contains about 6 per cent tannins.

Health Effects of Tannins

Health Effects of Natural Phenols and Polyphenols

Because of the large structural diversity of dietary polyphenols, it is difficult to assert specific health effects from such ubiquitous substances. Their antioxidant activities in chemical and biological assays are undisputed and many are associated with the health-promoting effects of fruits and vegetables, but to what extent these effects apply to entire organisms and clinical outcomes in human disease in particular, remains a controversially discussed topic in nutrition science and disease prevention.

Cardiovascular Health

A review published in 2012 found growing consensus for the hypothesis that the specific intake of food and drink containing relatively high concentrations of flavonoids may play a meaningful role in reducing the risk of cardiovascular disease (CVD). The reviewers stated that research to date had been of poor quality and the large and rigorous trials are needed better to study the science and to investigate possible adverse effects associated with excessive polyphenol intake: currently a lack of knowledge about safety suggests that polyphenol levels should not exceed that which occurs in a normal diet.

Toxicity

Toxicological concerns from dietary polyphenols have been voiced. They are unrelated to the acute toxicity of the phenols used in chemical industries. They are based on a number of *in-vitro* assays on the mutagenic and genotoxic properties of flavonols, such as quercetin. However, natural phenols and polyphenols are not classified as carcinogens. Their acute toxicity in humans and herbivores is generally very low due to their poor bioavailability. Tannins can have anti-feeding effects in livestock and interfere with nutrient absorption. This applies in particular to the astringent polyphenols and much less to the cinnamic and caffeic acid derivatives.

Bioavailability

Questions on the relationship between health benefits and polyphenols generally revolve around bioavailability. Gallic acid and isoflavones are the most well-absorbed phenols, followed by catechins (flavan-3-ols), flavanones and quercetin glucosides, but with different kinetics. The least well-absorbed phenols are the proanthocyanidins, galloylated tea catechins and anthocyanins.

Antioxidant Activity

As interpreted by the Linus Pauling Institute and the European Food Safety Authority (EFSA), dietary flavonoids have little or no direct antioxidant food value following digestion. Unlike controlled test tube conditions, the fate of natural phenols *in vivo* shows they are poorly conserved (less than 5 per cent), with most of what is absorbed existing as chemically-modified metabolites destined for rapid excretion.

Neonatal Effects

Many natural phenols, like the flavonoids, were found to be strong topoisomerase inhibitors *in vitro*, some of them were tested *in vivo* with similar results. Those substances share the property with some chemotherapeutic anticancer drugs such etoposide and doxorubicin. When tested some natural phenols induced DNA mutations in MLL gene, which are common findings in neonatal acute leukemia. The DNA changes were highly increased by treatment with flavonoids in cultured blood stem cells. Maternal high flavonoid content diet is suspected to increase risk of particularly acute myeloid leukemia in neonates. Natural phenols have both anticarcinogenic -proapoptotic effect and a carcinogenic, DNA damaging, mutagenic potential. Adults seem to rapidly metabolize most of phenols, so toxic, mutagenic effects may not be pronounced in regular low doses intaken with food. Some natural phenols - EGCG, for example - were found to rapidly induce detoxyfying Nrf2 transcription factor activity, which seems to be responsible for observed beneficial, antioxidative effects of the substances and which also leads to rapid degradation of the phenolic molecules. However, the human embryos detoxification system is not mature enough to deal with phenols, which can cross the placenta barrier. High intake of flavonoid compounds during pregnancy is suspected to increase risk of neonatal leukemia. Therefore "bioflavonoid" supplements should be not used by pregnant women.

Carcinogenicity

The Carcinogenic Potency Project, which is a part of the US EPA's Distributed Structure-Searchable Toxicity (DSSTox) Database Network, has been systemically testing the carcinogenicity of chemicals, both natural and synthetic and building a publicly available database of the results since about 1980. At this time, none of the dietary phenols and polyphenols are classified as carcinogens. Some polyphenols, particularly from the flavan-3-ol (catechin-type), have both anticarcinogenic-proapoptotic and mutagenic effects. The DNA changes were increased by treatment with flavonoids in cultured blood stem cells. Some natural polyphenols share the properties of some anticancer drugs such as etoposide and doxorubicin while other polyphenols may induce DNA mutations in the MLL gene, which are common findings in neonatal acute leukemia.

Tannin Market

Tannin production began at the beginning of the 19th century with the industrial revolution, to produce tanning material for the need for more leather. Before that time, processes used plant material and were long (up to six months). There was a collapse in the vegetable tannin market in the 1950s–1960s, due to the appearance of synthetic tannins, which were invented in response to a scarcity of vegetable tannins during World War II. At that time, many small tannin industry sites closed. Vegetable tannins are estimated to be used for the production of 10–20 per cent of the global leather production. The cost of the final product depends on the method used to extract the tannins, in particular the use of solvents, alkali and other chemicals used (for instance glycerin). For large quantities, the most cost-effective method is hot water extraction. Tannic acid is used worldwide as clarifying agent in alcoholic drinks and as aroma ingredient in both alcoholic and soft drinks or juices. Tannins from different botanical origins also find extensive uses in the wine industry.

Uses

Tannins are an important ingredient in the process of tanning leather. Tanbark from oak, mimosa, chestnut and quebracho tree has traditionally been the primary source of tannery tannin, though inorganic tanning agents are also in use today and account for 90 per cent of the world's leather production. Tannins produce different colours with ferric chloride (either blue, blue black, or green to greenish-black) according to the type of tannin. Iron gall ink is produced by treating a solution of tannins with iron (II) sulfate. Tannin is a component in a type of industrial particleboard adhesive developed jointly by the Tanzania Industrial Research and Development Organization and Forintek Labs Canada. *Pinus radiata* tannins has been investigated for the production of wood adhesives. Condensed tannins, *i.e.* quebracho tannin and Hydrolyzable tannins, *i.e.*, chestnut tannin, appear to be able to substitute a high proportion of synthetic phenol in phenol-formaldehyde resins for wood particle board. Tannins can be used for production of anti-corrosive primer, sold under brand name-Nox Primer for treatment of rusted steel surfaces prior to painting, rust converter to transform oxidized steel into a smooth sealed surface and rust inhibitor. The use of resins made of tannins has been investigated to remove mercury and methyl mercury from solution. Immobilized tannins have been tested to recover uranium from seawater.

Medical Uses and Potential

When incubated with red grape juice and red wines with a high content of condensed tannins, the poliovirus, herpes simplex virus and various enteric viruses are inactivated. In tissue-cultured cell assays tannins have shown antiviral, antibacterial and antiparasitic effects.

Tannins isolated from the stem bark of *Myracrodruon urundeuva* may offer protection against 6-hydroxydopamine-induced toxicity. Souza *et al.*, discovered that the tannins isolated from the stem bark also have anti-inflammatory and antiulcer activity in rodents, showing a strong antioxidant property with possible therapeutic applications. Foods rich in tannins can be used in the treatment of HFE hereditary hemochromatosis, a hereditary disease characterized by excessive absorption of dietary iron, resulting in a pathological increase in total body iron stores.

Wine and Health (Health Effects of Wine)

Wine and health is an issue of considerable discussion and research. Wine has a long history of use as an early form of medication, being recommended variously as a safe alternative to drinking water, an antiseptic for treating wounds, a digestive aid and as a cure for a wide range of ailments including lethargy, diarrhea and pain from child birth.

Ancient Egyptian Papyri and Sumerian tablets dating back to 2200 BC detail the medicinal role of wine, making it the world's oldest documented man-made medicine.[3] Wine continued to play a major role in medicine until the late 19th and early 20th century, when changing opinions and medical research on alcohol and alcoholism cast doubt on the role of wine as part of a healthy lifestyle and diet. In the late 20th and early 21st century, fueled in part by public interest in reports by the United States news broadcast 60 Minutes on the so-called "French Paradox", the medical establishment began to re-evaluate the role of moderate wine consumption in health.

Historical Role of Wine in Medicine

Early medicine was intimately tied with religion and the supernatural, with early practitioners often being priests and magicians. Wine's close association with ritual made it a logical tool for these early medical practices. Tablets from Sumerian culture and papyri from Ancient Egypt dating to 2200 BC include recipes for wine based medicines, making wine the oldest documented man made medicine. When the ancient Greeks introduced a more systematized approach to medicine, wine still retained its prominent role. The Greek physician Hippocrates considered wine a part of a healthy diet and advocated its use as a disinfectant for wounds, as well as a medium in which to mix other drugs for easier consumption by the patient. He also prescribed wine as a cure for various ailments ranging from diarrhea and lethargy to pain during childbirth. The medical practices of the ancient Romans involved the use of wine in a similar manner. In his 1st-century work *De Medicina*, the Roman encyclopedist Aulus Cornelius Celsus detailed a long list of Greek and Roman wines used for medicinal purposes. While treating gladiators in Asia Minor, the Roman physician Galen would use wine as a disinfectant for all types of wounds and even soaked exposed bowels before returning them to the body. During his four years tending to the gladiators, only five deaths occurred, compared to sixty deaths under the watch of the physician before him.

Religion still played a significant role in promoting wine's use for health benefit. The Talmud noted wine to be *"the foremost of all medicines: wherever wine is lacking, medicines become necessary."* In his first epistle to Timothy, Paul the Apostle recommended that his young colleague drink a little wine every now and then for the benefit of his stomach and digestion. While the Islamic Korancontained restrictions on all alcohol, Islamic doctors such as the Persian Avicenna in the 11th century AD noted that wine was an efficient digestive aid but, because of Islamic laws, were limited to only using it as a disinfectant while dressing wounds. Catholic monasteries during the middle Ages would also regularly use wine for various medical treatments.[3] So closely tied was the role of wine and medicine that the first printed book on the subject of wine was written in the 14th century by a physician, Arnaldus de Villa Nova, with lengthy essays on wine's suitability for treatment of a variety of medical ailments such dementia and sinus problems.

Risks of Consumption

The lack of safe drinking water during much of history may have been one reason for wine's popularity in medicine. Wine was still being used to sterilize water as late as the Hamburg cholera epidemic of 1892 in order to control the spread of the disease. However, the late 19th century and early 20th century ushered in a period of changing views on the role of alcohol and, by extension, wine in health and society. The Temperance movement began to gain steam by touting the ills of alcoholism, which was eventually defined by the medical establishment as a disease. Studies of the long and short-term effects of alcohol caused many in the medical community to reconsider the role of wine in medicine and diet.[3] Public opinion turned against consumption of alcohol in any form, leading to Prohibition in the United States and other countries. In some areas, wine was able to maintain a limited role, such as an exemption in the United States for "therapeutic wines" that were sold legally in drug stores. These wines were marketed for their medicinal benefits, but some wineries used this measure as a loophole to sell large quantities of wine for recreational consumption. In response, the United States government issued a mandate requiring producers to include an emetic additive that would induce vomiting above the consumption of a certain dosage level.

Throughout the mid to early 20th century, health advocates pointed to the risk of alcohol consumption and the role it played in a variety of ailments such as blood disorders, high blood pressure, cancer, infertility, liver damage, muscle atrophy, psoriasis, skin infections, strokes and long term brain damage. Studies showed a connection between alcohol consumption among pregnant mothers and an increased risk of mental retardation and physical abnormalities in what became known as fetal alcohol syndrome, prompting the use of warning labels on alcohol-containing products in several countries.

French Paradox and the Benefits of Consumption

The 1990s and early 21st century saw a renewed interest in the health benefits of wine, ushered in by increasing research suggesting that moderate wine drinkers have lower mortality rates than heavy drinkers or teetotalers. In November 1991, the U.S. news program *60 Minutes* aired a broadcast on the so-called "French Paradox". Featuring the research work of Bordeaux scientist Serge Renaud, the broadcast dealt with the seemingly paradoxical relationship between the high fat/high dairy diets of French people and the low occurrence of cardiovascular disease among them. The broadcast drew parallels to the American and British diets which also contained high levels of fat and dairy but which featured high incidences of heart disease. One of the theories proposed by Renaud in the broadcast was that moderate consumption of red wine was a risk-reducing factor for the French and that wine could have more positive health benefits yet to be studied. Following

Figure 17.11: The French have a Diet that is High in Full-Fat Dairy Products such as Cheeses and also have Low Rates of Heart Disease. One possible factor contributing to this "French paradox" is the regular consumption of red wine.

the *60 Minutes* broadcast, sales of red wine in the United States jumped 44 per cent over previous years.

This changing view of wine can be seen in the evolution of the language used in the U.S. Food and Drug Administration Dietary Guidelines. The 1990 edition of the guidelines contained the blanket statement that "wine has no net health benefit". By 1995, the wording had been changed to allow moderate consumption with meals providing the individual had no other alcohol-related health risk. From a research perspective, scientists began differentiating alcohol consumption among the various classes of beverages wine, beer and spirits. This distinction allowed studies to highlight the positive medical benefits of wine apart from the mere presence of alcohol. However wine drinkers tend to share similar lifestyle habits better diets, regular exercise, non-smoking that may in themselves be a factor in the supposed positive health benefits compared to drinkers of beer and spirits or those who abstain completely.

Moderate Consumption

Nearly all research into the positive medical benefits of wine consumptions make a distinction between moderate consumption, heavy and binge drinking. What constitutes a moderate, healthy level of consumption will vary by individual according to age, gender, genetics, weight and body stature as well as the situation- *i.e.* is food being consumed as well, are any other drugs currently in the individual's system, etc. Women, in general, tend to absorb alcohol quicker than men due to their lower body water content and difference in levels of stomach enzyme so their moderate levels of consumption tend to be lower than a male of equal age and weight.[6] Some doctors define "moderate consumption" as one 5-US-fluid-ounce

Figure 17.12: Some Doctors Define "Moderate" Consumption as one 5oz (150 ml) Glass of Wine per day for Women and Two Glasses per Day for Men.

(150 ml) glass of wine per day for women and two glasses per day for men. The view of consuming wine in moderation has a history almost as long as that of wine's role in medicine. The Greek poet Eubulus believed that three bowls (kylix) were the ideal amount of wine to consume. The number of three bowls for moderation is a common theme throughout Greek writing; today the standard 750 mL wine bottle contains roughly the volume of three Kylix cups (250 ml or 8 fl oz each). However, the Kylix cups would have contained a diluted wine, at a 1:2 or 1:3 dilutions with water. In his circa 375 BC play *Semele or Dionysus*, Eubulus has Dionysus say:

> "*Three bowls do I mix for the temperate: one to health, which they empty first, the second to love and pleasure, the third to sleep. When this bowl is drunk up, wise guests go home. The fourth bowl is ours no longer, but belongs to violence; the fifth to uproar, the sixth to drunken revel, the seventh to black eyes, the eighth is the policeman's, the ninth belong to biliousness and the tenth to madness and hurling the furniture*".

Chemical Composition

Natural Phenols and Polyphenols

Although red wine contains many other chemicals which may have health benefits, resveratrol has been studied the most. Resveratrol and other such compounds mainly fall in the category of phenolics. Cinnamates have been shown to have more antioxidant activity when exposed *in vitro* to the Fenton reaction (catalytic Fe (II) with hydrogen peroxide) than the other natural phenols present in wine.

Resveratrol

Research on potential health effects of resveratrol is in its infancy and the long-term effects of supplementation in humans are not known. Resveratrol is a stilbenoid phenolic compound found in wine produced in the grape skins and leaves of grape vines. It has received a lot of attention in both the media and medical research community for its potential health benefits. The production and concentration of resveratrol is not equal among all the varieties of wine grapes. Differences in clones, rootstock, *Vitis* species as well as climate conditions can affect the production of resveratrol. Also, because resveratrol is part of the defence mechanism in grapevines against attack by fungi or grape disease, the degree of exposure to fungal infection and grape diseases also appear to play a role. The Muscadinia family of vines, which has adapted over time through exposure to North American grape diseases such as phylloxera, has some of the highest concentrations of resveratrol among wine grapes. Among the European *Vitis vinifera*, grapes derived from the Burgundian Pinot family tend to have substantially higher amounts of resveratrol than grapes derived from the Cabernet family of Bordeaux. Wine regions with cooler, wetter climates that are more prone to grape disease and fungal attacks such as Oregon and New York tend to produce grapes with higher concentrations of resveratrol than warmer, dry climates like California and Australia. Although red wine and white vine varieties produce similar amounts of resveratrol, red wine contains more than white, since red wines are produced by maceration (soaking the grape skins in the mash). Other winemaking techniques, such as the use of certain strains of yeast during fermentation or lactic acid bacteria during malolactic fermentation, can have an influence on the amount of resveratrol left in the resulting wines. Similarly the use of certain fining agents during the clarification and stabilization of wine can strip the wine of some resveratrol molecules.

The prominence of resveratrol in the news and its association with positive health benefits has encouraged some wineries to highlight it in their marketing. In the early 21st century, the Oregon producer Willamette Valley Vineyards sought approval from the Alcohol and Tobacco Tax and Trade Bureau (TTB) to state on their wine labels the resveratrol levels of their wines which ranged from 19 to 71 micromoles per liter (higher than the average 10 micromoles per liter in most red wines). The TTB gave preliminary approval to the winery, making it the first to use such information on its labels. While resveratrol is the most widely publicized, there are other components in wine that have been the focus of medical research into potential health benefits. These include the compounds catechin and quercetin.

Anthocyanins

Red grapes are high in anthocyanins which are the source of the colour of various fruits, such as red grapes. The darker the red wine, the more anthocyanins present. Following dietary ingestion, anthocyanins undergo rapid and extensive metabolism that makes the biological effects presumed from *in vitro* studies unlikely to apply *in vivo*. Although anthocyanins are under basic and early-stage clinical research for a variety of disease conditions, there exists no sufficient evidence that they have any beneficial effect in the human body. The US FDA has issued warning letters, *e.g.*, to emphasize that anthocyanins are not a defined nutrient, cannot be assigned a dietary content level and are not regulated as a drug to treat any human disease.

Effect on the Body

Bones

Heavy alcohol consumption has been shown to have a damaging effect on the cellular processes that create bone tissue and long term alcoholic consumption at high levels increases the frequency of fractures. Epidemiological studies (studies done by interviewing subjects and studying their health records) have found a positive association between moderate alcohol consumption and increased bone mineral density (BMD). Most of this research has been conducted with postmenopausal women, but one study in men concluded that moderate consumption of alcohol may also be beneficial to BMD in men.

Cancer

The International Agency for Research on Cancer of the World Health Organization has classified alcohol as a Group 1 carcinogen. Research is ongoing with no conclusive results about the effect of alcohol consumption and cancer, though some studies suggest that moderate wine consumption may actually lower the risk for lung, ovarian and prostate cancer. In early 2009, three independent studies published in *Gastroenterology* suggest that moderate wine consumption may reduce the risk of certain forms of esophageal cancers such as esophageal adenocarcinoma and the precancerous condition Barrett's esophagus. In one study, conducted by Kaiser Permanente in California, respondents who reported drinking no more than 1 glass of wine a day had a 56 per cent decrease in the risk for developing Barrett's esophagus a rate lower than that of heavy and non-drinkers. Research conducted at the Yale School of Public Health in 2009; suggest that wine may have some protective benefits against some forms of cancer. Women diagnosed with non-Hodgkin's lymphoma were questioned about their alcohol consumption patterns and followed for an 8 to 12-year period. Compared to non-drinkers, women who had been drinking wine for at least 25 years prior were 33 per cent less likely to die over the five-year period following diagnosis and 26 per cent less likely to experience a relapse or develop a secondary cancer during that same five-year period. Of all the women in the study, 75 per cent of those who drank at least 12 glasses of wine over the course of their lifetime were alive after five years compared to 66 per cent of the women who never drank any wine. Women who drank beer and alcohol spirits showed no differences.

Cardiovascular System

Studies have shown that heavy drinkers put themselves at greater risk for

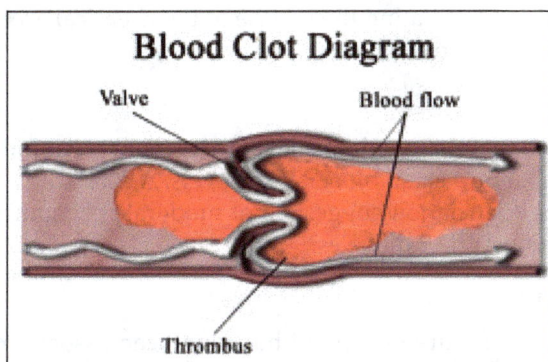

Figure 17.13: The Anticoagulant Properties of Wine may have the Potential Benefits of Reducing the Risk of Blood Clots that can Lead to Heart Disease.

heart disease and developing potentially fatal cardiac arrhythmias. Excessive alcohol consumption can cause higher blood pressure, increase cholesterol levels and weakened heart muscles. For moderate drinkers, medical research indicates moderate wine consumption may lower the mortality rate and risk of heart disease. Studies have shown that moderate wine drinking can improve the balance of low-density lipoprotein (LDL or "bad" cholesterol) to high-density lipoprotein (HDL "good" cholesterol), which has been theorized as to clean up or remove LDL from blocking arteries. The main cause of heart attacks and the pain of angina is the lack of oxygen caused by blood clots and atheromatous plaque build up in the arteries. The alcohol in wine has anticoagulant properties that limits blood clotting by making the platelets in the blood less prone to stick together and reducing the levels of fibrin protein that binds them together. However these anticoagulant properties of wine only stay in the system for a maximum of 24 hours after consumption. While having a glass of wine the night before may lower the risk of having a heart attack the next day, there is still the potential of long-term effects of alcohol. These anticoagulant properties can also be amplified adversely by binge drinking, with the individual becoming over-anticoagulated and at increase risk of a stroke or heart attack.

Dementia and Mental Functions

One of the short-term effects of alcohol is impaired mental function, which can cause behavioral changes and memory impairment. Long term effects of heavy drinking can inhibit new brain cell development and increase the risk for developing major depressive disorders. Studies have linked moderate alcohol consumption to lower risk of developing Alzheimer's and dementia though wine's role in this link is not yet fully understood. A 2009 study by Wake Forest University School of Medicine suggest that moderate alcohol consumption may help healthy adults ward off the risks of developing dementia but can accelerate declining memory for those already suffering from cognitive impairment. The reason for the potential positive benefit of moderate consumption is not yet identified and may even be unrelated to the alcohol but rather other shared lifestyle factors of moderate drinkers (such as exercise or diets). If it is the moderate consumption, researchers theorize that it may be alcohol's role in promoting the production of "good cholesterol" which prevents blood platelets from sticking together. Another potential role of alcohol in the body may be in stimulating the release of the chemical acetylcholine which influences brain function and memory.

Diabetes

Research has shown that moderate levels of alcohol consumed with meals do not have a substantial impact on blood sugar levels. A 2005 study presented to the American Diabetes Association suggests that moderate consumption may lower the risk of developing Type 2 diabetes.

Digestive System

The anti-bacterial nature of alcohol has long been associated with soothing stomach irritations and ailments like traveler's diarrhea where it was a preferred treatment to the less palatable bismuth treatments. The risk of infection from the bacterium *Helicobacter pylori*, strongly associated with causing gastritis and peptic ulcers as well as being closely linked to stomach cancer appears to lessen with moderate alcohol consumption. A German study conducted in the late 1990s showed that non-drinkers had slightly higher infection rates of *Helicobacter pylori* than moderate wine and beer drinkers. Wine's positive effects on the metabolism of cholesterol has been suggested as a link to lower occurrences of gallstones among moderate drinkers since cholesterol is a major component of gallstones.

Headaches

There are several potential causes of so-called "red wine headaches", including histamines/tyramines and the breakdown of some phenolic compounds in wine that carry the chemical messenger for serotonin. One culprit that is regularly dismissed by allergists as an unlikely cause of red wine headaches is sulfites which are used as a preservative in wine. Wine, like other alcoholic beverages, is a diuretic which promotes dehydration that can lead to headaches (such as the case often experienced with hangovers). In 2006, researchers from the University of California, Davis announced finding from genetic mapping that amino acids in wine that have been slightly modified by the fermentation process may be the cause of wine related headaches. The research suggests changes in fermentation techniques may help alleviate the risk for wine drinkers sensitive to these amino acids.

Calorie Intake

Compared to many beers and non-diet sodas, a serving of wine has a moderate amount of calories. A standard 5 oz serving of red wine (based on an average alcohol content of 13 per cent) contains approximately 106 calories and 2.51 g of carbohydrates. A similar serving of white wine contains approximately 100 calories and 1.18g of carbohydrates.

Psychological and Social

Danish epidemiological studies suggest that a number of psychological health benefits are associated with drinking wine. In a study testing this idea, Mortensen *et al.* (2001) measured socioeconomic status, education, IQ, personality, psychiatric symptoms and health related behaviors, which included alcohol consumption. The analysis was then broken down into groups of those who drank beer, those who drank wine and then those who did and did not drink at all. The results showed that for both men and women drinking wine was related to higher parental social status, parental education and the social status of the subjects. When the subjects were given an IQ test, wine drinkers consistently scored higher IQs than their counterpart beer drinkers. The average difference of IQ between wine and beer drinkers was 18 points. In regards to psychological functioning, personality and other health-related behaviors, the study found wine drinkers to operate at optimal levels while beer drinkers performed below optimal levels. As these social and psychological factors also correlate with health outcomes, they represent a plausible explanation for at least some of the apparent health benefits of wine.

Heavy Metals in Wine Controversy

In 2008, researchers from Kingston University in London discovered red wine to contain high levels of toxic metals relative to other beverages in the sample. Although the metalions, which included chromium, copper, iron, manganese, nickel, vanadium and zinc, were also present in other plant-based beverages, the sample wine tested significantly higher for all metal ions, especially vanadium. Risk assessment was calculated using "target hazard quotients" (THQ), a method of quantifying health concerns associated with lifetime exposure to chemical pollutants. Developed by the Environmental Protection Agency in the US and used mainly to examine seafood, a THQ of less than 1 represents no concern while, for example, mercury levels in fish calculated to have THQs of between 1 and 5 would represent cause for concern. The researchers stressed that a single glass of wine would not lead to metal poisoning, pointing out that their THQ calculations were based on the average person drinking one-third of a bottle of wine (250 ml) every day between the ages of 18 and 80. However the "combined THQ values" for metal ions in the red wine they analyzed were reported to be as high as 125. A subsequent study by the same university using a meta analysis of data based on wine samples from a selection of mostly European countries found equally high levels of vanadium in many red wines, showing combined THQ values in the range of 50 to 200, with some as high as 350. The findings sparked immediate controversy due to several issues: the study's reliance on secondary data; the assumption that all wines contributing to that data were representative of the countries stated; and the grouping together of poorly understood high-concentration ions, such as vanadium, with relatively low-level, common ions such as copper and manganese. While some publications printed lists of countries showing the "worst offenders" and reported that wine from other countries did not pose a health risk, others pointed out that the lack of identifiable wines and grape varieties, specific producers or even wine regions, provided only misleading generalizations that should not be relied upon in choosing wines.

In a news bulletin following the widespread reporting of the findings, the UK's National Health Service (NHS) were also concerned that "the way the researchers added together hazards from different metals to produce a final score for individual wines may not be particularly meaningful". Commentators in the US questioned the relevance of seafood-based THQ assessments to agricultural produce, with the TTB, responsible for testing imports for metal ion contamination, have not detected an increased risk. George Solas, quality assessor for the Canadian Liquor Control Board of Ontario (LCBO) claimed that the levels of heavy metal contamination reported were within the permitted levels for drinking water in tested reservoirs. Whereas the NHS also described calls for improved wine labeling as an "extreme response" to research which provided "few solid answers", they acknowledged the authors call for further research to investigate wine production, including the influence that grape variety, soil type, geographical region, insecticides, containment vessels and seasonal variations may have on metal ion uptake.

References

Abrahams, S., Tanner, G.J., Larkin, P.J., Ashton, A.R. (2002). Identification and biochemical characterization of mutants in proanthocyanidin pathway in Arabidopsis. *Plant Physiol*, 130: 561-576.

Abrahams, S., Tanner, G.J., Larkin, P.J., Ashton, A.R. (2002). Identification and biochemical characterization of mutants in proanthocyanidin pathway in Arabidopsis. *Plant Physiol*. 130: 561-576.

Achmadi, S., Syahbirin, G., Choong, E.T., Hemingway, R.W. (1994). Catechin-3-O-rhamnoside chain extender units in polymeric procyanidins from mangrove bark. *Phytochemistry* 35: 217-219.

Adam, W. B., Hardy, F., Nierenstein, M. (1931). "The Catechin of the Cacao Bean". *Journal of the American Chemical Society* 53 (2): 727.

Ageorges, A., Fernandez, L., Vialet, S., Merdinoglu, D., Terrier, N., Romieu, C. (2006). Four specific isogenes of the anthocyanin metabolic pathway are systematically co-expressed with the red colour of grape berries. *Plant Sci.* 170: 372-383.

Akinjogunla O. J., Yah C. S., Eghafona N. O. and Ogbemudia F. O. (2010). Antibacterial activity of leave extracts of *Nymphaea lotus* (Nymphaeaceae) on Methicillin resistant *Staphylococcus aureus* (MRSA) and Vancomycin resistant*Staphylococcus aureus* (VRSA) isolated from clinical samples". *Annals of Biological Research*, 1 (2), pages 174–184

Akiyama H., Fujii K., Yamasaki O., Oono T., Iwatsuki K. (2001). "Antibacterial action of several tannins against *Staphylococcus aureus*". *J. Antimicrob. Chemother.*48 (4): 487–91.

Alcohol drinking. Lyon: World Health Organization, International Agency for Research on Cancer. 1988. pp. 2–3.

Ashutosh Kar (2003). *Pharmacognosy And Pharmacobiotechnology*. New Age International. pp. 44.

Athanasiadou S. L., Kyriazakis, I., Jackson, F. and Coop, R. L. (2001). Direct anthelminthic effects of condensedtannins towards different gastrointestinal nematodes of sheep: *in vitro* and *in vivo* studies. *Vet. Parasitol.* (99),205-219.

Augustin Scalbert, (1992). Quantitative Methods for the Estimation of Tannins in Plant Tissues. Plant Polyphenols, Basic Life Sciences, Volume 59, pages 259-280.

Ayabe, S., Akashi, T. (2006). Cytochrome P450s in flavonoid metabolism. *Phytochem. Rev.* 5: 271- 282.

Bais, H. P., Vepachedu, R., Gilroy, S., Callaway, R. M., Vivanco, J. M. (2003). Allelopathy and exotic plant invasion: from molecules and genes to species interactions. *Science* 301: 1377-1380.

Bajaj, Y. P. S. (1988). *Medicinal and aromatic plants*. Biotechnology in agriculture and forestry 24. Berlin: Springer-Verlag.

Barahona R., Lascano, C. E., Cochran, R. C., Morril, J. L. and Titgemeyer, E. C. (1997). Intake, digestion andnitrogen utilization by sheep fed tropical legumes with contrasting tannin concentration and astringency. *J.Anim. Sci.* (75), 1633-1640.

Barry T. N. and Forss, D. A. (1983). The condensed tannin content of vegetative Lotus pedunculatus, its regulation byfertilizer application and effect upon protein solubility. *J. Sci. Food and Agric.* (34), 1047-1056.

Barry T. N. (1989). Condensed tannins: their role in ruminant protein and carbohydrate digestion and possibleeffects upon the rumen ecosystem. In: J. V. Nolan, R. A. Leng, D. I. Demeyer (eds.). *The Roles of Protozoa and Fungi in Ruminant Digestion.* Armidale NSW 2351, Australia: Penambul Books.

Barry T. N. and McNabb, W. C. (1999). The implications of condensed tannins on the nutritive value of temperateforages fed to ruminants. *Br. J. Nutr.* (81), 263-272.

Bate-Smith and Swain (1962). "Flavonoid compounds". In Florkin M., Mason H. S.*Comparative biochemistry* III. New York: Academic Press. pp. 75–809.

Baudry, A., Caboche, M., Lepiniec, L. (2006). TT8 controls its own expression in a feedback regulation involving TTG1 and homologous MYB and bHLH factors, allowing a strong and cell-specific accumulation of flavonoids in *Arabidopsis thaliana. Plant J.* 46: 768-779.

Baudry, A., Heim, M.A., Dubreucq, B., Caboche, M., Weisshaar, B., Lepiniec L. (2004). TT2, TT8 and TTG1 synergistically specify the expression of BANYULS and proanthocyanidin biosynthesis in *Arabidopsis thaliana. Plant J.* 39: 366-380.

Bavage, A.D., Davies, I.G., Robbins, M.P., Morris, P. (1997). Expression of an Antirrhinum dihydroflavonol reductase gene results inchanges in condensed tannin structure and accumulation in root cultures of *Lotus corniculatus* (bird's foot trefoil). *Plant Mol. Biol.* 35: 443-458.

Bavaresco L., Fregoni M., Trevisan M., Mattivi F., Vrhovsek U, Falchetti R. (2002). "The occurrence of piceatannol in grape". *Vitis* 41 (3): 133–6.

Baxter, I.R., Young, J.C., Armstrong, G., Foster, N., Bogenschutz, N., Cordova, T., Peer, W.A., Hazen, S.P., Murphy, A.S., Harper, J.F. (2005). A plasma membrane H+ -ATPase is required for the formation of proanthocyanidins in the seed coat endothelium of *Arabidopsis thaliana. Proc. Natl. Acad. Sci. USA* 102: 2649-2654.

Bednar, R.A., Hadcock, J.R. (1988). Purification and Characterizationo of Chalcone Isomerase from Soybeans. *J. Biol. Chem.* 263: 9582-9588.

Bisanda E. T. N., Ogola W. O., Tesha J. V. (2003). "Characterisation of tannin resin blends for particle board applications". *Cement and Concrete Composites* 25 (6): 593–8.

Bogs, J., Downey, M.O., Harvey, J.S., Ashton, A.R., Tanner, G.J., Robinson, S.P. (2005). Proanthocyanidin synthesis and expression of genes encoding leucoanthocyanidin reductase and anthocyanidin reductase in developing grape berries and grapevine leaves. *Plant Physiol.* 139: 652-663.

Bogs, J., Jaffé, F.W., Takos, A.M., Walker, A.R., Robinson, S.P. (2007). The grapevine transcription factor VvMYBPA1 regulates proanthocyanidin synthesis during fruit development. *Plant Physiol.* 143: 1347-1361.

Boralle N, Gottlieb H.E, Gottlieb O.R, Kubitzki K, Lopes L.M.X, Yoshida M and Young M.C.M. (1993). Oligostilbenoids from *Gnetum venosum*. *Phytochemistry*, vol. 34, no 5, pages 1403-1407.

Borevitz, J., Xia, Y., Blount, J.W., Dixon, R.A., Lamb, C. (2000). Activation tagging identifies a conserved MYB regulator of phenylpropanoid biosynthesis. *Plant Cell* 12: 2383-2393.

Botha, J.J., Ferreira, D., Roux, D.G. (1981). Synthesis of condensed tannins. Part 4. A direct biomimetic approach to [4,6]- and [4,8]- biflavonoids. *J. Chem. Soc., Perkin Trans.* 1, 1235-1245.

Braidot, E., Petrussa, E., Bertolini, A., Peresson, C., Ermacora, P., Loi, N., Terdoslavich, M., Passamonti, S., Macrì, F., Vianello, A. (2008). Evidence for a putative flavonoid translocator similar to mammalian bilitranslocase in grape berries (*Vitis vinifera* L.) during ripening. *Planta* 228: 203-213.

Brenner, Hermann, Berg, Gabriele, Lappus, Nicole, Kliebsch, Ulrike, Bode, Günter, Boeing, Heiner (1999). "Alcohol Consumption and Helicobacter pylori Infection: Results from the German National Health and Nutrition Survey". *Epidemiology* 10 (3): 214–8.

Britsch, L., Dedio, J., Saedler, H., Forkmann, G. (1993). Molecular characterization of flavanone 3β- hydroxylases. Consensus sequence, comparison with related enzymes and the role of conserved histidine residues. *Eur. J. Biochem.* 217: 745-754.

Brouillard R, S. Chassaing and A. Fougerousse, (2003). Why are grape/fresh wine anthocyanins so simple and why is it that red wine colour lasts so long? *Phytochemistry*, Volume 64, Issue 7, Pages 1179-1186.

Broun, P. (2005). Transcriptional control of flavonoid biosynthesis: a complex network of conserved regulators involved in multiple aspects of differentiation in Arabidopsis. *Curr. Opin. Plant Biol.* 8: 272-279.

Brown DE, Rashotte AM, Murphy AS *et al.* (2001). "Flavonoids act as negative regulators of auxin transport *in vivo* in arabidopsis". *Plant Physiol.* 126 (2): 524–35.

Bruyne, T.D., Pieters, L., Deelstra, H., Vlietinck, A. (1999). Condensed vegetable tannins: Biodiversity in structure and biological activities. *Biochem. Syst. Ecol.* 27: 445-459.

Buzzini, P., Turchetti, B., Ieri, F., Goretti, M., Branda, E., Mulinacci, N., Romani, A. (2007). Catechins and proanthocyanidins: naturally occurring O-heterocycles with antimicrobial activity. *Top Heterocycl. Chem.* 10: 239-263.

Cain, C.C., Saslowsky, D.E., Walker, R.A., Shirley, B.W. (1997). Expression of chalcone synthase and chalcone isomerase proteins in Arabidopsis seedlings. *Plant Mol. Biol.* 35: 377-381.

Carcinogenic Potency Project Official Website

Cardoso Do Vale, (1962). Chemical study of bark from *Commiphora angolensis* Engl. *J. Bol Escola Farm Univ Coimbra Edicao Cient*, volume 3, page 128 (abstract).

Carey, C.C., Strahle, J.T., Selinger, D.A., Chandler, V.L. (2004). Mutations in the pale aleurone colour1 regulatory gene of the Zea mays anthocyanin pathway have distinct phenotypes relative to the functionally similar TRANSPARENT TESTA GLABRA1 gene in *Arabidopsis thaliana*. *Plant Cell*, 16: 450-464.

Carron, T.R., Robbins, M.P., Morris, P. (1994). Genetic modification of condensed tannin biosynthesis in *Lotus corniculatus*. 1. *Heterologous antisense* dihydroflavonol reductase down-regulates tannin accumulation in 'hairy root' cultures. *Theor. Appl. Genet.* 87: 1006-1015.

Castellarin, S.D., Matthews, M.A., Gaspero, G.D., Gambetta, G.A., (2007). Water deficits accelerate ripening and induce changes in gene expression regulating flavonoid biosynthesis in grape berries. *Planta* 227: 101-112.

Chafe S. C. and Durzan, D. J. (1973). The development of the secretory cells of Ricinus and the problem of cellular differentiation. *Planta*. (113), 251-262

ChaichiSemsari M., MaheriSis, N., Sadaghian, M., Eshratkhah, B. and Hassanpour S. (2011). Effects of administration of industrial tannins on nutrient excretion parameters during naturally acquired mixed nematodeinfections in Moghani sheep. *J. Amer. Sci.*, 7(6), 245-248.

Chen Xiangming,Chen Heru and Li Weibin, (2006). Study on Fast Determination Content of Condensed Tannin Using Stiasny Method. Guangdong Chemical Industry, (abstract).

Chen, Z.-H., Nimmo, G.A., Jenkins, G.I., Nimmo, H.G. (2007). BHLH32 modulates several biochemical and morphological processes that respond to Pi starvation in Arabidopsis. *Biochem. J.* 405: 191-198.

Cheynier V, Duenas-Paton M, Salas E, Maury C, Souquet JM, Sarni-Manchado P, Fulcrand H (2006). "Structure and properties of wine pigments and tannins". *American Journal of Enology and Viticulture* 57: 298–305.

Cheynier, V. (2005). Polyphenols in foods are more complex than often thought. *Am. J. Clin. Nutr.* 81: 223S-229S.

Claire Dufour and Claude L. Bayonove, (1999). Interactions between Wine Polyphenols and Aroma Substances. An Insight at the Molecular Level. *J. Agric. Food Chem.*, 47 (2), pages 678–684.

Clifford M. N., Ramirez-Martinez J. R. (1991). "Tannins in wet-processed coffee beans and coffee pulp". *Food Chemistry* 40 (2): 191–200.

Cominelli, E., Gusmaroli, G., Allegra, D., Galbiati, M., Wade, H.K., Jenkins, G.I., Tonelli, C. (2008). Expression analysis of anthocyanin regulatory genes in response to different light qualities in *Arabidopsis thaliana*. *J. Plant Physiol.* 165: 886-894.

Conde, Elvira, Cadahía, Estrella, García-Vallejo, María Concepción, Fernández De Simón, Brígida (1998). "Polyphenolic Composition ofQuercus suberCork *from Different* spanish Provenances". *Journal of Agricultural and Food Chemistry* 46 (8): 3166.

Condensed tannins. Porter L. J. (1989). in *Natural Products of Woody Plants I*, Rowe J. W. (ed), Springer-Verlag: Berlin, Germany, pages 651–690

Cone, K.C. (2007). Anthocyanin synthesis in maize aleurone tissue. *Plant Cell Monogr.* 8: 121-139.

Corder R, Mullen W, Khan NQ (2006). "Oenology: red wine procyanidins and vascular health". *Nature* 444 (7119): 566.

Corder R., Mullen, W. and Khan, N. Q. (2006). Oenology: red wine procyanidins and vascular health. *Nature*, 444(7119): 566- 570. doi: 10.1038/444566a. PMID 17136085.

Creasey, L.L., Swain, T. (1965). Structure of condensed tannins. *Nature.* 208: 151-153.

Debeaujon, I., Nesi, N., Perez, P., Devic, M., Grandjean, O., Caboche, M., Lepiniec, L. (2003). Proanthocyanidin-accumulating cells in Arabidopsis testa: regulation of differentiation and role in seed development. *Plant Cell 15*: 2514-2531.

Debeaujon, I., Peeters, A.J.M., Léon-Kloosterziel, K.M., Koornneef, M. (2001). The TRANSPARENT TESTA 12 gene of Arabidopsis encodes a multidrug secondary transporter-like protein required for flavonoid sequestration in vacuoles of seed coat endothelium. *Plant Cell* 13: 853-871.

Degen A. A., Becker, K., Makkar, H. P. S. and Borowy, N. (1995). Acacia saligna as a fodder for desert livestock andthe interaction of its tannins with fiber fractions. *J. Sci.of Food Agric.* (68), 65-71.

Degradation of oligomeric procyanidins and anthocyanins in a Tinta Roriz red wine during maturation. C. Dallas, J.M. Ricardo-Da-Silva and Olga Laureano, Vitis, 1995, volume 34, issue 1, pages 51-56.

Deijs, W. B. (1939). Catechins isolated from tea leaves. Recueil des Travaux Chimiques des Pays-Bas, Volume 58, Issue 9, pages 805–830.

Delcour, J.A., Ferreira, D., Roux, D.G. (1983). Synthesis of condensed tannins. Part 9. The condensation sequence of leucocyanidin with (+)-catechin and with the resultant procyanidins. *J. Chem. Soc., Perkin Trans.* 1, 1711-1717.

Deluc, L., Barrieu, F., Marchive, C., Lauvergeat, V., Decendit, A., Richard, T., Carde, J.P., Mérillon, J.M., Hamdi, S. (2006). Characterization of a grapevine R2R3-MYB transcription factor that regulates the phenylpropanoid pathway. *Plant Physiol.* 140: 499-511.

Devic, M., Guilleminot, J., Debeaujon, I., Bechtold, N., Bensaude, E., Koornneef, M., Pelletier, G., Delseny, M. (1999). The BANYULS gene encodes a DFR-like protein and is a marker of early seed coat development. *Plant J.* 19: 387-398.

Dixon, R.A., Xie, D.Y., Sharma, S.B. (2005). Proanthocyanidins - a final frontier in flavonoid research? *New Phytol.* 165: 9-28.

Drabble, E., Nierenstein, M. (1907). "On the Rôle of Phenols, Tannic Acids and Oxybenzoic Acids in Cork Formation". *Biochemical Journal* 2 (3): 96–102.1.

Dubos C., Le Gourrierec, J., Baudry, A., Huep, G., Lanet, E., Debeaujon, I., Routaboul, J.-M., Alboresi, A., Weisshaar, B., Lepiniec, L. (2008). MYBL2 is a new regulator of flavonoid biosynthesis in *Arabidopsis thaliana*. *Plant J.* 55: 940-953.

El-Kereamy, A., Chervin, C., Roustan, J.P., Cheynier, V., Souquet, J.M., Moutounet, M., Raynal, J., Ford, C., Latché, A., Pech, J.C., Bouzayen, M. (2003). Exogenous ethylene stimulates the long-term expression of genes related to anthocyanin biosynthesis in grape berries. *Physiol. Plant.* 119: 175-182.

Elsevier Science Ltd. All rights reserved. PII: S0169-5347(00)01861-9.

Endt, D.V., Kijne, J.W., Memelink, J. (2002). Transcription factors controlling plant secondary metabolism: what regulates the regulators? *Phytochemistry* 61: 107-114.

Eric Verkaik, Anne G. Jongkindet, Frank Berendse, (2006). Short-term and long-term effects of tannins on nitrogen mineralisation and litter decomposition in kauri (*Agathis australis* (D. Don) Lindl.) forests". *Plant And Soil*, Volume 287, Numbers 1–2, pages 337–345.

Escribano-Bailon, T., Dangles O., Brouillard, R. (1996). Coupling reactions between flavylium ions and catechin. *Phytochemistry* 41: 1583-1592.

Escudier J.L., Kotseridis Y. and Moutounet M. (2002). Flash release and wine quality. Progrès Agricole et Viticole, 2002 (French)

Fahey Jr. G. C. and Jung, H.-J. G. (1989). Phenolic compounds in forages and fibrous feedstuffs. In: P.R. Cheeke(ed.). *Toxic. Plant Origin*, CRC Press, Inc. Boca Raton, Fla. (4), 123-127.

Feeny P. P. (1970). Seasonal changes in oak-leaf tannins and nutrients as a cause of spring feeding by winter mothcaterpillars. *Ecol.* (51), 565-581.

Ferreira, D., Bekker, R. (1996). Oligomeric proanthocyanidins: naturally occurring O-heterocycles. *Nat. Prod. Rep.* 13: 411-433.

Ferreira, D., Guyot, S., Marnet, N., Delgadillo, I., Renard, C.M., Coimbra, M.A. (2002). Composition of phenolic compounds in a Portuguese pear (*Pyrus communis* L. var. *S. Bartolomeu*) and changes after sun-drying. *J. Agric. Food Chem.* 50: 4537-4544.

Ferreira, D., Li, X.C. (2000). Oligomeric proanthocyanidins: naturally occurring O-heterocycles. *Nat. Prod. Rep.* 17: 193-212.

Ferreira, D., Slade, D. (2002). Oligomeric proanthocyanidins: naturally occurring O-heterocycles. *Nat. Prod. Rep.* 19: 517-541.

Flavonol profiles of Vitis vinifera red grapes and their single-cultivar wines. Castillo-Munoz Noelia, Gomez-Alonso Sergio, Garcia-Romero Esteban and Hermosin-Gutierrez Isidro (2007). *Journal of agricultural and food chemistry*, Vol. 55, no3, pp. 992-1002.

Fletcher, A.C., Porter, L.J., Haslam, E., Gupta, R.K. (1977). Plant proanthocyanidins. Part 3. Conformational and configurational studies of natural procyanidins. *J. Chem. Soc., Perkin Trans.* 1, 1628-1637.

Foo L.Y., Jones, W.T., Porter, L. J. and Williams, V. M. (1982). Proanthocyanidin polymers of fodder legumes. *Phytochemis.* (21), 933-935.

Ford, C.M., Boss, P.K., Høj, P.B. (1998). Cloning and characterization of Vitis vinifera UDPglucose: flavonoid 3-O-glucosyltransferase, a homologue of the enzyme encoded by the maize Bronze-1 locus that may primarily serve to glucosylate anthocyanidins *in vivo. J. Biol Chem.* 273: 9224-9233.

Forrest, G.I., Bendall, D.S. (1969). The distribution of polyphenols in the tea plant (*Camellia sinensis* L.). *Biochem J.* 113: 741-755.

Franz, M. (2005). "Merlot's Bad Press". *Washington Post.*

Fulcrand H, Duenas M, Salas E, Cheynier V (2006). "Phenolic reactions during winemaking and aging". *American Journal of Enology and Viticulture* 57: 289–297.

Gakh andrei A., Anisimova, Natalia Yu, Kiselevsky, Mikhail V., Sadovnikov, Sergey V., Stankov, Ivan N., Yudin, Mikhail V., Rufanov, Konstantin A., Krasavin, Mikhail Yu, Sosnov andrey V. (2010). "Dihydro-resveratrol A potent dietary polyphenol". *Bioorganic and Medicinal Chemistry Letters* 20 (20): 6149.

Gatto P., Vrhovsek U., Muth J., Segala C., Romualdi C., Fontana P., Pruefer D., Stefanini M., Moser C., Mattivi F. and Velasco R. (2008). "Ripening and genotype control stilbene accumulation in healthy grapes". *J Agric Food Chem.* 56 (24): 11773–85.

Gehm, B. D., McAndrews, JM, Chien, PY, Jameson, JL (1997). "Resveratrol, a polyphenolic compound found in grapes and wine, is an agonist for the estrogen receptor". *Proceedings of the National Academy of Sciences* 94 (25): 14138–43.

Geiss, F., Heinrich, M., Hunkler, D., Rimpler, H. (1995). Proanthocyanidins with (+)-epicatechin units from *Byrsonima crassifolia* bark. *Phytochemistry* 39: 635-643.

Gislason, Nick Emil, Currie, Bruce Lamonte, Waterhouse andrew Leo (2011). "Novel Antioxidant Reactions of Cinnamates in Wine". *Journal of Agricultural and Food Chemistry* 59 (11): 6221–6.

González S., Pabón M. L., Carulla J. (2002). "Effects of tannins on *in vitro* ammonia release and dry matter degradation of soybean meal". *Arch. Latinoam. Prod. Anim.* 10(2): 97–101.

Gonzalez, A., Zhao, M., Leavitt, J.M. Lloyd, A.M. (2008). Regulation of the anthocyanin biosynthetic pathway by the TTG1/bHLH/Myb transcriptional complex in Arabidopsis seedlings. *Plant J.* 53: 814-827.

Gonzalo H., Frutos, P., Giráldez, F. J., Mantecón, Á. R. and Álvarez Del Pino, M. C. (2003). Effect of different dosesof quebracho tannins extract on rumen fermentation in ewes. *Anim. Feed Sci. Technol.* (109), 65-78.

Goplen, B.P., Howarth, R.E., Sarkar, S.K., Lesins, K. (1980). A search for condensed tannins in annual and perennial species of *Medicago, Trigonella* and *Onobrychis. Crop Sci.* 20: 801-804.

Gottlieb O. R. (1990). Phytochemical differentiation and function. _Phytochemis._ (29), 1715- 1724.

Grasser, Georg (1922). _Synthetic Tannins._ F. G. A. Enna. (trans.).

Gross P (2009), _New Roles for Polyphenols. A 3-Part report on Current Regulations and the State of Science,_ Nutraceuticals World

Gu, L., Kelm, M.A., Hammerstone, J.F., Beecher, G., Holden, J., Haytowitz, D., Prior, R.L. (2003). Screening of foods containing proanthocyanidins and their structural characterization using LCMS/MS and thiolytic degradation. _J. Agric. Food Chem._ 51: 7513-7521.

Gu, L., Kelm, M.A., Hammerstone, J.F., Zhang, Z., Beecher, G., Holden, J., Haytowitz, D., Prior, R.L. (2003). Liquid chromatographic/electrospray ionization mass spectrometric studies of proanthocyanidins in foods. J. Mass Spectrom. 38: 1272-1280.

Guilford, J. M., Pezzuto, J. M. (2011). "Wine and Health: A Review". _American Journal of Enology and Viticulture_ 62 (4): 471–86.

Guyot, S., Marnet, N., Drilleau, J. (2001). Thiolysis-HPLC characterization of apple procyanidins covering a large range of polymerization states. _J. Agric. Food Chem._ 49: 14-20.

Guyot, S., Vercauteren, J., Cheynier, V. (1996). Structural determination of colourless and yellow dimers resulting from (+)-catechin coupling catalyzed by grape polyphenoloxidase. _Phytochemistry._ 42: 1279-1288.

Habauzit, V., Morand, C. (2011). "Evidence for a protective effect of polyphenols-containing foods on cardiovascular health: An update for clinicians". _Therapeutic Advances in Chronic Disease_ 3 (2): 87–106.

Hagerman A. E. and Buttler, L. G. (1981). The specificity of proanthocyanid in-protein interactions. _J. Bio. Chem.,_(256), 4494-4497.

Hague, Theresa, Petroczi andrea andrews, Paul LR, Barker, James, Naughton, Declan P (2008). "Determination of metal ion content of beverages and estimation of target hazard quotients: A comparative study". _Chemistry Central Journal_ 2: 13.

Halliwell B (2007). "Dietary polyphenols: Good, bad, or indifferent for your health?". _Cardiovasc Res_ 73 (2): 341–347.

Harding, G. (2005). _A Wine Miscellany._ New York: Clarkson Potter. pp. 66–70, 90 and 108.

Harry Snyder, (1893). Notes on Löwenthal's method for the determination of tannin. _J. Am. Chem. Soc.,_ volume 15, issue 10, pages 560–563.

Hartmann, U., Sagasser, M., Mehrtens, F., Stracke, R., Weisshaar, B. (2005). Differential combinatorial interactions of cis-acting elements recognized by R2R3-MYB, BZIP and BHLH factors control light-responsive and tissue-specific activation of phenylpropanoid biosynthesis genes. _Plant Mol. Biol._ 57: 155-171.

Haslam E. (1974). The Shikimate Pathway. New York: John Wiley and Sons.Haslam E. (1986). Hydroxybenzoic acid and the enigma of gallic acid. In: Conn, E.E.

(ed.). *The Shikimic Acid Pathway. Rec. Adv. Phytochem.*, Plenum Press, New York, (20), 163-200.

Haslam E. (1989). *Plant Polyphenols- Vegetable Tannins Revisited*. Camb. University Press, Cambridge, UK.

Haslam, Edwin (2007). "Vegetable tannins – Lessons of a phytochemical lifetime". *Phytochemistry* 68 (22–24): 2713–2721.

Hassanpour S., Sadaghian, M., MaheriSis, N., Eshratkhah, B. and ChaichiSemsari, M. (2011). Effect of condensedtannin on controlling faecal protein excretion in nematode-infected sheep: *in vivo* study. *J. Amer. Sci.*, 7(5), 896-900.

Hättenschwiler S. and Vitousek, P. M. (2000). The role of polyphenols in terrestrial ecosystem nutrient cycling. *Trends Ecol. Evol.*, 15(6), 238- 243.

He, Fei, Mu, Lin, Yan, Guo-Liang, Liang, Na-Na, Pan, Qiu-Hong, Wang, Jun, Reeves, Malcolm J., Duan, Chang-Qing (2010). "Biosynthesis of Anthocyanins and Their Regulation in Coloured Grapes". *Molecules* 15 (12): 9057–91.

Hemingway, R.W., Foo, L.Y., Porter, L.J. (2008). Linkage isomerism in trimeric and polymeric 2,3-cisprocyanidins. J. Chem. Soc., Perkin Trans. 1 1982, 1209-1216. *Molecules* 13: 2695

Hemingway, R.W., Laks, P.E. (1985). Condensed tannins: a proposed route to 2R,3R-(2,3-cis)- proanthocyanidins. J. *Chem. Soc., Chem. Commun.* 746-747.

Henry Richardson Procter, Edmund Stiasny and Harold Brumwel, E. and F.N. (1912). *Leather Chemists' Pocket-Book: A Short Compendium of Analytical Methods*. Spon, Limited, 1912 - 223 pages (book at google books)

Hirner, A.A., Veit, S., Seitz, H.U. (2001). Regulation of anthocyanin biosynthesis in UV-A-irradiated cell cultures of carrot and in organs of intact carrot plants. *Plant Sci.* 161: 315-322.

Holton, T.A., Brugliera, F., Lester, D.R., Tanaka, Y., Hyland, C.D., Menting, J.G.T., Lu, C.Y., Farcy, E., Stevenson, T.W., Cornish, E.C. (1993). Cloning and expression of cytochrome P450 genes controlling flower colour. *Nature* 366: 276-279.

Hunt, A.G., Xu, R., Addepalli, B., Rao, S., Forbes, K.P., Meeks, L.R., Xing, D., Mo, M., Zhao, H., Bandyopadhyay, A., Dampanaboina, L., Marion, A., Von Lanken, C., Li, Q.Q. (2008). Arabidopsis mRNA polyadenylation machinery: Comprehensive analysis of protein-protein interactions and gene expression profiling. *BMC Genomic.* 9: Article number 220.

Iida, N., Inatomi, Y., Murata, H., Inada, A., Murata, J., Lang, F.A., Matsuura, N., Nakanishi, T. (2007). A new flavone xyloside and two new flavan-3-ol glucosides from *Juniperus communis* var. *depressa. Chem. Biodivers.* 4: 32-42.

Ishida, T., Hattori, S., Sano, R., Inoue, K., Shirano, Y., Hayashi, H., Shibata, D., Sato, S. Kato, T., Tabata, S., Okada, K., Wadaa, T. (2007). Arabidopsis TRANSPARENT TESTA GLABRA2 is directly regulated by R2R3 MYB transcription factors and is involved in regulation of GLABRA2 transcription in epidermal differentiation. *Plant Cell* 19: 2531-2543.

Izabela Woc³awek-Potocka, Chiara Mannelli, Dorota Boruszewska, Ilona Kowalczyk-Zieba, Tomasz Waœniewski and Dariusz J. Skar¿yñski, (2013). "Diverse Effects of Phytoestrogens on the Reproductive Performance: Cow as a Model," (and references therein) *International Journal of Endocrinology*, Article ID 650984, 15 pages.

Jackson F. S., Barry, T. N., Lascano, C. E. and Palmer, B. (1996). The extractable and bound condensed tannincontent of leaves from tropical tree, shrub and forage legumes. *J. Sci. Food Agric.* (71), 103-110.

JáNváRy, LáSzló, Hoffmann, Thomas, Pfeiffer, Judith, Hausmann, Ludger, TöPfer, Reinhard, Fischer, Thilo C., Schwab, Wilfried (2009). "A Double Mutation in the Anthocyanin 5-O-Glucosyltransferase Gene Disrupts Enzymatic Activity in Vitis vinifera L". *Journal of Agricultural and Food Chemistry* 57 (9): 3512–8.

Jean-Marc Brillouet, Charles Romieu, Benoît Schoefs, Katalin Solymosi, Véronique Cheynier, Hélène Fulcrand, Jean-Luc Verdeil and Geneviève Conéjéro, (2013). The tannosome is an organelle forming condensed tannins in the chlorophyllous organs of Tracheophyta". *Annals of Botany*, September 11.

Jende-Strid, B. (1993). Genetic control of flavonoid biosynthesis in barley. *Hereditas* 119: 187-204.

Jeong, S.T., Goto-Yamamoto, N., Kobayashi, S., Esaka, M. (2004). Effects of plant hormones and shading on the accumulation of anthocyanins and the expression of anthocyanin biosynthetic genes in grape berry skins. *Plant Sci.* 167: 247-252.

Jez, J.M., Bowman, M.E., Dixon, R.A., Noel, J.P. (2000). Structure and mechanism of the evolutionarily unique plant enzyme chalcone isomerase. *Nat. Struct. Biol.* 7: 786-791.

Jez, J.M., Noel, J.P. (2002). Reaction mechanism of chalcone isomerase. *J. Biol.Chem.* 277: 1361- 1369.

Jin, H., Cominelli, E., Bailey, P., Parr, A., Mehrtens, F., Jones, J., Tonelli, C., Weisshaar, B., Martin, C. (2000). Transcriptional repression by AtMYB4 controls production of UV-protecting sunscreens in Arabidopsis. *EMBO J.* 19: 6150-6161.

Jingren He, Celestino Santos-Buelga, Nuno Mateus and Victor de Freitas, (2006). Isolation and quantification of oligomeric pyranoanthocyanin-flavanol pigments from red wines by combination of column chromatographic techniques. *Journal of Chromatography* A, Volume 1134, Issues 1-2, 17, Pages 215-225.

Johnson H. (1989). *Vintage: The Story of Wine* pg 35-46 Simon and Schuster.

Johnson, C.S., Kolevski, B., Smyth, D.R. (2002). TRANSPARENT TESTA GLABRA2, a trichome and seed coat development gene of Arabidopsis, encodes a WRKY transcription factor. *Plant Cell* 14: 1359-1375.

Johnson, Hugh (1989). *Vintage: The Story of Wine.* Simon and Schuster. p. 126.

Jonathan Yisa, (2009). Phytochemical Analysis and Antimicrobial Activity of *Scoparia dulcis* and *Nymphaea lotus*". *Australian Journal of Basic and Applied Sciences*, 3(4): pages 3975–3979

Jung H. G. and Fahey Jr. G. C. (1983). Nutritional implications of phenolic monomers and lignin: a review. *J. Anim. Sci.* (57), 206-219.

Kadam, S. S., Salunkhe, D. K., Chavan, J. K. (1990). *Dietary tannins: consequences and remedies.* Boca Raton: CRC Press. p. 177.

Katie E. Ferrell, Thorington, Richard W. (2006). *Squirrels: the animal answer guide.* Baltimore: Johns Hopkins University Press. p. 91.

Kelm, M.A., Hammerstone, J.F., Schmitz, H.H. (2005). Identification and quantitation of flavanols and proanthocyanidins in foods: How good are the datas? *Clin. Dev. Immunol.* 12: 35-41.

Kennedy JA, Matthews MA, Waterhouse AL (2002). "Effect of Maturity and Vine Water Status on Grape Skin and Wine Flavonoids". *Am. J. Enol. Vitic.* 53 (4): 268–74.

Kennedy James A. and Hayasaka Yoji, (2004). *Compositional investigation of pigmented tannin.* A.C.S. symposium series, vol. 886, pp. 247-264.

Kennedy, J.A., Jones, G.P. (2001). Analysis of proanthocyanidin cleavage products following acidcatalysis in the presence of excess phloroglucinol. *J. Agric. Food Chem* 49: 1740-1746.

Kennedy, J.A., Powell, K.J. (1985). Polyphenol interactions with aluminium (III) and iron (III): their possible involvement in the podzolization process. *Aust. J. Chem.* 38: 879-888.

Kim, J.H., Kim, B.G., Ko, J.H., Lee, Y., Hur, H.G., Lim, Y., Ahn, J.H. (2006). Molecular cloning, expression and characterization of a flavonoid glycosyltransferase from *Arabidopsis thaliana. Plant Sci.* 170: 897-903.

Kitamura, S., Shikazono, N., Tanaka, A. (2004). TRANSPARENT TESTA 19 is involved in the accumulation of both anthocyanins and proanthocyanidins in Arabidopsis. *Plant J.* 37: 104- 114.

Kobayashi, S., Goto-Yamamoto, N., Hirochika, H. (2005). Association of VvmybA1 gene expression with anthocyanin production in grape (*Vitis vinifera*) skin-colour mutants. *J. Japan. Soc. Hort. Sci.* 74: 196-203.

Kobayashi, S., Ishimaru, M., Hiraoka, K., Honda, C. (2002). Myb-related genes of the Kyoho grape (*Vitis labruscana*) regulate anthocyanin biosynthesis. *Planta,* 215: 924-933.

Koes, R., Verweij, W., Quattrocchio, F. (2005). Flavonoids: a colourful model for the regulation and evolution of biochemical pathways. *Trends Plant Sci.* 10: 236-242.

Kolodziej H., Kiderlen A. F. (September 2005). "Antileishmanial activity and immune modulatory effects of tannins and related compounds on *Leishmania* parasitised RAW 264.7 cells". *Phytochemistry* 66 (17): 2056–71.

Koppes, L. L.J., Dekker, J. M., Hendriks, H. F.J., Bouter, L. M., Heine, R. J. (2005). "Moderate Alcohol Consumption Lowers the Risk of Type 2 Diabetes: A meta-analysis of prospective observational studies". *Diabetes Care* 28 (3): 719–25.

Kortekamp, A. (2006). Expression analysis of defence-related genes in grapevine leaves after inoculation with a host and a non-host pathogen. *Plant Physiol. Biochem.* 44: 58-67.

Kreuzaler, F., Hahlbrock, K. (1972). Enzymatic synthesis of aromatic compounds in higher plants. Formation of naringenin (5,7,4'-trihydroxyflavanone) from p-coumaroyl coenzyme A and malonyl coenzyme A. *FEBS Lett.* 28: 69-72.

Kubo, H., Kishi, M., Goto, K. (2008). Expression analysis of ANTHOCYANINLESS2 gene in Arabidopsis. Plant Sci. doi: 10.1016/j.plantsci.2008.08.006.

Laino, C. (2009). "Moderate Drinking May Cut Dementia Risk". WebMD.

Laino, C. (April 21, 2009). "Study Shows Wine Drinkers With Non-Hodgkin's Lymphoma Less Likely to Die or Have Relapse". WebMD.

Lalusin, A.G., Nishita, K., Kim, S.H., Ohta, M. Fujimura, T. (2006). A new MADS-box gene (IbMADS10) from sweet potato (*Ipomoea batatas* (L.) Lam) is involved in the accumulation of anthocyanin. *Mol. Gen. Genomics.* 275: 44-54.

Lea, U.S., Slimestad, R., Smedvig, P, Lillo, C. (2007). Nitrogen deficiency enhances expression of specific MYB and bHLH transcription factors and accumulation of end products in the flavonoid pathway. *Planta* 225: 1245-1253.

Lee, C.Y., Jaworski, A. (1987). Phenolic compounds in white grapes grown in New York. *Am. J. Enol. Vitic.* 38: 277-281.

Lee, C.Y., Jaworski, A. (1990). Identification of some phenolics in white grapes. *Am. J. Enol. Vitic.* 41: 87-89.

Lees G. L., Gruber, M. Y. and Suttill, N. H. (1995). Condensed tannin in sainfoin. II. Occurrence and changes duringleaf development. *Can. J. Bot.* (73), 1540-1547.

Lees G.L. (1992). Condensed tannins in some forage legumes: their role in the prevention of ruminant pasture bloat. *Basic Life Sci.* 59: 915-934.

Li, Jingge, Maplesden, Frances (1998). "Commercial production of tannins from radiata pine bark for wood adhesives" (PDF). *IPENZ Transactions* 25 (1/EMCh).

Löwe, Zeitschrift für Chemie, 1868, 4, 603

Lü L., Liu S. W., Jiang S. B., Wu S. G. (2004). "Tannin inhibits HIV-1 entry by targeting gp41". *Acta Pharmacol. Sin.* 25 (2): 213–8.

Lukaèin, R., Urbanke, C., Gröning, I., Matern, U. (2000). The monomeric polypeptide comprises the functional flavanone 3β-hydroxylase from Petunia hybrida. *FEBS Lett.* 467: 353-358.

Luz Sanz M., Martinez-Castro Isabel and Moreno-Arribas M. Victoria, (2008). Identification of the origin of commercial enological tannins by the analysis of monosaccharides and polyalcohols. *Food chemistry*, vol. 111, no3, pp. 778-783.

Lvarez-Rodrã-Guez, Marã-a Luisa, Belloch, Carmela, Villa, Mercedes, Uruburu, Federico, Larriba, Germãjn, Coque, Juan-Josã© R (2003). "Degradation of vanillic acid and production of guaiacol by microorganisms isolated from cork samples". *FEMS Microbiology Letters* 220 (1): 49–55.

MacNeil, K. (2001). *The Wine Bible* pg 34 Workman Publishing.

Maggiolini, M, A G Recchia, D Bonofiglio, S Catalano, A Vivacqua, A Carpino, V Rago, R Rossi and S Andò, (2005). The red wine phenolics piceatannol and myricetin act as agonists for estrogen receptor in human breast cancer cells. *Journal of Molecular Endocrinology.* 35 269-281.

Maheri-Sis N., Chaichi Semsari, M., Eshratkhah, B., Sadaghian, M., Gorbani, A. and Hassanpour, S. (2011).Evaluation of the effects of Quebracho condensed tannin on faecal egg counts during naturally acquired mixed nematode infections in Moghani sheep. *Annals Biol. Res.*, 2 (2), 170-174.

Malan, J.C.S., Young, D.A., Steynberg, J.P., Ferreira, D. (1990). Oligomeric flavanoids. Part 10. Structure and synthesis of the first tetrahydropyrano[3,2-g]chromenes related to (4,6)-bis-(-)- fisetinidol profisetinidins. *J. Chem. Soc., Perkin Trans.* 1. 227-234.

Manach, C, Williamson, G, Morand, C, Scalbert, A, Rémésy, C (2005). "Bioavailability and bioefficacy of polyphenols in humans. I. Review of 97 bioavailability studies". *The American journal of clinical nutrition* 81 (1 Suppl): 230S–242S.

Mangan J. L. (1988). Nutritional effects of tannins in animal feeds. *Nutr. Res. Reviews.* (1), 209-231.

Mansoon, P. (1994). "The Father of the French Paradox". *Wine Spectator.*

Mansson, P. (2001). "Eat Well, Drink Wisely, Live Longer". *Wine Spectator.*

Marie-hélène Salagoity-Auguste, Christian Tricard, Frédéric Marsal and Pierre Sudraud, (1986). Preliminary Investigation for the Differentiation of Enological Tannins According to Botanical Origin: Determination of Gallic Acid and Its Derivatives. *Am. J. Enol. Vitic.* 37: 4: 301-303.

Marinova, K., Pourcel, L., Weder, B., Schwarz, M., Barron, D., Routaboul, J.M., Debeaujon, I., Klein, M. (2007). The Arabidopsis MATE transporter TT12 acts as a vacuolar Flavonoid/H+ -Antiporter active in proanthocyanidin-accumulating cells of the seed coat. *Plant Cell* 19: 2023-2038.

Marion Kite, Roy Thomson (2006). *Conservation of leather and related materials.* Butterworth-Heinemann. p. 23.

Marles, M.A.S., Ray, H., Gruber, M.Y. (2003). New perspectives on proanthocyanidin biochemistry and molecular regulation. *Phytochemistry* 64: 367-383.

Marrs, K.A., Alfenlto, M.R., Lloyd, A.M., Walbot, V. (1995). A glutathione S-transferase involved in vacuolar transfer encoded by the maize gene Bronze-2. *Nature* 375: 397-400.

Martens, S., Teeri, T., Forkmann, G. (2002). Heterologous expression of dihydroflavonol 4-reductases from various plants. *FEBS Lett.* 531: 453-458.

Marti´nez T. F., Moyano, F. J., Di´az, M., Barroso, F. G. and Alarco´ n, F. J. (2004). Ruminaldegradation of tannin-treated legume meals. *J. Sci. Food Agric.* (84), 1979-1987.

Mathieu F., Jouany J. P. (1993). "Effect of chestnut tannin on the fermentability of soyabean meal nitrogen in the rumen". *Ann Zootech* 42 (2): 127.

Matthews, S., Mila, I., Scalbert, A., Pollet, B., Lapierre, C., Hervé du Penhoat, C.L.M., Rolando, C., Donnelly, D.M.X. (1997). Method for estimation of proanthocyanidins based on their acid depolymerization in the presence of nucleophiles. *J. Agric. Food Chem.* 45: 1195-1201.

Mattivi F. (1993). "Solid phase extraction of trans-resveratrol from wines for HPLC analysis". *Zeitschrift für Lebensmittel- Untersuchung und Forschung* 196 (6): 522–5.

Mattivi F., Guzzon R., Vrhovsek U., Stefanini M. and Velasco R. (2006). "Metabolite Profiling of Grape: Flavonols and Anthocyanins". *J Agric Food Chem.* 54 (20): 7692–7702.

Mattivi F., Vrhovsek U., Masuero D., Trainotti D. (2009). "Differences in the amount and structure of extractable skin and seed tannins amongst red grape cultivars". *Australian Journal of Grape and Wine Research* 15: 27–35.

Max R. A. (2010). Effect of repeated wattle tannin drenches on worm burdens, faecal egg counts and egghatchability during naturally acquired nematode infections in sheep and goats. *Vet. Parasitol.* (169), 138-143.

McGee, Harold (2004). *On food and cooking: the science and lore of the kitchen.* New York: Scribner. p. 714.

McLernon, D. J., Powell, J. J., Jugdaohsingh, R., MacDonald, H. M. (2012). "Do lifestyle choices explain the effect of alcohol on bone mineral density in women around menopause?" *American Journal of Clinical Nutrition* 95 (5): 1261–9.

McSweeney C. S., Palmer, B., McNeill, D. M. and Krause, D. O. (2001). Microbial interactions with tannins: nutritional consequences for ruminants. *Anim. Feed Sci. Technol.* (91), 83-93.

Min B. R. and Hart, S. P. (2003). Tannins for suppression of internal parasites. J. Anim. Sci. (81), 102-109.Minson D. J. and McLeod, M. N. (April 13-23, 1970). The digestibility of temperate and tropical legumes. In: Norman, M.J.T. (ed.). *Proceedings of the XI International Grasslands Congress. Surfer's Paradise,* Queensland, Australia, University of Queensland Press, St. Lucia,

Morel-Salmi, Cécile, Souquet, Jean-Marc, Bes, Magali, Cheynier, Véronique (2006). "Effect of Flash Release Treatment on Phenolic Extraction and Wine Composition". *Journal of Agricultural and Food Chemistry* 54 (12): 4270–6.

Mori, K., Sugaya, S., Gemma, H. (2005). Decreased anthocyanin biosynthesis in grape berries grown under elevated night temperature condition. *Sci. Hortic.* 105: 319-330.

Mortensen, Erik L., Jensen, HH, Sanders, SA, Reinisch, JM (2001). "Better Psychological Functioning and Higher Social Status May Largely Explain the Apparent Health Benefits of Wine: A Study of Wine and Beer Drinking in Young Danish Adults".*Archives of Internal Medicine* 161 (15): 1844–8.

Mueller, L.A., Goodman, C.D., Silady, R.A., Walbot, V. (2000). AN9, a petunia glutathione Stransferase required for anthocyanin sequestration, is a flavonoid-binding protein. Plant Physiol, 123: 1561-2000.

Muller-Harvey I., McAllan A. B. (1992). "Tannins: Their biochemistry and nutritional properties". *Adv. Plant Cell Biochem. and Biotechnol.* 1: 151–217.

Muñoz-Espada, A. C., Wood, K. V., Bordelon, B., Watkins, B. A. (2004). "Anthocyanin Quantification and Radical Scavenging Capacity of Concord, Norton and Marechal Foch Grapes and Wines". *Journal of Agricultural and Food Chemistry* 52 (22): 6779–86.

Nakajima, J., Tanaka, Y., Yakushiji, H., Saito, K. (2001). Reaction mechanism from leucoanthocyanidin to anthocyanidin 3-glucoside, a key reaction for colouring in anthocyanin biosynthesis. *J. Biol. Chem.* 276: 25797-25803.

NAS (National Academy of Sciences). (1984). Leucaena: promising forage and tree crop for the tropics. NAS,Washington, D.C., USA.Patraa A. K. and J. Saxena. (2010). Exploitation of dietary tannins to improve rumen metabolism and ruminantnutrition. *J. Sci. Food Agric.* (91), 24-37.

National Center for Computational Toxicology (NCCT) DSSTox Official Website.

Naughton, Declan P, Petróczi andrea (2008). "Heavy metal ions in wines: Meta-analysis of target hazard quotients reveal health risks". *Chemistry Central Journal* 2: 22.

Nave F1, Petrov V, Pina F, Teixeira N, Mateus N and de Freitas V, J Phys Chem B. (2010). Thermodynamic and kinetic properties of a red wine pigment: catechin-(4,8)-malvidin-3-O-glucoside. Volume 114, issue 42, pages 13487-13496.

Navia, Jeanette. "Could Tannins Explain Classic Migraine Triggers?" 1988

Nesi, N., Debeaujon, I., Jond, C., Pelletier, G., Caboche, M., Lepiniec, L. (2000). The TT8 gene encodes a basic helix-loop-helix domain protein required for expression of DFR and BAN genes in *Arabidopsis siliques. Plant Cell.* 12: 1863-1878.

Nesi, N., Debeaujon, I., Jond, C., Stewart, A.J., Jenkins, G.I., Caboche, M., Lepiniec, L. (2002). The TRANSPARENT TESTA16 locus encodes the Arabidopsis bisister MADS domain protein and is required for proper development and pigmentation of the seed coat. *Plant Cell* 14: 2463- 2479.

Nesi, N., Jond, C., Debeaujon, I., Caboche, M., Lepiniec, L. (2001). The Arabidopsis TT2 gene encodes an R2R3 MYB domain protein that acts as a key determinant for proanthocyanidin accumulation in developing seed. *Plant Cell*, 13: 2099-2114.

Nierenstein, M. (1915). "The Formation of Ellagic Acid from Galloyl-Glycine by*Penicillium*". *The Biochemical Journal* 9 (2): 240–244.

Nierenstein, M. (1932). "A biological synthesis of m-digallic acid". *The Biochemical Journal* 26 (4): 1093–1094.

Nierenstein, M., Potter, J. (1945). "The distribution of myrobalanitannin". *The Biochemical Journal* 39 (5): 390–392.

Nobre-Junior, Helio V. (2007). "Neuroprotective Actions of Tannins from*Myracrodruon urundeuva* on 6-Hydroxydopamine-Induced Neuronal Cell Death". *Journal of Herbs, Spices and Medicinal Plants* (Haworth Press) 13 (2). Retrieved 8 November 2007.

Nutton, Vivian (1973). "The Chronology of Galen's Early Career". *The Classical Quarterly* 23 (1): 158–71.

Okamoto T. (2005). *Int J Mol Med.*16 (2): 275-8.

Ono, E., Hatayama, M., Isono, Y., Sato, T., Watanabe, R., Yonekura-Sakakibara, K., FukuchiMizutani, M., Tanaka, Y., Kusumi, T., Nishino, T., Nakayama, T. (2006). Localization of a flavonoid biosynthetic polyphenol oxidase in vacuoles. *Plant J.* 45: 133-143.

Oszmianski, J., Cheynier, V., Moutounet, M. (1996). Iron-catalyzed oxidation of (+)-catechin in model systems. *J. Agric. Food Chem.* 44: 1712-1715.

Oszmianski, J., Lee, C.Y. (1990). Enzymatic oxidative reaction of catechin and chlorogenic acid in a model system. *J. Agric. Food Chem.* 38: 1202-1204.

Pang, Y., Peel, G.J., Wright, E., Wang, Z., Dixon, R.A. (2007). Early steps in proanthocyanidin biosynthesis in the model legume *Medicago truncatula*. *Plant Physiol.* 145: 601-615.

Paolini, M, Sapone andrea, Valgimigli, Luca (2003). "Avoidance of bioflavonoid supplements during pregnancy: a pathway to infant leukemia?". *Mutation Research/Fundamental and Molecular Mechanisms of Mutagenesis* 527 (1–2): 99–101.

Paolocci, F., Bovone, T., Tosti, N., Arcioni, S., Damiani, F. (2005). Light and an exogenous transcription factor qualitatively and quantitatively affect the biosynthetic pathway of condensed tannins in *Lotus corniculatus* leaves. *J. Exp. Bot.* 56: 1093-1103. *Molecules* 2008, 13 2703

Paolocci, F., Robbins, M.P., Madeo, L., Arcioni, S., Martens, S., Damiani, F. (2007). Ectopic expression of a basic helix-loop-helix gene transactivates parallel pathways of proanthocyanidin biosynthesis.Structure, expression, analysis and genetic control of leucoanthocyanidin 4- reductase and anthocyanidin reductase genes in Lotus corniculatus. *Plant Physiol.* 143: 504-516.

Passos, C.P., Cardoso, S.M., Domingues, M.R.M., Domingues, P., Silva, C.M., Coimbra, M.A. (2007). Evidence for galloylated type-A procyanidins in grape seeds. *Food Chem.* 105: 1457-1467.

Peleg, H., Gacon, K., Schlich, P., Noble, A. C. (1999). Bitterness and astringency of flavan-3-ol monomers, dimers and trimers. *J. Sci. Food Agric.* 79: 1123-1128.

Pelletier, M.K., Murrell, J.R., Shirley, B.W. (1997). Characterization of flavonol synthase and leucoanthocyanidin dioxygenase genes in Arabidopsis. Further evidence for differential regulation of "Early" and "Late" genes. *Plant Physiol.* 11: 1437-1445.

Peltonen, P.A., Vapaavuori, E., Julkunen-Tiitto, R. (2005). Accumulation of phenolic compounds in birch leaves is changed by elevated carbon dioxide and ozone. *Global Change Biol.* 11: 1305-1324.

Perkin, A. G., Nierenstein, M. (1905). "CXLI. Some oxidation products of the hydroxybenzoic acids and the constitution of ellagic acid. Part I". *Journal of the Chemical Society, Transactions* 87: 1412.

Petit, P., Granier, T., d'Estaintot, B.L., Manigand, C., Bathany, K., Schmitter, J.M., Lauvergeat, V., Hamdi, S., Gallois, B. (2007). Crystal structure of grape dihydroflavonol 4-reductase, a key enzyme in flavonoid biosynthesis. *J. Mol. Biol.* 368: 1345-1357.

Pollnitz, Alan P, Pardon, Kevin H, Sefton, Mark A (2000). "Quantitative analysis of 4-ethylphenol and 4-ethylguaiacol in red wine". *Journal of Chromatography A* 874 (1): 101–9.

Porter, L.J. (1988). The Flavonoids. In *Flavans and Proanthocyanidins*, Harborne, J.B., Ed., Chapman and Hall: London, UK, pp. 21-62.

Porter, L.J. (1993). The Flavonoids: Advances in Research Since 1986. In *Flavans and Proanthocyanidins*, Harborne, J.B., Ed., Chapman and Hall: London, UK, pp. 23-55.

Pourcel, L., Routaboul, J.M., Kerhoas, L., Caboche, M., Lepiniec, L., Debeaujon, I. (2005). TRANSPARENT TESTA10 Encodes a laccase-like enzyme involved in oxidative polymerization of flavonoids in Arabidopsis seed coat. *Plant Cell.* 17: 2966-2980.

Prieur, C., Rigaud, J., Cheynier, V., Moutounet, M. (1994). Oligomeric and polymeric procyanidins from grape seeds. *Phytochemistry* 36: 781-784.

Prior, R. L., Gu, L. (2005). Occurrence and biological significance of proanthocyanidins in the American diet. *Phytochemistry* 66: 2264-2280.

Punyasiri, P.A., Tanner, G.J., Abeysinghe, I.S.B., Kumar, V., Campbell, P.M., Pradeepa, N.H.L. (2004). *Exobasidium vexans* infection of *Camellia sinensis* increased 2,3-cis isomerisation and gallate esterification of proanthocyanidins. *Phytochemistry* 65: 2987-2994.

Puupponen-Pimiä R., Nohynek L., Meier C. (2001). "Antimicrobial properties of phenolic compounds from berries". *J. Appl. Microbiol.* 90 (4): 494–507.

Queensland.Mooney, H. A., Harrison, A. T. and Morrow, P. A. (1975). Environmental limitations of photosynthesis on aCalifornia evergreen srub (*Heteroeles erbutifolia*). *Oecologia.* (19), 293-302.

Quideau, Stéphane (2009). "Why bother with Polyphenols". *Groupe Polyphenols.* Retrieved 21 August 2012.

Ramachandra Rao, S., Ravishankar G. A. (2002). Plant cell cultures: Chemical factories of secondary metabolites. *Biotechnol. Adv.* (20), 101-153.

Ramsay, N.A., Walker, A.R., Mooney, M., Gray, J.C. (2003). Two basic-helix-loop-helix genes (MYC- 146 and GL3) from Arabidopsis can activate anthocyanin biosynthesis in a white-flowered Matthiola incana mutant. *Plant Mol. Biol.* 52: 679-688.

Rao, L. J. M., Yada, H., Ono, H., Ohnishi-Kameyama, M., Yoshida, M. (2004). Occurrence of antioxidant and radical scavenging proanthocyanidins from the Indian minor spice nagkesar (*Mammea longifolia* planch and triana syn). *Bioorg. Med. Chem.* 12: 31-36.

Rasmussen, S.E., Frederiksen, H., Krogholm, K.S., Poulsen, L. (2005). Dietary proanthocyanidins: occurrence, dietary intake, bioavailability and protection against cardiovascular disease. *Mol. Nutr. Food Res.* 49: 159-174.

Ray, H., Yu, M., Auser, P., Blahut-Beatty, L., Mckersie, B, Bowley, S., Westcott, N., Coulman, B., Lloyd, A., Gruber, M.Y. (2003). Expression of anthocyanins and proanthocyanidins after transformation of alfalfa with maize Lc. *Plant Physiology* 132: 1448-1463.

Redco Foods, Inc. Wagner, RF, Inspections, Compliance, Enforcement and Criminal Investigations, US FDA. 22 February 2010. Retrieved 10 November 2014.

Reed J. D. (1995). Nutritional toxicology of tannins and related polyphenols in forage legumes. *J. Anim. Sci.* (73), 1516-1528.

Remy-Tanneau, S., Guernevé, C.L., Meudec, E., Cheynier, V. (2003). Characterization of a colourless anthocyanin-flavan-3-ol dimer containing both carbon-carbon and ether interflavanoid linkages by NMR and mass spectrometry. *J. Agric. Food Chem.* 51: 3592-3597.

Ribéreau-Gayon, P. (1972). *Plant Phenolics*, Hafner Publishing Company: New York, p. 254.

Robbins, C.T., Mole, S., Hagerman, A.E., Hanley, T.A. (1987). Role of tannins in defending plants against ruminants: reduction in dry matter digestion? *Ecologl.* 68: 1606-1615.

Robbins, M.P., Paolocci, F., Hughes, J.W., Turchetti, V., Allison, G., Arcioni, S., Morris, P., Damiani, F., (2003). Sn, a maize bHLH gene, modulates anthocyanin and condensed tannin pathways in Lotus corniculatus. *J. Exp. Bot.* 54: 239-248.

Robert L. Wolke, Marlene Parrish (2005). *What Einstein told his cook 2: the sequel: further adventures in kitchen science.* W. W. Norton and Company. p. 433.

Robinson J. (ed) (2006). "*The Oxford Companion to Wine*" Third Edition pg 680 Oxford University Press.

Robinson J. (ed) (2006). "*The Oxford Companion to Wine*" Third Edition pg 569 Oxford University Press.

Robinson J. (ed) (2006). "*The Oxford Companion to Wine*" Third Edition pg 144 Oxford University Press.

Robinson j. (ed) (2006). "*The Oxford Companion to Wine*" Third Edition, page 492, Oxford University Press.

Robinson J. (ed) 92006). "*The Oxford Companion to Wine*" Third Edition pg 727 Oxford University Press.

Robinson J. (ed) (2006). "*The Oxford Companion to Wine*" Third Edition pg 517-518 Oxford University Press.

Robinson J. (ed) (2006). *"The Oxford Companion to Wine"* Third Edition pg 273-274 Oxford University Press.

Robinson J. (ed) (2006). *"The Oxford Companion to Wine"* Third Edition pg 24 Oxford University Press.

Robinson, J., ed. (2006). *The Oxford Companion to Wine* (3rd ed.). Oxford University Press. p. 433.

Robinson, J., ed. (2006). *The Oxford Companion to Wine* (3rd ed.). Oxford University Press. pp. 341–2.

Robinson, J., ed. (2006). *The Oxford Companion to Wine* (3rd ed.). Oxford University Press. p. 569.

Ross, J. A. (2000). "Dietary flavonoids and the MLL gene: A pathway to infant leukemia?". *Proceedings of the National Academy of Sciences* 97 (9): 4411–3.

Ross, JA (1998). "Maternal diet and infant leukemia: a role for DNA topoisomerase II inhibitors?" *International journal of cancer. Supplement* 11: 26–8.

Ryan, K.G., Swinny, E.E., Markham, K.R., Winefield, C. (2002). Flavonoid gene expression and UV photoprotection in transgenic and mutant Petunia leaves. *Phytochemistry* 59: 23-32.

Sadaghian M., Hassanpour, S., Maheri-Sis, N., Eshratkhah, B., Gorbani, A. and Chaichi-Semsari, M. (2011). Effectsof different levels of wattle tannin drenches on faecal egg counts during naturally acquired mixed nematodeinfections in Moghani sheep. *Annals Biol. Res.*, 2 (1), 226-230.

Sagasser, M., Lu, G.H., Hahlbrock, K., Weisshaar, B. (2001). A. thaliana TRANSPARENT TESTA 1 is involved in seed coat development and defines the WIP subfamily of plant zinc finger proteins. *Gene. Dev.* 16: 138-149.

Saito, K., Kobayashi, M., Gong, Z, Tanaka, Y., Yamazaki, M. (1999). Direct evidence for anthocyanidin synthase as a 2-oxoglutarate-dependent oxygenase molecular cloning and functional expression of cDNA from a red forma of Perilla frutescens. *Plant J.* 17: 181-189.

Saito, K., Yamazaki, M. (2002). Tansley review no. 138 Biochemistry and molecular biology of the latestage of biosynthesis of anthocyanin: lessons from Perilla frutescens as a model plant. *New Phyto.* 155: 9-23.

Sano, T., Oda, E., Yamashita, T., Naemura, A., Ijiri, Y., Yamakoshi, J., Yamamoto, J. (2005). Antithrombotic effect of proanthocyanidin, a purified ingredient of grape seed. *Thromb. Res.* 115: 115-121.

Santos-Buelga, C., Kolodzieij, H., Treutter, D. (1995). Procyanidin trimers possessing a doubly linked structure from Aesculus hippocastanum. *Phytochemistry* 38: 499-504.

Santos-Buelga, C., Scalbert, A. (2000). Proanthocyanidins and tannin-like compounds – nature, occurrence, dietary intake and effects on nutrition and health. *J. Sci. Food Agric.* 80: 1094- 1117.

Sarni, Pascale, Fulcrand, Hélène, Souillol, Véronique, Souquet, Jean-Marc, Cheynier, Véronique (1995). "Mechanisms of anthocyanin degradation in grape must-like model solutions".*Journal of the Science of Food and Agriculture* 69 (3): 385.

Sarni-Manchado, P., Cheynier, V., Moutounet, M., (1999). Interactions of grape seed tannins with salivary proteins. *J. Agric. Food Chem.* 47: 42-47.

Sarni-Manchado, Pascale, Cheynier, Véronique, Moutounet, Michel (1999). "Interactions of Grape Seed Tannins with Salivary Proteins". *Journal of Agricultural and Food Chemistry* 47(1): 42–7.

Sarni-Manchado, Pascale, Cheynier, Véronique, Moutounet, Michel (1997). "Reactions of polyphenoloxidase generated caftaric acid o-quinone with malvidin 3-O-glucoside". *Phytochemistry* 45 (7): 1365.

Sarwar, G., Bell, J.M., Sharby, T.F., Jones, J.D. (1981). Nutritional evaluation of meals and meal fractions derived from rape and mustard seed. *Can. J. Anim. Sci.* 61: 719-733.

Saslowsky, D., Winkel-Shirley, B. (2001). Localization of flavonoid enzymes in Arabidopsis roots. *Plant J.* 27: 37-48.

Sato, M., Maulik, G., Ray, P.S., Bagchi, D., Das, D.K. (1999). Cardioprotective effects of grape seed proanthocyanidin against ischemic reperfusion injury. *J. Mol. Cell. Cardiol.* 31: 1289-1297.

Scalbert A., Morand, C., Manach, C. and Rémésy, C. (2002). Absorption and metabolism of polyphenols in the gutand impact on health. *Biomed. Pharma.*, 56 (6), 276-82.

Scalbert, A. (1991). Antimicrobial properties of tannins. *Phytochemistry.* 30: 3875-3883.

Schiavone A., Guo K., Tassone S. (2008). "Effects of a natural extract of chestnut wood on digestibility, performance traits and nitrogen balance of broiler chicks". *Poult. Sci.* 87 (3): 521–7.

Schofield, P., Mbugua, D.M., Pell, A.N. (2001). Analysis of condensed tannins: a review. *Anim. Feed Sci. Tech.* 91, 21-40.

Sharma, S.B., Dixon, R.A. (2005). Metabolic engineering of proanthocyanidins by ectopic expression of transcription factors in *Arabidopsis thaliana. Plant J.* 44: 62-75.

Sharma, S.D., Meeran, S.M., Katiyar, S.K. (2007). Dietary grape seed proanthocyanidins inhibit UVBinduced oxidative stress and activation of mitogenactivated protein kinases and nuclear factor-êB signaling *in vivo* SKH-1 hairless mice. *Mol. Cancer. Ther.* 6: 995-1005.

Shinya Kanzaki, Keizo Yonemori and Akira Sugiura, (2001). Identification of Molecular Markers Linked to the Trait of Natural Astringency Loss of Japanese Persimmon (*Diospyros kaki*) Fruit. *J. Amer. Soc. Hort. Sci.*, 126(1), pages 51–55 (article)

Shuman, Tracy C., ed. (2005). "Alcohol and Heart disease". WebMD.

Simon Mole (1993). "The Systematic Distribution of Tannins in the Leaves of Angiosperms: A Tool for Ecological Studies". *Biochemical Systematics and Ecology* 21(8): 833–846.

Skirycz, A., Jozefczuk, S., Stobiecki, M., Muth, D., Zanor, M.I., Witt, I., Mueller-Roeber, B. (2007). Transcription factor AtDOF4,2 affects phenylpropanoid metabolism in *Arabidopsis thaliana. New Phytol.* 175: 425-438.

Slade, D., Ferreira, D., Marais, J.P.J. (2005). Circular dichroism, a powerful tool for the assessment of absolute configuration of flavonoids. *Phytochemistry* 66: 2177-2215.

Souza, S. M. C., Aquino, L. C., Milach Jr, A. C., Bandeira, M. A., Nobre, M. E., Viana, G. S. (2006). "Antiinflammatory and antiulcer properties of tannins from*Myracrodruon urundeuva* Allemão (Anacardiaceae) in Rodents". *Phytotherapy Research* (John Wiley and Sons) 21 (3): 220–225.

Spector, L. G., Xie, Y, Robison, LL, Heerema, NA, Hilden, JM, Lange, B, Felix, CA, Davies, SM *et al.* (2005). "Maternal Diet and Infant Leukemia: The DNA Topoisomerase II Inhibitor Hypothesis: A Report from the Children's Oncology Group". *Cancer Epidemiology Biomarkers and Prevention* 14 (3): 651–5.

Spiers, C. W. (1914). The Estimation of Tannin in Cider. *The Journal of Agricultural Science*, Volume 6, Issue 01, pages 77-83.

Springob, K., Nakajima, J., Yamazaki, M., Saito, K. (2003). Recent advances in the biosynthesis and accumulation of anthocyanins. *Nat. Prod. Rep.* 20: 288-303.

Stafford H. A. (1990). *Flavanoid Metabolism.* CRC Press, Boca Raton, Fla., USA, pp. 63-99.

Stafford, H.A. (2008). Proanthocyanidins and the lignin connection. Phytochemistry 1988, 27, 1-6. *Molecules* 13: 2694

Steynberg, P.J., Steynberg, J.P., Bezuidenhoudt, B.C.B., Ferreira, D. (1994). Cleavage of the interflavanyl bond in 5-deoxy (A ring) proanthocyanidins. *J. Chem. Soc., Chem. Commun.* 31-32.

Strick, R., Strissel, PL, Borgers, S, Smith, SL, Rowley, JD (2000). "From the Cover: Dietary bioflavonoids induce cleavage in the MLL gene and may contribute to infant leukemia". *Proceedings of the National Academy of Sciences* 97 (9): 4790–5.

Subarnas, A., Wagner, H. (2000). Analgesic and anti-inflammatory activity of the proanthocyanidin shellegueain A from Polypodium feei METT. *Phytomedicine* 7: 401-405.

Sun, W., Miller, J.M. (2003). Tandem mass spectrometry of the B-type procyanidins in wine and B-type dehydrodicatechins in an autoxidation mixture of (+)-catechin and (-)-epicatechin. *J. Mass Spectrom.* 38: 438-446.

Tabacco E., Borreani G., Crovetto G. M., Galassi G., Colombo D., Cavallarin L. (2006). "Effect of chestnut tannin on fermentation quality, proteolysis and protein rumen degradability of alfalfa silage". *J. Dairy Sci.* 89 (12): 4736–46.

Takahata, Y., Ohnishi-Kameyama, M., Furuta, S., Takahashi, M., Suda, I. (2001). Highly polymerized procyanidins in brown soybean seed coat with a high radical-scavenging activity. *J. Agric. Food Chem.* 49: 5843-5847.

Takashi Sakaguchia, Akira Nakajimaa (1987). "Recovery of Uranium from Seawater by Immobilized Tannin". *Separation Science and Technology* 22 (6): 1609–23.

Tanner, G.J., Francki, K.T., Abrahams, S., Watson, J.M., Larkin, P.J., Ashton, A.R. (2003). Proanthocyanidin Biosynthesis in Plants. Proanthocyanidin biosynthesis in plants. Purification of legume leucoanthocyanidin reductase and molecular cloning of its cDNA. *J. Biol Chem.* 278: 31647-31656.

Tannin in Tropical Woods. Doat J, Bois. For Tmp. (1978). volume 182, pages 34-37

Tarascou, I., Barathieu, K., Simon, C., Ducasse, M.A. andré, Y., Fouquet, E., Dufourc, E.J., Freitas, V., Laguerre, M., Pianet, I. (2006). A 3D structural and conformational study of procyanidin dimers in water and hydro-alcoholic media as viewed by NMR and molecular modeling. *Magn. Reson. Chem.* 44: 868-880.

Taylor, L.P., Briggs, W.R. (1990). Genetic Regulation and Photocontrol of Anthocyanin Accumulation in Maize Seedlings. *Plant Cell*, 2: 115-127.

The Tannin Handbook, Ann E. Hagerman, 1988 (book).

Thirman MJ, Gill HJ, Burnett RC, Mbangkollo D, McCabe NR, Kobayashi H. (1993). "Rearrangement of the MLL gene in acute lymphoblastic and acute myeloid leukemias with 11q23 chromosomal translocations". *N Engl J Med* 329 (13): 909–14.

Thompson, R.S., Jacques, D., Haslam, E., Tanner, R.J.N. (1972). Plant proanthocyanidins -Part 1- Introduction: the isolation, structure and distribution in nature of plant procyanidins. *J. Chem. Soc., Perkin Trans.* 1, 1387-1399.

Tian, L., Kong, W.F., Pan, Q.H., Zhan, J.C., Wen, P.F., Chen, J.Y., Wan, S.B., Huang, W.D. (2006). Expression of the chalcone synthase gene from grape and preparation of an anti-CHS antibody. *Protein Expres. Purif.* 50: 223-228.

Tian, L., Pang, Y., Dixon, R.A. (2007). Biosynthesis and genetic engineering of proanthocyanidins and (iso)flavonoids. *Phytochem. Rev.* 1-21.

Tian, L., Wan, S.B., Pan, Q.H., Zheng, Y.J., Huang, W.D. (2008). A novel plastid localization of chalcone synthase in developing grape berry. *Plant Sci.* 175: 431-436.

Torres J., Olivares S., De La Rosa D., Lima L., Martínez F., Munita C. S., Favaro D. I. T. (1999). "Removal of mercury(II) and methylmercury from solution by tannin adsorbents". *Journal of Radioanalytical and Nuclear Chemistry* 240 (1): 361–5.

Uber die Bestimmung des Gerbstoffs. J. Lowenthal, Z. *Anal. Chem*, 1877, volume 16, pages 33-48,

Ubi, B.E., Honda, C., Bessho, H., Kondo, S., Wada, M., Kobayashi, S., Moriguchi, T. (2006). Expression analysis of anthocyanin biosynthetic genes in apple skin: effect of UV-B and temperature. *Plant Sci.* 170: 571-578.

Van der Linden MH *et al.* (2012) Diagnosis and management of neonatal leukaemia. *Semin Fetal Neonatal Med* 17(4): 192-5.

Van Soest P. J. (1982). Nutritional Ecology of the Ruminant. Corvallis, Oregon: O and B Books, Inc. Waghorn G. C. (2008). Beneficial and detrimental effects of dietary condensed tannins for sustainable sheep and goat production-Progress and challenges. *Anim. Feed Sci. Technol.* (147), 116-139.

Van Waalwijk Van Doorn-Khosrovani, S. B., Janssen, J., Maas, L. M., Godschalk, R. W.L., Nijhuis, J. G., Van Schooten, F. J. (2007). "Dietary flavonoids induce MLL translocations in primary human CD34+ cells". *Carcinogenesis* 28 (8): 1703–9.

Varea s., García-Vallejo, M, Cadahía, E, De Simón, FernáNdez (2001). "Polyphenols susceptible to migrate from cork stoppers to wine". *European Food Research and Technology*213: 56.

Vattem D. A., Ghaedian R., Shetty K. (2005). "Enhancing health benefits of berries through phenolic antioxidant enrichment: focus on cranberry". *Asia Pac J Clin Nutr* 14(2): 120–30.

Vaughn, K.C., Lax, A.R., Duke, S.O. (1988). Polyphenol oxidase: the chloroplast enzyme with no established function. *Physiol. Plant.* 72: 659-665.

Vessela Atanasova, Hélène Fulcrand, Véronique Cheynier and Michel Moutounet, (2002). Effect of oxygenation on polyphenol changes occurring in the course of wine-making. *Analytica Chimica Acta*, Volume 458, Issue 1, 29 Pages 15-27.

Vidal, S., Cartalade, D., Souquet, J.M., Fulcrand, H., Cheynier, V. (2002). Changes in proanthocyanidin chain length in winelike model solutions. *J. Agric. Food Chem.* 50: 2261-2266.

Vrhovsek U. (1998). "Extraction of Hydroxycinnamoyltartaric Acids from Berries of Different Grape Varieties". *J Agric Food Chem.* 46 (10): 4203–8.

Walker, A.R., Davison, P.A., Bolognesi-Winfield, A.C., James, C.M., Srinivasan, N., Blundell, T.L., Esch, J.J., Marks, M.D., Gray, J.C. (1999). The TRANSPARENT TESTA GLABRA1 locus, which regulates trichome differentiation and anthocyanin biosynthesis in Arabidopsis, encodes a WD40 repeat protein. *Plant Cell* 11: 1337-1349.

Walker, A.R., Lee, E., Bogs, J., McDavid, D.A.J., Thomas, M.R., Robinson, S.P. (2007). White grapes arose through the mutation of two similar and adjacent regulatory genes. *Plant J.* 49: 772- 785.

Walker-Bone, Karen (2012). "Recognizing and treating secondary osteoporosis". *Nature Reviews Rheumatology* 8 (8): 480–92.

Wang, Xilong, Sato, T., Xing, Baoshan, Tao, S. (2005). "Health risks of heavy metals to the general public in Tianjin, China via consumption of vegetables and fish". *Science of the Total Environment* 350 (1–3): 28–37.

Wang, Y., Frutos, P., Gruber, M.Y., Ray, H., McAllister, T.A. (2006). *In vitro* ruminal digestion of anthocyanidin-containing alfalfa transformed with the maize Lc regulatory gene. *Can. J. Anim. Sci.* 86: 1119-1130.

Weber, H.A., Hodges, A.E., Guthrie, J.R., O'Brien, B.M., Robaugh, D., Clark, A.P., Harris, R.K., AlgaierL, J.W., Smith, C.S. (2007). Comparison of proanthocyanidins in commercial antioxidants: grape seed and pine bark extracts. *J. Agric. Food Chem.* 55: 148-156.

Weinges, K., Freundenberg, K. (1965). Condensed proanthocyanidins from cranberries and cola nuts. *Chem. Commun.* 2: 220-222.

Welford, R.W.D., Turnbull, J.J., Claridge, T.D.W., Prescott, A.G., Schofield, C.J. (2001). Evidence for oxidation at C-3 of the flavonoid C-ring during anthocyanin biosynthesis. *Chem. Commun.* 1828-1829.

Wellmann, F., Griesser, M., Schwab, W., Martens, S., Eisenreich, W., Matern, U., Lukaèin R. (2006). Anthocyanidin synthase from Gerbera hybrida catalyzes the conversion of (+)-catechin to cyanidin and a novel procyanidin. *FEBS Lett.* 580: 1642-1648.

Wilbert, C. (2008). "Red Wine May Cut Risk of Lung Cancer". WebMD.

Williams RJ, Spencer JP, Rice-Evans C (2004). "Flavonoids: antioxidants or signalling molecules?" *Free Radical Biology and Medicine* 36 (7): 838–49.

Williams, Robert J, Spencer, Jeremy P.E, Rice-Evans, Catherine (2004). "Flavonoids: Antioxidants or signalling molecules?". *Free Radical Biology and Medicine* 36 (7): 838–49.

Wirth, J., Morel-Salmi, C., Souquet, J.M., Dieval, J.B., Aagaard, O., Vidal, S., Fulcrand, H., Cheynier, V. (2010). "The impact of oxygen exposure before and after bottling on the polyphenolic composition of red wines". *Food Chemistry* 123: 107.

Xie, D.Y., Dixon, R.A. (2005). Proanthocyanidin biosynthesis – still more questions than answers? *Phytochemistry* 66: 2127-2144.

Xie, D.Y., Sharma, S.B., Dixon, R.A. (2004). Anthocyanidin reductases from Medicago truncatula and *Arabidopsis thaliana*. *Arch. Biochem. Biophys.* 422: 91-102.

Xie, D.Y., Sharma, S.B., Paiva, N.L., Ferreira, D., Dixon, R.A. (2003). Role of anthocyanidin reductase, encoded by BANYULS in plant flavonoid biosynthesis. *Science* 299: 396-399.

Xie, D.Y., Sharma, S.B., Wright, E., Wang, Z.Y., Dixon, R.A. (2006). Metabolic engineering of proanthocyanidins through co-expression of anthocyanidin reductase and the PAP1 MYB transcription factor. *Plant J.* 45: 895-907.

Ya-Ling Hsu, Hsin-Lin Liang, Chih-Hsing Hung and Po-Lin Kuo, (2011). Syringetin, a flavonoid derivative in grape and wine, induces human osteoblast differentiation through bone morphogenetic protein-2/extracellular signal-regulated kinase 1/2 pathway. *Molecular Nutrition and Food Research*, Volume 53 Issue 11, Pages 1452-1461.

Yamakoshi, J., Saito, M., Kataoka, S., Kikuchi, M. (2002). Safety evaluation of proanthocyanidin-rich extract from grape seeds. *Food Chem. Toxicol.* 40: 599-607.

Yokotsuka, K., Singleton, V.L., (1997). Disappearance of anthocyanins as grape juice is prepared and oxidized with PPO and PPO substrates. *Am. J. Enol. Vitic.* 48: 13-25.

Zelman, Kathleen M. (2005). "Wine: How Much Is Good for You?". WebMD.

Zhang, F., Gonzalez, A., Zhao, M., Payne, C.T. Lloyd, A. (2003). A network of redundant bHLH proteins functions in all TTG1-dependent pathways of Arabidopsis. *Development*, 130: 4859-4869.

Zhao, M., Yang, B., Wang, J. S., Liu, Y., Yu, L. M., Jiang, Y. M. (2007). Immunomodulatory and anticancer activities of flavonoids extracted from litchi (*Litchi chinensis* Sonn.) pericarp. *Int. Immunopharmacol.* 7: 162-166.

Zheng G.C., Lin Y.L. and Yazaki Y. (1991). Tannin analysis of *Acacia mearnsii* bark - a comparison of the hide-powder and Stiasny methods. *ACIAR Proceedings Series*, No. 35, pages 128-131 (abstract)

Zheng Guangcheng, Lin Yunlu and Y. Yazaki, (1991). Bark tannin contents of *Acacia mearnsii* provenances and the relationship between the hide-powder and the Stiasny methods of estimation. *Australian Forestry*, Volume 54, Issue 4, pages 209-211.

Zimmermann, I.M., Heim, M.A., Weisshaar, B., Uhrig, J.F. (2004). Comprehensive identification of *Arabidopsis thaliana* MYB transcription factors interacting with R/B-like BHLH proteins. *Plant J.* 40: 22-34.

Zucker, W.V. (1983). Tannins: does structure determine function? An ecological perspective. *Am. Nat.* 121: 335-365.

Chapter 18

Terpenoids

The terpenoids sometimes called isoprenoids, are a large and diverse class of naturally occurring organic chemicals similar to terpenes, derived from five-carbon isoprene units assembled and modified in thousands of ways. Most are multicyclic structures that differ from one another not only in functional groups but also in their basic carbon skeletons. These lipids can be found in all classes of living things and are the largest group of natural products. Plant terpenoids are used extensively for their aromatic qualities. They play a role in traditional herbal remedies and are under investigation for antibacterial, antineoplastic and other pharmaceutical functions. Terpenoids contribute to the scent of eucalyptus, the flavors of cinnamon, cloves and ginger, the yellow colour in sunflowers and the red colour in tomatoes. Well-known terpenoids include citral, menthol, camphor, salvinorin A in the plant *Salvia divinorum*, the cannabinoids found in cannabis, ginkgolide and bilobalide found in *Ginkgo biloba* and the curcuminoids found in turmeric and mustard seed. The steroids and sterols in animals are biologically produced from terpenoid precursors. Sometimes terpenoids are added to proteins, *e.g.*, to enhance their attachment to the cell membrane; this is known as isoprenylation.

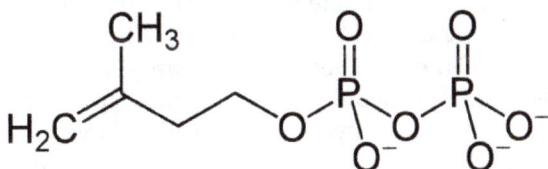

Chemical Structure of the Terpenoid isopentenyl Pyrophosphate

Structure and Classification

Terpenes are hydrocarbons resulting from the combination of several isoprene units. Terpenoids can be thought of as modified terpenes, wherein methyl groups have been moved or removed, or oxygen atoms added. (Some authors use the term "terpene" more broadly, to include the terpenoids.) Just like terpenes, the terpenoids can be classified according to the number of isoprene units used:

☆ Hemiterpenoids, 1 isoprene unit (5 carbons)

☆ Monoterpenoids, 2 isoprene units (10C)

☆ Sesquiterpenoids, 3 isoprene units (15C)

☆ Diterpenoids, 4 isoprene units (20C) (*e.g.* ginkgolides)

☆ Sesterterpenoids, 5 isoprene units (25C)

☆ Triterpenoids, 6 isoprene units (30C) (*e.g.* sterols)

☆ Tetraterpenoids, 8 isoprene units (40C) (*e.g.* carotenoids)

☆ Polyterpenoid with a larger number of isoprene units

Terpenoids can also be classified according to the number of cyclic structures they contain. Terpenoids can be tested for using the Salkowski test. Meroterpenes are any compound, including many natural products, having a partial terpenoid structure.

Biosynthesis

There are two metabolic pathways of creating terpenoids:

Mevalonic Acid Pathway

Many organisms manufacture terpenoids through the HMG-CoA reductase pathway, the pathway that also producescholesterol. The reactions take place in the cytosol. The pathway was discovered in the 1950s.

MEP/DOXP Pathway

The 2-C-methyl-D-erythritol 4-phosphate/1-deoxy-D-xylulose 5-phosphate pathway (MEP/DOXP pathway), also known as [non-mevalonate pathway] or mevalonic acid-independent pathway, takes place in the plastids of plants and apicomplexan protozoa, as well as in many bacteria. It was discovered in the late 1980. Pyruvate and glyceraldehyde 3-phosphate are converted by DOXP synthase (Dxs) to 1-deoxy-D-xylulose 5-phosphate and by DOXP reductase (Dxr, IspC) to 2-C-methyl-D-erythritol 4-phosphate (MEP). The subsequent three reaction steps catalyzed by 4-diphosphocytidyl-2-C-methyl-D-erythritol synthase (YgbP, IspD), 4-diphosphocytidyl-2-C-methyl-D-erythritol kinase (YchB, IspE) and 2-C-methyl-D-erythritol 2, 4-cyclodiphosphate synthase (YgbB, IspF) mediate the formation of 2-C-methyl-D-erythritol 2,4-cyclopyrophosphate (MEcPP). Finally, MEcPP is converted to (*E*)-4-hydroxy-3-methyl-but-2-enyl pyrophosphate (HMB-PP) by HMB-PP synthase (GcpE, IspG) and HMB-PP is converted to isopentenyl pyrophosphate (IPP) and dimethylallyl pyrophosphate (DMAPP) by HMB-PP reductase (LytB, IspH). IPP and DMAPP are the end-products in either pathway and are the precursors of isoprene, monoterpenoids (10-carbon), diterpenoids (20-carbon), carotenoids (40-carbon), chlorophylls and plastoquinone-9 (45-carbon). Synthesis of all higher terpenoids proceeds via formation of geranyl pyrophosphate (GPP), farnesyl pyrophosphate (FPP) and geranylgeranyl pyrophosphate (GGPP).

Although both pathways, MVA and MEP, are mutually exclusive in most organisms, interactions between them have been reported in plants and few bacteria species.

Organism	Pathways
Bacteria	MVA or MEP
Archaea	MVA
Green Algae	MEP
Plants	MVA and MEP
Animals	MVA
Fungi	MVA

References

Ayoola, GA (2008). "Phytochemical Screening and Antioxidant Activities of Some Selected Medicinal Plants Used for Malaria Therapy in Southwestern Nigeria". *Tropical Journal of Pharmaceutical Research* **7** (3): 1019–1024. Retrieved 18 September 2014.

Michael Specter (2009). "A Life of Its Own". *The New Yorker*.

Chapter 19

Trypsin Inhibitor

A trypsin inhibitor is a type of serine protease inhibitor that reduces the biological activity of trypsin. Trypsin is an enzyme involved in the breakdown of many different proteins, including as part of digestion in humans and other animals. As a result, protease inhibitors that interfere with its activity can have an antinutritional effect.

Major Commercial Sources

Source	Inhibitor	Molecular Weight	Inhibitory Power	Details
Blood plasma	α_1-antitrypsin	52 kDa		Also known as serum trypsin inhibitor
Lima beans		8–10 kDa	2.2 times weight	A mixture of six different inhibitors
Bovine pancreas and lung	Aprotinin	6.5 kDa	2.5 times weight	Also known as BPTI (basic pancreatic trypsin inhibitor) and Kunitz inhibitor. Best-known pancreatic inhibitor. Inhibits several different serine proteases
Raw avian egg white	Ovomucin	8–10 kDa	1.2 times weight	The ovomucoids are a mixture of several different glycoprotein protease inhibitors
Soybeans		20.7–22.3 kDa	1.2 times weight	A mixture of several different inhibitors. All also bind chymotrypsin to a lesser degree.

A study revealing that a protease inhibitor from the eggs of the freshwater snail *Pomacea canaliculata*, interacting as a trypsin inhibitor with the protease of potential predators, was reported in 2010, the first direct evidence for this mechanism in the animal kingdom.

Clinical Significance

The peptide tumor-associated trypsin inhibitor (TATI) has been used as a marker of mucinous ovarian carcinoma, urothelial carcinoma and renal cell carcinoma. TATI is metabolised by the kidneys and is, thus, elevated in patients with renal failure. It may be elevated in nonneoplastic processes such as pancreatitis and can be used as a prognostic marker in this setting (levels above 70 micrograms/L are associated with poor prognosis).

Fifty percent of stage I mucinous ovarian carcinomas are associated with elevated TATI and nearly 100 per cent of stage IV tumors show elevated TATI. Eighty-five to 95 per cent of pancreatic adenocarcinomas are associated with increased TATI (but elevation in pancreatitis limits the clinical utility of TATI in this setting; see above). Sixty percent of gastric adenocarcinomas show elevated TATI, in particular tumors of diffusely infiltrative/signet ring type. TATI, thus, complements CEA, which is elevated exclusively in intestinal type adenocarcinoma of the stomach. In urothelial carcinoma, TATI expression varies with stage, ranging from 20 per cent in low-stage tumors to 80 per cent of high-stage tumors. TATI sensitivity in the setting of renal cell carcinoma is approximately 70 per cent. Elevated TATI is more likely to be seen in patients with advanced-stage disease. In nearly all tumor types studied, TATI is a marker of poor prognosis.

Protease Inhibitor

In biology and biochemistry, protease inhibitors are molecules that inhibit the function of proteases. Many naturally occurring protease inhibitors are proteins. In medicine, *protease inhibitor* is often used interchangeably with alpha 1-antitrypsin (A1AT, which is abbreviated PI for this reason). A1AT is indeed the protease inhibitor most often involved in disease, namely in alpha 1-antitrypsin deficiency.

Classification

Protease inhibitors may be classified either by the type of protease they inhibit, or by their mechanism of action. In 2004 Rawlings and colleagues introduced a classification of protease inhibitors based on similarities detectable at the level of amino acid sequence. This classification initially identified 48 families of inhibitors that could be grouped into 26 related super family (or clans) by their structure. According to the MEROPS database there are now 85 families of inhibitors. These families are named with an I followed by a number; for example, I14 contains hirudin-like inhibitors.

By Protease

Classes of *proteases* are:

- ☆ Aspartic protease inhibitors
- ☆ Cysteine protease inhibitors
- ☆ Metalloprotease inhibitors
- ☆ Serine protease inhibitors (serpins)
- ☆ Threonine protease inhibitors

☆ Trypsin inhibitors

☆ Kunitz STI protease inhibitor

By Mechanism

Classes of *inhibitor mechanisms of action* are:

☆ Suicide inhibitor

☆ Transition state inhibitor

☆ Protein protease inhibitor (see serpins)

☆ Chelating agents

Families

Inhibitor I9

Proteinase propeptide inhibitors (sometimes referred to as activation peptides) are responsible for the modulation of folding and activity of the peptidase pro-enzyme or zymogen. The pro-segment docks into the enzyme, shielding the substrate binding site, thereby promoting inhibition of the enzyme. Several such propeptides share a similar topology, despite often low sequenceidentities. The propeptide region has an open-sandwich antiparallel-alpha/antiparallel-beta fold, with two alpha-helices and fourbeta-strands with a (beta/alpha/beta)x2 topology. The peptidase inhibitor I9 family contains the propeptide domain at the N-terminusof peptidases belonging to MEROPS family S8A, subtilisins. The propeptide is removed by proteolytic cleavage; removal activating the enzyme.

Inhibitor I10

This family includes both microviridins and marinostatins. It seems likely that in both cases it is the C-terminus which becomes the active inhibitor after post-translational modifications of the full length, pre-peptide. it is the ester linkages within the key, 12-residue, region that circularise the molecule giving it its inhibitory conformation.

Inhibitor I24

This family includes PinA, which inhibits the endopeptidase La. It binds to the La homotetramer but does not interfere with the ATPbinding site or the active site of La.

Inhibitor I29

The inhibitor I29 domain, which belongs to MEROPS peptidase inhibitor family I29, is found at the N-terminus of a variety of peptidaseprecursors that belong to MEROPS peptidase subfamily C1A; these include cathepsin L, papain and procaricain. It forms an alpha-helical domain that runs through the substrate-binding site, preventing access. Removal of this region by proteolytic cleavage results in activation of the enzyme. This domain is also found, in one or more copies, in a variety of cysteine peptidase inhibitors such as salarin.

Inhibitor I34

The saccharopepsin inhibitor I34 is highly specific for the aspartic peptidase saccharopepsin. In the absence of saccharopepsin it is largely unstructured, but in its presence, the inhibitor undergoes a conformational change forming an almost perfect alpha-helixfrom Asn2 to Met32 in the active site cleft of the peptidase.

Inhibitor I36

The peptidase inhibitor family I36 domain is only found in a small number of proteins restricted to *Streptomyces* species. All have fourconserved cysteines that probably form two disulphide bonds. One of these proteins from *Streptomyces nigrescens*, is the well characterised metalloproteinase inhibitor SMPI. The structure of SMPI has been determined. It has 102 amino acid residues with two disulphide bridges and specifically inhibits metal loproteinases such as thermolysin, which belongs to MEROPS peptidase family M4. SMPI is composed of two beta-sheets, each consisting of four antiparallel beta-strands. The structure can be considered as two Greek key motifs with 2-fold internal symmetry, a Greek key beta-barrel. One unique structural feature found in SMPI is in its extension between the first and second strands of the second Greek key motif which is known to be involved in the inhibitory activity of SMPI. In the absence of sequence similarity, the SMPI structure shows clear similarity to both domains of the eye lens crystallins, both domains of the calcium sensor protein-S, as well as the single-domain yeast killer toxin. The yeast killer toxin structure was thought to be a precursor of the two-domain beta gamma-crystallin proteins, because of its structural similarity to each domain of the beta gamma-crystallins. SMPI thus provides another example of a single-domain protein structure that corresponds to the ancestral fold from which the two-domain proteins in the beta gamma-crystallin superfamily are believed to have evolved.

Inhibitor I42

Inhibitor family I42 includes chagasin, a reversible inhibitor of papain-like cysteine proteases. Chagasin has a beta-barrel structure, which is a unique variant of the immunoglobulin fold with homology to human CD8alpha.

Inhibitor I48

Inhibitor family I48 includes clitocypin, which binds and inhibits cysteine proteinases. It has no similarity to any other known cysteine proteinase inhibitors but bears some similarity to a lectin-like family of proteins from mushrooms.

Inhibitor I53

Members of this family are the peptidase inhibitor madanin proteins. These proteins were isolated from tick saliva.

Inhibitor I67

Bromelain inhibitor VI, in the Inhibitor I67 family, is a double-chain inhibitor consisting of an 11-residue and a 41-residue chain.

Inhibitor I68

The Carboxypeptidase inhibitor I68 family represents a family of tick carboxypetidase inhibitors.

Inhibitor I78

The peptidase inhibitor I78 family includes *Aspergillus* elastase inhibitor.

Compounds

   ☆ Aprotinin

   ☆ Bestatin

   ☆ Calpain inhibitor I and II

   ☆ Chymostatin

   ☆ E-64

   ☆ Leupeptin (N-acetyl-L-leucyl-L-leucyl-L-argininal)

   ☆ alpha-2-Macroglobulin

   ☆ Pefabloc SC

   ☆ Pepstatin

   ☆ PMSF (phenylmethanesulfonyl fluoride)

   ☆ TLCK

   ☆ Trypsin inhibitors

Chemical and Biological Synthesis of Trypsin Inhibitor

Diverse gene-encoded cyclic peptides are produced by species from all three domains of life. The 14-residue, head-to-tail cyclized plant peptide, SFTI-1 (Sunflower Trypsin Inhibitor 1) (Luckett, *et al.*, 1999). Figure 19.1 has attracted much attention due to its great stability and capability to potently inhibit trypsin (Ki 0.1 nM) as well as the epithelial serine protease matriptase (Ki 0.92 nM) (Long, *et al.* (1999) giving it exciting promise as a drug lead and a protein engineering scaffold (Long, *et al.* (1999). Although peptides such as SFTI-1 are routinely produced by chemical synthesis, the biological mechanisms that enable biosynthesis of ribosomally synthesized cyclic peptides are largely unknown. Recently Mylne *et al.*, described the biosynthetic origin of SFTI-1 and a related peptide SFT-L1 (Mylne, *et al.*, 2011). Both emerge from seed storage protein precursors PawS1 and PawS2, respectively, using each seed protein's own maturing protease for their release (Mylne, *et al.*, 2011). We used transgenic constructs in the model plant Arabidopsis thaliana combined with proteomics and MALDI mass spectrometry to study this unusual dual-fate for PawS1 and identified the residues that are critical for SFTI-1 maturation and cyclisation (Mylne, *et al.*, 2011). Here we compare best practice for chemical synthesis of SFTI-1 with how plants biologically create the same product.

There is a long history of the synthesis of cyclic peptides by manual solid-phase peptide synthesis (White and Yudin, 2011). Here we describe the optimized synthesis of the 14 residue cyclic peptide, SFTI-1 using BOC (t-butoxycarbonyl) chemistry with

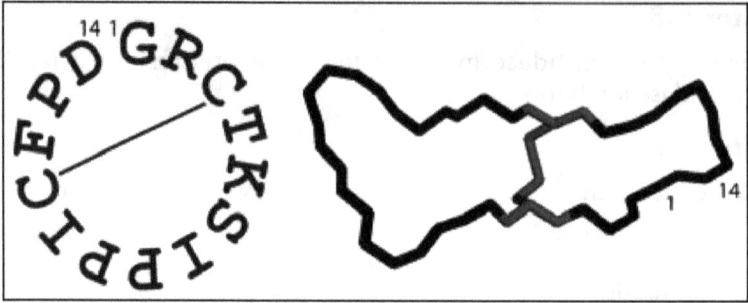

Figure 19.1: Sequence of SFTI-1 (Left) and a Model of its Backbone Structure (PDB 1SFI). When chemically synthesized, SFTI-1 ends with an N-terminal Cys, but its biological ligation point is between Gly1 and Asp14.

in situ neutralization, HBTU [2-(1-Hbenzotriazol-1-yl)-1, 1, 3, 3-tetramethyluronium hexafluorophosphate] (Schnolzer, *et al.*, 1992) and a C-terminal thioester linker. Peptides were assembled on a PAM-Gly-Boc (phenylacetamidomethyl-glycine-t-Boc) resin on a 0.5-mmol scale and designed to contain the C-terminal thioester linker, S-trityl-β-mercaptopropionic acid and an N-terminal cysteine. Peptides were subsequently cleaved from the resin by HF cleavage with p-cresol as the scavenger (9:1 HF:p-cresol by volume). The cleavage reaction was incubated at <5 to 0°C for 90 min; HF was removed under vacuum and the peptide precipitated with diethyl ether, then filtered and re-dissolved in 50 per cent acetonitrile 0.045 per cent TFA and lyophilised. The crude peptide was subsequently purified by RP-HPLC using a gradient of 0–80 per cent acetonitrile 0.045 per cent TFA over 80 min. Analytical HPLC and ESI-MS confirmed peptide purity and mass. Importantly, the folding for SFTI-1 is a two-step process. The two steps ensure that the more favorable reaction, disulfide bond formation occurs secondarily to cyclization, the less favorable reaction, as the free sulfur donated from the cysteine is required to form a thiol for cyclization. Firstly, the peptides were reduced and cyclized in 0.1 M ammonium bicarbonate pH 8.2 with 0.5 mg/mL TCEP overnight at room temperature, followed by RP-HPLC purification. Secondly, the peptides were oxidized in 0.1 M ammonium bicarbonate pH 8.2 overnight at room temperature, followed by RP-HPLC purification as above to yield fully folded and pure peptides. From ~200 mg of crude cleaved peptide ~40 mg of over 98 per cent pure fully folded SFTI-1 may be obtained using this method. Sunflowers produce SFTI-1 with ease. We cloned the precursor gene for SFTI-1, *PawS1* and were surprised to find that the sequence for SFTI-1 was encoded along with a much larger protein of entirely different function. PawS1 is a preproalbumin, which encodes seed storage albumin protein as well as SFTI-1. Seed storage albumins are matured from proalbumins by the action of a Cys-protease called asparaginyl endopeptidase (AEP, aka legumain, vacuolar processing enzyme). AEP usually cleaves at Asn and, to a lesser extent, at Asp. Within PawS1, SFTI-1 ends with Asp and is preceded by Asn suggesting AEP matures both the albumin and SFTI-1. We used an *Arabidopsis* mutant lacking AEP to confirm that AEP was required to release SFTI-1 at both proto-termini and is the best candidate for the ligation reaction. Furthermore, an in-depth mutagenesis of *PawS1* and subsequent testing of these constructs *in vivo* revealed residues essential for SFTI-1 processing. These experiments combined with the wealth of information about albumin maturation (Hara-Hishimura, *et al.*, 1993) allowed us to propose a model for the processing of SFTI-1 from within PawS1 (Figure 19.2). PawS preproalbumin is sent to the ER where its signal peptide is cleaved and with the aid of hairpin formation from hydrophobic clustering within SFTI-1 the Cys38-Cys46 disulfide bond is formed. The proalbumin is then matured at several points by AEP in multivesicular bodies as they traffic to protein storage vacuoles. During this processing SFTI-1 is released at the N-terminus (PawS1 Gly36) and during the final cleavage at its C-terminal P1 aspartic acid, a reactive, thioester acyl-intermediate is created. Instead of the typical attack by water and bond hydrolysis, we propose this reactive intermediate is instead attacked at its carbonyl carbon by the unmasked amino terminus of the glycine, held in proximity to the thioester by the disulfide bond. This attack results in peptide bond formation and reconstitutes the AEP active site thiol.

Solid phase peptide synthesis is the best way to study peptides for applications in drug design. However, to produce them on a large-scale by chemical synthesis becomes costly. We propose to use plants as a cost effective manufacturing process for producing cyclic peptide therapeutics based on molecules produced naturally by plants. To do this, requires an in-depth knowledge of their *in planta* processing so that we may manipulate this processing to vary the peptides produced. To enhance this understanding we will continue to approach this problem by combining synthetic chemistry with plant genetic engineering.

Alpha 1-antitrypsin

Alpha-1 antitrypsin or α_1-antitrypsin (A1AT) is a protease inhibitor belonging to the serpin superfamily. It is generally known as serum trypsin inhibitor. Alpha 1-antitrypsin is also referred to as alpha-1 proteinase inhibitor (A1PI) because it inhibits a wide variety of proteases. It protects tissues from enzymes of inflammatory cells, especially neutrophil elastase and has a reference range in blood of 1.5 - 3.5 gram/liter (in US the reference range is generally expressed as mg/dL or micromoles), but the concentration can rise manyfold upon acute inflammation. In its absence, neutrophil elastase is free to break down elastin, which contributes to the elasticity of the lungs, resulting in respiratory complications such asemphysema, or COPD (chronic obstructive pulmonary disease) in adults and cirrhosis in adults or children.

Function

A1AT is a 52-kDa serpin and, in medicine, it is considered the most prominent serpin; the terms $\alpha 1$-*antitrypsin* and *protease inhibitor* (P_i) are often used interchangeably. Most serpins inactivate enzymes by binding to them covalently, requiring very high levels to perform their function. In theacute phase reaction, a further elevation is required to "limit" the damage caused by activated neutrophil granulocytes and their enzyme elastase, which breaks down the connective tissue fiber elastin. Like all serine protease inhibitors, A1AT has a characteristic secondary structure of beta sheets and alpha helices. Mutations in these areas can lead to

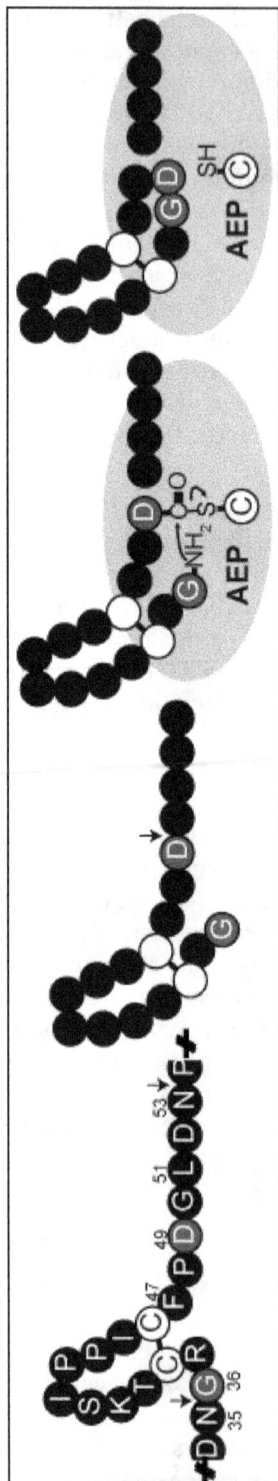

Figure 19.2: Proposed Model for SFTI-1 Biosynthesis (adapted from Mylne *et al.*, 2011) [3]. AEP cleavage occurs firstly at N35 and N53, later at D49 followed by ligation of G36 and D49 (grey) to form the cyclic peptide, SFTI-1.

non-functional proteins that can polymerise and accumulate in the liver (infantile hepatic cirrhosis).

Role in Disease

Disorders of this protein include alpha 1-antitrypsin deficiency, an autosomal codominant hereditary disorder in which a deficiency of alpha 1-antitrypsin leads to a chronic uninhibited tissue breakdown. This causes the degradation especially of lung tissue and eventually leads to characteristic manifestations of pulmonary emphysema. Evidence has shown that cigarette smoke can lead to oxidation of methionine 358 of α_1-antitrypsin (382 in the pre-processed form containing the 24 amino acid signal peptide), a residue essential for binding elastase; this is thought to be one of the primary mechanisms by which cigarette smoking (or second-hand smoke) can lead to emphysema. Because A1AT is expressed in the liver, certain mutations in the gene encoding the protein can cause misfolding and impaired secretion, which can lead to liver cirrhosis. An extremely rare form of P_i, termed $P_{iPittsburgh}$, functions as an antithrombin (a related serpin), due to a mutation (Met358Arg). One person with this mutation has been reported to have died of a lethal bleeding diathesis.

Liver biopsy will show abundant PAS-positive globules within periportal hepatocytes.

Nomenclature

The protein was originally named "antitrypsin" because of its ability to covalently bind and irreversibly inactivate the enzymetrypsin *in vitro*. Trypsin, a type of peptidase, is a digestive enzyme active in the duodenum and elsewhere. The term *alpha-1* refers to the protein's behavior on protein electrophoresis. On electrophoresis, the protein component of the blood is separated by electric current. There are several *clusters*, the first being albumin, the second being the *alpha*, the third *beta* and the fourth *gamma* (immunoglobulins). The non-albumin proteins are referred to as globulins. The *alpha* region can be further divided into two sub-regions, termed "1" and "2". Alpha 1-antitrypsin is the main protein of the alpha-globulin 1 region. Another name used is *alpha-1 proteinase inhibitor* (α_1-PI).

Genetics

The gene is located on the long arm of the fourteenth chromosome (14q32.1). Over 100 different variants of α_1-antitrypsin have been described in various populations. North-Western Europeans are most at risk for carrying one of the most common mutant forms of A1AT, the Z mutation (Glu342Lys on M1A, rs28929474).

Biochemical Properties

A1AT is a single-chain glycoprotein consisting of 394 amino acids in the mature form and exhibits a number of glycoforms. The three N-linked glycosylations sites are mainly equipped with so-called diantennary N-glycans. However, one particular site shows a considerable amount of heterogeneity since tri- and even tetraantennary N-glycans can be attached to the Asparagine 107 (ExPASy amino acid nomenclature). These glycans carry different amounts of negatively-charged sialic acids, this causes the heterogeneity observed on normal A1AT when analysed by isoelectric focussing. In addition, the fucosylated triantennary N-glycans were shown to have the fucose as part of a so-called Sialyl Lewis x epitope, which could confer this protein particular protein-cell recognition properties. The single cysteine residue of A1AT in position 256 (ExPASy nomenclature) is found to be covalently linked to a free single cysteine by a disulfide bridge.

Therapeutic Use

Recombinant alpha 1-antitrypsin is not yet commercially available, but is under investigation as a therapy for alpha 1-antitrypsin deficiency. Therapeutic concentrates are prepared from the blood plasma of blood donors. The US FDA has approved the use of three alpha 1-antitrypsin products derived from a human plasma: Prolastin, Zemaira and Aralast. These products for intravenous augmentation A1AT therapy can cost up to $100,000 per year per patient. They are administered intravenously at a dose of 60 mg/kg once a week. A recent study analyzed and compared the three FDA-approved products in terms of their primary structure and glycosylation. All three products showed minor differences compared to the normal human plasma A1AT and are introduced during the specific purifications procedures. However, these detected differences are not believed to have any negative implications to the patients. Aerosolized-augmented A1AT therapy is under study. This involves inhaling purified human A1AT into the lungs and trapping the A1AT into the lower respiratory tract. This method proves more successful than intravenous-augmented A1AT therapy because intravenous use of A1AT results in only 10 per cent -15 per cent of the A1AT reaching the lower respiratory tract, whereas 25 per cent -45 per cent of A1AT can reach the lower respiratory tract through inhalation. However, inhaled A1AT may not reach the elastin fibers in the lung where elastase injury actually occurs. Further study is currently underway.

Kunitz STI Protease Inhibitor

Trypsin and Protease Inhibitor

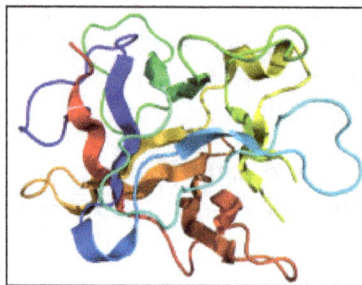

Structure of a Kunitz-type trypsin inhibitor

Identifiers

☆ Symbol   Kunitz_legume
☆ Pfam   PF00197

☆ InterPro   IPR002160

☆ PROSITE   PDOC00255

☆ SCOP   1tie

☆ SUPERFAMILY   1tie

Kunitz soybean trypsin inhibitor is a type of protein contained in legume seeds which functions as a protease inhibitor. Kunitz-type Soybean Trypsin Inhibitors are usually specific for either trypsin or chymotrypsin. They are thought to protect seeds against

Background

Two types of trypsin inhibitors are found in soy: the Kunitz trypsin inhibitor (KTI) and the Bowman-Birk inhibitor (BBI). KTI is a large (20,100 daltons), strong inhibitor of trypsin, while BBI is much smaller (8,000 daltons) and inhibits both trypsin and chymotrypsin. Both inhibitors have significant anti-nutritive effects in the body, affecting digestion by hindering protein hydrolysis and activation of other enzymes in the gut. In soy, KTI is found in much larger concentrations than BBI is soy, however, to achieve the highest nutritional value from this ingredient, both of these inhibitors must be denatured in some way. Whole soybeans have been reported to contain 17–27 mg of trypsin inhibitor per gram.

Structure

Proteins from the Kunitz family contain from 170 to 200 amino acid residues and one or two intra-chain disulfide bonds. The best conserved region is found in their N-terminalsection. The crystal structures of soybean trypsin inhibitor (STI), trypsin inhibitor DE-3 from the Kaffir tree *Erythrina caffra* (ETI) and the bifunctional proteinase K/alpha-amylase inhibitor from wheat (PK13) have been solved, showing them to share the same 12-stranded beta sheet structure as those of interleukin 1 and heparin-binding growth factors. The beta-sheets are arranged in 3 similar lobes around a central axis, 6 strands forming an anti-parallel beta barrel. Despite the structural similarity, STI shows no interleukin-1 bioactivity, presumably as a result of their primary sequence disparities. The active inhibitory site containing the scissile bond is located in the loop between beta-strands 4 and 5 in STI and ETI. The STIs belong to a superfamily that also contains the interleukin-1 proteins, heparin binding growth factors (HBGF) and histactophilin, all of which have very similar structures, but share no sequence similarity with the STI family.

Action and Consequences of Trypsin Inhibitors

Trypsin inhibitors require a specific three-dimensional structure in order to follow through with inactivation of trypsin in the body. They bind strongly to trypsin, blocking its active site and instantly forming an irreversible compound and halting digestion of certain proteins. Trypsin, a serine protease, is responsible for cleaving peptide bonds containing carbonyl groups from arginine or lysine. After a meal, trypsin is stimulated by cholecystokinin and undergoes specific proteolysis for activation. Free trypsin is then able to activate other serine proteases, such as chymotrypsin, elastase and more trypsin (by autocatalysis), or continue breaking down proteins. However, if trypsin inhibitors (specifically KTI) are present, the majority of trypsin in the cycle of digestion is inactivated and ingested proteins remain whole. Effects of this occurrence include gastric distress and pancreatich yperplasia (proliferation of cells) or hypertrophy (enlargement of cells). The amount of soy inhibitors is directly related to the amount of trypsin it will inhibit, therefore a product with high concentration of soy is suspect to produce large values of inhibition. In a rat model, animals were fed either soy protein concentrate or direct concentrate of the Kunitz trypsin inhibitor. In both instances, after a week the rats showed a dose-related increase in pancreas weight due to both hyperplasia and hypertrophy. This indicates that long-term consumption of a diet high in soy with strong trypsin inhibitor activity may produce unwanted effects in humans as well.

Inactivation of Trypsin Inhibitors

A significant amount of research is being done to determine the best method of inhibitor inactivation. The most successful methods found so far include:

- ☆ Heat
- ☆ Freezing
- ☆ Addition of Sulfites

Cancer Research

While trypsin inhibitors have been widely regarded as anti-nutritive factors in soy, research is currently being done on the inhibitors' possible anti-carcinogenic characteristics. Some research has shown that protease inhibitors can cause irreversible suppressive effect on carcinogenic cell growth. However, the mechanism is still unknown. The cancers showing positive results for this new development are colon, oral, lung, liver and esophageal cancers. Further research is still necessary to determine things such as the method of delivery for this natural anti-carcinogen, as well as performing extensive clinical trials in this area.

References

Brzin J, Rogelj B, Popovic T, Strukelj B, Ritonja A (2000). "Clitocypin, a new type of cysteine proteinase inhibitor from fruit bodies of mushroom clitocybe nebularis". *J. Biol. Chem.* 275 (26): 20104–9.

Daly, N., *et al.*, *J. Biol. Chem.* 281, 23668-23675 (2006).

De Mais, Daniel. *ASCP Quick Compendium of Clinical Pathology*, 2nd Ed. ASCP Press 2009.

DiPietro CM, Liener IE (1989). Soybean Protease Inhibitors in Foods. *J Food Sci.* 35: 535-540.

Dreon M. S., Ituarte S. and Heras H. (2010). "The Role of the Proteinase Inhibitor Ovorubin in Apple Snail Eggs Resembles Plant Embryo Defense against Predation". PLoS ONE 5(12): e15059.

Figueiredo da Silva AA, de Carvalho Vieira L, Krieger MA, Goldenberg S, Zanchin NI, GuimarÃ£es BG (2007). "Crystal structure of chagasin, the endogenous cysteine-protease inhibitor from Trypanosoma cruzi". *J. Struct. Biol.* 157 (2): 416–23.

Green TB, Ganesh O, Perry K, Smith L, Phylip LH, Logan TM, Hagen SJ, Dunn BM, Edison AS (2004). "IA3, an aspartic proteinase inhibitor from Saccharomyces cerevisiae, is intrinsically unstructured in solution". *Biochemistry* 43 (14): 4071–81.

Groves MR, Taylor MA, Scott M, Cummings NJ, Pickersgill RW, Jenkins JA (1996). "The prosequence of procaricain forms an alpha-helical domain that prevents access to the substrate-binding cleft". *Structure* 4 (10): 1193–203.

Gruis, D., *et al., Plant Cell* 16, 270-290 (2004).

Hara-Hishimura, I., *et al., Plant J.* 4, 793-800 (1993)

Hiraiwa, N., *et al., FEBS Lett.* 447, 213-216 (1999)

Iwanaga S, Okada M, Isawa H, Morita A, Yuda M, Chinzei Y (2003). "Identification and characterization of novel salivary thrombin inhibitors from the ixodidae tick, *Haemaphysalis longicornis*". *Eur. J. Biochem.* 270 (9): 1926–34.

Jain SC, Shinde U, Li Y, Inouye M, Berman HM (1998). "The crystal structure of an autoprocessed Ser221Cys-subtilisin E-propeptide complex at 2.0 A resolution". *J. Mol. Biol.* 284 (1): 137–44.

Long, Y., *et al., Bioorg. Med. Chem. Lett.* 11, 2515-2519 (1999).

Luckett, S., *et al., J. Mol. Biol.* 290, 525-533 (1999).

Monteiro AC, Abrahamson M, Lima AP, Vannier-Santos MA, Scharfstein J (2001). "Identification, characterization and localization of chagasin, a tight-binding cysteine protease inhibitor in *Trypanosoma cruzi*". *J. Cell. Sci.* 114 (Pt 21): 3933–42.

Murai H, Hara S, Ikenaka T, Oda K, Murao S (1985). "Amino acid sequence of Streptomyces metallo-proteinase inhibitor from Streptomyces nigrescens TK-23". *J. Biochem.* 97 (1): 173–80.

Murzin AG, Lesk AM, Chothia C (1992). "Beta-Trefoil fold. Patterns of structure and sequence in the Kunitz inhibitors interleukins-1 beta and 1 alpha and fibroblast growth factors". *J. Mol. Biol.* 223 (2): 531–43.

Mylne, J., *et al., Nat. Chem. Biol.* 7, 257-259 (2011).

Ohno A, Tate S, Seeram SS, Hiraga K, Swindells MB, Oda K, Kainosho M (1998). "NMR structure of the *Streptomyces metalloproteinase* inhibitor, SMPI, isolated from *Streptomyces nigrescens* TK-23: another example of an ancestral beta gamma-crystallin precursor structure". *J. Mol. Biol.* 282 (2): 421–33.

Olonen A, Kalkkinen N, Paulin L (2003). "A new type of cysteine proteinase inhibitor–the salarin gene from Atlantic salmon (*Salmo salar* L.) and Arctic charr (*Salvelinus alpinus*)". *Biochimie* 85 (7): 677–81.

Onesti S, Brick P, Blow DM (1991). "Crystal structure of a Kunitz-type trypsin inhibitor from *Erythrina caffra* seeds". *J. Mol. Biol.* 217 (1): 153–76.

Otegui, M., *et al., Plant Cell* 18, 2567-2581 (2006).

Rawlings ND, Tolle DP, Barrett AJ (2004). "Evolutionary families of peptidase inhibitors". *Biochem. J.* 378 (Pt 3): 705–16.

Rawlings ND, Tolle DP, Barrett AJ (2004). "Evolutionary families of peptidase inhibitors". *Biochem. J.* 378 (Pt 3): 705–16.

Schnolzer, M., *et al., Int. J. Pept. Protein Res.* 40, 180-193 (1992).

Swedberg, J., *et al., Chem. Biol.* 16, 633-643 (2009).

Tanaka K, Aoki H, Oda K, Murao S, Saito H, Takahashi H (1990). "Nucleotide sequence of the gene for a metalloproteinase inhibitor of *Streptomyces nigrescens* (SMPI)". *Nucleic Acids Res.* 18 (21): 6433.

Tangrea MA, Bryan PN, Sari N, Orban J (2002). "Solution structure of the pro-hormone convertase 1 pro-domain from Mus musculus". *J. Mol. Biol.* 320 (4): 801–12.

Wang SX, Pandey KC, Scharfstein J, Whisstock J, Huang RK, Jacobelli J, Fletterick RJ, Rosenthal PJ, Abrahamson M, Brinen LS, Rossi A, Sali A, McKerrow JH (2007). "The structure of chagasin in complex with a cysteine protease clarifies the binding mode and evolution of an inhibitor family". *Structure* 15 (5): 535–43.

White, C., Yudin, A. *Nat. Chem.* 3, 509-524 (2011).

Chapter 20

Potential Health Benefits and Problems

Secondary metabolites and antinutrients commonly found in plant foods have both adverse effects and health benefits. For example. phytic acid, lectins, phenolic compounds (tannins), saponins and enzyme (amylase and protease) inhibitors have been shown to reduce the availability of nutrients and cause growth inhibition, while phytoestrogens and lignans have been linked with infertility problems. However. phytic acid, lectins, phenolic compounds. amylase inhibitors and saponins have also been sihown to reduce the blood glucose and insulin responses to starchy foods and/or the plasma cholesterol and triglycerjdes. In addition, phytic acid. phenolics, saponins, protease inhibitors, phytoeslrogens and lignans have been related to reduced cancer risks. Because antinutrients can also be mitigating agents, they need reevaluation and perhaps a change in name in the future.

Current recommendations suggest that the intake of grains, fruits and vegetables be increased for better health and management of chronic diseases such as cardiovascular disease, diabetes and cancer (Committee on Diet and Health, 1989; Scientific Review Committee 1990). However, there are concerns that the high intake of these foods may also increase the intake of antinutrients which are present in them (Liener, 1986; Roebuck, 1987; Morgan and Fenwick, 1990). At the same time, the health beneficial effects of grains, fruits and vegetables have been attributed in part to some of their naturally occurring antinutrients (Wattenberg, 1983; Stich and Rosin, 1984; Thompson, 1986, 1988, 1989; Adlercreutz, 1988, 1990, 1991; Cheng and Ho, 1988; Hayatsu *et al.,* *1988;* Setchell and Adlercreutz, 1988; Troll and Kennedy, 1989; Namiki, 1990; Oakenful and Sidhu, 1990; Adlercreutz *et al.,* 1991; Messina and Barnes, 1991; Messina and Messina, 1991; Caragay, 1992). Although not an exhaustive review, this paper aims to describe some of the problems and benefits associated with the antinutrients and, hopefully, to provide a more balanced view of these substances in health and disease. A number of antinutrients are present in foods but this paper will describe only a few of those in which health benefits have also been reported.

The Problems

Phytic Acid

Phytic acid (myoinositol 1,2,3,4,5,6, hexakis-dihydrogen phosphate; PA) is present in foods in concentrations ranging from 0-1 to 6-0 per cent (Reddy *et al.*, 1982; Yoon *et al.*, 1983; Harland and Oberleas, 1987; Ferguson *et al.*, 1988). It is found as crystalline globoid inside protein bodies in the cotyledon of legumes or oilseeds or in the bran region of the cereal grains (Reddy *et al.*, 1982). Its highly ne atively charged structure at a wide range of pH values makes it very reactive with other positively charged ions such as,minerals, forming insoluble complexes which are less available for digestion and absorption in the small intestine. This is the main reason why PA has traditionally been considered as an antinutrient.

The adverse effect of PA in mineral availability depends on a number of factors including the concentration of PA and the strength of its binding with different minerals. For example, !inc (Zn) forms one of the strongest mineral complexes with PA (Evans and Martin, 1988); when the PN/Zn molar ratio in the diet exceeds 10, the Zn status of the individual is likely to be compromised (Morris and Ellis, 1989). However, since calcium (Ca) forms an even stronger and more insoluble complex with Zn and PA, a dietary PA x Ca/Zn molar ratio exceeding 3-5 mols/kg in rats or 100 mmols/1000 kcal (1 cal = 4·2 J) in humans is thought to be a better indicator of adverse effect on Zn status (Fordyce *et al.*, 1987; Morris and Ellis, 1989; Thompson, 1989). Other factors influencing mineral availability in the presence of PA include the method of food processing, type of PA (added on endogenous), presence of other mineral binding agents such as dietary fiber) oxalic acid and oxalic acid and tannins which may compete with PA for binding with minerals, dietary protein concentration, presence of dietary, intestinal or bacterial phytase, whether the PA is taken in the same meal as the mineral source or in separate meals, presence of other minerals and metabolic adaptation of the individual to a high PA diet. Several recent reviews have discussed their effects in detail as well as the effect of PA on other minerals. (Cheryan,.1980; Cosgrove, 1980; Davies, 1982; Maga, 1982; Reddy *et al.*, 1982; Kratzer and Vohra, 1986; Morris, 1986; Hallberg, 1987; Morris and Ellis, 1989; Spivey-Fox and Tao, 1989; Thompson, 1989; Torre *et al.*, 1991).

PA can also react directly with the positively charged group or indirectly with the negatively charged group of the proteins mediated by a positively charged mineral ion such as calcium. It can bind with starch either directly by hydrogen bonding with the phosphate group or indirectly through the proteins to which it is associated with. The formation of these complexes is likewise thought to reduce the solubility and digestibility of the proteins or starch and several ill-vitro (Barre and van Huot, 1965a, b; Camus and Laporte, 1976; Singh and Krikorian, 1982; Knuckles *et al.*, 1985; Carnovale *et al.*, 1988; Knuckles *et al.*, 1989) and *in vivo* (Atwal *et al.*, 1980) studies have in-deed shown reductions in protein digestibility by PA. However, other in-vitro (Serraino *et al.*, 1985; Reddy *et al.*, 1988) and *in vivo* studies in animals (McDonald *et al.*, 1978; Yoshida *et al.*, 1982; Thompson and Serraino, 1986) and humans (Reinhold *et al.*, 1973) have shown the contrary and thus the evidence that PA affects protein digestibility remains inconclusive.

Protease Inhibitors

Protease inhibitors are abundant in raw cereals and legumes, particularly soybeans. Because of their protein nature, they can be easily denatured and inactivated by heat although about 5-20 per cent of the activity may still remain in commercially available soya products (Rackis and Gumbman, 1981; Hathcock, 1991). Protease inhibitors have been associated with growth inhibition and pancreatic hypertrophy in some experimental animals (Hathcock, 1991). These effects are thought to be related not to the reduced digestibility of the protein since the addition of trypsin inhibitors to free amino acid diets has been shown to also inhibit the rat growth, but to the negative feedback mechanism that exists in the small intestine (Fushiki and Iwai, 1989; Hathcock, 1991). In the presence of trypsin inhibitors, inactivation and small intestinal loss of trypsin can take place. These can trigger the release of the hormone cholecystokinin from the intestinal mucosa which then can induce the pancreas to synthesise more tryesin. Because trypsin is rich in sulphur containing amino acids, Its preferential synthesIs can cause an increased requirement for sulphur-containing amino acids for the synthesis of other body tissues; this, in turn, can contribute to the loss in body weight (Liener and Kakade, 1980). At the same time, the stress on the pancreas can cause hypertrophy and hyperplasia of the exocrine (acinar) cells (Liener, 1986; Roebuck, 1987) which then can lead to the formation of adenomatous nodules (McGuiness *et al.*, 1980). The feeding of purified trypsin inhibitor or raw soya flour containing protease inhibitors has been shown topotentiate the effects of pancreatic carcinogens (Morgan *et al.*, 1977; Roebuck, 1987; Hathcock, 1991).

The adverse effects of protease inhibitors have been observed primarily in animals with a high weight of pancreas, expressed as a percentage of body weight (0·29-0·8), such as rats, mice, chickens, hamsters and young guinea pigs (Liener and Kakade, 1980). No pancreatic enlargement has been seen in large animals with a small pancreas (0·06-0.24) such as dogs, pigs, calves and monkeys. Thus it has been suggested that no effect should be expected in humans because of their small sized pancreases (0·09-0·12). However, the provision of raw soya flour or purified trypsin inhibitor directly to the duodenum of human subjects caused an increased secretion of pancreatic enzymes and serum levels of cholecystokinin (Liener *et al.*, 1988) suggesting that a negative feedback mechanism also exists in humans. Nonetheless, because the incidence of pancreatic cancer is lower than normal frequency in vegetarians (Seventh Day Adventists) and in the Japanese population where the intake of soya beans and other foods rich in protease inhibitors is high (Kennedy and Billings, 1987; Troll and Kennedy, 1989), the effect of protease inhibitors on the pancreas of humans remains unclear and needs further elucidation.

Phytoestrogens and Lignanas

Phytoestrogens and lignans are widely distributed in the plant kingdom including cereals, legumes, oilseeds, fruits and vegetables. The two main classes of phytoestrogens are the isoflavones and the coumestans, the former having attracted greater attention (Price and Fenwick, 1985; Setchell and Adlercreutz, 1988). The important coumestans include coumestrol, 4'-methyoxycoumestrol, sativol, trifoliol and repensol. Of the isoflavones, the most abundant are the glycosides of genistein (4', 5, 7 -trihydroxy-isoflavone) and daidzein (4',7-dihy-droxy isoflavone) and their 4-methyl ether derivatives formononetin (7-hydroxy-4'-methoxy isoflavone) and Biochanin A (5,7-dihydroxy-4'-methoxyiso-flavanone). Lignans are a group of diphenolic compounds with dibenzylbutane skeleton structures and similar characteristics to the phytoestrogens (Setchell and Adlercreutz, 1988). Many different plant lignans have been identified and/or isolated but recently the mammalian lignans enterolactone (*trans*- 2, 3- bis-(3- hydroxy benzy lyl)- bu tyrolactone) and enterodiol (2,3-bis (3-hydroxybenzyl) butan-1,4-diol) have gained much attention. They are produced by the bacterial flora in the colon from the plant lignans matairesinol and secoisolari-ciresinol, respectively and have been detected in the biological fluids of man and animals. The phytoestrogens, plant lignans and their bacterial products undergo enterohepatic circulation and some have been detected in the urine.

Concern for the presence of dietary phytoestrogens started in the 1940s with reports of infertility in sheep in Western Australia which decimated the sheep-breeding industry (Bennetts *et al.*, 1946). The infertility syndrome, referred to as Clover disease and characterised by a cystic condition of the ovaries, an irreversible endometriosis and failure to conceive, appear to have been caused by grazing in pasture having a high content of clover (*Trifolium subterraneum*) which is rich in phytoestrogens (Bradbury and White, 1954). Upon ingestion, formononetin and daidzein can be converted to equol (4',7-dihydroxyisoflavan) by the bacteria in the colon. It is equol that is thought to be responsible for the infertility (Lindsay and Kelly, 1970). More recently, jnfertility and liver disease in captive cheetahs were linked to the intake of phytoestrogen containing soya bean protein (Setchell *et al.*, 1987). With structural similarities to phytoestrogens, lignans are also thought to have oestrogenic and antifertility effects. In addition, there are concerns that phytoestrogens and lignans may enhance tumor growth since oestrogens have growth-stimulatory effects (Miller, 1990). Their aromatic structure suggests that they may-mso be carcinogenic. However, based on binding inhibition studies using 3H -12-0-tetradecanoylphorbol-13-acetate with a mouse skin particulate fraction, the mammalian lignans and phytoestrogens at 10-4 mol/litre level did not appear to have tumour promoting effects (Adlercreutz, 1988). The phytoestrogens daidzein, genistein, formononetin and biochanin A are also non-mutagenic based on the *Salmonella/mammalian* microsome assay. Nevertheless, some purified lignans are toxic and, for example, in high doses, the plant lignan podophyllum can produce renal impairment, nausea, vomiting, delirium, stupor and coma (Ayres, 1990).

The endogenous oestrogens such as the steroids oestrone, oestradiol and oestriol, generally act by binding to the oestrogen receptor in the cell nucleus; the 'transformed' steroid-oestrogen receptor complex then binds to specific sites of the DNA, resulting in transcription of specific genes and the known oestrogenic responses, *e.g.* increased growth rate, hypertrophy and hyperplasia throughout the reproductive tract (Adams, 1989). Because of their similarity in shape and structure to the endogenous steroids, non-steroidal compounds such as diethylstilbisterol, phytoestrogens and lignans can also bind to the oestrogen receptor, although with different binding affinities (Jordan *et al.*, 1985). The oestrogenic effects of phytoestrogens are generally weaker than the steroidal oestrogens due to the lower binding affinity of phytoestrogens to the oestrogen receptor, the lower stability of the phytoestrogen-oestrogen receptor complex that is fanned and the lesser ability of the complex to be transformed (Martin *et al.*, 1978; Tang and Adams, 1980; Kitts *et al.*, 1984). Phytoestrogens may act as oestrogens or antioestrogens depending on the concentration of the steroid oestrogens (Folman and Pope, 1966). Whether phytoestrogens have beneficial or adverse effect appears to depend on what capacity they are acting.

Saponins

Saponins are a diverse group of compounds commonly found in legumes, *e.g.* chick peas, soya beans, lentils, peanuts, lentils, *Phaseoius* beans and alfalfa sprouts; and in some plants commonly used as flavourings, herbs or spices, *e.g.* ginseng, fenugreek, sage, quillaja bark, thyme, sarsaparilla and nutmeg (Oakenful and Sidhu, 1990). Their structures are characterised by the presence of a steroid or triterpene group, referred to as the aglycone, linked to one or more sugar molecules. The presence of both polar (sugar) and nonpolar (steroid or triterpene) groups provide saponins with strong surface-active properties which then are responsible for many of its adverse and beneficial biological effects.

A well-known toxic effect of saponin is its ability to lyse erythrocytes,this, in general, being due to its' Interaction with the cholesterol in the erythrocyte membrane (Birk and Peri, 1980). If provided intravenously to mammals, saponins can cause local inflammation and in large doses, can result in death due to massive release of erythrocyte debris and reduction in the oxygen-carrying capacity of the blood (Scott *et al.*, 1985). Saponins can also lyse other cells such as those found in the intestinal mucosa and consequently affect nutrient Decreased weight gain has been observed with high saponin intake due to a number of reasons including reduced food intake attributable to the bitter taste of saponin,(Birk and Petri, 1980), or decreased absorption and utilisation of nutrients caused either by (a) the inhibition of metabolic and digestive enzymes, *e.g.* protease, amylase, lipase and cholinesterase inhibition by soya saponin (Cheeke, 1971) and chymotrypsin, protease and succinoxidase inhibition by alfalfa saponins (Birk and Petri, 1980); or (b) binding with nutrients, *e.g.* Zn binding by alfalfa saponins (West and Greger, 1978). More information on the toxic effects of saponins are available in several reviews (Birk and Petri, 1980; Oakenful, 1981; Price *et al.*, 1987; Oakenful and Sidhu, 1989, 1990).

Phenolic Compounds

Phenolic compounds encompass a wide variety of compounds characterised by the presence of an aromatic ring with one or more hydroxyl groups and a variety of substituents. Although those found in the plant kingdom can be subdivided into many groupings, this paper will be limited to a few groups which are commonly found in foods and have potential health benefits. Phytoestrogens (isoflavones) and lignans are one group of phenolic compounds already described. Other groups include the phenolic acids and flavonoids. Many of the phenolic acids are eIther denvatives of benzoic acid (C6-Cl structure), _e.g._ phydroxybenzoic, gallic, syringic, vanillic protocatechuic acid, or derivates of cinnamic acid (C6-C3 structure), _e.g._ caffeic, ferulic, sinapic, pcoumaric which are commonly found as esters of quinic or sugars (Deshpande _et al._, 1984). Chlorogenic acid is an ester of caffeic and quinic acids. The flavonoids have a basic C6-C3-C6 structure and include the anthocyanin pigments, flavonols, flavariols and isoflavones. They occur mostly as glycosides except the liavanols which tend to polymerise to condensed tannins. The tannins could be classified either as condensed or hydrolysable. Most condensed tannins are polymers of flavan-3-ols (catechins) or flavan 3: 4 diols (leucoanthocyanidins) while most wrolysable tannins are glucose or poly hydric alcohol esterified with gallic acid (gallotannins) or hexahydrodiphenic acid (ellagitannins). The stable dilactone of the latter is ellagic acid.

The antinutritional and toxic effects of phenolic compounds, particularly the tannins, have been categorised as depression in food/feed intake, formation of the less digestible tannin-dietary protein complexes, inhibition of digestive enzymes, increased excretion of endogenous protein, malfunctions in digestive tract and toxicity of absorbed tannin or its metabolites (Price and Butler, 1980; Singleton, 1981; Deshpande _et al._, 1984; Salunkhe _et al._, 1990). Increased risks of cancer of the mouth and oesophagus have been linked todietary tannins in some epidemiological studies (Morton, 1970, 1972; Kapadin _et al._, 1983). Oral ingestion of plant extracts containing tannins did not induce tumours but did produce sarcoma at the injection site when injected subcutineously to rats (O'Gara _el al._, 1971). Early studies reported carcinogenic effects of quercetin, a major flavonoid in foods, although more recent work suggest the contrary (Deschner _et al._, 1991; Sahu and Washington, 1991).

Although the depressed food intake has been related in part to the astringent taste of the phenolic compounds, the adverse effects of tannins have traditionally been attributed to their ability to bind with proteins and hence inhibit their digestion land absorption. However, this mechanism has been challenged for a number of reasons (Butler, 1989a, b). First, the affinity of tannins to various proteins differs, with the proteins rich in the amino acid proline such as the collagen from animal tissues and the alcohol-soluble storage proteins of cereal seed being strongly bound. Since these proteins are deficient in essential amino acids, the tannin effect on protein digestion is not as adverse as when tannins bind non-preferentially or preferentially to those protein rich in essential amino acids. Secondly, although tannins may also bind to the digestive enzymes (Griffiths, 1986) this is more often seen in _in vitro_ studies where the enzymes are the only proteins present. In some cases where the substrate is present in "a larger concentration than the enzymes, enhancement rather than inhibition of enzyme activity has been observed. Thirdly, _in vivo_ (animal) studies have demonstrated no change, increase or decrease in digestive enzyme activities with high tannin diets. The increased nitrogen output in the faeces after high tannin diet has been attributed not to undigested dietary protein but to endogenous protein.

Tannins are thought to not severely affect the protein digestion but rather inhibit the utilization of the digested and absorbed proteins (Butler, 1989a). This was based on the observation in rats that the presence of tannins almost completely prevented the rat growth while the digestibility of the proteins did not differ much (80 and 88 per cent in the high tannin and tannin-free diets, respectively). On the other hand, the efficiency with which the digested food was converted to biomass was depressed by tannin to 10 per cent of that in a tannin-free diet. It also has been proposed that the body develops certain defence mechanisms against tannins, one of which is the induction of proline- rich, tannin-binding salivary proteins; this may explain the survival of animals who normally consume high-tannin-containing foods. The observations suggest that the antinutritive effect of tannins is more complex than previously thought and needs further elucidation.

Amylase Inhibitor

Inhibitors of a-amylase were first reported in buck-wheat and later in wheat, rye, beans, mangoes, taro root, acorns, legumes, potatoes, sorghum and oats. The characteristics of many of these inhibitors have been reviewed (Buonocore and Silano, 1986; Gallaher and Schneeman, 1986; Whitaker, 1989; Belitz and Weder, 1990). With a few exceptions, most amylase inhibitors from plants are active against animal amylases but inactive against bacterial, fungal and plant enzymes. The inhibitor forms a complex with the amylase, the extent of which depends on a number of factors including the pH, ionic strength, temperature, time of interaction and inhibitor concentration. The complex formation can inactivate the amylase and in turn cause reduction in starch digestion.

Growth reduction might be expected when the intake of starch is limited and its availability is further decreased in inefficient digestion. However, the data on the effect of amylase inhibitors in growth as well as the size of the pancreas are conflicting. When fed to rats, amylase inhibitors derived from wheat reduced the growth rate in one study (Lang *et al.*, 1974) but not in another, although pancreas enlargement and hyperplasia were observed (Granum and Eskeland, 1981). In chicks, wheat amylase inhibitors caused pancreatic hypertrophy, increased pancreatic amylase activity and reduced initial growth rate when given encapsulated to prevent gastric destruction but has no effect when given without encapsulation (Macri *et al.*, 1977). Nevertheless, after 4 weeks, the chicks fed the encapsulated inhibitor had growth rates which were similar to those of the control, suggesting their ability to adapt. Amylase inhibitors derived from kidney beans did not affect the weight gain in rats (Savaiano *et al.*, 1977) while white bean inhibitor retarded growth and caused liver and kidney changes (Maranesi *et al.*, 1984). These conflicting data have been attributed to differences in the pH sensitivity (Gallaher and Schneeman, 1986) and the sensitivity of pancreatic amylase to different amylase inhibitors (Ho *et al.*, 1981). In humans, the intake of amylase inhibitors in the form of 'starch blockers' prepared from *Phaseolus* beans has been reported to cause gastrointestinal; symptoms, *e.g.* diarrhoea, nausea and vomiting; however, many of the 'starch blocker' preparations were not pure amylase inhibitors and likely that other antinutritional factors, *e.g.* lectins, protease inhibitors, contributed to these acts as well (Kilpatrick *et al.*, 1983; Liener *et al.*, 1984).

Lectins or Haemagglutinins

Lectinss or haemagglutinins are sugar-binding proteins which are able to bind and agglutinate red blood cells. They are found in most plant including those that maybe eaten without at treatment or processing (Nachbar and Oppenm, 1980). Lectins are specific not only in the gars that they bind to on the cell membranes. It also in their toxicity. The lectins from jack beans, horse gram, lima beans, kidney beans, mung beans, winged beans and castor beans are I toxic when taken orally but the lectins from soya beans and peanut are not (Liener, 1989).

The toxicity of lectins is characterised by owth inhibition in experimental animals and dirrhoea, nausea, bloating and vomiting in humans (Liener, 1989). When injected, lectins can agglutinate red blood cells followed by haemolysis and heath in extreme cases. Several outbreaks in Eng-and after the intake of improperly cooked beans aye been related to the presence of lectins in the beans although the role of other antinutrients uch as enzyme inhibitors in these cases has 'not been ruled out (Noah *et al.*, 1980; Bender and Readi, 1982; Bender, 1983). Heat processing can reduce the toxicity of lectins as it can be denatured by heat, but low temperature or slow cooking may not be enough to completely eliminate its toxicity (Thompson *et al.*, 1983).

The toxic effects of lectins relate to their binding with the specific receptor sites on the epithelial cell of the intestinal mucosa which then causes lesion, disruption and abnormal development of the microvillae (Liener, 1989). Consequently, absorption of nutrients is impaired. The intake of raw beans or purified lectins from beans has been shown to decrease the absorption of sugars (Jaffe and Cornejo, 1961; Santidrian, 1981; Ishiguro *et al.*, 1984; Donatucci *et al.*, 1987), amino acids (Kawatra and Bhatia, 1979), lipid, nitrogen and vitamin B_{12} (Banwell *et al.*, 1983; Dobbins *et al.*, 1986). Lectins can also impair the nutrient absorption by reducing the activity of brush border enzymes, *e.g.* ptidases, disaccharidases, alkaline phosphatase, utamyl transferase (Kim *et al.*, 1976; Triadou and,udran, 1983; Jindal *et al.*, 1984; Rouanet *et al.*, 988; Liener, 1989), pepsin and/or pancreatin fhompson *et al.*, 1986) and pancreatic and saliary amylase (Rea *et al.*, 1985; Thompson and Gabon, 1987; Fish and Thompson, 1991). The increased secretion of mucin and the increased weight and number of intestinal mucosal cells in the presence of lectins (Pusztai *et al.*, 1986) have also been thought to lead to endogenous loss of litrogen and aggravated toxic effects of lectin with respect to protein utilization. The carbohydrates and proteins that are undigested and unabsorbed in the small intestines reach the colon where they are fermented by the bacterial flora to short-chain fatty-acids and gases. This increased fermentation contributes to some of the gastrointestinal symptoms associated with the intake of raw beans or purified lectins.

Bacterial overgrowth or colonization of coliform bacteria in the small intestine has been observed upon feeding raw beans or purified lectins and this has also been suggested to contribute to lectin toxicity (Liener, 1989). Thinning of the mucous layer in the jejunum which was extensively populated by bacteria and the damage to the microvilli of the duodenum has been thought to be related to the bacterial overgrowth. How lectin increases bacterial colonisation is unclear but it may be related to the polyvalent nature of lectin; this allows lectin to bind to both the mucosal cells and bacteria at the same time and fix the bacteria close to the intestinal mucosa.

The lectin induced disruption of the intestinal mucosa may allow entrance of the bacteria and their endotoxins to the blood stream and cause toxic response (Jayne Williams, 1973; Banwell *et al.*, 1985). Lectins, themselves, may also be internalized and cause systemic effects such as increased protein catabolism and breakdown of stored fat and glycogen and disturbance in mineral metabolism (Pusztai, 1986; Liener, 1989).

The Health Benefits

Antinutrients also have some health benefits which, interestingly, appear to be similar to those suggested for the dietary fibers in fruits, vegetables and grains, *e.g.*, lower blood glucose and hormonal responses to starchy foods, decrease blood lipids and decrease cancer risks.

Lower Blood Glucose and Hormonal Responses

Phytic Acid, Lectins, Phenolics and Amylase Inhibitors

Starchy foods which are digested slowly and result in lower blood glucose response have been suggested to be more beneficial to health and in the management of diabetes and hyperlipidaemia (Wolever, 1990). Hence, factors causing such effects may be considered desirable and current evidence suggests that antinutrients are one of them (Thompson, 1988).

Of the different carbohydrate foods tested for starch digestibility and blood glucose response, those which are rich in antinutrients, *e.g.* legumes, showed the lowest values (Jenkins *el al.*, 1986; Thompson, 1988). A significant negative relationship was observed between the blood glucose response to different starchy foods, expressed as glycemic index and the intake of PA (Yoon *et al.*, 1983), lectins or haemagglutinins (Rea *et al.*, 1985), or polyphenolic compounds or tannins (Thompson *el al.*, 1984) in these foods. Similarly, the concentration of these anti nutrients in the starchy food negatively related to the *in vitro* rate of starch digestion (Y oon *et al.*, 1983; Thompson *et al.*, 1984; Rea *et al.*, 1985). When navy bean was fractionated to starch, protein, fiber and soluble whey fractions and the digestibility of the starch fraction was determined alone or in combination with the protein, fiber or soluble whey fractions, the greatest lowering of the starch digestion rate was observed when the starch was combined with the soluble whey fraction, the fraction which is the richest in antinutrients such as tannins, lectins and PA (Thompson, 1988). The removal of PA from navy beans caused an increase in blood glucose response while the addition of PA back to the beans flattened the response (Thompson *et al.*, 1987). When added to wheat flour in unleavened bread preparations, PA (2 per cent, starch basis) also reduced the starch digestion rate and glycaemic response (Yoon *et al.*, 1983). The same trends were seen when tannins (Thompson and Yoon, 1984), lectins (Rea *et al.*, 1985; Thompson and Gabon, 1987; Fish and Thompson, 1991) or amylase inhibitor (Boivin *et al.*, 1987, 1988) were added to the starch or diet. All these observations suggest that PA, lectins, tannins or amylase inhibitors may be reducing the blood glucose response by influencing the rate of starch digestion. However, blood glucose response to a test meal of 50 g glucose was significantly reduced when PA at a 0·8 per cent level was also present (Demjen and Thompson, 1991), indicating that PA may also be exerting its effect by mechanisms not involving starch digestion.

Antinutrients such as PA, tannins, lectins and amylase inhibitors may lower the rate of starch digestion and hence blood glucose response by the same mechanisms which make them anti nutrients (Thompson, 1988). They can bind directly with the amylase enzyme, thus inactivating it. PA or tannins may also bind with Ca which is needed to stabilise amylase activity, or with the starch to influence its degree of gelatinisation or its accessibility to the digestive enzymes. Independent of starch digestion, PA may influence blood glucose response by slowing down gastric emptying (Sognen, 1965a, b). Lectins may also influence blood -glucose response by binding with the intestinal mucosal cell, causing disruption and interference in the absorption of nutrients. An insulin-lowering effect of lectin' has also been reported (Pusztai, 1986).

The observation that the addition of amylase inhibitor from wheat or microbial source can reduce the blood glucose and raise insulin levels1–after the raw starch intake by rats, dogs and humans (Puis and Keup, 1973) stimulated interest in the use of this antinutrient for therapeutic purposes in diabetes and for the control of obesity. Several companies commercialised, for caloric control amylase inhibi to preparations referred to as starch blockers. However, clinical studies showed that they do not influence the fecal calorie excretion (Bo-Linn *et al.*, 1983), the post-prandial concentrations of plasma glucose and insulin and breath hydrogen (Carlson *et al.*, 1983; Hollenbeck *et al.*, 1983) and ^{13}C-starch metabolism (Garrow *et al.*, 1983). These are thought to be related to the inactivation of the inhibitors by the gastrointestinal secretions and low pH, the excess amylase in the small intestine, starch digestion by mucosal glucoamylase and/or the low purity and hence low anti-amylase activity of the starch blocker (Bo-Lin *et al.*, 1982; Rosenberg, 1982; Carlson *et al.*, 1983; Hollenbeck *et al.*, 1983; Layer *et al.*, 1985). Subsequent studies suggest that it was likely due to the latter since partial purification of the amylase inhibitors caused inhibition of amylase activity both *in vitro* and *in vivo* (Layer *et al.*, 1985, 1986a). As well, significant reductions in post-prandial plasma glucose, insulin, C-peptide and gastric inhibitory polypeptide were observed when amylase inhibitors were given with a starch meal or mixed meal to healthy or diabetic individuals (Layer *et al.*, 1986a,b; Boivin *et al.*, 1987) and when given to non-insulin-dependent diabetics in a weight maintenance diet for 2 weeks (Boivin *et al.*, 1988). While mild diarrhoea occurred on the first day of treatment, this subsided even though carbohydrate malabsorption persisted, suggesting a certain degree of human adaptation to antinutrients such as the amylase inhibitor.

Reduction of Blood Lipids

Saponins

Of the antinutrients, saponins have been studied the most extensively for the hypocholesterolaemic effect. Decreases in cholesterol have been observed when chicks (Newman *et al.*, 1958; Morgan *et al.*, 1972), rats (Cayen, 1971; Yamamoto *et al.*, 1975; Malinow *et al.*, 1977; Oakenful *et al.*, 1979, 1984; Oakenful and Sidhu, 1984; Rao and Kendall, 1986; Sharma, 1986), mice (Reshef *et al.*, 1976) or monkeys (Malinow *et al.*, 1978, 1981) were fed diets containing either purified quillaja (Newman *et al.*, 1958; Oakenful and Sidhu, 1984; Rao and Kendall, 1986), fenugreek (Sharma, 1986), tomatine (Cayen, 1971), gypsophilla (Morgan *et al.*, 1972), digitonin (Morgan *et al.*, 1972; Malinow *et al.*, 1978), saikosaponin (Yamamoto *et al.*, 1975), saponaria (Oakenftil *et al.*, 1979), soya (Oakenful *et al.*, 1984), chickpea (Malinow *et al.*, 1977) or alfalfa (Reshef *et al.*, 1976; Malinow *et al.*, 1977, 1980a,b, 1981) saponins with (all references above) or without (Rao and Kendall, 1986) the presence of cholesterol. Without dietary cholesterol, saponaria saponin did not exhibit a hypocholesteroiemic effect in nits (Sautier *et al.*, 1979; Gibney and Taylor, 1980), pigs (Topping *et al.*, 1980), or rabbits (Pathirana *et al.*, 1981).

While in animal studies the evidence for hypo-cholesterolaemic effect of saponins is very strong particularly when fed in the presence of cholesterol, the evidence in human studies is less conclusive. Diets containing foods rich in saponins (300–500 mg/day) such as soya, alfalfa, chickpea, bean meal and yucca (Mathur *et al.*, 1968; Sirtori *et al.*, 1977; Bingham *et al.*, 1978; Malinow *et al.*, 1980b; Bingwen *et al.*, 1981; Molgaard *et al.*, 1987) reduced the plasma cholesterol by 16-24 per cent. However, such foods also contain other substances which may contribute to cholesterol lowering, *e.g.* phytosterols, Isoflavones, protein, fiber, PA, etc. and it is unclear how much of the lowering can be directly attributed to the saponin content. Feeding purified saponins, as was done in animal studies, has been easy to do in humans due to the unknown longterm toxicity of saponins.

Saponins may lower cholesterol by binding with dietary cholesterol (Coulston and Evans, 1960; Gestener *et al.*, 1972) and preventing its absorption and/or by binding with bile acids, interfering with its enterohepatic circulatl and increasing its faecal excretion (Oakenul and Sidhu, 1983; Oakenful, 1986; Sidhu and Oakenful, 1986). Increased bile acid excretion may cause compensatory increase in bile acid synthesis from cholesterol in the liver and consequent lowering of the plasma cholesterol. These mechanisms are feasible due to the amphiphilic structure of saponin. Increased bile acid excretion has been seen in cases where plasma cholesterol is decreased by high saponin diet (Mathur *et al.*, 1968; Malinow *et al.*, 1980b; Oakenful and Sidhu, 1990).

Phytic Acid, Amylase Inhibitors, Phenolics, Phytoestrogens and Lignans

Other antinutrients have also demonstrated some hypocholesterolaemic effect. The addition of PA (0·2 - 9 per cent) to the diet of rats significantly reduced the plasma cholesterol and triglyceride levels (Klevay, 1977; Sharma, 1980, 1984; Jariwalla *et al.*, 1990). This was suggested to be related to the ability of PA to bind to Zn and thus lower the plasma Zn to copper (Cu) ratio; lower ratios tend to pre-dispose humans to cardiovascular disease (Klevay, 1977). In addition, the effect may be related to the ability of PA to reduce the plasma glucose and in-sulin concentrations which, in turn, may lead to reduced stimulus for hepatic lipid synthesis (Thompson, 1988; Wolever, 1990). Amylase inhibitors, with their ability to reduce plasma glucose and insulin (Boivin *et al.*, 1987, 1988), may also lower the plasma cholesterol and triglyceride. Phenolic compounds such as ferulic and *p*–coumaric acid, at low levels found in bengal gram, lowered the blood lipid levels in rats; they are thought to contribute to the hypocholesterolaemic effect of bengal gram (Sharma, 1980, 1984). Because the plasma sex hormone binding globulin (SHBG) and liver 5-alpha-reductase concentrations have positive relationships with plasma HDL (high density lipoprotein) cholesterol and apolipoprotein A-1 and because the urinary excretion of the mammalian lignans enterolactone and enterodiol and the phytoestrogens also have positive relationship with the SHBG concentration, these antinutrients have been thought as well to influence blood lipid levels and coronary heart disease (Adlercreutz, 1990). Interestingly, tamoxifen, wd antioestrogen used as adjuvant therapy in breast cancer patients and with close structural and biological similarities with phytoestrogen and lignans, has been shown to inhibit cholesterol synthesis and lower the plasma cholesterol in some studies (Schapira *et al.*, 1990).

Reduction in Cancer Risk

Recently, the US National Cancer Institute identified a number of foods which are thought to be protective against cancer and which research is currently being encouraged (Troll and Kennedy, 1989; Messina and Barnes, 1991; Messina and Messina, 1991; Caragay, 1992). Several substances in these foods which may be responsible for the cancer protective effect have been suggested and interestingly, many of them are antinutrients including protease inhibitors, PA, phytoestrogens and lignans, saponins and phenolic compounds. Some of the supporting evidences are described below.

Phytic Acid

When PA was added to the diet of rats, a significant negative relationship was observed between the PA levels (0·6-2·0 per cent) and the epithelial cell proliferation (expressed as labeling index) in the ascending and the descending colon (Nielsen *et al.*, 1987). Because carcinogenic initiation involves the alteration of DNA molecules by carcinogens and this oncogenic response is usually greater in highly proliferating than slowly proliferating cells, this suggests.reducing colon cancer risk in the presence of PA. Indeed when PA (1 or 2 per cent) was given to the drinking water either 1 week before or 2 weeks or 5 months after the induction of carcinogen, a significant reduction in tumour number and tumour volume in the colon of rats or mice were observed (Shamsuddin *et al.*, 1988, 1989); Shamsuddin and Ullah, 1989; Ullah and Shamsuddin, 1990). The tumor-suppressing effect of magnesium oxide was suppressed by PA (Jariwalla *et al.*, 1988). Mammary cancer risk may also be reduce2. by PA as indicated by the decreased cell proliferation, nuclear aberration and intraductal proliferation in the mammary gland of mice fed PA particularly when the concentration of iron (Fe) or Ca is also high (Thompson and Zhang, 1991). A significant relationship was observed between the epithelial cell proliferation in the colon and in the mammary gland suggesting a similar effect of PA and minerals in these tissues. In a wide spectrum organ carcinogenesis model using three different types of carcinogens, however, no significant effect of PA on the colon, lung, oesophagus, for stomach, small intestine, kidney and thyroid gland were seen (Horose *et al.*, 1991). On the other hand, reduction in the incidence of liver hyper plastic nodules, hepatocellular carcinoma and eosinophilic focus in the pancreas and increase in bladder papillomas were observed, suggesting differences in the effect of PA on the various organs.

PA may reduce cancer risk by a number of mechanisms (Shamsuddin *et al.*, 1988, 1989; Shamsuddin and Ullah, 1989; Thompson, 1989; Graf and Eaton, 1990; Thompson and Zhang, 1991). PA may bind Fe, a catalyst of lipid peroxidation; free-radical formation during lipid oxidation can cause breakdown of cellular membranes and thus encourage cell proliferation. PA may also bind the Zn needed for DNA synthesis and, again, indirectly result in reduced cell proliferation. The slowing down of starch digestion by PA can cause some of the starch to escape the colon. The starch then is fermented by the colonic bacteria producing short-chain fatty acids (SCF A), gases and more bacterial cells. The SCF A lowers the pH to a level which may be protective; low pH may protect by reducing the solubility of bile acids and neutralizing the ammonia which have been suggested as tumour promoters (Newmark and Lupton, 1990). Furthermore, butyric acid produced in high concentrations from starch fermentation has been shown to be a primary respiratory fuel and modulator of gene expression and cell growth in the colon (Cummings and Macfarlane, 1991). PA may also interact with bacterial enzymes such as, a-glu curonidase and mucinase in the colon; reduction of the, a-glucuronidase activity reduces the conversion of primary to secondary bile acids which are thought to promote tumourigenesis. Mechanisms involving the inositol triphosphate which is produced upon PA hydrolysis (Shamsuddin and Ullah, 1989; Shamsuddin *et al.*, 1989, Ullah and Shamsuddin, 1990) and the enhancement of the baseline natural killer cell activity by PA (Baten *et al.*, 1989) have also been suggested.

Protease Inhibitors

While the protease inhibitors have been linked with pancreatic cancer in animal studies, they may also act as anticarcinogenic agents as suggested by animal studies, *in vitro* cell culture work and epidemiological data which show low cancer mortality rate in human population with high dietary intake of protease inhibitors. These were discussed in recent workshops (Troll and Kennedy, 1989; Messina and Barnes, 1991; Messina and Messina, 1991). *In vitro*, protease inhibitors have been shown to suppress the malignant transformation of cells induced by different types of carcinogens, *e.g.* ionizing radiation, UV light, chemical carcinogens, steroid hormones (Kennedy, 1984; Kennedy and Billings, 1987). Of the protease inhibitors, the most effective are those with chymotrypsin inhibitor activity such as those found in soya bean, chick pea and potato. The Bowman-Birk inhibitor (BBI) derived from soya bean has been shown to either inhibit or prevent the development of chemically induced cancer of the liver (St. Clair *et al.*, 1990), lung (Witschi and Kennedy, 1989), colon (St. Clair *et al.*, 1990), oral (Messadi *et al.*, 1986) and oesophagus (Messina and Barnes, 1991).

Several mechanisms whereby protease inhibitors inhibit carcinogenesis have been hypothesised (Kennedy and Billings, 1987; Troll *et al.*, 1987, Troll, 1989). The inhibitor may reduce the digestion of proteins and thus the amino acids available to the growing cancer cells; deprivation of amino acids particularly leucine, phenylalanine and tyrosine has been shown to prevent the growth of mouse hepatoma and mammary adeno-carcinoma (Troll *et al.*, 1987). The inhibitors may stop the ongoing cellular process begun by carcinogen exposure by reversing the carcinogen-induced change in oncogene expression. They may inhibit the formation of superoxide anion radicals (O_2^-) and hydrogen peroxide (H_2O_2) by neutrophils induced by tumour promoters; these oxygen-reactive species can damage or modify cellular DNA. The inhibitors may inhibit the oxygen radical-induced DNA polymerase, the enzyme involved in the formation of poly (ADP-ribose).

Phytoestrogens and Lignans

Epidemiological data and biological properties of phytoestrogens and lignans suggest that they may also be important in the prevention and control of cancer, particularly the hormone-dependent ones. The urinary excretion of lignans (enterolactone and enterodiol) and the isoflavonic phytoestrogens (equol, daidzein, O-desmethylangolensin) was significantly lower in breast cancer patients and omnivores than in the vegetarians with a lower risk of cancer (Adlercreutz *et al.*, 1986). Japanese women on traditional diets and thought to be of lower cancer risk excrete 140 per cent higher amounts of phytoestrogens than the Finnish women of high cancer risk (Adlercreutz, 1988; Adlercreutz *et al.*, 1988). Non-human primates excreting very high levels of lignans are very resistant to the carcinogenic effect of oestrogens even in the presence of carcinogens. Many lignans have antimitotic properties and have been shown to prevent the growth of various tumours studied in the chemotherapy programs of the US National Cancer Institute (Hartwell, 1976). Antiproliferative effects of synthetic and natural flavonoids, *e.g.* daidzein, on tumour cells of the human breast cancer line ZR–75-1, as well as on human peripheral blood lymphocytes have also been reported (Hirano *et al.*, 1989a, b).

Support for the hypothesis that phytoestrogens have anticarcinogenic effects also comes from studies which showed strong relationships between cancer risk and soya bean intake in the Japanese population and between soya bean intake and urinary phytoestrogen excretion (Adlercreutz *et al.*, 1988). The feeding of soya bean either as chips or protein isolates also caused inhibition of mammary tumours induced by N-methyl-N-nitrourea (MNU) or 7,12 dimethylbenz(a)-anthracene (DMBA) (Barnes *et al.*, 1990), an effect which was attributed to the phytoestrogens rather than the protease inhibitors since autoclaving the soya to inactivate the protease inhibitors produced similar tumor inhibitory effect.

In short-term studies, diet supplementation with flax seed, the richest source of mammalian lignan precursors (100-800 times higher than 66 common plant foods) (Axelson *et al.*, 1982; Thompson *et al.*, 1991), lowered the epithelial cell proliferation and aberrant crypt foci formation in the colon (Serraino and Thompson, 1992a) and the epithelial cell proliferation and nuclear aberration in the mammary gland of rats (Serraino and Thompson, 1991), suggesting a reduction in colon and mammary cancer risks, respectively. In long-term studies, significant reduction in the tumour size was observed when flax seed was fed at the promotion stage of tumourigenesis (Serraino and Thompson, 1992b). A tendency for the number of tumours to be lower in the rats fed the flax seed at the initiation and at both the initiation and promotion stage of tumourigenesis was also observed. These effects are thought to be related to the flax seed lignans although the role of the other components in flax seed have not been ruled out (Serraino and hompson, 1991, 1992a, b; Serraino *et al.*, 1992; hompson *et al.*, 1992).

How phytoestrogens and lignans may influence lrcinogenesis is still unclear but several mechaisms have been postulated in several reviews Setchell and Adlercreutz, 1988; Adlercreutz, 1990, 991; Adlercreutz *et al.*, 1991) based on current evidences.

(a) Since the urinary excretion of the total lignans and total phytoestrogens in omnivores, lactovegetarians and breast cancer patients correlated positively to the plasma SHBG and negatively to the percentage of free oestradiol and percentage of free testosterone, the phytoestrogens and lignans may stimulate the SHBG synthesis in the liver and reduce the metabolic clearance rate, bioavailability and adverse effects of the steroids.

(b) The lignans and phytoestrogens have weak oestrogenic/antioestrogenic effects; they compete with oestrogens for the oestrogen receptors and hence prevent the growth of tumour cells. For example, enterolactone alone can stimulate the growth of MCF breast cancer cells but in the presence of oestradiol the growth either was less or did not differ from the control.

(c) As was observed *in vitro* in the case of enterolactone and an intermediate between matairesinol and enterodiol, lignans and phytoestrogens may have arornatase inhibitory effects; they prevent the production of oestrone from androstenedione and consequently deny the tumour a source of endogenous oestrogen.

(d) They may inhibit the tyrosine-specific protein kinase and DNA topoisomerase as was observed for genistein (Markovits *et al.*, 1989). As protein kinase activity is associated with cellular receptors for several growth factors (Akiyama *et al.*, 1978; Ogawara *et al.*, 1989; Teraoka *et al.*, 1989) and also with oncogene products of retroviral src gene family, suggesting roles in cell proliferation and transformation, the reduction in its activity may represent reduced cancer risk.

(e) Typical of phenolic compounds, lignans and phytoestrogens may act as antioxidants. For example, nordihydroguaiaretic acid (NDGA), a lignan used as a food antioxidant, has been shown to protect from TPA -induced mouse skin tumour (Nakadate *et al.*, 1982a), ornithine decarboxylase (Nakadate *et al.*, 1982b) and protein kinase C activities (Nakadate, 1989) and inflammation (Young *et al.*, 1983); these have been linked to its ability to inhibit lipoxygenase activity and to act as antioxidant (Agarwal *et al.*, 1991). Furthermore, NDGA inhibits the cytochrome PASO-mediated monooxygenase activity in epidermal and hepatic microsomal preparations suggesting that it may inhibit the activation of promutagens and procarcinogens since they require metabolic activation catalysed by the cytochrome P-450 enzymes (Agarwal *et al.*, 1991).

(f) Phytoestrogens and lignans may exert their effects by influencing cholesterol homeostasis. Enterolactone and enterodiol have been shown *in vitro* to inhibit the activity of cholesterol-7-a-hydroxylase, the rate-limiting enzyme in the formation of primary bile acids from cholesterol (Sanghvi *et al.*, 1984). Since primary bile acids are converted by the colonic bacterial flora to secondary bile acids such as deoxycholic acid which has been correlated with increased colon cancer risk, the enzyme inhibition may represent a reduction in cancer risk.

Saponins

The biological properties of saponins suggest that they may also have some anticarcinogenic effect (Messina and Barnes, 1991) through a number of mechanisms. Saponins are able to bind cholesterol and bile acids (Oakenful and Sidhu, 1983; Oakenful, 1986; Sidhu and Oakenful, 1986). The binding of the primary bile acids by saponins would reduce their bacterial conversion to secondary bile acids and consequently reduce their ability to promote tumourigenesis. Saponins have been shown to enhance the immune response and thus have been used as adjuvants for oral vaccines (Chavali, 1978, 1987; Bomford, 1982; Maharaj *et al.*, 1986; Newmark, 1988; Heath *et al.*, 1991; Kensil *et al.*, 1991; Mowat *et al.*, 1991; Oosterlaken *et al.*, 1991). Astragalus saponinl induced the accumulation of cAMP in rabbit plasma and since increased levels of cAMP inhibit the DNA synthesis in tumour cells *in vitro*, they may also inhibit tumor growth *in vivo* (Yindi, 1984). Saponins have been shown to decrease the growth of human epidermoid carcinoma cells (Aswal *et al.*, 1984) and human cervical carcinoma cells (Sati *et al.*, 1985) and to inhibit Epstein-Barr virus genome expression (Tokuda, 1988). Saponins from the Dolichos-Fal-calUs-Klein root were cytotoxic to Sarcoma 37 cells (Huang *et al.*, 1982).

Phenolic Compounds

The carcinogenesis process involves two major stages: initiation and promotion. The initiation stage involves the intracellular formation of reactive electrophiles that can attack chemical nucleophiles in the cell, including DNA, thus causing its chemical or structural alteration. The promotion stage is the malignant transformation of the damaged cells leading to mutagenesis or carcinogenesis (Newmark, 1984; Namiki, 1990). The initiator may be the agent itself or a product of its metabolic activation by mixed function oxidases. Thus, to block carcinogenesis, a substance may act as follows: (a) prevent the formation of carcinogen (*e.g.* N-nitroso compounds, active oxygen species) in foods or endogenously, (b) alter the enzymic activation of the potential initiator to ultimate carcinogen, (c) stimulate carcinogen detoxifying systems, or (d) block or trap the ultimate carcinogen electrophiles in a nucleophilic chemical reaction to form innocuous products, or inhibit by arachidonic acid metabolites via the lipoxygenase or cyclooxygenase pathways (Wattenberg, 1983; Newmark, 1984).

Some recent reviews have identified phenolic compounds in foods as antimutagens and anticarcinogens, perhaps functioning in one or more of the above ways (Wattenberg, 1983; Newmark, 1984; Stich and Rosin, 1984; Cheng and Ho, 1988; Hayatsu *et al.*, 1988; Williams *et al.*, 1989; Namiki, 1990; Caragay, 1992). Several phenolics (*e.g.* chlorogenic acid, gallic acid, caffeic acid, tannic acid, catechin) have been shown to inhibit the mutagenic effects of both directacting carcinogens (*e.g.* benzo(a)pyrene diol epoxide) and carcinogens that require metabolic activation (*e.g.* aflatoxin B_1) and to trap nitrite, thus reducing the nitrosating species and preventing the endogenous formation of carcinogenic nitroseamines (Stich and Rosin, 1984). In mice fed diets containing 1·1 per cent caffeic or ferulic acid, about 40 per cent reduction in benzo(a)pyrene induced stomach cancer was observed (Wattenberg *et al.*, 1980). Ellagic acid produced significant reduction in mice skin tumour risk promoted by 12-0-tetradecanoyl-phorbol-13-acetate (Chang *et al.*, 1985) or induced by 3-methylcholanthrene (Mukhtar *et al.*, 1986). Ellagic acid, quercetin or robinetin injection also inhibited the incidence of lung tumours in mice (Lesca, 1983; Chang *et al.*, 1985) although other studies showed no significant effect of ellagic acid on lung or skin tumours (Smart *et al.*, 1986). Quercetin and rutin caused reduction in azoxymethanolin-duced colon cancer (Deschner *et al.*, 1991). Green tea polyphenols, particularly epigallocatechin gallate, have been shown to reduce the mutagenicity of AFB1, 4-NQO, BaP, MelQ and cigarette smoke condensate, to induce the activity of glutathione-S-transferase and superoxide dismutase and to reduce the development of skin, liver and lung cancer (Cheng and Ho, 1988).

Other Benefits

Lignan-rich plant products are components of many Chinese and Japanese folk medicines (Ayres, 1990). Rheumatoid arthritis, gastric and duodenal ulcers, scrofula (tuberculous-like disease of the lymph glands generally localised in the neck), venereal wart, nasal papillomas and psoriasis are diseases reported to be treated by lignan-rich plant products (Ayres, 1990) although more studies are needed to confirm that lignan is the active ingredient in these products. Other reported effects of lignans include diuretic action (Prestegane B) (Plante *et al.*, 1987), digitalis-like action (enterolactone) (Fagoo *et al.*, 1986) antagonistic action of platelet-activating factor (PAF) receptor (Braquet *et al.*, 1985) and priming action on superoxide production on human neutrophiles (2,3-dibenzylbutane-IA-diol, DBB) (Morikawa *et al.*, 1990).

PA has been suggested to have a role in the prevention of dental caries and platelet aggregation, in the treatment of hypercalciuria and kidney stones and as antidote against acute lead poisoning, primarily due to its mineral-binding ability (Graf, 1986; Graf and Eaton, 1990).

Saponins are present in a number of herbal remedies, *e.g.* ginseng, liquorice, sarsaparilla, which appear to have expectorant and antiflammatory effects (Shibata, 1977). Several phenolic compounds, *e.g.* monomeric hydrolysable tannins, oligomeric ellagitannins and condensed tannins, having galloyl groups or hexahydroxydiphenoyl groups, have potent inhibitory effect on herpes simplex virus types 1 and 2 (HSV-1, HSV-2) infection (Fukuchi *et al.*, 1989). These viruses are linked to several human diseases including gingivos-tomatitis, stomatitis, meningitis and venereally transmitted genital disease. Three hydroxyl groups in the benzene ring and high molecular weight of the phenolic compounds are needed for the inhibition of the virus to take place. Interestingly, protein binding and antitumour activities also are related to molecular weight and dimeric ellagitannins with high molecular weight also have strong anti-HIV activity (Asanaka *et al.*, 1988; Fukuchi *et al.*, 1989).

Conclusion

It is evident that both adverse and health benefits may be attributed to antinutrients in foods. It is also evident that, in many cases, the same interactions that make them ntinutritive also are responsible for their beneficial effects. However, health benefits are possible at a certain level of antinutrient intake without causing much adverse effects. For example, Oriental women who consume more soya products (containing phytoestrogens and protease inhibitors) than American women and have lower cancer risks do not seem to have reproductive and pancreatic cancer problems (Messina and Barnes, 1991). Lower blood glucose response to antinutrient-containing test meals (*e.g.* cooked cereals, legumes or starchy meal with added PA or amylase inhibitors) have also been observed without causing much gastrointestinal discomfort (Boivin *et al.*, 1987, 1988; Thompson, 1988).

Undoubtedly, the physiological effects of anti-nutrients are related to their level of intake and the conditions they are taken, *e.g.* presence of other dietary constituents, nutritional and health status of the individual. Thus, if we are to balance their risks with benefits, there is a need to obtain more information on the concentration of antinutrients in foods and their level of intake. More dose- response studies to determine the level that will produce health benefits without causing any adverse effect should be conducted in the future. Since several anti nutrients appear to have the same mitigating effects, *e.g.* reduce cancer risk or cardiovascular disease, studies to determine the synergistic or antagonistic effects of mixtures of these antinutrients are also of interest. The interaction of antinutrients with other components of the diet and the mechanism of their action need to be further addressed if we are to understand the role of antinutrients in health and disease. Because of the potential health benefits of antinutrients, it is probably inappropriate to refer to these substances as 'antinutrients' or 'natural toxins'. The time has come for us to reevaluate not only their presence in our diet but also their name.

References

Adams, R. (1989). Phytoestrogens. In *Toxicants of Plant Origin. IV. Phenolics*, ed. P. R. Cheeke. CRC Press. Boca Raton, FL, USA, pp. 24-51.

Adlercreutz, H. (1988). Lignans and phytoestrogens. Possible preventive role in cancer. In *Frontiers of Gastrointestinal Research*, ed. P. Rozen. S. Karger, Basel. Switzerland, pp. 165-76.

Adlercreutz, H. (1990). Western diet and Western disease: some hormonal and biochemical mechanisms and associations. *Scand. 1. Clin. Lab Invest.* 50 (suppl. 201). 3-23.

Adlercreutz, H. (1991). Diet and sex hormone metabolism. In *Nutrition, Toxicity and Cancer*, ed. J. R. Rowland. CRC Press, Boca Raton, FL, USA, pp. 137-93.

Adlercreutz, H., Fotsis, T., Bannwart, C, Wahala, K., Makela, T., Brunow, G. and Haase, T. (1986). Determina-tion of urinary lignans and phytoestrogens metabolites. po-tential antiestrogen and anticarcinogens, in urine of women on various habitual diets. *J. Steroid Biochem.*, 25, 791-7.

Adlercreutz, H., Honjo, H., Higashi, A., Fotsis, T., Hamalainen, E., Hasegawa, T. and Okada, H. (1988). Lignan and phytoestrogen excretion in Japanese consuming traditional diet. *Scand. J. Clin. Lab. Invest.*, 48 (suppl. 190), 190.

Adlercreutz, H., Mousavi, Y., Loukpvaara, M. and Hamalainen, E. (1991). Lignans, isoflavones, sex hormone metabolism and breast cancer. In *The New Biology of Steroid Hormones*, ed. R. Hochberg and F. Naftolin. Raven Press, New York, USA, pp. 145-54.

Agarwal, R., Wang, Z., Bik, D. P. and Mukhtar, H. (1991). Nordihydroguairetic acid, an inhibitor of lipoxygenase, also inhibits cytochrome P-450-mediated monooxygenase activity in rat epidermal and hepatic microsomes. *Drug Metab. Disp.*, 19,620–4.

Akiyama, T., Ishida, J., Nakagawa, S., Ogawara, H., Watan-abe, S., Itoh, N., Shibuya, M. and Fukami, Y. (1987). Genistein, a specific inhibitor of tyrosine-specific protein kinases. *J. Bioi. Chern.*, 262, 5592-5.

Asanaka, M., Kurimura, T., Koshiura, R., Okuda, T., Mori, M. and Yokoi, H. (1988). Tannin as candidate for an anti -HIV drug. In *Proc. Fourth Internat. Conf. on Immunopharmacology*, Osaka, p. 35.

Aswal, B. S., Bhakuni, A. K. and Kar, A. (1984). Screening of Indian plants for biological activity. Part X. Indian. *J. Exper. Bioi.*, 22,312-32.

Atwal, A., Eskin, N., McDonald, B. and Vaisey-Genser, M. (1980). The effects of phytase on nitrogen utilization and zinc metabolism in young rats. *Nutr. Rep. Int.*, 21, 257-67, Axelson, M., Sjovall, J., Gustafsson, B. E. and Setchell, K. D. R. (1982). Origin of lignan in mammals and identification of precursors from plants. *Nature*, 298, 659-60.

Ayres, L. (1990). *Lignans-Chemical, Biological and Clinical Properties*. Cambridge University Press, Cambridge, UK.

Banwell, J. G., Boldt. D., Meyers, J., Weber, Jr, F. L., Miller, B. and Howard, R. (1983). Phytohemagglutinin derived from red kidney beans (*P. vulgaris*). A cause for intestinal mal-absorption wi th bacterial overgrowth in the rat. *Gastroenlerology*, 84, 506-15.

Banwell, J. G., Howard, R., Cooper, D. and Costerton, J. W. (1985). Intestinal microbial flora after feeding phytohemag-glutinin to rats. *Appl. Environ. Mircobiol.*, 50, 68-80.

Barnes, S., Grubbs, c., Setchell, K. D. R. and Carlson, J. (1990). Soybeans inhibit ammary tumours in models of breast cancer. In *Mutagens and Carcinogens in the Diet*, ed. (M. W. Pariza, H. Aeschbacher, J. S. Felton and S. Sato. Wiley-Liss, New York, USA, pp. 239-53.

Barre, R. and van Huot, N. (1965a). Etude de la combinaison de I' acjde phytique avec la serum albumine humaine native, acetylee et des amine. *Bull. Soc. Chim. Biol.*, 47, 1399-409.

Barre, R. and van Huot, N. (1965b). Etude de la combinaison de l'ovalbumine avec les acids phosphorique, u-glycero-phosphorique et phytique. *Bull. Soc. Chim. Biol.*, 47, 1419-27.

Baten, A., Ullah, A., Tomazic, V. J. and Shamsuddin, A. M. (1989). Inositol phosphate induced enhancement of natural killer cell activity correlates with tumor suppression. *Carcinogenesis*, 10, 1595-8.

Belitz, A. D. and Weder, J. K. P. (1990). Protease inhibitors of hydrolases in plant foodstuffs. *Food Rev. Int.*, 6, 151-211.

Bender, A. E. (1983). Hemagglutinins (lectins) in beans. *Food Chem.*, II, 309-20.

Bender, A. E. and Readi, G. B. (1982). Toxicity of kidney beans (*Phaseolus vulgaris*) with special reference to lectins. *1. Plant Foods*, 4, 15-22.

Bennetts, H. W., Underwood, E. J. and Shier, F. L. (1946). A specific breeding problem of sheep on subterranean clover pastures in Western Australia. *Aust. J. Agric. Res.*, 22, 131-8.

Bingham, R., Harris, D. H. and Laga, T. (1978). Yucca plant saponin in the treatment of hypertension and hyper-cholesterolemia. *J. Appl. Nutr.*, 30, 127-32.

Bingwen, L., Zhaofeng, W., Wanzhen, L. and Rongjue, Z. (1981). Effects of bean meal on serum cholesterol and triglycerides. *Chin. Med. J.*,94,455-8.

Birk, Y. and Peri, I. (1980). In *Toxic ConstilUents of Plant Foodstuffs*, ed. J. E. Liener. Academic Press, New York, USA, pp. 161-82.

Boivin, M., Zinsmeister, A. R., Go, V. L. and DiMagno, E. P. (1987). Effect of purified amylase inhibitor on carbohydrate metabolism after a mixed meal in healthy humans. *Mayo Clinic Proc.*, 62, 249-55.

Boivin, M., Flourie, B., Rizza, R. A., Go, V. L. and DiMagno, E. P. (1988). Gastrointestinal and metabolic effects of amylase inhibition in diabetics. *Gastroenterology*, 94, 387-94.

Bo·Linn, G. W., Santa Ana, C. A., Morawski, S. G. and Ford-tran, J. S. (1982). Starch blockers - their effect on calorie absorption from high starch meal. *New England J. Med.*, 307,1413-16.

Bomford, R. (1982). Studies on the cellular site of action of the adjuvant activity of saponin for sheep erythrocytes. *Arch. Allergy Appl. Immunol.*, 67, 127-31.

Bradbury, R. B. and White, D. E. (1954). Oestrogens and related substances in plants. *Vitam Horm.* (NY), 12, 207-33.

Braquet, P., Robin, J. P., Esanu, A., Landais, Y., Vilain, B., Baroggi, N., Touvay, C. and Etienne, A. (1985). Isolation and characterization of endogenous PAF (platelet activat-ing factor)-inhibiting factors (EPIFs) in human and mon-key urine. *Prostaglandins*, 30, 692.

Buonocore, V. and Silano, V. (1986). Biochemical, nutritional and toxicological aspects of alpha amylase inhibitors from plant foods. In *Nutritional and Toxicological Significance of Enzyme Inhibitors in Foods,* ed. M. Friedman. Plenum Press, New York, USA, pp. 483-508.

Butler, L. G. (1989a). New perspectives on the antinutritive effects of tannins. In *Food Proteins,* ed. J. E. Kinsella and W. G. Soucie, AOCS, Champaign, IL, USA, pp. 402-9.

Butler, L. G. (1989b). Sorghum polyphenols. [n *Toxicants of Plant Origin,* ed. P. R. Cheeke. CRC Press, Boca Raton, FL, USA, pp. 95-122.

Camus, M. C. and Laporte, J. C. (1976). Inhibition de la prote-olyse pepsique en vitro par de ble. Role de l'acide phytique des issues. *Ann. Bioi. Biochem. Biophys.,* 16,719-29.

Caragay, A. B. (1992). Cancer protective foods and ingredi-ents. *Food Technol.,* 46, 65-8.

Carlson, G. L., Li, B., Bass, P. and Olsen, W. A. (1983). A bean alpha amylase inhibitor formulation (starch blocker) is ineffective in man. *Science,* 219, 393-5.

Carnovale, E., Lugaro, E. and Lombardi-Boccia, G. (1988). Phytic acid in fava beans and peas: effect on protein availability. *Cereal Chern.,* 65, 114-17.

Cayen, M. N. (1971). Effect of dietary tomatine on choles-terol metabolism in the rat. *J. Lipid Res.,* 12,482-90.

Chang, R. L., Huang, M.-T., Wood, A. W., Wong, C.-Q., Newmark, H. L., Yagi, H., Sayer, J. M., Jerina, D. M. and Conney, A. H. (1985). Effect of ellagic acid and hydroxy-lated fiavonoids in the tumorigenicity of benzo(a)pyrene and (+)-7,8-dihydroxy-9,I 0-epoxy-7,8,9,I O-tetrahydrobenzo-(a)pyrene on mouse skin and in the newborn mouse. *Carcinogenesis,* 6, 1127-33.

Chavali, S. R. (1978). An *in vitro* study of immunomodu-latory effects of some saponins. *Int. J. Immunopharmacol.,* 9,675-9.

Chavali, S. R. (1987). Adjuvant effects of orally administered saponins on humoral and cellular immune responses in mice. *Immunobiology,* 174, 347-52.

Cheeke, P. R. (1971). Nutritional and physiological implica-tion of saponin. A review. *Can. J. Anim. Sci.,* 51, 621.

Cheng, S. and Ho, C. (1988). Mutagens, carcinogens and in-hibitors in Chinese foods. *Food Rev. Int.,* 4,353-74.

Cheryan, M. (1980). Phytic acid interactions in food systems. *S,RC Crit. Rev. Food Sci. Nutr.,* 13,297-335.

Committee on Diet and Health (1989). *Diet and Health. Im-plications for Reducing Chronic Disease Risk.* National Academy Press, Washington DC, USA.

Cosgrove, D. J. (1980). *Inositol Phosphates: Their Chemistry. Biochemistry and Physiology.* Elsevier Publishing Co., New York, USA.

Coulson, C. B. and Evans, R. A. (1960). Effect of saponin, sterols and linoleic acid on the weight increase of growing rats. *Br. 1. Nutr.,* 14, 121-34.

Cummings, J. H. and Macfarlane, G. T. (1991). The control and consequences of bacterial fermentation in the human colon. *J. Appl. Bacteriol.,* 70, 443-59.

Davies, N. T. (1982). Effects of phytic acid on mineral avail-ability. In *Dietary Fiber in Health and Disease,* ed. G. Va-houny and D. Kritchevsky. Plenum Press, New York, USA, pp. 105-16.

Demjen, A. and Thompson, L. U. (1991). Calcium and phytic acid independently lower the glycemic response to a glu-cose load. In *Proc. 34th Can. Fed. 'Bioi. Soc. Meeting.* Canadian Federation of Biological Societies, Ottowa, Canada, p. 53 (abst)

Deschner, E. E., Ruperto, J., Wong, G. and Newmark, H. L. (1991). Quercetin and rutin as inhibitors of azoxymethanol-induced colonic neoplasia. *Carcinogenesis,* 12, 1193.

Deshpande, S. S., Sathe, S. K. and Salunkhe, D. K. (1984). Chemistry and safety of plant polyphenols. In *Nutritional and Toxicological Aspects of Food Safety,* ed. M. Friedman. Plenum Press, New York, USA, pp. 457-95.

Dobbins, J. W., Lurenson, J. P., Gorelick. F. S. and Banwell, J. G. (1986). Phytohemagglutinin from red kidney beans (*P. vulgaris*) inhibits sodium and chloride absorption in the rabbit ileum. *Gastroenterology,* 90, 1907-13.

Donatucci. D. A., Liner. I.E. and Gross. C. J. (1987). Binding of navy beans (*P. vulgaris*) lectin to the intestinal cells of the rat and its effect on the absorption of glucose. *J. Nutr.,* 117,2154.

Evans. W. J. and Martin, C. J. (1988). Interactions of Mg(II), Co(Il), Ni(II) and Zn(II) with phytic acid. VIII. A calori-metric study. *J. Inorg. Biochem.,* 32, 259-67.

Fagoo. M. Braquet, P., Robin, J. P., Esanu, A. and God-fraind, T. (1986). Evidence that mammalian lignans show endogenous digitalis-like activities. *Biochem. Biophys. Res. Commun.,* 134. 1064-70.

Ferguson, E. L., Gibson, R. S., Thompson, L. U., Ounpuu, S. and Berry, M. (1988). Phytase, zinc and colon contents of 30 East African foods and their calculated Phytate: Zn, Ca:

Phytate and [Ca][Phytate]/[Zn] molar ratios. *J. Food Compo Anal.,* 1, 316–25.

Fish, B. and Thompson, L. U. (1991). Lectin-tannin interac-tions and their influence on pancreatic amylase activity and starch digestion. 1. *Agric. Food Chem.,* 39, 727-31.

Folman, Y. and Pope, G. S. (1966). The interaction in the im-mature mouse of potent oestrogens with coumestrol, genis-tein and other uterovaginotrophic compounds of low potency. 1. *Endocrinol.,* 34, 215-22.

Fordyce, E. J., Forbes, R. M., Robbins, K. R. and Erdman, Jr. 1. W. (1987). Phytate x calcium/zinc molar ratios: Are they predictive of zinc bioavailability? 1. *Food Sci.,* 52, 440-4.

Fukuchi, K., Sakagami, H., Okuda, T., Hatano, T., Tanima, S., Kitajima, K., Inoue, Y., Inoue, S., Ichikawa, S., Nonoyama, M. and Konno, K. (1989). Inhibition of herpes simplex virus infection by tannins and related compounds. *Antiviral Res.,* 11, 285-98.

Fushiki, T. and Iwai, K. (1989). Two hypotheses on the feed-back regulation of pancreatic enzyme secretion. *F ASEB* 1., 3,121-6.

Gallaher, D. and Schneeman, B. O. (1986). Nutritional and metabolic response to plant inhibitors of digestive enzymes. [n *Nutritional and Toxicological Significance of Enzyme Inhibitors in Foods,* ed. M. Friedman. Plenum Press, New York, USA, pp. 167-84.

Garrow, J. S., SOlt, P. F., Heels, S., Nair, K. S. and Halliday, D, (1983). Starch blockers are ineffective in man. *Lancet, 1,* W-61.

Gestener, B., Assa, Y., Henis, Y., Tencer, Y., Royman, M. Birk, Y. and Bondi, A. (1972). Interaction of lucerne saponins with sterols. *Biochem. Biophys. Acta.,* 270, 181-7.

Gibney, M. J. and Taylor, T. G. (1980). Effects of soy protein and saponin on serum and liver lipids in rats. *Atherosclerosis,* 36, 595-6.

Graf, E. (1986). Chemistry and applications of phytic acid: An overview. In *Phytic Acid Chemistry and Applications,* ed. E. Graf. Pilat us Press, Minneapolis, MN, USA, pp. 1-21.

Graf, E. and Eaton, J. W. (1990). Antioxidant functions of phytic acid. *Free Rad. Biol. Med.,* 8, 61-9.

Granum, P. E. and Eskeland, B. (1981). Nutritional signifi-cance of alpha-amylase inhibitors from wheat. *Nut. Rep. Int.,* 23, 155-62.

Griffiths, D. W. (1986). The inhibition of digestive enzymes by polyphenolic compounds. In *Nutritional and Toxicological Significance of Enzyme Inhibitors in Foods,* ed. M. Friedman. Plenum Press, New York, USA, pp. 509-16.

Hallberg, L. (1987). Wheat fiber, phytate and iron absorption. *Scand. J. Gastroenterol.,* 22 (supp\. 129). 73-9.

Harland, B. and Oberleas, D. (1987). Phytate in foods. *World Rev. NUlr. Diet,* 52, 235-58.

Hartwell, J. L. (1976). Types of anticancer agents isolated from plants. *Cancer Treat. Rep.,* 60, 1031-67.

Hathcock, J. N. (1991). Residue trypsin inhibitor: Data needs for risk assessment. In *Nutritional and Toxicological Conse-quences of Food Processing,* ed. M. Friedman. Plenum Press, New York, USA, pp. 273-9.

Hayatsu, H., Arimoto, S. and Negishi, T. (1988). Dietary in-hibitors of mutagenesis and carcinogenesis. *MUl. Res.,* 202, 429-46.

Heath, A. W., Nyan, 0., Richards, C. E. and Playfair, J. H. (1991). Effects of interferon gamma and saponin on lym-phocyte traffic are inversely related to adjuvanticity and enhancement of MHC II expression. *Int. Immunol.,* 3, 285-92.

Hirano, T., Oka, K. and Akiba, M. (l989a). Antiproliferative effects of synthetic and naturally occurring flavonoids on tumor cells of the human breast carcinoma cell line ZR-75-1. *Res. Comm. Chern. Pathol. Pharmacol.,* 64, 69-78.

Hirano, T., Oka, K., Kawashima, E. and Akiba, M. (1989b). Effects of synthetic and naturally occurring flavonoids on mitogen-induced proliferation of human peripheral-blood lymphocytes. *Life Sci.*, 45, 1407–41.

Ho, R., Aranda, C. and Venico, J. (1981). Species differences in response to two naturally occurring alpha amylase inhibitors. *J. Pharm. Pharmacol.*, 33, 351-8.

Hollenbeck, C. B., Coulston, A. M., Quan, R., Becker, T. R., Vreman, H. J., Stevenson, D. K. and Reaven, G. M. (1983). Effects of a commercial starch blocker preparation on car-bohydrate digestion and absorption: *in vivo* and *in vitro* studies. *Am. J. Clin. Nutr.*, 38, 498-503.

Horose, M., Ozaki, K., Takaba, K., Fukushima, S., Shirai, T. and Ito, N. (1991). Modifying effects of the naturally occurring antioxidants, gamma oryzanol, phytic acid. tannic acid and n-triacontan-16, 18-dione in a rat wide spectrum organ carcinogenesis model. *Carcinogenesis*, 12, 1917-21.

Huang, H.-P., Cheng, C.-F. and Lin, W. Q. (1982). Antitumor activity of total saponins from Dolichos-Falactus-Klein. *Acta Pharmacal. Sinica*, 3, 386.

Ishiguro, M., Harada, H., Ichiki, O., Sekine, I., Nishmni, I. and Kikutani, M. (1984). Effects of ricin, a protein toxin, on glucose absorption by rat small intestine. *Chern. Pharmacol. Bull.*, 32,3141-7.

Jaffe, W. G. and Cornejo, G. (1961). La accion de una proteina toxica, aislada de caraotas negras (*P. vulgaris*), sobre la absorcion intestinal en ratas. *Acta Ciet. Venez.*, 12, 59-61 (in French).

Jariwalla, R. J., Sabin, R., Lawson, S., Bloch, D. A., Pren-der, M. andrews, V. and Herman, Z. S. (1988). Effects of dietary phytic acid (phytate) on the incidence and growth rate of tumours promoted in Fisher rats by magnesium supplement. *Nutr. Res.*, 8, 813-27.

Jariwalla, R. J., Sabin, R., Lawson, S. and Herman, Z. S. (1990). Lowering of serum cholesterol and triglycerides and modulation of divalent cation by dietary phytase. *J. Appl. Nutr.*,42, 18-28.

Jayne-Williams, D. J. (1973). The influence of dietary jack bean and of concanavalin A on the growth of conventional and gnotobiotic Japanese quail. *Nat. New Bioi*, 243, 150-1.

Jenkins, D. J. A., Wolever, T. M. S., Jenkins, A. L., Thomp-son, L. U., Rao, A. V. and Francis, T. (1986). The glycemic index. Blood glucose response- to foods. In *Basic and Chemical Aspects of Dietary Fiber*, ed. G. V. Vahouny and D. Kritchevsky. Alan R. Liss, Inc., New York, USA, pp. 167-79.

Jindhal, S., Soni, G. L. and Singh, R. (1984). Biochemical and histopathological studies in albino rats fed on soybean lectin. *Nutr. Rep. Int.*, 29, 95-106.

Jordan, V. C., Mittal, S., Gosden, B., Koch, R. and Lieberman, M. E (1985). Structure-activity relationships of estrogen. *Environ. Health Perspecl.*, 61, 97-110.

Kapadin, D. J., Subba Rao, G. and Morton, J. F. (1983). Herbal tea consumption and esophageal cancer. In *Carcinogens and Mutagens in the Environmental* (Vol

3 - Naturally Occurring Compounds: Epidemiology and Distribution), ed. H. F. Stich. CRC Press, Boca Raton, FL, USA, pp.3-12.

Kawatra, B. L. and Bhatia, I. S. (1979). Isolation and charac-terization of a glycoprotein from field peas (*Pisum arvense* L.). *Biochem. Physiol. Pflanz.*, 174,283-8.

Kennedy, A. R. (1984). Promotion and other interactions be-tween agents in the induction of transformations *in vitro* in fibroblasts. In *Mechanisms of Tumor Promotion* (Vol II -Tumor Promotion and Carcinogenesis *in vitro*), ed. T. J. Siaga. CRC Press, Boca Raton, FL, USA, pp. 13-55.

Kennedy, A. R. and Billings, P. C. (1987). Anticarcinogenic ac-tions of protease inhibitors. In *Anticarcinogenesis and Radiation Protection*, ed. P. A. Corutti, O. F. Nygaard and M. G. Simic. Plenum Press, New York, USA, pp. 285-95.

Kensil, C. R., Patel, U., Lennick, M. and Marciani, D. (1991). Separation and characterization of saponins with adjuvant activity from *Quillaja saponaria* Molina Cortex. *J. Im-munol.*, 146, 431-7.

Kilpatrick, D. c., Green, C. and Yap, P. L. (1983). Lectin con-tent of slimming pills. *Br. M ed. J.*, 286, 305-10.

Kim, S., Brophy, E. J. and Nicholson, J. A. (1976). Rat intesti-nal brush border membrane peptidases II. Enzymatic properties, immunochemistry and interaction with lectins of two different forms of the enzyme. *J. Bioi. Chem.*, 251, 3206-12.

Kitts, W. D., Newsome, F. and Runeckles, V. (1984). The estrogenic and antiestrogenic effects of coumestrol and zeralonone on the immature rat uterus. *Can. J. Anim. Sci.*, 63,823-30.

Klevay, L. M. (1977). Hypocholesterolemia due to sodium phytate. *Nutr. Rep. Int.*, 15, 587-93.

Knuckles, B. E., Kuzmicky, D. and Betschart, A. (1985). Effect of phytate and partially hydrolyzed phytate on *in vitro* protein digestibility. *J. Food Sci.*, 50, 1080–2.

Knuckles, B. E., Kuzmicky, D. D., Gumbman, M. R. and Betschart, A. A. (1989). Effect of myoinositol phosphate esters and *in vitro* and *in vivo* digestibility of proteins. *J. Food Sci.*, 54, 1348-50.

Kratzer, F. H. and Vohra, P. (1986). *Chelates in Nutrifion*. CRC Press, Boca Raton, FL, USA. Lang, J. A., Chang-Hum, L. E., Reyes, P. S. and Briggs, G. M. (1974). Interference of starch metabolism by alpha-amylase inhibitors. *FASEB J.*, 33,718.

Layer, P., Carlson, G. L. and DiMagno, E. P. (1985). Partially purified white bean amylase inhibitor reduces starch digestion *in vitro* and inactivates intraduodenal amylase in humans. *Gastroenterology*, 88, 1895-902.

Layer, P., Zinsmeister, A. R. and DiMagno, E. P. (1986a). Effects of decreasing intraluminal amylase activity on starch digestion and postprandial gastrointestinal function in humans. *Gastroenterology*, 91, 41-8.

Layer, P., Rizza, R. A., Zinsmeister, A. R., Carlson, G. L. and DiMagno, E. P. (1986b). Effect of purified amylase inhibitor on carbohydrate tolerance in normal subjects and patients with diabetes melitus. *Mayo C/in. Proc.*, 61,442-7.

Lesca, P. (1983). Protective effects of ellagic acid and other plant phenols on benzo[a] pyrene-induced neoplasm in mice. *Carcinogenesis*, 4, 1651-3.

Liener, I.E. (1986). Trypsin inhibitors: Concern for human nutrition or not? *J. Nutr.*, 116,920–3.

Liener, I.E. (1989). The nutritional significance of lectins. In *Food Proteins*, ed. J. E. Kinsella and W. G. Soucie. AOCS, Champaign, IL, USA, pp. 329-53.

Liener, I.E. and Kakade, M. L. (1980). Protease inhibitors. In *Toxic Constituents of Plant Foodstuffs*, ed. I.E. Liener. Academic Press, New York, USA, pp. 7-71.

Liener, I.E., Donatucci, D. A. and Tarcza, J. C (1984). Starch blocker: a potential source of trypsin inhibitors and lectin. *Am. J. Clin. Nutr.*, 39, 196–200.

Liener, I.E., Goodale, R. L., Desmukh, A., Satterberg, T. L., Ward, G., DiPietro, eM., Bankey, P. E. and Bomer, J. W. (1988). Effect of a trypsin inhibitor from soybeans (Bow-man-Birk) on the secretory activity of the human pancreas. *Gastroenterology*, 94, 419-27.

Lindsay, D. R. and Kelly, R. W. (1970). The metabolism of phytoestrogens in sheep. *Aust. Vet. J.*, 46, 219-22.

Macri, A., Parlamenti. R., Siolano, V. and Valfre, F. (1977). Adaption of the domestic cereal *Gallus domesticus* to continuous feeding of albumin amylase inhibitor from wheat as gastro-resistant microgranules. *Poultry Sci.*, 56, 434–41.

Maga, J. A. (1982). Phytate: its chemistry, occurrence, food interactions, nutritional significance and method of analysis. *J. Agric. Food Chern.*, 30, 1-9.

Maharaj, 1., Froh, K. J. and Campbell, J. B. (1986). Immune response of mice to inactivated rabies vaccine administered orally: potentiation by *Quillaja saponin*. *Can. J. Microbiol.*, 32,414-20.

Malinow, M. R., Mclaughlin, P., Papworth. L., Stafford, C, Livingston, A. L. and Cheeke, P. R. (1977). Effect of alfalfa saponins on intestinal cholesterol absorption in rats. *Am. J. Clin. Nutr.*, 30, 2061-7.

Malinow, M. R., Mclaughlin, P. and Stafford, C (1978). Pre-vention of hypercholesterolemia in monkeys (*Macaca fascicularis*) by digitonin. *Am. J. Clin. Nutr.*, 31, 814-18.

Malinow, M. R., McLaughlin, P. and Stafford, C (1980a). Al-falfa seeds: effects on cholesterol metabolism. *Experientia*, 36, 562–4.

Malinow, M. R., McLaughlin, P., Stafford, C, Livingston, A. L. and Kohler, G. O. (I 980b). Alfalfa saponins and alfalfa seeds. Dietary effects in cholesterol-fed rabbits. *A therosclerosis*, 37, 433-8.

Malinow, M. R., McLaughlin, P., Stafford, C., Lin, D. S., Livingston, A. L., Kohler, G. O. and McNulty, W. P. (1981). Cholesterol and bile acid balance in *Maca fascicularis*: effects of alfalfa saponins. *J. Clin. Invest.*, 67, 156–62.

Maranesi, M., Carenini, G. and Gentili, P. (1984). Nutritive studies on anti-amylase: 1. Influence of the growth rate, blood picture and biochemistry and histological parameters in rats. *Acta Vitaminol. Enzymol., 6,* 259-70.

Markovits, J., Linassier, C, Fosse, P., Couprie, J., Pierre, J., Jacquemin-Sablon, A., Saucier, J.-M., Le Pecq, J-B. and Larsen, A. K. (1989). Inhibitory effects of the tyrosine kinase inhibitor genistein on mammalian DNA to poisomerase II. *Cancer Res., 49,* 5111-7.

Martin, P. M., Horwitzkiz, K. B., Ryan, D. S. and McGuire, W. L. (1978). Phytoestrogens interaction with estrogen re-ceptors in human breast cancer cells. *Endocrinology, 103,* 1860-7.

Mathur, K. S., Khan, M. A. and Sharma, R. D. (1968). Hypocholesterolemic effect of Bengal gram: a long-term study in man. *Br. Med. J., 1,* 30-1.

McDonald, B., Lieden, S. and Hambraeus, K. (1978). Evalu-ation of the protein quality of rapeseed meals, flours and isolates. *Nutr. Rep. Int., 17,* 49-56.

McGuiness, E. E., Morgan, G. H., Levison, D. A., Frape, D. L., Hopwood, D. and Wormsley, K. G. (1980). The effects of long term feeding of soya flour on the ιɑl pancrcao. *Scand J. Gastroenterol., 15,* 497-502.

Messadi, D. V., Billings, P., Shklar, G. and Kennedy, A. R. (1986). Inhibition ot oral carcinogenesis by a protease in-hibitor. *J. Nat. Cancer Inst., 76,* 447-52.

Messina. M. and Barnes. S. (1991). The role of soy products in reducing risk of cancer. J. *Nat. Cancer Inst., 83.* 541-6. vM'essina. M. and Messina. V. (1991). Increasing use of soybean and their potential role in cancer prevention. J. *Am. Diet. Assoc.* 91. 836–40.

Miller. W. R. (1990). Endocrine treatment for breast cancer: Biological rationale and current progress. J. *Steroid Biochem. Molec. Biol.* 37. 467-80.

Molgaard. 1., von Schenck, H. and Olsson, A. G. (1987). Al-falfa seeds lower low density lipoprotein cholesterol and apoprotein B concentrations in patients with type II hyper-liporoteinemia. *Atherosclerosis, 65,* 173-9.

\Morgan, M. and Fenwick, G. (1990). Natural foodborne toxicants. *Lancet, 336,* 1492-5.

Morgan, B. Heald, M., Brooks, S. G., Tee, J. L and Green. J. (1972). The interactions between dietary saponin. cholesterol and related sterols in the chick. *Poultry Sci., 51,* 677-82.

Mo: gan. R. G. H., Levison, D. A., Hopwood, D., Saunders, J. H. B. and Worms ley, K. G. (1977). Potentiation of the action of azaserine on the rat pancreas by raw soya bean flour. *Cancer Lell., 3,* 87-90.

Morikawa, M., Abe, M., Yamauchi, Y., Inoue, M. and Tsuboi, M. (1990). Priming effect of 2,3-dibenzylbutane-I,4-dol (mammalian lignan) on superoxide production in human neutrophils. *Biochem. Biophys. Res. Comm., 168,* 194-9.

Morris, E. R. (1986). Phytate and dietary mineral bioavail-ability. In *Phytic Acid: Chemistry and Application,* ed. E. Graf. Pilatus Press, Minneapolis. MN, USA, pp. 57-76.

Morris, E. R. and Ellis, R. (1989). Usefulness of the dietary phytic acidlzinc molar ratio as an index of zinc bioavail-ability to rats and humans. *Biol. Trace Elem. Res., 19,* 107-17

Morton, 1. F. (1970). Tentative correlation of plant usage and esophageal cancer zones. *Econ. Bot., 24,* 217-26.

Morton, 1. F. (1972). Further association of plant tannin and human cancer. *Q.J. Crude Drug Res., 12,* 1829–41.

Mowat, A. M., Donachie, A. M., Reid, G. and 1arrett, O. (1991). Immune stimulating complexes containing Quil A and protein antigen prime class I MHC-restricted T lym-phocytes *in vitro* are immunogenic by the oral route. *Immunology, 72,* 317-22.

Mukhtar. H., Das, M. and Bickers, D. R. (1986). Inhibition of 3-methyl cholanthrene-induced skin tumorigenicity in BALB/c mice by chronic oral feeding of trace amounts of ellagic acid in drinking water. *Cancer Lell., 46,* 2262-5.

Nachbar, M. S. and Oppenheim, J. D. (1980). Lectins in the United States diet. A survey of lectins in commonly con-sumed foods and a review of the literature. *Am. J. Clin. Nutr., 33,* 2338–45.

Nakadate, T. (1989). The mechanism of skin tumor promo-tion caused by phorbol esters: possible involvement of arachidonic acid/cascade/lipoxygenase, protein kinase C and calcium/calmodulin systems. *Japanese Phamacol., 49,* 1-9.

akadate. T., Yamamoto, S., Iseki, H., Sonoda, S., Take-mura, A., Ura, A., Hosoda, Y. and Kato, R. (1982a). Inhibi-tion of TPA induced tumor promotion by NDGA, a lipoxygenase inhibitor and p-bromophenacyl bromide, a phospholipase A inhibitor. *Gann, 73,* 841-3.

Nakadate, T., Yamamoto, S., Ishii, M. and Kato, R. (1982b). Inhibition of 12-0-tetradecanoylphorbol-13-acetate-in-duced epidermal ornithine decarboxylase activity by phos-pholipase A2 inhibitors and lipoxidase inhibitor. *Cancer Res., 42,* 2841-5.

Namiki, M. (1990). Antioxidants/antimutagens in food. *Crit. Rev. Food Sci. Nutr., 29,* 273-99.

Newman, H. A. I., Kummerow, F. A. and Scott, H. M. (1958). Dietary saponin, a factor which may influence liver and serum cholesterol levels. *Poultry Sci.* 38. 42-6.

Newmark, H. L. (1984). A hypothesis for dietary components as blocking agents of chemical carcinogenesis: plant phenolics and pyrrole pigments. *Nutr. Cancer.* 6. 58-70.

Newmark. P. (1988). Bark extract amplifies vaccines. *Biotechnology, 6,* 23.

Newmark, H. L. and Lupton. J.R. (1990). Determinants and consequences of colonic pH: Implications for colon cancer. *Nutr. Cancer, 14,* 161-73.

Nielsen, B. K., Thompson, L. U. and Bird. R. P. (1987). Effect of phytic acid on colonic epithelial cell proliferation. *Cancer Lell.*, 37, 317-25.

Noah, N. D., Bender. A. E. Readi. G. B. and Gilbert. R. (1980). Food poisoning from raw red kidney beans. *Br. Med J*, 281, 236-7.

Oakenful, D. (1981). Saponins in food - A review. *Food Chern.*, 6, 19-40.

Oakenful, D. (1986). Aggregation of saponins and bile acids in aqueous solution. *Aust. J Chern.*, 39. 1671-3.

Oakenful, D. G. and Sidhu, G. S. (1983). A physico-chemical explanation for the effects of dietary saponins on choles-terol and bile salt metabolism. *Nutr. Rep. Int.*, 27. 1253-9.

Oakenful, D. G. and Sidhu, G. S. (1984). Prevention of dietary hypercholesterolemia by chickpea saponins and navy beans. *Proc. Nutr. Soc. Aust.*, 9, 104.

Oakenful, D. and Sidhu, G. S. (1989). Saponins. In *Toxicants of Plant Origin* (Vol. 2), ed. P. Cheeke. CRC Press, Boca Raton, FL, USA.

Oakenful, D. and Sidhu, G. S. (1990). Could saponins be a useful treatment for hypercholesterolemia? *Eur. J Clin. Nutr.*, 44, 79-88.

Oakenful, D. G., Fenwick, D. E., Topping, D. L. Illman, R. J. and Storer, G. B. (1979). Effects of saponins on bile acids and plasma lipids in the rat. *Br. J. Nutr.*, 42, 209-16.

Oakenful, D. G., Topping, D. L., Illman, R. J. and Fenwick, P. E. (1984). Prevention of dietary hypercholesterolemia in the rat by soya bean and quillaja saponin. *Nutr. Rep. Int.*, 29, 139-46.

O'Gara, R. W., Lee, C and Morton, J. F. (1971). Carcino-genicity of extracts of selected plants from Curacao after oral and subcutaneous administration to rodents. *J Nat. Cancer Inst.*, 46, 1131-7.

Ogawara, H., Akiyama, T, Watanabe, S., Ito, N., Kobori, M. and Sedoa, Y. (1989). Inhibition of tyrosine protein kinase activity by synthetic isoflavones and flavones. *J. Antibiotics*, XLI, 340-3.

Oosterlaken, T A., Brandenburg, A., Schelen, P. and Fransen, R. (1991). Efficient induction of Semliki Forest virus and mumps virus neutralizing anti-idiotypic antibodies using Quil A as adjuvant. *J lrnrnunol. Meth.*, 136, 169-75.

Pathirana, C, Gibney, M. J. and Taylor, T G. (1981). The effect of dietary protein source and saponins on serum lipids and excretion of bile acids and neutral sterols in rab-bits. *Br. J. Nutr.*, 46, 421-30.

Plante, G. E., Prevost, C., Chainey, A., Braquet, P. and Siros, P. (1987). Diuretic and natriuretic properties of prestegone B, a mammalian lignan. *Am. J. Physiol.*, 253, R375-8.

Price, K. R. and Fenwick, G. R. (1985). Naturally occurring oestrogens in foods - A review. *Food AddU. Contarn.*, 2, 73.

Price, K. R., Johnson, I. T and Fenwick, G. R. (1987). The chemistry and biological significance of saponins in foods and feedingstuffs. *CRC Crit. Rev. Food Sci. Nutr.*, 26, 27-135.

Price, M. L. and Butler, L. G. (1980). *Tannins and Nutrition.* Purdue University Agricultural Experimental Station Bul-letin No. 272, Purdue, W. Lafayette, IN, USA.

Puis, W. and Keup U. (1973). Influence of an alpha-amylase inhibitor (Bay-d7791) on blood glucose, serum insulin and EFA in starch loading tests in rats, dogs and man. *Diabetology*, 9, 97-101.

Pusztai, A. (1986). The biological effects of lectins in the diet of animals and man. In *Lectins. Biology. Biochemistry. Clinical Biochemistry* (Vol. 5), ed. T. C. Bog-Hansen and E. Van Driesche. Walter de Gruyter and Co., ew York, USA, pp.317-26.

Rackis, J. J. and Gumbman, M. (1981). Protease inhibitors: physiologic~1 properties and nutritional significance. In *Antinutrient and Natural Toxicants in Foods*, ed. R. L. Ory. Food and Nutrition Press, Westport, CT, USA, pp. 203-37.

Rao, A. V. and Kendall, C. W. (1986). Dietary saponins and serum lipids. *Food Chem. Toxicol.*, 24, 441.

Rea, R., Thompson, L. U. and Jenkins, D. J. A. (1985). Lectin in foods and their relation to starch digestibility. *Nutr. Res,5*, 919-29.

Reddy, N. R., Sathe, S. K. and Salunkhe, D. K. (1982). Phytates in legumes and cereals. *Adv. Food Res.*, 28, 1-92.

Reddy, N. R., Sathe, S. and Pierson, M. (1988). Removal of phytate from Great Northern bean (*P. vulgaris*) and its combined density fraction. *J. Food Sci.*, 53, 107-10.

Reinhold, J., Nasr, K., Lahimgarzadeh, A. and Hedayati, H. (1973). Effects of purified phytate and phytate-rich bread upon metabolism of zinc, calcium, phosphorus and nitrogen. *Lancet*, i, 283-8.

Reshef, G., Gestetner, B., Birk, Y. and Bondi, A. (1976). Effect of alfalfa saponins on the growth and some aspects of lipid metabolism of mice and quails. *J. Sci. Food Agric.*, 27, 63-72.

Roebuck. B. D. (1987). Trypsin inhibitors: Potential concern for humans. *J. Nutr.*, 117, 398-400.

Rosenberg, I. H. (1982). Starch blockers-still no calorie-free lunch. *New England J. Med.*, 307, 1444-5.

Rouanet, J. M., Lafont, J., Zambionino-Infante, J. L. and Besan ion, P. (1988). Selective effects of PHA on rat brush border hydro lases along the crypt-villus axis. *Experientia*, 44,340-1.

Sahu, S. C. and Washington, M. C. (1991). Effects of antioxidants on quercetin-induced nuclear DNA damage and lipid peroxidation. *Cancer Lett.*, 60, 259-64.

Salunkhe, D. K., Chavan, J. K. and Kadam, S. S. (1990). *Dietary Tannins: Consequences and Remedies.* CRC Press, Boca Raton, FL, USA.

Sanghvi, A., Diven, W. F., Seltman, H., Warty, V., Rizk, M., Kritchevsky, D. and Setchell, K. (1984). Inhibition of rat liver cholesterol 7-alpha hydroxylase and acyl-CoA: cholesterol acyl transferase activities by enterodiol and enterolactone. In *Drugs Affecting Lipid Metabolism,* ed. D. Kritchevsky, R. Paoletti and W. L. Holmes. Plenum Press, New York, USA, pp. 311-22.

Santidrian, S. (1981). Intestinal absorption of D-glucose, D-galactose and L-Ieucine in male growing rats fed raw field beans (*Vicia faba* L.) *J. Anim. Sci.*, 53, 414-19.

Sati, O. P., Pnant, G. and Nohara, T. (1985). Cytotoxic saponin from asparagus and agave. *Pharmazie*, 40, 586.

Sautier, c., Doucet, c., Flament, C. and Lemonnier, D. (1979). Effect of soy protein and saponins on serum, tissue and feces steroids in the rat. *Atherosclerosis*, 34, 233-41.

Savaiano, D. A., Powers, J. R., Costello, M. J., Whitaker, J. R. and Clifford, A. J. (1977). The effect of an alpha amylase inhibitor on the growth rate of weanling rats. *Nutr. Rep. Int.*, 15,443-9.

Scbapira, D. V., Kumar, N. B. and Lyman, G. H. (1990). Serum cholesterol reduction with tamoxifen. *Breast Cancer Res. Treat.*, 17, 3-7.

Scientific Review Committee (1990). *Nutrition Recommendations.* Health and Welfare Canada, Supply and Services Canada, Ottawa, Canada.

Scott, M. T., Gross-Sampson, M. and Bomford, R. (1985). Adjuvant activity of saponin: antigen localization studies. *Int. Arch. Allergy. Appl. Irnrnunol.*, 77, 409-12.

Serraino. M. and Thompson, L. U. (1991). The effect of flaxseed supplementation on early risk markers for mammary carcinogenesis. *Cancer Leu.*, 60, 135-42.

Serraino, M. and Thompson, L. U. (1992a). Flaxseed supplementation and early markers of colon carcinogenesis. *Can-er Leu.*, 63, 159-65.

Serraino, M. and Thompson, L. U. (1992b). The effect of flaxseed supplementation on the initiation and promotional stages of mammary tumorigenesis. *Nutr. Cancer*, 17, 153-9.

Serraino, M., Thompson, L. U., Savoie, L. and Parent, G. (1985). Effect of phytic acid on in-vitro digestibility of rapeseed protein and amino acids. 1. *Food Sci.*, 50, 1689-92.

Serraino, M., Thompson, L. U. and Cunnane, S. C. (1992). Effect of low level flaxseed supplementation on the fatty acid composition of mammary glands and tumors in rats. *Nutr. Res.*, 12, 767-72.

Setchell, K. D. R. and Adlercreutz, H. (1988). Mammalian lig-nans and phytoestrogens. Recent studies on the formation, metabolism and biological role in health and disease. In *Role of the Gut Flora in Toxicity and Cancer,* ed. I. R. Rowland, Academic Press, London, UK, pp. 315-45.

Setchell, K. D. R., Gosselin, S. J., Welsh, M. B., Johnston, J. O., Balistreri, W. F., Kramer, L. W., Dresser, B. L. and Tarr, M. J. (1987). Dietary estrogens - a probable cause of infertility and liver disease in captive cheetah. *Gastro-enterology*, 93, 225-33.

Shamsuddin, A. M. and Ullah, A. (1989). Inositol hexaphosphate inhibits large intestinal cancer in F-344 rats 5 months following induction by azoxyrnethane. *Carcinogenesis*, 10, 625-6.

Shamsuddin, A. M., Elsayed, A. M. and Ullah, A. (1988). Sup-pression of large intestinal cancer in F-344 rats by inositol hexaphosphate. *Carcinogenesis*, 9, 577-80.

Shamsuddin, A. M., Ullah, A. and Chakvarthy, A. (1989). In-ositol and inositol hexaphosphate suppress cell prolifera-tion and tumor formation in CD-I mice. *Carcinogenesis*, 10, 1461-3.

Sharma, R. D. (1980). Effect of hydroxy acids on hypocholesterolemia in rats. *Atherosclerosis*, 37, 463-8.

Sharma, R. D. (1984). Hypocholesterolemic effect of hydroxy acid components of Bengal gram. *Nutr. Rep. Int.*, 29, 1315-22.

Sharma, R. D. (1986). An evaluation of hypocholesterolemic factor of fenugreek seeds (*T. foenurn graecurn*) in rats. *Nutr. Rep. Int.*, 33, 669-77.

Shibata, S. (1977). Saponins with biological and pharmacological activity. In *New Natural Products Plant Drugs with Pharmacological. Biological or Therapeutical Activity*, ed. H. Wagner and F. Wolf. Springer-Verlag, Berlin, Germany, pp. 177-96.

Sidhu, G. S. and Oakenful, D. G. (1986). A mechanism for the hypocholesterolemic activity of saponins. *Br. J. Nutr.*, 55, 643-9.

Singh, M. and Krikorian, A. (1982). Inhibition of trypsin activity *in vitro* by phytate. *J. Agric. Food Chern.*, 30, 799-800.

Singleton, V. L. (1981). Naturally occurring food toxicants. Phenolic substances of plant origin common in foods. *Adv. Food Res.*, 27, 149-242.

Sirtori, C. R., Agradi, E., Conti, F., Mantero, I. and Gatti, E. (1977). Soybean-protein diet in the treatment of type II hyperlipoproteinaemia. *Lancet*, i, 275-7.

Smart, R. C., Huang, M.-C., Chang, R. L., Sayer, J. M., Je-rina, D. M., Wood, A. W. and Conney, A. H. (1986). Effect of ellagic and 3-0-decylellagic acid on the formation of benzo(a)pyrene-derived DNA adducts *in vivo* and on the tumorigenicity of 3-methylchloranthracene in mice. *Carcinogenesis*,7, 1669-76.

Sognen, E. (1965a). Apparent depression in the absorption of strychnine, alcohol, sulphanilamide after oral administra-tion of sodium fluoride sodium oxalate, tetracemin and sodium phytate. *Acta Pharmacol. Toxicol.*, 22, 8-18.

Sognen, E. (1965b). Effects of calcium binding substances on gastric emptying as well as intestinal transit and absorption in intact rats. *Acta Pharmacal. Toxical.*, 22, 31-48.

Spivey-Fox, M. R. and Tao, S.-H. (1989). Antinutritive effect of phytate and other phosphorylated derivatives. *Nutr. Toxicol.*, 3, 59-96.

St. Clair, ·W. H., Billings, P. C. and Kennedy, A. R. (1990). The effects of the Bowman-Birk protease inhibitor on c-myc expression and cell proliferation in the unirradiated and irradiated mouse colon. *Cancer Leu.*, 52, 145-52.

Stich, H. F. and Rosin, M. P. (1984). Naturally occurring phe-nolics as antimutagenic and anticarcinogenic agents. In *Nutritional and Toxicological Aspects of Food Safety*, ed. M. Friedman. Plenum Press, New York, USA. pp. 1-29.

Tang, B. Y. and Adams, N. R. (1980). Effect of equol on oestrogen receptors and on synthesis of DNA and protein in the immature rat uterus. *J. Endocrinol.*, 85, 291-7.

Teraoka, H., Ohmura, Y. and Tsukada, K. (1989). The nuclear matrix from rat liver is capable of phosphorylating exogenous tyrosine-containing substrates. *Biochem. Int.*, 18, 1203-10.

Thompson, L. U. (1986). Phytic acid. A factor influencing starch digestibility and blood glucose response. In *Phytic Acid: Chemistry and Applications*, ed. E. Graf. Pilatus Press, Minneapolis, MN, USA, pp. 173-94.

Thompson, L. U. (1988). Antinutrients and blood glucose. *Food Technol.*, 42, 123-32.

Thompson, L. U. (1989). Nutritional and physiological effects of phytic acid. In *Food Proteins*, ed. J. E. Kinsella and W. G. Soucie, AOCS, Champaign, IL, USA, pp. 410-31.

Thompson, L. U. and Gabon, J. E. (1987). Effect of lectins on salivary and pancreatic amylase activities and the rate of starch digestion. *J. Food Sci.*, 52, 1050-3.

Thompson, L. U. and Serraino, M. (1986). Effect of phytic acid reduction on rapeseed protein digestibility and amino acid absorption. *J. Agric. Food Chern.*, 34, 468-9.

Thompson, L. U. and Yoon, J. H. (1984). Starch digestibility as affected by polyphenols and phytic acid. *J. Food Sci.*, 49, 1228-9.

Thompson, L. U. and Zhang, L. (1991). Phytic acid and miner-als: effect on early markers of risk for mammary and colon carcinogenesis.

Thompson, L. U., Rea, R. and Jenkins, D. (1983). Effect of heat processing on hemagglutinin in red kidney beans. *J. Food Sci.*, 48,235-6.

Thompson, L. U., Yoon, J. H., Jenkins, D. J. A., Wolever, T. M. S. and Jenkins, A. L. (1984). Relationship between polyphenol intake and blood glucose response of normal and diabetic individuals. *Am. J. Clin. Nutr.*, 39, 745-51.

Thompson, L. U., Tenebaum, A. and Hui, H. (1986). Effect of lectins and the mixing of proteins on rate of protein digestibility. *J. Food Sci.*, 51, 150-3.

Thompson, L. U., Button, C. L. and Jenkins, D. J. A. (1987). Phytic acids and calcium effect on starch digestibility and glucose response to navy bean flour. *Am. 1. Clin. Nutr.*, 38, 481-S.

Thompson, L. U., Robb, P., Serraino, M. and Cheung, F. (1991). Mammalian lignan production from various foods. *Nutr. Cancer*, 16,43-52.

Thompson, L. U., Serraino, M., Cunnane, S. C., Barnes, S. and Mobbs, B. (1992). Effect of flaxseed supplementation on carcinogen- and non-carcinogen-treated rats. *Proc. US. Flax Institute Meeting*, 54, 32-6.

Tokuda, H. (1988). Inhibitory effects of 12-0-tetrade-canoylphorbol-13 acetate and teleocidin- B-induced Epstein-Barr virus by saponin and its related compounds. *Cancer Lett.*, 40, 309-17.

Topping, D. L., Storer, G. B., Calbert, G. D., Illman, R. J., Oak-enful, D. G. and Weller, R. A. (1980). Effect of dietary saponins on fecal bile acids and neutral sterols, plasma lipids and lipoprotein turn over in the pig. *Am. J. Clin. Nutr.*, 33, 78H.

Torre, M., Rodriguez, A. R. and Saura-Calixto, F. (1991). Effects of dietary fiber and phytic acid on mineral availability. *Crit. Rev. Food Sci. Nutr.*, 1, 1-22.

Triadou, N. and Audran, E. (1983). Interaction of the brush border hydro lases of the human small intestine with lectins. *Digestion*, 27, 1-7.

Troll, W. (1989). Protease inhibitors interfere with the necessary factors of carcinogenesis. *Environ. Health Perspect.*, 81, 59-62.

Troll, W. and Kennedy, A. R. (1989). Workshop report for the Division of Cancer Etiology, National Cancer Institute, National Institute of Health. Protein inhibitors as cancer chemipreventive agents. *Cancer Res.*, 49, 499-502.

Troll, W., Wiesner, R. and Frenkel, K. (1987). Anticarcinogenic action of protease inhibitors. In *Advances in Cancer Research* (Vol. 49). Academic Press, New York, USA, pp. 265-83.

Ullah, A. and Shamsuddin, A. M. (1990). Dose-dependent inhi-bition of large intestinal cancer by inositol hexaphosphate in F344 rats. *Carcinogenesis*, 11, 2219-22.

Wattenberg, L. W. (1983). Inhibition of neoplasia by minor dietary constituents. *Cancer Res.*, 43, 2448s-53s.

Wattenberg, L. W., Coccia, J. B. and Lam, L. K. T. (1980). Inhibitory effects of phenolic compounds on benzo(a)pyrene-induced neoplasia. *Cancer Res.*, 40, 2820–3.

West, L. G. and Greger, J. L. (1978). *In vitro* studies of saponin-vitamin complexation. *J. Food Sci.*, 43, 1340–5.

Whitaker, J. (1989). Alpha amylase inhibitors of higher plants and microorganisms. In *Food Proteins*, ed. J. E. Kinsella and W. G. Soucie. AOCS, Champaign, IL, USA, pp. 354-80.

Williams, D. E., Dashwood, R. H., Hendricks, J. D. and Bailey, G. S. (1989). Anticarcinogens and tumor promoters in foods. In *Food Toxicology. A Perspective on Relative Risks*, ed. S. L. Taylor and R. A. Scanlan. Marcel Dek.ker, New York, USA, pp. 101-50.

Witschi, H. and Kennedy, A. R. (1989). Modulation of lung tumor development in mice with the soybean-derived Bow-man-Birk protease inhibitor. *Carcinogenesis*, 10,2275-7.

Wolever, T. M. S. (1990). The glycemic index. *World Rev.Nutr. Diet.*, 62, 120–5.

Yamamoto, M., Kumagai, A. and Yamamura, Y. (1975). Structure and action of saikosaponins isolated from *Bupleurum falcaturn* L. *Arzneimittel-Forsch*, 25, 1240–3.

Yindi, z. (1984). Effects of astragulus, saponin-Ion cAMP and cGMP levels in plasma and DNA synthesis in regenerating liver. *Yao Hsueh Pao,* 19,619.

Yoon, J. H., Thompson, L. U. and Jenkins, D. J. A. (1983). The effect of phytic acid on in-vitro rate of starch digestibility and blood glucose response. *Am. J. Clin. Nutr.*, 38, 835–42.

Yoshida, T., Shinoda, S., Matsumoto, T. and Watarai, S. (1982). Feed digestibility and mineral balance of the diet of young mice kept in mouse cages inside or outside an isola-tor using varied concentrations of sodium phytate. *J. Nutr. Sci. Vitaminol.*, 28, 401-10.

Young, J. M., Wagner, B. M. and Spires, D. A. (1983). Tachy-phylaxis in 12-0-tetradecanoylphorbol-acetate- and arachidonic acid-induced ear edema. *J. Invest. Dermatol.*, 80, 48-52.

Part–II

Plant Pigments

Chapter 21

General Biological Pigment

Photosynthetic pigment or antenna pigment is a pigment that is present in chloroplasts or photosynthetic bacteria and captures the light energy necessary for photosynthesis. Green plants have five closely-related photosynthetic pigments (in order of increasing polarity):

☆ Carotene - an orange pigment

☆ Xanthophyll - a yellow pigment

☆ Chlorophyll a - a blue-green pigment

☆ Chlorophyll b - a yellow-green pigment

☆ Phaeophytin a[1] - a gray-brown pigment

☆ Phaeophytin b[1] - a yellow-brown pigment

Chlorophyll a is the most common of the six, present in every plant that performs photosynthesis. The reason that there are so many pigments is that each absorbs light more efficiently in a different part of the spectrum. Chlorophyll a absorbs well at a wavelength of about 400-450 nm and at 650-700 nm; chlorophyll b at 450-500 nm and at 600-650 nm. Xanthophyll absorbs well at 400-530 nm. However, none of the pigments absorbs well in the green-yellow region, which is responsible for the abundant green we see in nature. Well photosynthetic pigments have to do with the different component of light (*i.e.* the colours - ROYGBIV) and their specific wavelenght. Photosynthetic pigments based on their colours can capture the energy from that wavelenght. For example fucoxanthin can capture the brown colour of the light just like phycocyanin can capture the blue colour of the light. These photosynthetic pigments are usually the accessory pigments of Chlorophyll A which is the main pigments of most of the photosynthetic organisms in the earth.

The process of photosynthesis is crucial to the existence of life. In this process plants harness incoming light energy from the sun and convert it into a form of chemical energy that all organisms need in order to fuel body functions. In order for this process to occur, however, it is necessary that plants have a way to actively harness incoming light of all different wavelengths. This is done by a series of photosynthetic pigments that allow the plant to utilize as much of the light reaching the plant and increase the amount of chemical energy output. This website is designed to introduce you to the basics of the primary pigment (chlorophyll), group of secondary pigments (carotenoids) and a special type of water-soluble pigments (phycobilins) and how they are crucial to the process of photosynthesis.

Biological pigments, also known simply as pigments or biochromes are substances produced by living organisms that have a colourresulting from selective colour absorption. Biological pigments include plant pigments and flower pigments. Many biological structures, such as skin, eyes, feathers, fur and hair contain pigments such as melanin in specialized cells called chromatophores. Pigment colour differs from structural colour in that it is the same for all viewing angles, whereas structural colour is the result of selectivereflection or iridescence, usually because of multilayer structures. For example, butterfly wings typically contain structural colour, although many butterflies have cells that contain pigment as well.

Biological Pigments

Conjugated Systems for electron bond chemistry that causes these molecules to have pigment

- ☆ Heme/porphyrin-based: chlorophyll, bilirubin, hemocyanin, hemoglobin, myoglobin
- ☆ Light-emitting: luciferin
- ☆ Carotenoids:
- ☆ Hematochromes (algal pigments, mixes of carotenoids and their derivates)
- ☆ Carotenes: alpha and beta carotene, lycopene, rhodopsin
- ☆ Xanthophylls: canthaxanthin, zeaxanthin, lutein
- ☆ Proteinaceous: phytochrome, phycobiliproteins
- ☆ Polyene enolates: a class of red pigments unique to parrots
- ☆ Other: melanin, urochrome, flavonoids

Pigments in Plants

The primary function of pigments in plants is photosynthesis, which uses the green pigment chlorophyll along with several red and yellow pigments that help to capture as much light energy as possible. Other functions of pigments in plants include attracting insects to flowers to encourage pollination. Plant pigments include a variety of different kinds of molecule, including porphyrins, carotenoids, anthocyanins and betalains. All biological pigments selectively absorb certain wavelengths of light while reflecting others. The light that is absorbed may be used by the plant to power chemical reactions, while the reflected wavelengths of light determine the colour the pigment will appear to the eye.

The principal pigments are:

☆ **Chlorophyll** is the primary pigment in plants; it is a chlorin that absorbs yellow and blue wavelengths of light while reflecting green. It is the presence and relative abundance of chlorophyll that gives plants their green colour. All land plants and green algae possess two forms of this pigment: chlorophyll a and chlorophyll b. Kelps, diatoms and other photosynthetic heterokonts contain chlorophyll cinstead of b, while red algae possess only chlorophyll a. All chlorophylls serve as the primary means plants use to intercept light in order to fuel photosynthesis.

☆ **Carotenoids** are red, orange, or yellow tetraterpenoids. They function as accessory pigments in plants, helping to fuel photosynthesis by gathering wavelengths of light not readily absorbed by chlorophyll. But they also fulfill other important functions in plants as preventing photooxidative damage or by being precursors to the plant hormone abscisic acid. Carotenoids have been shown to act asantioxidants and to promote healthy eyesight in humans. Plants, in general, contain six ubiquitous carotenoids: neoxanthin, violaxanthin, antheraxanthin, zeaxanthin, lutein and ß-carotene, together with the main two chlorophylls (Chl), Chl a and Chl b. Indeed, the most familiar carotenoids in plants are ß carotene (an orange pigment), lutein (a yellow pigment found in fruits and vegetables and the most abundant carotenoid in plants) and lycopene (the red pigment responsible for the colour of tomatoes). However, this composition can be supplemented in some species and under some circumstances by the presence of other "non-ubiquitous" carotenoids as Lutein epoxide (in many woody species; lactucaxanthin (found in lettuce) or alpha carotene (in carrots). In general, photosynthetic pigment composition is affected by external and internal factors in plants.

☆ **Anthocyanins** (literally "flower blue") are water-soluble flavonoid pigments that appear red to blue, according to pH. They occur in all tissues of higher plants, providing colour in leaves, plant stem, roots, flowers and fruits, though not always in sufficient quantities to be noticeable. Anthocyanins are most visible in the petals of flowers, where they may make up as much as 30 per cent of the dry weight of the tissue.They are also responsible for the purple colour seen on the underside of tropical shade plants such asTradescantia zebrina; in these plants, the anthocyanin catches light that has passed through the leaf and reflects it back towards regions bearing chlorophyll, in order to maximize the use of available light.

☆ **Betalains** are red or yellow pigments. Like anthocyanins they are water-soluble, but unlike anthocyanins they are synthesized fromtyrosine. This class of pigments is found only in the Caryophyllales (including cactus and amaranth) and never co-occur in plants with anthocyanins. Betalains are responsible for the deep red colour of beets and are used commercially as food-colouring agents.

A particularly noticeable manifestation of pigmentation in plants is seen with autumn leaf colour, a phenomenon that affects the normally green leaves of many deciduous trees and shrubs whereby they take on, during a few weeks in the autumn season, various shades of red, yellow, purple and brown. Chlorophylls degrade into colourless tetrapyrroles known as *nonfluorescent chlorophyll catabolites* (NCCs). As the predominant chlorophylls degrade, the hidden pigments of yellow xanthophylls and orange beta-carotene are revealed. These pigments are present throughout the year, but the red pigments, the anthocyanins, are synthesized *de novo* once roughly half of chlorophyll has been degraded. The amino acids released from degradation of light harvesting complexes are stored all winter in the tree's roots, branches, stems and trunk until next spring when they are recycled to releaf the tree.

Pigments in Animals

Pigmentation is used by many animals for protection, by means of camouflage, mimicry, or warning colouration. Some animals including fish, amphibians and cephalopods use pigmented chromatophores to provide camouflage that varies to match the background. Pigmentation is used in signalling between animals, such as in courtship and reproductive behavior. For example, some cephalopods use their chromatophores to communicate. The photopigment rhodopsin intercepts light as the first step in the perception of light. Skin pigments such as melanin may protect tissues from sunburn by ultraviolet radiation.

However, some biological structures in animals, such as heme groups that help to carry oxygen in the blood, are coloured as a result of their structure. Their colour does not have a protective or signalling function.

Diseases and Conditions

A variety of diseases and abnormal conditions that involve pigmentation are in humans and animals, either from absence of or loss of pigmentation or pigment cells, or from the excess production of pigment.

☆ Albinism is an inherited disorder characterized by total or partial loss of melanin. Humans and animals that suffer from albinism are called "albinistic" (the term "albino" is also sometimes used, but may be considered offensive when applied to people).

☆ Lamellar ichthyosis, also called "fish scale disease", is an inherited condition in which one symptom is excess production of melanin. The skin is darker than normal and is characterized by darkened, scaly, dry patches.

☆ Melasma is a condition in which dark brown patches of pigment appear on the face, influenced by hormonal changes. When it occurs during a pregnancy, this condition is called the mask of pregnancy.

☆ Ocular pigmentation is an accumulation of pigment in the eye and may be caused by latanoprost medication.

☆ Vitiligo is a condition in which there is a loss of pigment-producing cells called melanocytes in patches of skin.

Pigments in Marine Animals

Carotenoids and Carotenoproteins

Carotenoids are the most common group of pigments found in nature. Over 600 different kinds of carotenoids are found in animals and plants. In plants, carotenoids are responsible for photo-protection, light-harvesting and singlet oxygen scavenging in the process of photosynthesis. This pigment is usually found in the chloroplast of plants and other photosynthetic organism such as algae and some bacteria. On the other hand, animals are incapable of making their own carotenoids. Thus, they rely on plants for these pigments.

Carotenoids form complexes with proteins which are known as carotenoproteins. These complexes are common among marine animals. The carotenoprotein complexes are responsible for the various colours (red, purple, blue, green, etc.) to these marine invertebrates for mating rituals and camouflage. There are two main types of carotenoproteins: Type A and Type B. Type A has carotenoids (chromogen) which are stoichiometrically associated with a simple protein (glycoprotein). The second type, Type B, has carotenoids which are associated with a lipo protein and is usually less stable. While Type A is commonly found in the surface (shells and skins) of marine invertebrates, Type B is usually in eggs, ovaries and blood. The colours and characteristic absorption of these carotenoprotein complexes are based upon the chemical binding of the chromogen and the protein subunits.

For example, the blue carotenoprotein, linckiacyanin has about 100-200 carotenoid molecules per every complex. In addition, the functions of these pigment-protein complexes also change their chemical structure as well. Carotenoproteins that are within the photosynthetic structure are more common, but complicated. Pigment-protein complexes that are outside of the photosynthetic system are less common, but have a simpler structure. For example, there are only two of these blue astaxanthin-proteins in the jellyfish, Velella velella, contains only about 100 carotenoids per complex. The most common carotenoprotein is astaxanthin, which gives off a purple-blue and green pigment. Astaxanthin's colour is formed by creating complexes with proteins in a certain order. For example, the crustochrin has approximately 20 astaxanthin molecules bonded with protein. When the complexes interact by exciton-exciton interaction, it lowers the absorbance maximum, changing the different colour pigments. In lobsters, there are various types of astaxanthin-protein complexes present. The first one is crustacyanin (max 632 nm), a slate-blue pigment found in the lobster's carapace. The second one is crustochrin (max 409), a yellow pigment which is found on the outer layer of the carapace. Lastly, the lipoglycoprotein and ovoverdin forms a bright green pigment that is usually present in the outer layers of the carapace and the lobster eggs.

Tetrapyrroles

Tetrapyrroles are the next most common group of pigments. They have four pyrrole rings, each ring consisting of C_4H_4NH. The main role of the tetrapyrroles is their connection in the biological oxidation process. Tetrapyrroles has a major role in electron transport and acts as a replacement for many enzymes. In addition, they also have a role in the pigmentation of the marine organism's tissues.

Melanin

Melanin is a class of compounds that serves as a pigment with different structures responsible for dark, tan, yellowish/reddish pigments in marine animals. It is produced as the amino acid tyrosine is converted into melanin, which is found in the skin, hair and eyes. Derived from aerobic oxidation of phenols, they are polymers.

There are several different types of melanins considering that they are an aggregate of smaller component molecules, such as nitrogen containing melanins. There are two classes of pigments: black and brown insoluble eumelanins, which are derived from aerobic oxidation of tyrosine in the presence of tyrosinase and the alkali-soluble phaeomelanins which range from a yellow to red brown colour, arising from the deviation of the eumelanin pathway through the intervention of cysteine and/or glutathione. Eumelanins are usually found in the skin and eyes. Several different melanins include melanoprotein (dark brown melanin that is stored in high concentrations in the ink sac of the cuttlefish Sepia Officianalis), echinoidea (found in sand dollars and the hearts of sea urchins), holothuroidea (found in sea cucumbers) and ophiuroidea (found in brittle and snake stars). These melanins are possibly polymers which arise from the repeated coupling of simple bi-polyfunctional monomdric intermediates, or of high molecular weights. The compounds benzothiazole and tetrahydroisoquinoline ring systems act as UV-absorbing compounds. There are several different types of melanins considering that they are an aggregate of smaller component molecules, such as nitrogen containing melanins.

Bioluminescence

The only light source in the deep sea, marine animals give off visible light energy called bioluminescence, a subset of chemiluminescence. This is the chemical reaction in which chemical energy is converted to light energy. It is estimated that 90 per cent of deep-sea animals produce some sort of bioluminescence. Considering that a large proportion of the visible light spectrum is absorbed before reaching the deep sea, most of the emitted light from the sea-animals is blue and green. However, some species may emit a red and infrared light and there has even been a genus that is found to emit yellow bioluminescence. The organ that is responsible for the emission of bioluminescence is known as photophores. This type is only present in squid and fish and is used to illuminate their ventral surfaces, which disguise their silhouettes from predators. The uses of the photophores in the sea-animals differ, such as lenses for controlling intensity of colour and the intensity of the light produced. Squids have both photophores and chromatophores which controls both of these intensities. Another thing that is responsible for the emission of bioluminescence, which is evident in the bursts of light that jellyfish emit, start with a luciferin (a photogen) and ends with the light emitter (a photagogikon.) Luciferin, luciferase, salt and oxygen react and combine to create a single unit called photo-proteins, which can produce light when reacted with another molecule such as Ca+. Jellyfish use this as a defense mechanism; when a smaller predator is attempting to devour a jellyfish, it will flash its lights, which would therefore lure a larger predator and chase the smaller predator away. It is also used as mating behavior. In reef-building coral and sea anemones, they fluoresce; light is absorbed at one wavelength and re-emitted at another. These pigments may act as natural sunscreens, aid in photosynthesis, serve as warning colouration, attract mates, warn rivals, or confuse predators.

Chromatophores

Chromatophores are colour pigment changing cells that are directly stimulated by central motor neurons. They are primarily used for quick environmental adaptation for camouflaging. The process of changing the colour pigment of their skin relies on a single highly developed chromatophore cell and many muscles, nerves, glail and sheath cells. Chromatophores contract and contain vesicles that stores three different liquid pigments. Each colour is indicated by the three types of chromatophore cells: erythrophores, melanophores and xanthophores. The first type is the erythrophores, which contains reddish pigments such as carotenoids and pteridines. The second type is the melanophores, which contains black and brown pigments such as the melanins. The third type is the xanthophores which contains yellow pigments in the forms of carotenoids. The various colours are made by the combination of the different layers of the chromatophores. These cells are usually located beneath the skin or scale the animals. There are two categories of colours generated by the cell biochrome and schematochromes. Biochromes are colours chemically formed microscopic, natural pigments. Their chemical composition is created to take in some colour of light and reflect the rest. In contrast, schematochromes (structural colours) are colours created by light reflections from a colourless surface and refractions by tissues. Schematochromes act like prisms, refracting and dispersing visible light to the surroundings, which will eventually reflect a specific combination of colours. These categories are determined by the movement of pigments within the chromatophores. The physiological colour changes are short-term and fast, found in fishes and are a result from an animal's response to a change in the environment. In contrast, the morphological colours changes are long-term changes, occur in different stages of the animal and are due the change of numbers of chromatophores. To change the colour pigments, transparency, or opacity, the cells alter in form and size and stretch or contract their outer covering.

Photo-protective Pigments

Due to damage from UV-A and UV-B, marine animals have evolved to have compounds that absorb UV light and act as sunscreen. Mycosporine-like amino acids (MAAs) can absorb UV rays at 310-360 nm. Melanin is another well-known UV-protector. Carotenoids and photopigments both indirectly act as photo-protective pigments, as they quench oxygen free-radicals. They also supplement photosynthetic pigments that absorb light energy in the blue region.

Defensive Role of Pigments

It's known that animals use their colour patterns to warn off predators; however it has been observed that a sponge pigment mimicked a chemical which involved the regulation of moulting of an amphipod that was known to prey on sponges. So whenever that amphipod eats the sponge, the chemical pigments prevent the moulting and the amphipod eventually dies.

Environmental Influence on Colour

Colouration in invertebrates varies based on the depth, water temperature, food source, currents, geographic location, light exposure and sedimentation. For example, the amount of carotenoid a certain sea anemone decreases as we go deeper into the ocean. Thus, the marine life that resides on deeper waters is less brilliant than the organisms that live in well-lit areas due to the reduction of pigments. In the colonies of the colonial ascidian-cyanophyte symbiosis Trididemnum solidum, their colours are different depending on the light regime in which they live. The colonies that are exposed to full sunlight are heavily calcified, thicker and are white. In contrast the colonies that live in shaded areas have more phycoerythrin (pigment that absorbs green) in comparison to phycocyanin (pigment that absorbs red), thinner and are purple. The purple colours in the shaded colonies are mainly due to the phycobilin pigment of the algae, meaning the variation of exposure in light changes the colours of these colonies.

Adaptive Colouration

Aposematism is the warning colouration to signal potential predators to stay away. In many chromodrorid nudibranchs, they take in distasteful and toxic chemicals emitted from sponges and store them in their repugnatorial glands (located around the mantle edge). Predators of nudibranchs have learned to avoid these certain nudibranchs based on their bright colour patterns. Preys also protect themselves by their toxic compounds ranging from a variety of organic and inorganic compounds.

Physiological Activities

Pigments of marine animals sever several different purposes, other than defensive roles. Some pigments are known to protect against UV (see photo-protective pigments.) In the nudibranch Nembrotha Kubaryana, tetrapyrrole pigment 13 has been found to be a potent antimicrobial agent. Also in this creature, tamjamines A, B, C, E and F has shown antimicrobial, antitumor and immunosuppressive activities. Sesquiterpenoids are recognized for their blue and purple colours, but it has also been reported to exhibit various bioactivities such as antibacterial, immunoregulating, antimicrobial and cytotoxic, as well as the inhibitory activity against cell division in the fertilized sea urchin and ascidian eggs. Several other pigments have been shown to be cytotoxic. In fact, two new carotenoids that were isolated from a sponge called Phakellia stelliderma showed mild cytotoxicity against mouse leukemia cells. Other pigments with medical involvements include scytonemin, topsentins and debromohymenialdisine have several lead compounds in the field of inflammation, rheumatoid arthritis and osteoarthritis respectively. There's evidence that topsentins are potent mediators of immunogenic inflation and topsentin and scytonemin are potent inhibitors of neurogenic inflammation.

Uses

Pigments may be extracted and used as dyes.

Pigments (such as astaxanthin and lycopene) are used as dietary supplements.

References

Bandaranayake, Wickramasinghe. (2010). "The nature and role of pigments of marine invertebrates." Natural Products Report. Cambridge, n.d. Web. 25 May.

Biochrome - biological pigment". *Encyclopedia Britannica*. Retrieved 27 January 2010.

Chang, Kenneth (2005). "Yes, It's a Lobster and Yes, It's Blue." The New York Times - Breaking News, World News and Multimedia. NY Times, 15 Mar.

Esteban R, Barrutia O, Artetxe U, Fernández-Marín B, Hernández A, García-Plazaola JI. (2014). Internal and external factors affecting photosynthetic pigment composition in plants: a meta-analytical approach. *New Phytol*. DOI: 10.1111/nph.13186

Esteban R, García-Plazaola JI. (2014). Involvement of a second xanthophyll cycle in non-photochemical quenching of chlorophyll fluorescence: the lutein epoxide story. In: Demmig-Adams B, Adams WW III, Garab G, Govindjee, eds. Non-Photochemical Fluorescence Quenching and Energy Dissipation in Plants, Algae and Cyanobacteria. *Advances in Photosynthesis and Respiration*. Springer, pp. 277-295.

García-Plazaola JI, Matsubara S, Osmond CB. (2007). The lutein epoxide cycle in higher plants: its relationships to other xanthophyll cycles and possible functions. *Funct. Plant Biol.* 34: 759-773.

Hortensteiner, S. (2006). "Chlorophyll degradation during senescence". *Annual Review of Plant Biology* **57**: 55–77. doi: 10.1146/annurev.arplant.57.032905.105212.

Milicua, JCG. (1984). "Structural characteristics of the carotenoids binding to the blue carotenoprotein from Procambarus clarkii." Structural characteristics of the carotenoids binding to the blue carotenoprotein from *Procambarus clarkii.* N.p., 25 Oct. Web. 24 May 2010.

Nadakal A. M. (2010). "Carotenoids and Chlorophyllic Pigments in the Marine Snail, Cerithidea Californica Haldeman, Intermediate Host for Several Avian Trematodes." Marine Biological Laboratory. JSTOR, n.d. Web. 26 May.

Rang, H. P. (2003). *Pharmacology*. Edinburgh: Churchill Livingstone. ISBN 0-443-07145-4. Page 146

Schmidt-Danner, Claudia. (2010). "Biosynthesis of Porphyrin Compounds." Tetrapyrroles. N.p., n.d. Web. 25 May.

Webexhibits. "Bioluminescence/Causes of Colour." WebExhibits. Web. 2 June 2010.

Young AJ, Phillip D, Savill J. (1997). Carotenoids in higher plant photosynthesis. In: Pessaraki M, ed. *Handbook of Photosynthesis*, New York, Taylor and Francis, pp. 575-596.

Zagalsky, P. (2010). "Colouration in Marine Invertebrates" "A central role for astaxanthin complexes." Crustacean. N.p., n.d. Web. 25 May.

Zagalsky, Peter F. (2010). "The lobster carapace carotenoprotein, a-crustacyanin." A possible role for tryptophan in the bathochromic spectral shift of protein-bound astaxanthin. N.p., n.d. Web. 25 May.

Chapter 22

Adaptive Colouration

Nearly all animals have some sort of adaptive colouration or camouflage patterning that is often linked to a behavior to make it adaptive.

Some of the functions of adaptive colouration include:

☆ Defense against predators,

☆ Communication with conspecifics,

☆ Attracting or deceiving mates,

☆ Repelling or deceiving rivals,

☆ Signalling alarm to conspecifics and so forth,

☆ Protection from the environment (*e.g.*, ultraviolet radiation, cold) and

☆ Approaching prey

Colouration, in biology, the general appearance of an organism as determined by the quality and quantity of light that is reflected or emitted from its surfaces. Colouration depends upon several factors: the colour and distribution of the organism's biochromes (pigments), particularly the relative location of differently coloured areas; the shape, posture, position and movement of the organism; and the quality and quantity of light striking the organism. The perceived colouration depends also on the visual capabilities of the viewer. Colouration is a dynamic and complex characteristic and must be clearly distinguished from the concept of "colour," which refers only to the spectral qualities of emitted or reflected light.

Many evolutionary functions have been suggested for the effects of colouration on optical signaling. An organism with conspicuous colouration draws attention to itself, with some sort of adaptive interaction the frequent result. Such "advertising" colouration may serve to repel or attract other animals. While conspicuous colouration emphasizes optical signals and thereby enhances communication, colouration may, conversely, suppress optical signals or create incorrect signals and thereby reduce communication. This "deceptive" colouration serves to lessen detrimental or maladaptive interactions with other organisms.

Colouration may also affect an organism in ways other than its interaction with other organisms. Such nonoptical functions of colouration include physiological roles that depend on the molecular properties (*e.g.*, strength and type of chemical bonds) of the chemicals that create colour. For example, dark hair is mechanically stronger than light hair and dark feathers resist abrasion better than light feathers. Colouration may also play a part in the organism's energy budget, because biochromes create colour by the differential reflection and absorption of solar engery. Energy absorbed as a result of colouration may be used in biochemical reactions, such as photosynthesis, or it may contribute to the thermal equilibrium of the organism. Nonoptical functions of colouration also include visual functions in which colouration or its pattern affects an animal's own vision. Surfaces near the eye may be darkly coloured, for instance, to reduce reflectance that interferes with vision.

Emitted light, the product of bioluminescence, forms a portion of the colouration of some organisms. Bioluminescence may reveal an organism to nearby animals, but it may also serve as a light source in nocturnal species or in deepwater marine animals such as the pinecone fishes (*Monocentris*). These fishes feed at night and have bright photophores, or bioluminescent organs, at the tips of their lower jaws; they appear to use these organs much like tiny searchlights as they feed on planktonic (minute floating) organisms.

Because many pigments are formed as the natural or only slightly modified by-products of metabolic processes, some colouration may be without adaptive function. Nonfunctional colouration can, for example, be an incidental effect of a pleiotropic gene (a gene that has multiple effects), or it can result from pharmacological reaction (as when the skin of a Caucasian person turns blue in cold water) or from pure chance. It seems unlikely, however, that any apparently fortuitous colouration could long escape the process of natural selection and thus remain totally without function.

Regardless of its adaptive advantages, a particular colouration or pattern of colouration cannot evolve unless it is within the species' natural pool of genetic variability. Thus a species may lack a seemingly adaptive colouration because genetic variability has not included that colouration or pattern in its hereditary repertoire.

Because humans are highly visual animals, we are naturally interested in and attentive to biological colouration. Human attention to colouration ranges from the purely aesthetic to the rigidly pragmatic. Soft, pastel colourations aid in increasing work efficiency and contribute to tranquil moods; bright, strongly contrasting colours seem to contribute to excitement and enthusiasm. These phenomena may be extensions of the basic human response to the soft blue, green and brown backgrounds of the environment as opposed to sharply contrasting warning colourations found on many dangerous organisms. It is possible that much of the aesthetic value humans attach to colouration is closely related to its broad biological functions.

Human interest in colouration has led to biological studies. The classical work by the Moravian abbot Gregor Mendel on inherited characteristics, based largely on plant colouration, formed the foundation for modern genetics. Colouration also aids in the identification of organisms. It is an easily perceived, described and compared characteristic. Related species living in different habitats, however, frequently have strikingly different colourations. Since colouration is susceptible to alteration in various functional contexts, it usually lacks value as a conservative characteristic for determining systematic relationships between all but the most closely related species.

Structural and Biochemical Bases for Colour

Organisms produce colour physically, by submicroscopic structures that fractionate incident light into its component colours (schemochromes); or chemically, by natural pigments (biochromes) that reflect or transmit (or both) portions of the solar spectrum. Pigmentary colours, being of molecular origin, may be expressed independently of structural colour and are not altered by crushing, grinding, or compression. Structural colours are often reinforced by the presence of biochromes and are altered or destroyed by crushing, grinding, or compression.

Structural Colours (Schemochromes)

The physical principles of total reflection, spectral interference, scattering and, to some extent, polychromatic diffraction, all familiar in reference to inanimate objects, are also encountered among tissues of living forms, most commonly in animals. In plants these physical principles are exemplified only by the total reflection of white light by some fungi and bacteria and by the petals of some flowers and barks and by some spectral interference in certain sea plants.

Reflection

Total reflection of light which imparts whiteness to flowers, birds' feathers, mammalian hair and the wings of certain butterflies often results from the separation of finely divided materials by minute air spaces. Secretions or deposits in tissues may also contribute to the whiteness; for example, the fat and protein in mammalian milk and the calcium carbonate in the shells of mollusks, crustaceans, certain echinoderms, corals and protozoans.

Interference

Fractionation of white light into its components occurs in organisms (chiefly animals) through interference: the incident light penetrates the animal structure and is reflected back through successive ultrathinly layered films, giving striking iridescence, even in diffuse light, as a result of the asynchrony between the wavelengths of visible light that enter and those that return. Brilliant interference colours may display variety or be predominantly of one kind, depending upon the relative thicknesses of layers and interlaminar spaces giving rise to the colours. Such colours also are changeable with the angle of vision of the viewer. Purely prismatic refraction of light (sometimes confused with interference iridescence) is probably rare in animals and is limited to instances in which direct beams of light impinge upon certain microcrystalline deposits. Polychromatic diffraction *e.g.*, by natural, fine gratings or regular fine striations may be observed among certain insects, but, like prismatic refraction, it is conspicuous only when a direct beam of light strikes such structures and they are viewed at an angle.

Scattering

A special instance of diffraction, often referred to as the Tyndall effect (after its discoverer, the 19th-century British physicist John Tyndall), results in the presence of blue colours in many animals. The Tyndall effect arises from the reflection of the shorter (blue) waves of incident light by finely dispersed particles situated above the dark layers of pigment, commonly melanin deposits. In these blue-scattering systems, the reflecting entities whether very small globules of protein or lipid, semisolid substances in aqueous mediums, or very small vesicles of air are of such small size as to approximate the shorter wavelengths of light (about 0.4 micron). The longer waves, such as red, orange and yellow, pass through such mediums and are absorbed by the dark melanin below; the short waves, violet and blue, encounter bodies of approximately their own dimensions and consequently are reflected back. Two types of colouration may act in combination; in some instances, for example, structurally coloured and pigmented layers may be superimposed. Most of the greens found in the skin of fishes, amphibians, reptiles and birds do not arise from the presence of green pigments (although exceptions occur); rather, they result from the emergence of scattered blue light through an overlying layer of yellow pigment. Extraction of the yellow pigment from the overlying cuticle of a green feather or of a reptilian skin leaves the object blue.

Pigments (Biochromes)

Plants and animals commonly possess characteristic pigments. They range in plants from those that impart the brilliant hues of many fungi, through those that give rise to the various browns, reds and greens of species that can synthesize their food from inorganic substances (autotrophs), to the colourful pigments found in the flowers of seed plants. The pigments of animals are located in nonliving skin derivatives such as hair in mammals, feathers in birds, scales in turtles and tortoises and cuticles and shells in many invertebrates. Pigments also occur within living cells of the skin. The outermost skin cells may be pigmented, as in humans, or special pigment-containing cells, chromatophores, may occur in the deeper layers of the skin. Depending on the colour of their pigment, chromatophores are termed melanophores (black), erythrophores (red), xanthophores (yellow), or leucophores (white).

Phenoxazones and Sclerotins

Once confused with melanins, biochromes such as phenoxazones and sclerotins show a similar colour series (yellow, ruddy, brown, or black). Genetic research, notably with reference to eye pigments of the fruitfly, *Drosophila melanogaster*, has resulted in the description of a class of so-called ommochromes, which arephenoxazones. The ommochromes not only are conspicuous in the eyes of insects and crustaceans but have also been detected in the eggs of the echiurid worm*Urechis* *caupo* and in the changeable chromatophores in the skin of cephalopods. In addition to being responsible for the brown, vermilion, cinnabar and other colours of insect eyes, ommochromes are also sometimes present in the molting fluid and integument. They are distinguished from the melanins by solubility in formic acid and in dilute mineral acids, by manifestation of violet colours in concentrated sulfuric acid and by showing reversible colour changes with oxidizing and reducing agents. The ommochromes, which are derived from breakdown of the amino acid tryptophan, include ommatins and ommins. The ommatins, although complex in chemical structure, are relatively small molecules. The ommins are large molecules, in which the chromogenic moiety is seemingly condensed with longer chains, such as peptides (amino acids linked together). Clerotins arise as a result of an enzyme-catalyzed tanning of protein. Certain roaches secrete a phenolase enzyme, the glucoside of a dihydroxyphenol and a glycosidase. Mixing of these substances results in the release of the phenolic compound from glucose and its combination, via a reaction catalyzed by the phenolase, with protein; the products are pink, ruddy and ultimately dark-brown polymers that are incorporated into the insect's body cuticle and egg cases. Similar reactions take place in the carapace (the shell covering the body) of certain crustaceans.

Purines and Pterins

Although the purine compounds cannot be classed as true pigments they characteristically occur as white crystals they often contribute to the general colour patterns in lower vertebrates and invertebrates. That purines are excretory materials is illustrated by the uric acid (or urates) and guanine found in the excrement of birds and of uric acid found in that of reptiles. Uric acid has also been detected in the mucus excreted by sea anemones and urates are present in small amounts in the urine of higher apes and humans. The white, silvery, or iridescent chromatophores, both stationary iridocytes and changeable leucophores, of some fishes, amphibians, lizards and cephalopods contain microcrystalline aggregates of the purine guanine; a layer of white skin on the underside of many fishes, called the stratum arginatum, is particularly rich in guanine. Closely related to the purines and formerly classed among them are the pterins, so named from their notable appearance in and first chemical isolation from the wings of certain butterflies. Both purines and pterins contain a six-atom pyrimidine ring; in purines this ring is chemically condensed with an imidazole ring; pterins contain the pyrazine ring. Pterins occur as white, yellow, orange, or red granules in association with insect wing material.

Flavins (Lyochromes)

Flavins constitute a class of pale-yellow, greenly fluorescent, water-soluble biochromes widely distributed in small quantities in plant and animal tissues. The most prevalent member of the class is riboflavin (vitamin B_2). Flavins are synthesized by bacteria, yeasts and green plants; riboflavin is not manufactured by animals, which therefore are dependent upon plant sources. Riboflavin is a component of an enzyme capable of combining with molecular oxygen; the product, which is yellow, releases the oxygen in the cell with simultaneous loss of colour.

Miscellaneous Pigments

The chemical constitution of many pigments remains imperfectly known. Only a few of the more conspicuous examples are mentioned below.

Hemocyanins

Copper-containing proteins called hemocyanins occur notably in the blood of larger crustaceans and of gastropod and cephalopod mollusks. Hemocyanins are colourless in the reduced, or deoxygenated, state and blue when exposed to air or to oxygen dissolved in the blood. Hemocyanins serve as respiratory pigments in many animals, although it has not been established that they perform this function wherever they occur.

Hemerythins

Iron-containing, proteinaceous pigments, hemerythrins are present in the blood of certain bottom-dwelling marine worms (notably burrowing sipunculids) and of the brachiopod *Lingula*; the pigments serve as oxygen-carriers.

Hemovanadin

Pale-green pigment, hemovanadin, is found within the blood cells (vanadocytes) ofsea squirts (Tunicata) belonging to the families Ascidiidae and Perophoridae. The biochemical function of hemovanadin, a strong reducing agent, is unknown.

Actiniochrome

A relatively rare pigment, actiniochrome occurs in red or violet tentacle tips and in the stomodeum (oral region) of various sea anemones. The pigment plays no recognized physiological role.

Adenochrome

Adenochrome is a nonproteinaceous pigment that occurs as garnet-red inclusions at high concentrations in the glandular, branchial heart tissues of *Octopus bimaculatus*. The compound contains small amounts of ferric iron and some nitrogen and gives a positive reaction for pyrroles. It is believed to be an excretory product.

Control of Colouration

Genetic Control

Colouration is in large measure determined genetically. As mentioned earlier, the inheritance of colour in garden peas provided part of the basis for the pioneering studies of heredity by Mendel. These studies led Mendel to postulate the existence of discrete units of heredity that segregate independently of one another during the formation of reproductive cells. The studies also led to his discovery of the phenomenon of dominance. The basic units of heredity are now known as genes and the variant forms of a given gene are termed alleles. Among species that reproduce sexually, an individual normally possesses a pair of alleles for any gene one inherited from the female parent and one from the male parent. These two alleles are situated at corresponding loci on the paired chromosomes found in diploid cells *i.e.*, cells containing two similar sets of complementary chromosomes. Segregation of the alleles occurs during formation of reproductive cells, with the result that only one of the pairs enters each cell, which is called a haploid cell. In his experiments Mendel crossed purple-flowered peas with white-flowered ones. The plants he used in these crosses were true-breeding for flower colour, meaning that the purple-flowered plants were descended for generations from only other purple-flowered plants and that the white-flowered plants were likewise descended for generations from only other white-flowered plants. Because of these true-breeding characteristics, Mendel postulated that the original plants were homozygous for the trait of flower colour in other words, that each plant carried a pair of identical heredity units (*i.e.,* alleles) for this trait. When he crossed purple-flowered peas with white-flowered ones, he obtained a first filial (F$_1$) generation in which all the offspring had purple flowers. He therefore deduced that the unit for purple (usually designated R) was dominant over the unit for white (r). Thus in the parental generation the purple-flowered plants can be designated RR (indicating that they are homozygous for the dominant allele) and the white-flowered plants can be symbolized as rr. The F$_1$ plants were heterozygous for flower colour (Rr), but they expressed purple colour because of the complete dominance of the allele Rover r. Dominance may be incomplete, however; a crossing between homozygous red Japanese four-o'clocks (*Mirabilis*) and homozygous white ones yields heterozygous Rr offspring, which are all pink. A cross of the heterozygous pink generation of four-o'clocks with each other yields a second generation with the colour ratio of 25 percent red (RR), 50 percent pink (Rr) and 25 percent white (rr). This is because each of the parent (F$_1$) plants produces equal numbers of R- and r-containing reproductive cells through segregation and there is a random chance of either type of male haploid cell (gamete) fertilizing either of the two female types. For peas, on the other hand, the ratio resulting from a cross of parent (F$_1$) plants is three purple (one RR and two Rr) to one white (rr) because of the dominance of R. Although the principle of inheritance of colour and colouration patterns in all organisms is like that for the two plants described above, it is usually far more complex. Within the species population, a particular gene may have multiple alleles instead of two; thus numerous combinations within any individual may be possible; in addition, the colouration may depend upon genes at several sites. In this case either all pairs may segregate simultaneously and more or less independently into the gametes, or the genes may be linked in their inheritance by location on the same chromosome. Such possibilities, together with different degrees of dominance, result in tremendously complex hereditary bases for the genetic control of colour and colour patterns within many species.

Physiological Control

The development of colouration often depends upon regulatory substances (hormones) secreted by endocrine glands. In birds the level of the hormonethyroxine determines the colouration of feathers and bill, although specific seasonal biochromes are often laid down under the influence of sex hormones, as in the beak of the starling, which turns from black to yellow in early spring. The variability in control among bird species is so great, however, that generalizations are impossible. Hormonally controlled colour changes also occur in mammals; for example, swellings in the genital areas that become pink due to vascularization during the reproductive season. The species specificity of colouration patterns, however, always depends on a genetically determined responsiveness of various target tissues to certain hormones.

Chromatophores occur in cephalopods, crustaceans, insects, fishes, amphibians and lizards and are responsible for the most rapid colour changes. They allow conspicuous display of a biochrome by dispersing it in the chromatophore-bearing surface, or they conceal the biochrome by concentrating it into small areas. Chromatophores are of three kinds. The chromatophoric organs of cephalopodsconsist of an elastic sac filled with biochrome and controlled by a ring of radiating muscle fibres. These fibres contract in response to neural stimulation, thereby stretching the sac into a broad, thin disk. Chromatophoric syncytia occur incrustaceans, the movement of biochrome being due to the ebb and flow of cytoplasm through fixed tubular spaces that collapse when the cell is contracted and fill when the cell expands. Chromatophoric syncytia are hormonally controlled. Cellular chromatophores, the third kind, are found in vertebrates. In these cells melanic granules flow in stable cellular processes that maintain a fixed position, unlike the contracting and expanding processes of the syncytia. Control among vertebrates is varied: chromatophores of bony fishes are controlled by the autonomic nervous system; those of elasmobranch fishes (sharks and rays) and lizards are controlled by hormones and nerves; those of amphibians are regulated by hormones alone. One animal may contain biochromes of several colours, commonly red, yellow, black and reflecting white; prawns also have a blue biochrome. By appropriate migrations of biochromes, an animal can achieve substantial alterations in colour or shade for varying periods of time. In prawns, dispersion of blue and yellow yields green; unequal dispersion of biochromes over parts of the body produces patterns of colouration. Rapid physiological colour changes are supplemented by morphological ones, the animal either gradually synthesizing or destroying biochromes, usually in an adaptive manner.

The Adaptive Value of Biological Colouration

Colouration and the pattern of colouration play a central role in the lives of plants and animals even those species in which vision is lacking or not the dominant sense. For example, cryptic colouration often goes hand in hand with cryptic behaviour; nonreflective colours occur on the faces of birds that forage in bright sunlight; and abrasion-resistant colouration occurs more often among species that inhabit abrasive habitats than among species that inhabit nonabrasive habitats. The functions of biological colouration fall into three broad categories: (1) optical functions, in which colouration affects the animal's or plant's visibility to other animals; (2) visual functions, in which colouration affects the animal's own vision; and (3) physiological functions, in which the molecular properties of biochromes play a role unrelated to either optical signaling or vision.

Optical Functions: Deceptive Colouration

Deceptive colouration depends on four factors: the coloured organism, hereafter referred to as the organism; its model, which may be the background against which it is concealed; the spectral quality of the illumination; and the visual sensitivity and behaviour of the animal or animals that the organism is deceiving. To some extent the following discussion considers the relationships among the four factors separately; but in reality the deceptive, optical effect results from the interaction of all four factors. There are two basic types of deceptive colouration: (1) concealing colouration, or camouflage, in which the organism blends into its surroundings; and (2) mimicry, in which the organism is not hidden but rather presents a false identity by its resemblance to another species.

Camouflage

Background Matching

Background matching is probably the most common form of concealment. It makes little difference whether the background model is an animate or inanimate object since both involve the initial establishment and continued maintenance of the concealment. Not only colouration but also the form and the activities or behaviour of the organism in relation to its model are important. The simplest examples of background matching are provided by the fish eggs and planktonic (free-floating) larval fishes that exist in the uniformly blue environment of the open sea *i.e.*, those that are pelagic. They usually possess minimal pigmentation and are transparent. In other organisms and environments the behaviour and form of the organism become more important as adjuncts to colouration. Evidence of the importance of the choice of a proper background is provided by three differently coloured species of lizards of the genus *Anolis*, which form mixed hunting groups over the same background. Many of the individuals are easily perceived on this background, but, when disturbed, they conceal themselves by segregating according to species over the appropriately coloured backgrounds. Camouflage may also be accomplished through a change in colouration. Many flatfishes, for example, show a remarkable ability to match the pattern of the surface on which they are resting. Somenudibranchs, a group of marine gastropods, such as *Phestilla melanobrachia*, manage to establish and maintain their resemblance to the background by ingesting portions of their model, which is the living coral on which they live. The pigments in the coral polyps are deposited in diverticula (branches) of the gut and occasionally in the epidermis and show through as nearly perfect camouflage. The slow-moving nudibranchs are very difficult to see on their coral host and when they move to differently coloured coral, their colouration changes as their food source changes.

Some of the parasites that live on marine fishes conceal themselves in a similar manner. Flukes, or monogenean trematodes, gorge themselves on their hosts' tissues and biochromes and appear to remain within areas on the host that have similar pigmentation. The adaptive significance of the colouration is known to lie in escape from predation by the third party, cleaning organisms such as the fish*Labroides*, which feeds on the external parasites of other fishes. Several decoratorcrabs use portions of the model for concealment by picking up algae and sponges and placing them on the carapace (upper shell) to cover their own colouration; the algae and sponges continue to live as if in their normal habitat.

Visual Functions

Biological colouration can play a variety of roles in an animal's visual system. For example, facial colouration can help determine the amount of light that is reflected into the eyes. Among animals living in brightly lit habitats, too much reflected light could have undesirable effects on vision. It could, for example, produce blinding glare or dazzle; it might result in high luminance in parts of the visual field, thereby diminishing contrast in other parts of the field; or it could cause adaptation to a higher illuminance level than is appropriate for the remainder of the visual field. Birds that forage in sunlight for aerial insects a visually demanding task have bills that are black. Apparently the black colouration reduces reflectance that interferes with their vision. Vision itself depends on a biochrome that consists of a protein, opsin, attached to a chromophore. The chromophore may be either retinal (vitamin A_1), in which case the molecule is called rhodopsin; or 3-dehydroretinal (vitamin A_2), in which case the molecule is called porphyropsin. When light enters the eye and strikes the visual biochrome, the molecule undergoes a chemical change that stimulates the receptor nerve and thereby produces a visual stimulus.

In addition to the visual pigments, the eyes of many invertebrates contain biochromes that affect the spectrum of light that reaches the photoreceptors. Similarly, oil droplets in the retina and epithelium of vertebrate eyes contain carotenoids that may affect colour perception. More importantly, the epithelium contains melanin, which absorbs stray light that penetrates the retina without being absorbed by the visual pigments. In insect eyes a similar function is performed by ommochromes in secondary pigment cells surrounding the photoreceptors. Among many nocturnal vertebrates the white compound guanine is found in the epithelium or retina of the eye. This provides a mirrorlike surface, the tapetum lucidum, which reflects light outward and thereby allows a second chance for its absorption by visual pigments at very low light intensities. Tapeta lucida produce the familiar eyeshine of nocturnal animals.

Physiological Functions

The discussion of biochromes earlier in this article touched upon the many important physiological roles of biological pigments, including that of the chlorophylls in photosynthesis and of the hemoglobins in oxygen transport. This section provides examples of other physiological effects of biological colouration. Hair and feathers that contain melanin are more durable than those that lack this biochrome. Increased durability probably accounts for the dark, melanic wing tips of most birds. It may also be a contributing factor to the high proportion of black among birds that live in deserts, which are exceptionally abrasive habitats.

The absorption of solar energy by dark skin, scales, feathers, or hair is often associated with increased heat gain and reduced metabolic rates. Because birds lose a large amount of body heat through their uninsulated legs, dark leg colouration may help to warm the legs by absorbing solar energy, thereby reducing heat loss. Such reduced heat loss may explain why dark-legged North American woodwarblers (Parulidae) arrive in their northern breeding areas earlier than light-legged woodwarblers. Dark feathers, however, may actually reduce the amount of solar energy that penetrates to and is absorbed by a bird's skin. With fully erect plumage in moderate winds, a dark bird in full sunlight absorbs less heat into its body than a light bird does. This may also be a factor contributing to the high proportion of black among desert-dwelling birds.

Photoactivation of 7-dehydrocholesterol into vitamin D occurs throughout the epidermis of humans in the presence of ultraviolet light. The melanization of human skin may be an adaptation to optimize synthesis of vitamin D by permitting more or less ultraviolet radiation to penetrate the epidermis. A widespread response to increased light levels is the addition of melanin, or darkening of the body for example, tanning in humans. Such melanic shielding protects the tissues of the organism from potentially dangerous levels of ultraviolet radiation. Since the ultraviolet shield need protect only easily damaged cells in the nervous and reproductive systems, it does not necessarily have to lie in the skin but can instead be located internally, immediately around sensitive organs. When the ultraviolet shield is internal, external colouration may conform to other selection pressures.

Water is conserved by reducing evaporative loss and by reducing excretory water loss. Insects reduce evaporative water loss by adding melanin to the cuticle, melanin being more waterproof than other biochromes. The black-coloured beetle *Onymacris laeviceps* loses significantly less water than does the white-coloured beetle *O. brincki* when both species are kept without food under identical conditions. Quinones also darken insect exoskeletons and in *Drosophila* quinones contribute to the low permeability of the exoskeleton. Some insects avoid excretory water loss by depositing nitrogenous wastes in the exoskeleton, which is shed periodically. In these species external colouration is a consequence of nitrogen excretion.

Some arthropods produce offensive odours as a means of defense against predators. These odours derive from *p*-benzoquinones in the exoskeleton and are correlated with the chromatic properties of the molecules. Consequently, colouration in these species may be a consequence of selection for chemical defense.

Colouration Changes in Individual Organisms

Short-term Changes

Most rapid colour changes are chromatophoric ones that alter the colour of the organism through the dispersion or concentration of biochromes. Emotion plays a role in such changes among some cephalopods, fishes and horned lizards (*Phrynosoma*). When excited, certain fishes and horned lizards undergo a transient blanching that probably results from the secretion of adrenaline (epinephrine), a hormone known to concentrate the dark biochrome of vertebrates. Excitedcephalopods exhibit spectacular displays of colour, with waves of colour rippling across the body. Chromatophoric colour change is slower in vertebrates than in cephalopods. Although some fish may complete a colour change within a minute (compared to half a second or less for cephalopods), most vertebrates require several minutes to several hours. Colour changes extending over several hours are often entrained to external cycles.Fiddler crabs (*Uca*) that live in the intertidal zone show a complex pattern of cyclic chromatophoric colour change that is entrained not only to the local tidal cycle but also to the lunar and solar cycles. So important is this cyclic colour change that the response is innate to every part of the integument. The legs of a fiddler crab can be removed and sustained for a few days in saline solution; during this time melanophores in the legs continue to disperse and concentrate their melanin according to the cycle at the time they were removed from the body.

Changes in colour that extend over periods of several months may involve the synthesis or destruction of chromatophores or biochromes. The quantities of deposited guanine in some fishes vary in proportion to the relative lightness in colour of the background upon which they are living. Greenfish, or opaleye (*Girella nigricans*), kept in white-walled aquariums became very pale during a four-month period, storing about four times the quantity of integumentary guanine as was recoverable from the skins of individuals living in black-walled aquariums but receiving the same kind and amounts of food and the same overhead illumination. Some chromatophores respond directly to relevant environmental stimuli, independent of the nervous system. Such response occurs in the young of some fish and of the clawed frog (*Xenopus*); but in older individuals the nervous system, which is by this time fully developed, controls responsiveness. More typically the chromatophore response is mediated by the sensorimotor system from the start. The eye plays a major role in cephalopods and most vertebrates, particularly in animals capable of matching complex backgrounds, but the pineal organ (a light-sensitive organ on top of the brain) and a generalized dermal light sense may also mediate the chromatophore response.

Seasonal Changes

Seasonal changes of fields and forests include the annual colour changes involving foliage, flowers, fruits and seeds of plants. Many birds and mammals undergo seasonal molts, replacing their plumage or pelage with differently coloured feathers or hair. Winter whitening of the willow ptarmigan (*Lagopus lagopus*) andvarying hare (*Lepus*) are examples of a shift in camouflage coincident with a change in the background colouration. Many songbirds adopt a bright, contrasting nuptial plumage during the breeding season, reverting to a drabber winter plumage during the postnuptial molt.

Seasonal colour changes are usually regulated by light (mediated by the visual or pineal systems) or by temperature. Decreasing day lengths initiate whitening in the willow ptarmigan, whereas falling temperatures initiate whitening in the weasel (*Mustela erminea*). The spring molt of the varying hare is stimulated by the lengthening day, but the rate of molt depends on temperature. Seasonal changes in colouration may occur without a molt as a result of bleaching or wear, for example, the bleaching of human hair in the summer sun and birds that have bright colours based on carotenoids.

Age-related Changes

Colour changes during the life of an individual are common. Graying hair is a familiar badge of the elderly, both in humans and, to varying degrees, in other mammals. Among primate groups, particularly gorillas and chimpanzees, silver hairs indicate both age and dominance. Young birds of many species have a juvenile plumage that gives way to either an adult plumage in short-lived birds or a series of immature plumages in longer-lived species. Most gulls, for example, are deep gray or brown during their first year and become increasingly white thereafter. Changes of colour are also associated with age and size in many fish; for example, the blue parrot fish changes from a vertically barred pattern to all blue in association with increasing age and size.

Colouration Changes in Populations

Colouration changes occur not only in individuals but in populations as well. These latter result from evolutionary pressures *i.e.,* agents of natural selection that act upon the natural variations in colour types (morphs) found among the population. As a result of such pressures, certain colour morphs have increased odds of surviving and passing on their colouration pattern. Depending on the nature of the selection pressures, the population may come to include substantial numbers of individuals of different colour morphs; or one morph may become dominant, largely supplanting an earlier dominant colour form.

When individual colour variation is discontinuous within a species, that species is said to be polychromatic. The white-throated sparrow (*Zonotrichia albicollis*) of North America, for example, has individuals with white-and-black head stripes and other individuals with tan-and-brown head stripes. The different colourations are not associated with age, sex, or geographic region. Polychromatism may evolve in response to predation. A predator that successfully takes one prey type may then concentrate its search on others of this type and hence may overlook differently coloured prey of the same species. The phenomenon known as a perceptual set or a search image is exemplified by the predator of the European snail *Cepaea*. Predators encounter one morph and form a search image; they continue to hunt for that one form until its increasing rarity causes the predator to hunt randomly, encounter a different morph and form a new search image. In this way, oscillating selection pressures maintain several contemporaneous colour morphs among the snail population.

Evolutionary colour changes dictated by shifting selection are suggested by many populations that show geographical or temporal clines (graded series of morphological characters). For example, the common flicker (*Colaptes auratus*) has yellow markings in eastern North America and red markings in western North America, suggesting a change in selection pressure as one move from east to west. The best documented temporal shift in selection is the industrial melanism of noctuid and geometrid moths in England and Europe. The proportion of melanic, or darkly coloured, individuals in about 70 species of moths has increased dramatically since the 1850s. This increase correlates with the Industrial Revolution and the associated pollution of the countryside. Prior to that time, tree trunks, the normal daytime resting place of these nocturnal moths, had been covered by scattered whitish lichens. The trunks have turned dark in areas of industrial development because the lichens have been killed by pollutants and the trunks have been dirtied by soot. Blotched gray moths, previously protected from predation by birds, are now vulnerable, while the dark moths are less conspicuous. The shift to melanic populations in the United States lagged behind that in England and Europe, as did the industrialization process; but in Michigan by the early 1970s darkly coloured individuals formed up to 97 percent of some populations in regions where melanism was unknown before 1927. Since the 1970s in England there has been a reversal in the number of melanic individuals of some species, a sign that efforts to curb air pollution are having far-reaching effects. The many diverse functions discussed above lead to the inevitable conclusion that no single function can explain the colouration of living things. While biologists are far from a comprehensive theory that predicts the hues and patterns of colouration of plants and animals, such a theory will have to integrate the optical, visual and physiological functions of biological colouration.

Adaptive Colouration in Animals is a 500 page textbook about camouflage, warning colouration and mimicry by the University of Cambridge Zoologist Hugh Cott, first published during the Second World War in 1940; the book sold widely and made him famous. The book's general method is to present a wide range of examples from across the animal kingdom of each type of colouration, including marine invertebrates and fishes as well as terrestrial insects, amphibians, reptiles, birds and mammals. The examples are supported by a large number of Cott's own drawings, diagrams and photographs. This essentially descriptive natural history treatment is supplemented with accounts of experiments by Cott and others. The book is divided into three parts: concealment, advertisement and disguise. Part 1, concealment, covers the methods of camouflage, which are colour resemblance, countershading, disruptive colouration and shadow elimination. The effectiveness of these, arguments for and against them and experimental evidence, are described. Part 2, advertisement, covers the methods of becoming conspicuous, especially for warning displays in aposematic animals. Examples are chosen from mammals, insects, reptiles and marine animals and empirical evidence from feeding experiments with toads is presented. Part 3, disguise, covers methods of mimicry that provide camouflage, as when animals resemble leaves or twigs and markings and displays that help to deflect attack or to deceive predators with deimatic displays. Both Batesian mimicry and Müllerian mimicry are treated as adaptive resemblance, much like camouflage, while a chapter is devoted to the mimicry and behaviour of the cuckoo. The concluding chapter admits that the book's force is cumulative, consisting of many small steps of reasoning and being a wartime book, compares animal to military camouflage. Cott's textbook was at once well received, being admired both by zoologists and naturalists and among allied soldiers. Many officers carried a copy of the book with them in the field. Since the war it has formed the basis for experimental investigation of camouflage, while its breadth of coverage and accuracy have ensured that it remains frequently cited in scientific papers.

Camouflage

A primary defense against predators throughout the animal kingdom (and against the enemy during human warfare) is to avoid detection or recognition through camouflage. Achieving effective camouflage requires a suite of appropriate actions by an animal:

1. Sensing the local environment (including the animals in it),
2. Filtering the sensory input,
3. Using selected sensory input to make a behavior decision,
4. Directing the appropriate effectors (be they muscles/postures/colour patterns, etc.) to achieve some form of camouflage and
5. Implementing the appropriate behavior to render the camouflage effective.

How many Kinds of Camouflage are there?

There is still active debate on how to best "categorize" camouflage. Generally the tactics involve hindering or preventing *detection* or *recognition*. Some of the generally accepted mechanisms of camouflage include:

☆ General background resemblance (or "background matching,"

☆ Deceptive resemblance (or "masquerade" including mimicry),

☆ Disruptive colouration,

☆ Countershading/concealment of the shadow.

Other mechanisms include self-shadow concealment, obliterative shading, distractive markings, flicker-fusion camouflage, motion dazzle and motion camouflage. Ongoing research worldwide is occurring on many of these features.

Cephalopods: Ultimate Adaptive Colouration?

A distinguishing feature of cephalopods is that individual animals can change their appearance with a speed and diversity unparalleled in the animal kingdom: we term this "rapid, neutrally controlled polymorphism." Some squids, octopuses and cuttlefish can show 30-50 different appearances. In fact, these marine invertebrates manifest most aspects of their behavior through body patterning. An example of their versatility is that unlike other animals that use one or a few mechanisms of camouflage cephalopods use most of the mechanisms listed above.

Sensory/Motor Mechanisms of Achieving Adaptive Colouration in Cephalopods

Due to their sophisticated neural control of the skin, cephalopods can adapt to a wide range of backgrounds. What sensory cues do they use to achieve background matching? Vision is probably the main cue, but cephalopods do not seem to have colour vision. We are currently investigating the mechanisms and functions of polarization sensitivity in cephalopods. We have also begun to look at visual features of the background that cuttlefish use to switch on disruptive colouration. Tactile cues seem not to be used by cuttlefish for regulating their skin texture - vision seems to be used. Olfactory cues are used by cuttlefish females to choose mates, yet we do not know if/how this translates to certain body patterns for communication. Read more about skin ultra structure and neurobiology.

How do you Study Camouflage?

Field work is mandatory to experience and understand the dynamic features of light throughout each daily, lunar, seasonal and yearly cycle. Field work also enables observation and analysis of animal behavior under natural conditions. Laboratory experiments provide detailed testing of sensory cues, motor output and sequences of behavior. Computer simulations allow additional testing of hypotheses at multiple levels of analysis.

References

Adamo, S.A. and Hanlon, R.T. (1996). Do cuttlefish (Cephalopoda) signal their intentions to conspecifics during agonistic encounters? *Anim. Behav.* 52: 73-81.

Anderson, J.C., Baddeley, R.J., Osorio, D., Shashar, N., Tyler, C.W., Ramachandran, V.S., Cook A.C. and Hanlon, R.T. (2003). Modular organization of adaptive colouration in flounder and cuttlefish revealed by independent component analysis. *Network: Comput. Neural Syst.* 14: 321-333.

äthger, L.M., Denton, E.J., Marshall, J. and Hanlon, R.T. (2008). Mechanisms and behavioral functions of structural colouration in cephalopods. *Journal of the Royal Society Interface* 6: S149-S164.

äthger, L.M., Chiao, C-C., Barbosa, A, Buresch, K., Kaye, S. and Hanlon, R.T. (2007). Disruptive colouration elicited on controlled natural substrates in cuttlefish, *Sepia officinalis. J. Exp. Biol.* 210: 2657-2666.

äthger, L.M. and Cronin, T.W. (2007). Spectral and spatial properties of the polarized light reflections on the arms of squid (*Loligo pealeii*) and cuttlefish (*Sepia officinalis* L.). *Journal of Experimental Biology.* 210: 3624-3635.

Barbosa, A., Mäthger, L.M., Chubb, C., Florio, C., Chiao, C-C. and Hanlon, R.T. (2007). Disruptive colouration in cuttlefish: a visual perception mechanism that regulates ontogenetic adjustment of skin patterning. *Journal of Experimental Biology* 210: 1139-1147.

Chiao, C-C., Chubb, C., Buresch, K., Siemann, L. and Hanlon, R.T. (2009). The scaling effects of substrate texture on camouflage patterning in cuttlefish. *Vision Research* 49 (13): 1647-1656.

Chiao, C-C. and Hanlon, R.T. (2001). Cuttlefish camouflage: visual perception of size, contrast and number of white squares on artificial substrata initiates disruptive colouration. *J. Exp. Biol.* 204: 2119-2125.

Chiao, C-C. and Hanlon, R.T. (2001). Cuttlefish cue visually on area–not shape or aspect ratio–of light objects in the substrate to produce disruptive body patterns for camouflage. *Biol. Bull.* 201: 269-270.

Cooper, K.M., Hanlon, R.T. and Budelmann, B.U. (1990). Physiological colour change in squid iridophores II. Ultrastructural mechanisms in *Lolliguncula brevis. Cell Tissue Res.* 259: 15-24.

Cornwell, C.J., Messenger, J.B. and Hanlon, R.T. (1997). Chromatophores and body patterning in the squid *Alloteuthis subulata. J. Mar. Biol. Assoc. U.K.* 77: 1243-1246.

DiMarco, F.P. and Hanlon, R.T. (1997). Agonistic behavior in the squid *Loligo plei* (Loliginidae, Teuthoidea): Fighting tactics and the effects of size and resource value. *Ethology* 103(2): 89-108.

Forsythe, J.W. and Hanlon, R.T. (1997). Foraging and associated behavior by Octopus cyanea Gray, 1849 on a coral atoll, French Polynesia. *J. Exp. Mar. Biol. Ecol.* 209: 15-31.

Hanlon, R.T. (2005). Disruptive body patterning of cuttlefish (*Sepia officinalis*) requires visual information regarding edges and contrast of objects in natural substrate backgrounds. *Biological Bulletin.* 208: 7-11.

Hanlon, R.T. and Messenger, J.B. (1996). *Cephalopod Behaviour.* Cambridge University Press.

Hanlon, R.T., Conroy, L.-A. and Forsythe, J.W. (2008). Mimicry and foraging behaviour of two tropical sand-flat octopus species off North Sulawesi, Indonesia. *Biological Journal of the Linnean Society*, 93: 23-38.

Hanlon, R.T. (2007). Cephalopod dynamic camouflage. *Current Biology* 17 (11): 400-404.

Hanlon, R.T., Naud, M.-J., Forsythe, J.W., Hall, K., Watson, A.C. and McKechnie, J. (2007). Adaptable night camouflage by cuttlefish. *American Naturalist* 169 (4): 543-551.

Hanlon, R.T., Forsythe, J.W. and Joneschild, D.E. (1999). Crypsis, conspicuousness, mimicry and polyphenism as antipredator defences of foraging octopuses on Indo-Pacific coral reefs, with a method of quantifying crypsis from video tapes. *Biol. J. Linn. Soc.* 66: 1-22.

Hanlon, R.T., Maxwell, M.R., Shashar, N., Loew, E.R. and Boyle, K.-L. (1999). An ethogram of body patterning behavior in the biomedically and commercially valuable squid *Loligo pealei* off Cape Cod, Massachusetts. *Biol. Bull.* 197(1): 49-62.

Hanlon, R.T., Smale, M.J. and Sauer, W.H.H. (1994). An ethogram of body patterning behavior in the squid *Loligo vulgaris reynaudii* on spawning grounds in South Africa. *Biol. Bull.* 187(3): 363-372.

Hanlon, R.T., Cooper, K.M., Budelmann, B.U. and Pappas, T.C. (1990). Physiological colour change in squid*iridophores* I. Behavior, morphology and pharmacology in *Lolliguncula brevis*. *Cell Tissue Res.* 259: 3-14.

Hanlon, R.T. (1988). Behavioral and body patterning characters useful in taxonomy and field identification of cephalopods. *Malacologia* 29(1): 247-264.

Hanlon, R.T., Messenger, J.B. (1988). Adaptive colouration in young cuttlefish (*Sepia officinalis L.*): The morphology and development of body patterns and their relation to behaviour. *Phil. Trans. R. Soc. Lond. B* 320: 437-487.

Hanlon R.T., Chiao C-C., Mäthger, L.M., Barbosa A, Buresch K.C. and Chubb, C. (2009). Cephalopod dynamic camouflage: bridging the continuum between background matching and disruptive colouration. *Philosophical Transactions of the Royal Society* B 364: 429-437.

Mäthger, L.M. and Hanlon, R.T. (2006). Anatomical basis for camouflaged polarized light communication in squid. *Biology Letters* 2(4): 494-496.

Mäthger, L.M., Barbosa, A. and Hanlon, R.T. (2009). Cuttlefish use visual cues to control 3-dimensional skin papillae for camouflage. *Journal of Comparative Physiology* A 195: 547-555.

Mäthger, L.M., Shashar, N. and Hanlon, R.T. (2009). Do cephalopods communicate using polarized light reflections from their skin? *Journal of Experimental Biology* 212: 2133-2140.

Mäthger, L.M. and Hanlon, R.T. (2007). Malleable skin colouration in cephalopods: selective reflectance, transmission and absorbance of light by chromatophores and iridophores. *Cell and Tissue Research* 329: 179-186.

Novicki, A., Budelmann, B.U. and Hanlon, R.T. (1990). Brain pathways of the chromatophore system in the squid *Lolliguncula brevis*. *Brain Res.* 519(1-2): 315-323.

Shashar, N. and Hanlon, R.T. (1997). Squids (*Loligo pealei* and *Euprymna scolopes*) can exhibit polarized light patterns produced by their skin. *Biol. Bull.* 193(2): 207-208.

Williams, S.B., Pizarro, O., How, M., Mercer, D., Powell, G., Marshall, J. and Hanlon, R.T. (2009(. Surveying nocturnal cuttlefish camouflage behaviour using an AUV. *IEEE International Conference on Robotics and Automation*, p. 214-219.

Wolterding, M.R. (1989). Behavior, body patterning, growth and life history of Octopus *briareus*cultured in the laboratory. *Am. Malacol. Bull.* 7(1): 21-45.

Chapter 23

Anthocyanins

Anthocyanins are water-soluble vacuolar pigments that may appear red, purple, or blue depending on the pH. They belong to a parent class of molecules called flavonoids synthesized via the phenylpropanoid pathway; they are odorless and nearly flavorless, contributing to taste as a moderately astringent sensation. Anthocyanins occur in all tissues of higher plants, including leaves, stems, roots, flowers and fruits. Anthoxanthins are clear, white to yellow counterparts of anthocyanins occurring in plants. Anthocyanins are derived from anthocyanidins by adding sugars.

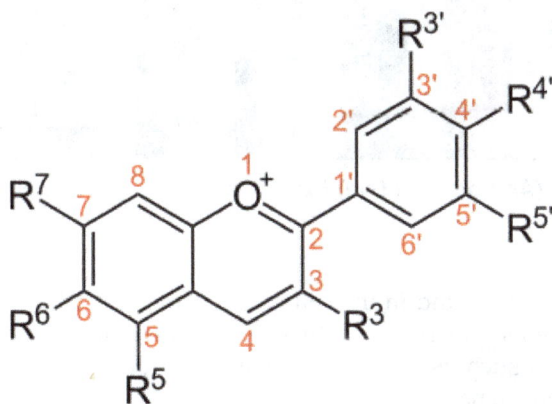

Anthocyanins are Glucosides of Anthocyanidins, the Basic Chemical Structure

Light Absorbance

The absorbance pattern responsible for the red colour of anthocyanins may be complementary to that of green chlorophyll in photosynthetically active tissues such as young *Quercus coccifera* leaves. It may protect the leaves from attacks by plant eaters that may be attracted by green colour.

pH

Anthocyanins are generally degraded at higher pH. However, some anthocyanins, such as petanin (petunidin 3-[6-*O*-(4-*O*-(*E*)-*p*-coumaroyl-*O*-α-L-rhamnopyranosyl)-β-D-glucopyranoside]-5-*O*-β-D-glucopyranoside), are resistant to degradation at pH 8 and can be used as a food colourant.

Use as pH Indicator

Anthocyanins can be used as pH indicators because their colour changes with pH; they are pink in acidic solutions (pH < 7), purple in neutral solutions (pH ~ 7), greenish-yellow in alkaline solutions (pH > 7) and colourless in very alkaline solutions, where the pigment is completely reduced.

Red Cabbage (Anthocyanin Dye) Extract at Low pH (Left) to High pH (Right)

Occurrence

Anthocyanins are found in the cell vacuole, mostly in flowers and fruits but also in leaves, stems and roots. In these parts, they are found predominantly in outer cell layers such as the epidermis and peripheral mesophyll cells. Most frequently occurring in nature are the glycosides of cyanidin, delphinidin, malvidin, pelargonidin, peonidin and petunidin. Roughly 2 per cent of all hydrocarbons fixed in photosynthesis are converted into flavonoids and their derivatives such as the anthocyanins. No fewer than 10^9 tons of anthocyanins are produced in nature per year. Not all land plants contain anthocyanin; in the Caryophyllales (including cactus, beets and amaranth), they are replaced by betalains. Anthocyanins and betalains have never been found in the same plant. Plants with abnormally high anthocyanin quantities are popular as ornamental plants.

In Food

Plants rich in anthocyanins are *Vaccinium* species, such as blueberry, cranberry and bilberry; *Rubus* berries, including black raspberry, red raspberry and blackberry; blackcurrant, cherry, eggplant peel, black rice, Concord grape, muscadine grape, red cabbage and violet petals. Anthocyanins are less abundant in banana, asparagus, pea, fennel, pear and potato and may be totally absent in certain cultivars of green gooseberries. Red-fleshed peaches are rich in anthocyanins. The highest recorded amount appears to be specifically in the seed coat of black soybean (*Glycine max* L. Merr.) containing some 2,000 mg per 100 g, in purple corn kernels and husks and in skins and pulp of black chokeberry (*Aronia melanocarpa* L.) (Table). Due to critical differences in sample origin, preparation and extraction methods determining anthocyanin content, the values presented in the adjoining table are not directly comparable. Nature, traditional agriculture and plant breeding have produced various uncommon crops containing anthocyanins, including blue- or red-flesh potatoes and purple or red broccoli, cabbage, cauliflower, carrots and corn. Garden tomatoeshave been subjected to a breeding program using introgression lines of genetically modified organisms (but not incorporating them in the final purple tomato) to define the genetic basis of purple colouration in wild species originally fromChile and the Galapagos Islands. The variety known as "Indigo Rose" became commercially available to the agricultural industry and home gardeners in 2012. Investing tomatoes with high anthocyanin content doubles their shelf-life and inhibits growth of a post harvest mold pathogen, *Botrytis cinerea*. Tomatoes have also been genetically modified with transcription factors from snapdragons to produce high levels of anthocyanins in the fruits. Anthocyanins can also be found in naturally ripened olives and are partly responsible for the red and purple colours of some olives.

Food Source	Anthocyanin Content in mg per 100 g
Açaí	320
Blackcurrant	190–270
Aronia	1,480[8]
Eggplant	750
Blood orange	~200
Marion blackberry	317
Black raspberry	589
Raspberry	365
Wild blueberry	558
Cherry	350–400
Redcurrant	80–420
Purple corn	1,642
Purple corn leaves	10x more than in kernels
Concord grape	326
Norton grape	888t

In Leaves of Plant Foods

Content of anthocyanins in the leaves of colourful plant foods, such as purple corn, blueberries or lingonberries, is about 10 times higher than in the edible kernels or fruit. The colour spectrum of grape berry leaves can be analysed to evaluate the amount of anthocianins. Fruit maturity, quality and harvest time can be evaluated on the basis of the spectrum analysis.

Functions

In flowers, bright-reds and -purples are adaptive for attracting pollinators. In fruits, the colourful skins also attract the attention of animals, which may eat the fruits and disperse the seeds. In photosynthetic tissues (such as leaves and sometimes stems), anthocyanins have been shown to act as a "sunscreen", protecting cells from high-light damage by absorbing blue-green and ultraviolet light, thereby protecting the tissues from photoinhibition, or high-light stress. This has been shown to occur in red juvenile leaves, autumn leaves and broad-leaf evergreen leaves that turn red during the winter. The red colouration of leaves has been proposed to possibly camouflage leaves from herbivores blind to red wavelengths, or signal unpalatability, since anthocyanin synthesis often coincides with synthesis of unpalatable phenolic compounds. In addition to their role as light-attenuators, anthocyanins also act as powerful antioxidants. However, it is not clear whether anthocyanins can significantly contribute to scavenging of free radicals produced through metabolic processes in leaves, since they are located in the vacuole and, thus, spatially separated from metabolic reactive oxygen species. Some studies have shown hydrogen peroxide produced in other organelles can be neutralized by vacuolar anthocyanin.

Autumn Leaf Colour

The reds, the purples and their blended combinations that decorate autumn foliage come from anthocyanins. Unlike the carotenoids, these pigments are not present in the leaf throughout the growing season, but are actively produced towards the end of summer. They develop in late summer in the sap of the cells of the leaf and this development is the result of complex interactions of many influences, both inside and outside the plant. Their formation depends on the breakdown of sugars in the presence of bright light as the level ofphosphate in the leaf is reduced. Anthocyanins are present in about 10 per cent of tree species in temperate regions, although in certain areas such as New England, up to 70 per cent of tree species may produce the pigment. Many science textbooks incompletely state that autumn colouration (including red) is the result of breakdown of green chlorophyll, which unmasks the already-present orange, yellow and red pigments (carotenoids, xanthophylls and anthocyanins, respectively). While this is indeed the case for the carotenoids and xanthophylls (orange and yellow pigments), anthocyanins are not synthesized until the plant has begun breaking down the chlorophyll.

Structure

Selected Anthocyanidins and their Substitutions

Basic Structure	Anthocyanidin	R_3'	R_4'	R_5'	R_3	R_5	R_6	R_7
	Aurantinidin	-H	-OH	-H	-OH	-OH	-OH	-OH
	Cyanidin	-OH	-OH	-H	-OH	-OH	-H	-OH
	Delphinidin	-OH	-OH	-OH	-OH	-OH	-H	-OH
	Europinidin	-OCH$_3$	-OH	-OH	-OH	-OCH$_3$	-H	-OH
	Pelargonidin	-H	-OH	-H	-OH	-OH	-H	-OH
	Malvidin	-OCH$_3$	-OH	-OCH$_3$	-OH	-OH	-H	-OH
	Peonidin	-OCH$_3$	-OH	-H	-OH	-OH	-H	-OH
	Petunidin	-OH	-OH	-OCH$_3$	-OH	-OH	-H	-OH
	Rosinidin	-OCH$_3$	-OH	-H	-OH	-OH	-H	-OCH$_3$

Glycosides of Anthocyanidins

The anthocyanins, anthocyanidins with sugar group(s), are mostly 3-glucosides of the anthocyanidins. The anthocyanins are subdivided into the sugar-free anthocyanidin aglycones and the anthocyanin glycosides. As of 2003, more than 400 anthocyanins had been reported while more recent literature (early 2006), puts the number at more than 550 different anthocyanins. The difference in chemical structure that occurs in response to changes in pH is the reason why anthocyanins are often used as pH indicators, as they change from red in acids to blue in bases.

Stability

Anthocyanins are thought to be subject to physiochemical degradation *in vivo* and *in vitro*. Structure, pH, temperature, light, oxygen, metal ions, intramolecular association and intermolecular association with other compounds (copigments, sugars, proteins, degradation products, etc.) are generally known to affect the colour and stability of anthocyanins.[31] B-ring hydroxylation status and pH have been shown to mediate the degradation of anthocyanins to their phenolic acid and aldehyde constituents. Indeed, significant portions of ingested anthocyanins are likely to degrade to phenolic acids and aldehyde *in vivo*, following consumption. This characteristic confounds scientific isolation of specific anthocyanin mechanisms *in vivo*.

Biosynthesis

1. Anthocyanin pigments are assembled like all other flavonoids from two different streams of chemical raw materials in the cell:
   ☆ One stream involves the shikimate pathway to produce the amino acid phenylalanine.
   ☆ The other stream produces three molecules of malonyl-CoA, a C3 unit from a C2 unit (acetyl-CoA).

2. These streams meet and are coupled together by the enzyme chalcone synthase, which forms an intermediate chalcone-like compound via a polyketide folding mechanism that is commonly found in plants.

3. The chalcone is subsequently isomerized by the enzyme chalcone isomerase to the prototype pigment naringenin.

4. Naringenin is subsequently oxidized by enzymes such as flavanone hydroxylase, flavonoid 3' hydroxylase and flavonoid 3' 5'-hydroxylase.

5. These oxidation products are further reduced by the enzyme dihydroflavonol 4-reductase to the corresponding colourless leucoanthocyanidins.

6. Leucoanthocyanidins were once believed to be the immediate precursors of the next enzyme, a dioxygenase referred to as anthocyanidin synthase or leucoanthocyanidin dioxygenase. Flavan-3-ols, the products of leucoanthocyanidin reductase (LAR), have been recently shown to be their true substrates.

7. The resulting unstable anthocyanidins are further coupled to sugar molecules by enzymes such as UDP-3-O-glucosyltransferase to yield the final relatively stable anthocyanins.

More than five enzymes are thus required to synthesize these pigments, each working in concert. Even a minor disruption in any of the mechanism of these enzymes by either genetic or environmental factors would halt anthocyanin production. While the biological burden of producing anthocyanins is relatively high, plants benefit significantly from environmental adaptation, disease tolerance and pest tolerance provided by anthocyanins. In anthocyanin biosynthetic pathway, L-phenylalanine is converted to naringenin by phenylalanine ammonialyase (PAL), cinnamate 4-hydroxylase (C4H), 4-coumarate CoA ligase (4CL), chalcone synthase (CHS) and chalcone isomerase (CHI). And then, the next pathway is catalyzed the formation of complex aglycone and anthocyanin composition by flavanone 3-hydroxylase (F3H), flavonoid 3'-hydroxylase (F3'H), dihydroflavonol 4-reductase (DFR), anthocyanidin synthase (ANS), UDP-glucoside: flavonoid glucosyltransferase (UFGT) and methyl transferase (MT). Among those, UFGT is divided into UF3GT and UF5GT, which are responsible for the glucosylation of anthocyanin to produce stable molecules. In *Arabidopsis thaliana*, two glycosyltransferases, UGT79B1 and UGT84A2, are involved in the anthocyanin biosynthetic pathway. The UGT79B1 protein converts cyanidin 3-O-glucoside to cyanidin 3-O-xylosyl(1→2)glucoside. UGT84A2 encodes sinapic acid: UDP-glucosyltransferase.

Genetic Analysis

The phenolic metabolic pathways and enzymes can be studied by mean of transgenesis of genes. The *Arabidopsis* regulatory gene in the production of anthocyanin pigment 1 (*AtPAP1*) can be expressed in other plant species.

Potential Food Value

Anthocyanins are considered secondary metabolites as a food additive with E number E163 (INS number 163); they are approved for use as a food additive in the EU, Australia and New Zealand. Although anthocyanins are powerful antioxidants *in vitro*, this antioxidant property is unlikely to be conserved after the plant is consumed. As interpreted by the Linus Pauling Institute and European Food Safety Authority, dietary anthocyanins and other flavonoids have little or no direct antioxidant food value following digestion. Unlike controlled test-tube conditions, the fate of anthocyanins *in vivo* shows they are poorly conserved (less than 5 per cent), with most of what is absorbed existing as chemically modified metabolites that are rapidly excreted. The increase in antioxidant capacity of blood seen after the consumption of anthocyanin-rich foods may not be caused directly by the anthocyanins, but instead may result from increased uric acid levels derived from metabolism of flavonoids.

Dye-Sensitized Solar Cells

Anthocyanins have been used in organic solar cells because of their ability to convert light energy into electrical energy. The many benefits to using dye-sensitized solar cells instead of traditional pn junction silicon cells include lower purity requirements and abundance of component materials, such as titania, as well as the fact they can be produced on flexible substrates, making them amenable to roll-to-roll printing processes.

Research on Health Benefits

General Research

Richly concentrated as pigments in berries, anthocyanins were the topics of research presented at a 2007 symposium on health benefits that may result from berry consumption. Anthocyanins also fluoresce, enabling a tool for plant cell research to allow live cell imaging for extended periods of time without a requirement for other fluorophores.

Cancer Research

According to the American Cancer Society, there have been no studies in humans showing that any phytochemical supplement can prevent or treat cancer.

Use as Visual Markers to Mark Genetically Modified Materials

Anthocyanin production can be engineered into genetically modified materials to enable their visual identification.

Specific Functions

Anthocyanins are flavonoids. A major function of anthocyanins is to provide colour to most flowers and fruits. The colours can help attract pollinating animals to flowers and animals that will help disperse seeds. Anthocyanins are also thought help protect leaves from ultraviolet radiation but some botanists think that may not be true for all plant species. Anthocyanins are also thought to deter herbivores in some species. Xanthophylls are carotenoids. The main xanthophyll in leaves is lutein. Xanthophylls are structural components of the light harvesting antenna in chloroplasts. They function as accessory pigments for harvesting light at wavelengths that chorophyll cannot and transfer the light energy to chlorophyll. They also absorb excess light energy and dissipate it in order to avoid damage in what is termed the Xanthophyll Cycle. The xanthophyll, zeaxanthin, appears to have a role in sensing blue light which is involved in stomatal opening and in phototropism.

Anthocyanins are water soluble strong colours and have been used to colour food since historical times. Extracts of berries have been used to colour drinks, pastries and other foods. There are, however, some drawbacks in the use of anthocyanins in food. Anthocyanins are water soluble, which restricts the use and are pH dependent. When the acidity changes, the colour changes. For example the colour of red cabbage is enhanced with the addition of vinegar or other acid. On the other hand, when cooked in aluminium pans, which cause a more alkaline environment, the colour changes to purple and blue. The colour is also susceptible towards temperature, oxygen, UV-light and different co-factors. Temperature may destroy the flavylium ion and thus causes loss of colour. Temperature also causes Maillard reactions, in which the sugar residues in the anthocyanins may be involved.

Light may have a similar effect. Oxygen may destroy the anthocyanins, as do other oxidizing reagents, such as peroxides and vitamin C. Many other components in plants and foods may interact with the anthocyanins and either destroy, change or increase the colour. Quinones in apples, for example, enhance the degradation of anthocyanins, whereas the addition of sugar to strawberries stabilises the colour. All these factors limit the use of anthocyanins in foods. Some loss of colour during storage may be prevented by storing at low temperatures, in dark containers or under oxygen-free packaging. In practice the pure colours are very hard to obtain and most often (crude) extracts are used as food colours. Grape peel (E163(i)) and black currant extract (E163(iii)) are the most widely used anthocyanin mixtures in foods.

Activity and Toxicity

Anthocyanins, when used as food colours, are not toxic and do not exceed the levels that may be ingested by consuming berries or other anthocyanin-coloured fruits. The toxicity of the highly concentrated anthocyanin mixtures that are presently sold as food supplements is not well investigated. There are indications those anthocyanins as antioxidants have health promoting effects, which include reduced risk of coronary heart disease, improved visual activity and antiviral activity. However, many of these claims are not proven scientifically. Anthocyanins are important antioxidants, but their effect on human health is still poorly understood.

References

Agati, Giovanni, Pinelli, Patrizia, Cortés Ebner, Solange *et al.* (2005). "Nondestructive evaluation of anthocyanins in olive (*Olea europaea*) fruits by in situ chlorophyll fluorescence spectroscopy". *Journal of Agricultural and Food Chemistry* 53 (5): 1354–63.doi: 10.1021/jf048381d.

Andersen, Øyvind M (2001). "Anthocyanins". *Encyclopedia of Life Sciences. Eds.* John Wiley and Sons, Ltd. doi: 10.1038/npg.els.0001909. ISBN 0470016175.

Andersen, Øyvind M., Jordheim, Monica (2008). "Anthocyanins- food applications". *5th Pigments in Food congress- for quality and health*. University of Helsinki. ISBN 978-952-10-4846-3.

Archetti, Marco, Döring, Thomas F., Hagen, Snorre B. *et al.* (2011). "Unravelling the evolution of autumn colours: an interdisciplinary approach". *Trends in Ecology and Evolution* 24(3): 166–73. doi: 10.1016/j.tree.2008.10.006.

Australia New Zealand Food Standards Code"Standard 1.2.4. Labelling of ingredients". Retrieved 27 October 2011.

Bramley, R.G.V., Le Moigne, M., Evain, S. *et al.* (2011). "On-the-go sensing of grape berry anthocyanins during commercial harvest: development and prospects". *Australian Journal of Grape and Wine Research* 17: 316–326.

Butelli, Eugenio, Titta, Lucilla, Giorgio, Marco *et al.* (2008). "Enrichment of tomato fruit with health-promoting anthocyanins by expression of select transcription factors". *Nature Biotechnology* 26 (11): 1301–8. doi: 10.1038/nbt.1506.

Cevallos-Casals, BA, Byrne, D, Okie, WR, Cisneros-Zevallos, L (2006). "Selecting new peach and plum genotypes rich in phenolic compounds and enhanced functional properties". *Food Chemistry* 96: 273–328. doi: 10.1016/j.foodchem.2005.02.03.

Cherepy, Nerine J., Smestad, Greg P., Grätzel, Michael, Zhang, Jin Z. (1997). "Ultrafast Electron Injection: Implications for a Photoelectrochemical Cell Utilizing an Anthocyanin Dye-Sensitized TiO_2 Nanocrystalline Electrode". *The Journal of Physical Chemistry B* 101 (45): 9342–51. doi: 10.1021/jp972197w.

Choung, Myoung-Gun, Baek, In-Youl, Kang, Sung-Taeg *et al.* (2001). "Isolation and determination of anthocyanins in seed coats of black soybean (*Glycine max* (L.) Merr.)". *J. Agric. Food Chem.* 49 (12): 5848–51. doi: 10.1021/jf010550w.

Da Qiu Zhao, Chen Xia Han, Jin Tao Ge, Jun Tao 2012). "Isolation of a UDP-glucose: Flavonoid 5-O-glucosyltransferase gene and expression analysis of anthocyanin biosynthetic genes in herbaceous peony (*Paeonia lactiflora* Pall.)". *Electronic Journal of Biotechnology* 15 (6). doi: 10.2225/vol15-issue6-fulltext-7.

Davies, Kevin M. (2004). *Plant pigments and their manipulation*. Wiley-Blackwell. p. 6.ISBN 1-4051-1737-0.

De Rosso, VV, Morán Vieyra, FE, Mercadante, AZ, Borsarelli, CD (2008). "Singlet oxygen quenching by anthocyanin's flavylium cations". *Free Radical Research* 42 (10): 885–91.doi: 10.1080/10715760802506349. PMID 18985487.

Fossen T, Cabrita L andersen OM (1998). "Colour and stability of pure anthocyanins influenced by pH including the alkaline region". *Food Chemistry* 63 (4): 435–440. doi: 10.1016/S0308-8146(98)00065-X.

Francis, F.J. (1999). *Colourants*. Egan Press. ISBN 1-891127-00-4.

Grätzel, Michael (2003). "Dye-sensitized solar cells". *Journal of Photochemistry and Photobiology* 4 (2): 145–53. doi: 10.1016/S1389-5567(03)00026-1.

Hosseinian FS, Beta T (2007). "Saskatoon and wild blueberries have higher anthocyanin contents than other Manitoba berries". *Journal of Agricultural and Food Chemistry* 55 (26): 10832–8. doi: 10.1021/jf072529m.

Jack Sullivan (1998). "Anthocyanin". *Carnivorous Plant Newsletter* (CPN) *September 1998*. Archived from the original on 1 November 2009. Retrieved 6 October 2009.

Karageorgou P, Manetas Y (2006). "The importance of being red when young: anthocyanins and the protection of young leaves of *Quercus coccifera* from insect herbivory and excess light". *Tree Physiol* 26 (5): 613–621. doi: 10.1093/treephys/26.5.613.

Kong, JM, Chia, LS, Goh, NK *et al.* (2003). "Analysis and biological activities of anthocyanins". *Phytochemistry* 64 (5): 923–33. doi: 10.1016/S0031-9422(03)00438-2.

Kovinich, N, Saleem, A, Arnason, JT, Miki, B (2010). "Functional characterization of a UDP-glucose: flavonoid 3-O-glucosyltransferase from the seed coat of black soybean (*Glycine max* (L.) Merr.)". *Phytochemistry* 71 (11–12): 1253–63. doi: 10.1016/j.phytochem.2010.05.009.

Kovinich, N, Saleem, A, Rintoul, TL *et al.* (2012). "Colouring genetically modified soybean grains with anthocyanins by suppression of the proanthocyanidin genes ANR1 and ANR2". *Transgenic Res.* 21 (4): 757–71. doi: 10.1007/s11248-011-9566-y.

Krenn, L, Steitz, M, Schlicht, C *et al.* (2007). "Anthocyanin- and proanthocyanidin-rich extracts of berries in food supplements–analysis with problems". *Pharmazie* 62 (11): 803–12.

Li, C. Y., Kim, H. W., Won, S. R. *et al.* (2008). "Corn husk as a potential source of anthocyanins". *Journal of Agricultural and Food Chemistry* 56 (23): 11413–6.doi: 10.1021/jf802201c.

Li, Xiang, Gao, Ming-Jun, Pan, Hong-Yu *et al.* (2010). "Purple canola: Arabidopsis PAP1 increases antioxidants and phenolics in *Brassica napus* leaves". *J. Agric. Food Chem.* 58 (3): 1639–1645. doi: 10.1021/jf903527y.

Lieberman S (2007). "The antioxidant power of purple corn: a research review". *Alternative and Complementary Therapies* 13 (2): 107–110. doi: 10.1089/act.2007.13210.

Lotito SB, Frei B, Frei (2006). "Consumption of flavonoid-rich foods and increased plasma antioxidant capacity in humans: cause, consequence, or epiphenomenon?". *Free Radic. Biol. Med.* 41 (12): 1727–46. doi: 10.1016/j.freeradbiomed.2006.04.033.

Michaelis, Leonor, Schubert M.P., Smythe C.V. (1936). "Potentiometric Study of the Flavins". *J. Biol. Chem* 116 (2): 587–607.

Muñoz-Espada, A.C., Wood, K. V., Bordelon, B., Watkins, B. A. (2004). "Anthocyanin Quantification and Radical Scavenging Capacity of Concord, Norton and Marechal Foch Grapes and Wines". *Journal of Agricultural and Food Chemistry* 52 (22): 6779–86.doi: 10.1021/jf040087y.

Nakajima, J, Tanaka, Y, Yamazaki, M, Saito, K (2001). "Reaction mechanism from leucoanthocyanidin to anthocyanidin 3-glucoside, a key reaction for colouring in anthocyanin biosynthesis". *The Journal of Biological Chemistry* 276 (28): 25797–803.doi: 10.1074/jbc.M100744200.

Phytochemicals. American Cancer Society. January 2013. Retrieved September 2013.

Purple as a tomato: towards high anthocyanin tomatoes. *Trends in Plant Science* Vol.14 No.5

Purple tomato 'may boost health', Health, BBC News online, 26 October 2008

Scientific Opinion on the substantiation of health claims related to various food(s)/food constituent(s) and protection of cells from premature aging, antioxidant activity, antioxidant content and antioxidant properties and protection of DNA, proteins and lipids from oxidative damage pursuant to Article 13(1) of Regulation (EC) No 1924/20061, EFSA Panel on Dietetic Products, Nutrition and Allergies (NDA)2, 3 European Food Safety Authority (EFSA), Parma, Italy, *EFSA Journal* 2010, 8(2): 1489.

Scott J (2012). "Purple tomato debuts as 'Indigo Rose'". Oregon State University Extension Service, Corvallis. Retrieved 9 Sep 2014.

Seeram, Navindra P. (2008). "Berry Fruits: Compositional Elements, Biochemical Activities and the Impact of Their Intake on Human Health, Performance and Disease". *Journal of Agricultural and Food Chemistry* 56 (3): 627–9. doi: 10.1021/ jf071988k.

Siriwoharn T, Wrolstad RE, Finn CE, Pereira CB (2004). "Influence of cultivar, maturity and sampling on blackberry (*Rubus* L. Hybrids) anthocyanins, polyphenolics and antioxidant properties". *Journal of Agricultural and Food Chemistry* 52 (26): 8021–30.doi: 10.1021/jf048619y.

Siriwoharn, T, Wrolstad, RE, Finn, CE, Pereira, CB (2004). "Influence of cultivar, maturity and sampling on blackberry (Rubus L. Hybrids) anthocyanins, polyphenolics and antioxidant properties". *J Agric Food Chem* 52 (26): 8021–30. doi: 10.1021/jf048619y.

Stafford, Helen A. (1994). "Anthocyanins and betalains: evolution of the mutually exclusive pathways". *Plant Science* 101 (2): 91–98. doi: 10.1016/0168-9452(94)90244-5.

Stan Kailis and David Harris (2007). "The olive tree *Olea europaea*". *Producing Table Olives*. Landlinks Press. pp. 17–66. ISBN 978-0-643-09203-7.

Studies force new view on biology of flavonoids", by David Stauth, *EurekAlert!*. Adapted from a news release issued by Oregon State University

UK Food Standards Agency: "Current EU approved additives and their E Numbers". Retrieved 27 October 2011.

Vyas, P, Kalidindi, S, Chibrikova, L. (2013). "Chemical analysis and effect of blueberry and lingonberry fruits and leaves against glutamate-mediated excitotoxicity". *Journal of Agricultural and Food Chemistry* 61 (32): 7769–76. doi: 10.1021/jf401158a.

Wada L, Ou B (2002). "Antioxidant activity and phenolic content of Oregon caneberries".*Journal of Agricultural and Food Chemistry* 50 (12): 3495–500. doi: 10.1021/jf0114051.

Williams RJ, Spencer JP, Rice-Evans C, Spencer, Rice-Evans (2004). "Flavonoids: antioxidants or signalling molecules?". *Free Radical Biology and Medicine* 36 (7): 838–49.doi: 10.1016/j.freeradbiomed.2004.01.001.

Wiltshire EJ, Collings DA, Collings (2009). "New dynamics in an old friend: dynamic tubular vacuoles radiate through the cortical cytoplasm of red onion epidermal cells". *Plant and Cell Physiology* 50 (10): 1826–39. doi: 10.1093/pcp/pcp124.

Woodward, G, Kroon, P, Cassidy, A, Kay, C (2009). "Anthocyanin stability and recovery: implications for the analysis of clinical and experimental samples". *J. Agric. Food Chem.* 57(12): 5271–8. doi: 10.1021/jf900602b.

Wu X, Gu L, Prior RL, McKay S (2004). "Characterization of anthocyanins and proanthocyanidins in some cultivars of Ribes, Aronia and Sambucus and their antioxidant capacity". *Journal of Agricultural and Food Chemistry* 52 (26): 7846–56. doi: 10.1021/jf0486850.

Yonekura-Sakakibara K, Fukushima A, Nakabayashi R *et al.* (2012). "Two glycosyltransferases involved in anthocyanin modification delineated by transcriptome independent component analysis in *Arabidopsis thaliana*". *Plant J.* 69 (1): 154–67.doi: 10.1111/j.1365-313X.2011.04779.x.

Zhang, Y., Butelli, E., De Stefano, R. *et al.* (2013). "Anthocyanins Double the Shelf Life of Tomatoes by Delaying Overripening and Reducing Susceptibility to Gray Mold". *Current Biology* 23 (12): 1094. doi: 10.1016/j.cub.2013.04.072.

Chapter 24

Betalain

Betalains are a class of red and yellow indole-derived pigments found in plants of the Caryophyllales, where they replace anthocyaninpigments. Betalains also occur in some higher order fungi.They are most often noticeable in the petals of flowers, but may colour the fruits, leaves, stems and roots of plants that contain them. They include pigments such as those found in beets.

The name "betalain" comes from the Latin name of the common beet (*Beta vulgaris*), from which betalains were first extracted. The deep red colour of beets, bougainvillea, amaranth and many cactuses results from the presence of betalain pigments. The particular shades of red to purple are distinctive and unlike that of anthocyanin pigments found in most plants.

There are two categories of betalains:

☆ Betacyanins include the reddish to violet betalain pigments. Among the betacyanins present in plants include betanin, isobetanin, probetanin and neobetanin.

☆ Betaxanthins are those betalain pigments which appear yellow to orange. Among the betaxanthins present in plants includevulgaxanthin, miraxanthin, portulaxanthin and indicaxanthin.

Plant physiologists are uncertain of the function that betalains serve in those plants which possess them, but there is some preliminary evidence that they may have fungicidal properties. Furthermore, betalains have been found in fluorescent flowers.

Chemistry

It was once thought that betalains were related to anthocyanins, the reddish pigments found in most plants. Both betalains and anthocyanins are water soluble pigments found in the vacuoles of plant cells. However, betalains are structurally and chemically unlike anthocyanins and the two have never been found in the same plant together. For example, betalains contain nitrogen whereas anthocyanins do not. It is now known that betalains are aromatic indole derivatives synthesized from tyrosine. They are not related chemically to the anthocyanins and are not even flavonoids. Each betalain is a glycoside and consists of a sugar and a coloured portion. Their synthesis is promoted by light. The most heavily studied betalain is betanin, also called beetroot red after the fact that it may be extracted from red beet roots. Betanin is a glucoside and hydrolyzes into the sugar glucose and betanidin. It is used as a food colouring agent and the colour is sensitive to pH. Other betalains known to occur in beets are isobetanin, probetanin and neobetanin. The colour and antioxidant capacity of betanin and indicaxanthin (betaxanthin derived of L-proline) are affected by dielectric microwave heating. Addition of TFE (2,2,2-trifluoroethanol) is reported to improve the hydrolytic stability of some betalains in aqueous solution. Furthermore, a betanin-europium(III) complex has been used to detect calcium dipicolinate in bacterial spores, including *Bacillus anthracis* and *B. cereus*. Other important betacyanins are amaranthine and isoamaranthine, isolated from species of *Amaranthus*.

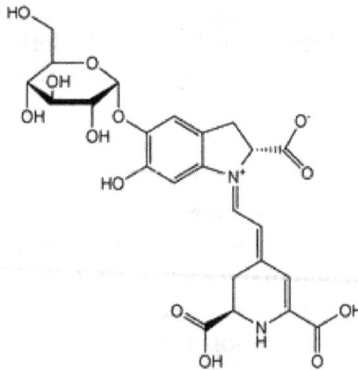

Chemical Structure of Betanin

Taxonomic Significance

Betalain pigments occur only in the Caryophyllales and some Basidiomycota (mushrooms). Where they occur in plants, they sometimes coexist with anthoxanthins (yellow to orange flavonoids), but never occur in plant species with anthocyanins. Among the flowering plant order Caryophyllales, most members produce betalains and lack anthocyanins. Of all the families in the Caryophyllales, only the Caryophyllaceae (carnation family) and Molluginaceae produce anthocyanins instead of betalains. The limited distribution of betalains among plants is a synapomorphy for the Caryophyllales, though their production has been lost in two families.

Economic Uses

Betanin is commercially used as a natural food dye. It can cause beeturia (red urine) and red feces in some people who are unable to break it down. The interest of the food industry in betalains has grown since they were identified by *in vitro* methods as antioxidants. which may protect against oxidation of low-density lipoproteins. The 'Hopi Red Dye' amaranth is rich in betacyanins and produces red flowers which the Hopi Amerindians used as the source of a deep red dye.

Semisynthetic Derivatives

Betanin extracted from the red beet was used as starting material for the semisynthesis of an artificial betalainic coumarin, which was applied as a fluorescent probe for the live-cell imaging of *Plasmodium*-infected erythrocytes.

Health Benefits

1. **Beets are widely known to help create red blood cells:** The red pigment colour of beets is from a group of phytonutrients known as ***betalains***. Betalains are antioxidant, anti-inflammatory and detoxifying agents that are richer in beets than other plant foods.

2. **Anti-Inflammatory:** The anti-inflammatory aspect of betalains helps prevent many chronic diseases and promotes cardiovascular health. Early research indicates the betalains support nerve and eye tissue better than most other antioxidants.

3. **Anti-Cancer:** Among other **health benefits of beetroot**, beet betalains provide some cancer prevention capacity, especially against colon cancer. *In vitro* (test tube or petri dish) lab tests have reported beet betalain suppression of human cancer cells. It has been observed that betalains provide a more varied and higher antioxidant value than most other vegetables containing beta-carotene.

4. **Rich in Vitamin C, Folate, Other Nutrients:** In addition to the high betalain phytonutrient content, beets are very high in vitamin C, folate (a *natural* source of folic acid), manganese, magnesium and potassium. These vitamins help bolster the immune system and support a healthy pregnancy for women, among other things. The fiber in beets is similar to carrots and supports gastrointestinal health as well.

5. **Detox:** When it comes to detoxing, certain enzymes in beet betalains stimulate glutathione production and connect toxins to glutathione molecules. Then the toxins are neutralized and excreted harmlessly.

6. **Improves Digestion:** A high fiber count makes beets a prime choice for those with digestive issues. Fiber helps to promote digestive and colon health by cleaning out the gastrointestinal system and making for regular, healthy bowel movements.

7. **Digestive markers:** Beeturia, pink or red urine occurs sometimes after consuming beets, beet soup, or beet juice. Beeturia could be an indicator of a low iron absorption at worst. A more likely occurrence is pink or red water surrounding one's stool after consuming beet preperations.

Healthy Reasons to Eat Beets

1. Weight Loss

Beets taste sweet, but a cup of cooked beet contains only 60 calories and is full of fiber. This is a perfect food for weight management. The sugars in beets are smart carbs, since they come in a natural whole food form. Unlike white sugar, the beet calories come with a lot of nutrients and phytochemicals.

2. Brain and Energy Boost

Beets are high in natural nitrates, which are converted to nitric oxide in the body. Nitric oxide is known to expand the walls of blood vessels so you can enjoy more oxygen, more nutrients and more energy. Studies have shown nitric oxide to increase the efficiency of the mitochondria (your energy powerhouses). The results of these studies were impressive.

A single small serving (70 ml) of beetroot juice reduced resting blood pressure by 2 per cent. A single small serving increased the length of time professional divers could hold their breath by 11 per cent. Cyclists who drank a single larger serving (500 ml) of beetroot juice were able to ride up to 20 per cent longer.

3. Super Antioxidant for a Long, Healthy, Pain-Free Life

Antioxidants help to reverse the daily accumulated wear and tear on the body, known as aging. Beets are a very good source of commonly known antioxidants like vitamin C and manganese, but it is their lesser-known antioxidants which give them their true value.

The blood-red colour of beets comes from a powerful group of antioxidants called betalains. There are hundreds of studies on the positive health benefits of betalains. A short summary of the results shows that they help in the areas of cancer, heart disease, diabetes and inflammation. If you go to Google Scholar and type in 'betalains,' you will see 3790 scholarly references on this subject.

4. Anti-Inflammatory Benefits

The inflammatory response is a natural function of the body which saves our lives when it responds to the acute stresses in our lives, like bacterial infection and injury. Due to the constant stress in our modern lives, however, this inflammation becomes chronic. It is as though our body is constantly in a battle. Inflammation has been linked to a number of symptoms and diseases including: Wrinkles; Susceptibility to infections; Cancer; Arthritis; Bronchitis; Chronic pain; Diabetes; High blood pressure; Osteoporosis; Heart disease; Candidiasis.

The blood-red betalains in beets have been shown to reduce chronic inflammation.

5. Cancer Prevention

Preliminary tests suggest that beetroot ingestion can be one of the useful means to prevent lung and skin cancer. Other studies have shown that beet juice inhibits the formation of cancer-causing compounds called nitrosamines.

6. Cell Detoxification and Cleansing

The antioxidants in beets have been shown to support what is called phase 2 cleansing. In phase 2 cleansing, unwanted toxic substances are chemically combined with a small nutrient group. This combination neutralizes the toxin and makes them sufficiently water-soluble so they can be excreted through the urine. This is therefore deep cleansing on a cellular level that may have long term health benefits.

7. Improved Mental Health

The betalains in beets has been used in certain treatments of depression. It also contains tryptophan, which relaxes the mind and creates a sense of well-being, similar to chocolate.

Promote Optimal Health

The pigments that give beets their rich colours are called *betalains*. There are two basic types of betalains: betacyanins and betaxanthins. Betacyanins are pigments are red-violet in colour. Betanin is the best studied of the betacyanins. Betaxanthins are yellowish in colour. In light or dark red, crimson, or purple coloured beets, betacyanins are the dominant pigments. In yellow beets, betaxanthins predominate and particularly the betaxanthin called vulgaxanthin. All betalains come from the same original molecule (betalamic acid). The addition of amino acids or amino acid derivatives to betalamic acid is what determines the specific type of pigment that gets produced. The betalain pigments in beets are water-soluble and as pigments they are somewhat unusual due to their nitrogen content. Many of the betalains function both as antioxidants and anti-inflammatory molecules. At the same time, they themselves are also very vulnerable to oxidation (change in structure due to interaction with oxygen). In addition to beets, rhubarb, chard, amaranth, prickly pear cactus and Nopal cactus are examples of foods that contain betalains.

It's interesting to note that humans appear to vary greatly in their response to dietary betalains. In the United States, only 10-15 per cent of adults are estimated to be "betalain responders." A betalain responder is a person who has the capacity to absorb and metabolize enough betalains from beet (and other foods) to gain full antioxidant, anti-inflammatory and Phase 2 triggering benefits. (Phase 2 is the second step in our cellular detoxification process).

Antioxidant Benefits

What's most striking about beets is not the fact that they are rich in antioxidants; what's striking is the unusual mix of antioxidants that they contain. We're used to thinking about vegetables as rich in antioxidant carotenoids and in particular, beta-carotene; among all well-studied carotenoids, none is more commonly occurring in vegetables than beta-carotene. When it comes to antioxidant phytonutrients that give most red vegetables their distinct colour, we've become accustomed to thinking about anthocyanins. (Red cabbage, for example, gets it wonderful red colour primarily from anthocyanins.) Beets demonstrate their antioxidant uniqueness by getting their red colour primarily from betalain antioxidant pigments (and not primarily from anthocyanins). Coupled with their status as a very good source of the antioxidant manganese and a good source of the antioxidant vitamin C, the unique phytonutrients in beets provide antioxidant support in a different way than other antioxidant-rich vegetables. While research is largely in the early stage with respect to beet antioxidants and their special benefits for eye health and overall nerve tissue health, we expect to see study results showing these special benefits and recognizing beets as a standout vegetable in this area of antioxidant support.

Anti-Inflammatory Benefits

Many of the unique phytonutrients present in beets have been shown to function as anti-inflammatory compounds. In particular, this anti-inflammatory activity has been demonstrated for betanin, isobetanin and vulgaxanthin. One mechanism allowing these phytonutrients to lessen inflammation is their ability to inhibit the activity of cyclo-oxygenase enzymes (including both COX-1 and COX-2). The COX enzymes are widely used by cells to produce messaging molecules that trigger inflammation. Under most circumstances, when inflammation is needed, this production of pro-inflammatory messaging molecules is a good thing. However, under other circumstances, when the body is undergoing chronic, unwanted inflammation, production of these inflammatory messengers can make things worse. Several types of heart disease–including atherosclerosis–are characterized by chronic unwanted inflammation. For this reason, beets have been studied within the context of heart disease and there are some encouraging although very preliminary results in this area involving animal studies and a few very small scale human studies. Type 2 diabetes–another health problem associated with chronic, unwanted inflammation–is also an area of interest in this regard, with research findings at a very preliminary stage.

In addition to their unusual betalain and carotenoid phytonutrients, however, beets are also an unusual source of betaine. Betaine is a key body nutrient made from the B-complex vitamin, choline. (Specifically, betaine is simply choline to which three methyl groups have been attached.) In and of itself, choline is a key vitamin for helping regulate inflammation in the cardiovascular system since adequate choline is important for preventing unwanted build-up of homocysteine. (Elevated levels of homocysteine are associated with unwanted inflammation and risk of cardiovascular problems like atherosclerosis.) But betaine may be even more important in regulation of our inflammatory status as its presence in our diet has been associated with lower levels of several inflammatory markers, including C reactive protein, interleukin-6 and tumor necrosis factor alpha. As a group, the anti-inflammatory molecules found in beets may eventually be shown to provide cardiovascular benefits in large-scale human studies, as well as anti-inflammatory benefits for other body systems.

Other Health Benefits

It's important to note two other areas of potential health benefits associated with beets: anti-cancer benefits and fiber-related benefits. The combination of antioxidant and anti-inflammatory molecules in beets makes this food a highly-likely candidate for risk reduction of many cancer types. Lab studies on human tumor cells have confirmed this possibility for colon, stomach, nerve, lung, breast, prostate and testicular cancers. Eventually, we expect to see large-scale human studies that show the risk-reducing effect of dietary beet intake for many of these cancer types.

Beet fiber has also been a nutrient of increasing interest in health research. While many people tend to lump all food fiber into one single category called "dietary fiber," there is evidence to suggest that all dietary fiber is not the same. Beet fiber (along with carrot fiber) are two specific types of food fiber that may provide special health benefits, particularly with respect to health of our digestive tract (including prevention of colon cancer) and our cardiovascular system. Some beet fiber benefits may be due to the pectin polysaccharides that significantly contribute to the total fiber content.

References

Augustsson K, Michaud DS, Rimm EB, (2003). A prospective study of intake of fish and marine fatty acids and prostate cancer. Cancer Epidemiol Biomarkers Prev. 2003 May,12(1)64-7.

Bartoloni, F. H., Gonçalves, L. C. P., Rodrigues, A. C. B., Dörr, F. A., Pinto, E. and Bastos, E. L. (2013). "Photophysics and hydrolytic stability of betalains in aqueous trifluoroethanol".*Monatshefte für Chemie - Chemical Monthly* 144 (4): 567–571. doi: 10.1007/s00706-012-0883-5.

Bobek P, Galbavy S, Mariassyova M. (2000). The effect of red beet (Beta vulgaris var. rubra) fiber on alimentary hypercholesterolemia and chemically induced colon carcinogenesis in rats. *Nahrung* 2000 Jun,44(3): 184-187.

Cronquist, Arthur (1981). *An Integrated System of Classification of Flowering Plants.* New York: Columbia University Press. pp. 235–9. ISBN 0-231-03880-1.

Elbandy MA and Abdelfadeil MG. (2010). Stability of betalain pigments from a red beetroot (Beta vulgaris). Poster Session Presentation. The First International Conference of Food Industries and Biotechnology and Associated Fair. Al-Baath University, North Sinai, Egypt. Available online at: www.albaath univ. edu.sy/foodex2010/connections/Posters/6.pdf.

Escribano, J., M. A. Pedreño, F. García-Carmona, R. Muñoz (1998). "Characterization of the antiradical activity of betalains from *Beta vulgaris* L. roots". *Phytochem. Anal.* 9 (3): 124–7.doi: 10.1002/(SICI)1099-1565(199805/06)9: 3<124: : AID-PCA401>3.0.CO,2-0.

Francis, F.J. (1999). *Colourants.* Egan Press. ISBN 1-891127-00-4.

Gandia-Herrero, F., Garcia-Carmona, F. and Escribano, J. (2005). Floral fluorescence effect, Nature 437, 334-334. doi: 10.1038/437334a

Gonçalves, LCP, Di Genova, BM, Dörr, FA, Pinto, E, Bastos, EL (2013). "Effect of dielectric microwave heating on the colour and antiradical capacity of betanin". *Journal of Food Engineering* 118 (1): 49–55. doi: 10.1016/j.jfoodeng. 2013.03.022.

Gonçalves, L.C.P., Da Silva, S. M., DeRose, P. ando, R. A., Bastos, E. L. (2013). "Beetroot-pigment-derived colourimetric sensor for detection of calcium dipicolinate in bacterial spores".*PLoS ONE* 8: e73701. doi: 10.1371/journal. pone.0073701.

Gonçalves, L. C. P., Trassi, M. A. D., Lopes, N. B., Dörr, F. A., Santos, M. T., Baader, W. J., Oliveira, V. X. and Bastos, E. L. (2012). "A comparative study of the purification of betanin". *Food Chem.* 131: 231–238. doi: 10.1016/j.foodchem. 2011.08.067.

Gonçalves LCP, Tonelli RR, Bagnaresi P, Mortara RA, Ferreira AG, Bastos EL (2013). Sauer, Markus, ed. "A Nature-Inspired Betalainic Probe for Live-Cell Imaging of Plasmodium-Infected Erythrocytes". *PLoS ONE* 8 (1): e53874. doi: 10.1371/ journal. pone. 0053874.

Kimler, L. M. (1975). "Betanin, the red beet pigment, as an antifungal agent". *Botanical Society of America, Abstracts of papers* 36.

Lee CH, Wettasinghe M, Bolling BW (2005). Betalains, phase II enzyme-inducing components from red beetroot (*Beta vulgaris* L.) extracts. *Nutr Cancer.* 53(1): 91-103.

Lucarini M, Lanzi S, D'Evoli L (2006). Intake of vitamin A and carotenoids from the Italian population–results of an Italian total diet study. *Int J Vitam Nutr Res.* May, 76(3): 103-9.

Raven, Peter H., Ray F. Evert, Susan E. Eichhorn (2004). *Biology of Plants* (7th ed.). New York: W. H. Freeman and Company. p. 465. ISBN 0-7167-1007-2.

Reddy MK, Alexander-Lindo RL and Nair MG. (2005). Relative inhibition of lipid peroxidation, cyclooxygenase enzymes and human tumor cell proliferation by natural food colours. *J Agric Food Chem.* Nov 16,53(23): 9268-73.

Renner-Nance J. (2009). Improving the stability and performance of naturally derived colour additives. DD Williamson Support Center Presentation, Louisville, KY, June 8. Available online at: www.naturalcolours.com/./File/ naturalcolours/./Session139-02.pdf.

Robinson, Trevor (1963). *The Organic Constituents of Higher Plants.* Minneapolis: Burgess Publishing. p. 292.

Salisbury, Frank B., Cleon W. Ross (1991). *Plant Physiology* (4th ed.). Belmont, California: Wadsworth Publishing. pp. 325–326. ISBN 0-534-15162-0.

Song W, Derito CM, Liu MK (2010). Cellular antioxidant activity of common vegetables. *J Agric Food Chem.* Jun 9, 58(11): 6621-9.

Stafford, Helen A. (1994). "Anthocyanins and betalains: evolution of the mutually exclusive pathways". *Plant Science* 101 (2): 91–98. doi: 10.1016/0168-9452(94)90244-5.ISSN 0168-9452. Retrieved 20 May 2013.

Strack D, Vogt T, Schliemann W (2003). "Recent advances in betalain research". *Phytochemistry* 62 (3): 247–69. doi: 10.1016/S0031-9422(02)00564-2.

Tesoriere, Luisa, Mario Allegra, Daniela Butera, Maria A. Livrea (2004). "Absorption, excretion and distribution of dietary antioxidant betalains in LDLs: potential health effects of betalains in humans". *American Journal of Clinical Nutrition* 80 (4): 941–5.

Chapter 25

Bioluminescence

Bioluminescence is the production and emission of light by a living organism. It is a form of chemiluminescence. Bioluminescence occurs widely in marine vertebrates and invertebrates, as well as in some fungi, microorganisms including some bioluminescent bacteria and terrestrial invertebrates such as fireflies. In some animals, the light is produced by symbiotic organisms such as *Vibrio* bacteria. The principal chemical reaction in bioluminescence involves the light-emitting pigment luciferin and the enzyme luciferase, assisted by other proteins such as aequorin in some species. The enzyme catalyzes the oxidation of luciferin. In some species, the type of luciferin requires cofactors such as calcium or magnesium ions and sometimes also the energy-carrying molecule adenosine triphosphate (ATP). In evolution, luciferins vary little: one in particular, coelenterazine, is found in nine different animals (phyla), though in some of these, the animals obtain it through their diet. Conversely, luciferases vary widely in different species. Bioluminescence has arisen over forty times inevolutionary history.

Both Aristotle and Pliny the Elder mentioned that damp wood sometimes gives off a glow and many centuries later Robert Boyle showed that oxygen was involved in the process, both in wood and in glow-worms. It was not until the late nineteenth century that bioluminescence was properly investigated. The phenomenon is widely distributed among animal groups, especially in marine environments wheredinoflagellates cause phosphorescence in the surface layers of water. On land it occurs in fungi, bacteria and some groups of invertebrates, including insects.

The uses of bioluminescence by animals include counter-illumination camouflage, mimicry of other animals, for example to lure prey and signallingto other individuals of the same species, such as to attract mates. In the laboratory, luciferase-based systems are used in genetic engineering and for biomedical research. Other researchers are investigating the possibility of using bioluminescent systems for street and decorative lighting and a bioluminescent plant has been created.

History

Before the development of the safety lamp for use in coal mines, dried fish skins were used in Britain and Europe as a weak source of light. This experimental form of illumination avoided the necessity of using candles which risked sparking explosions of firedamp. Another safe source of illumination in mines was bottles containing fireflies. In 1920, the American zoologist E. Newton Harvey published a monograph, *The Nature of Animal Light*, summarizing early work on bioluminescence. Harvey notes that Aristotle mentions light produced by dead fish and flesh and that both Aristotle and Pliny the Elder (in his *Natural History*) mention light from damp wood. He also records that Robert Boyle experimented on these light sources and showed that both they and the glow-worm require air for light to be produced. Harvey notes that in 1753, J. Baker identified the flagellate *Noctiluca* "as a luminous animal" "just visible to the naked eye" and in 1854 Johann Florian Heller (1813-1871) identified strands (hyphae) of fungi as the source of light in dead wood. While sailing in these latitudes on one very dark night, the sea presented a wonderful and most beautiful spectacle. There was a fresh breeze and every part of the surface, which during the day is seen as foam, now glowed with a pale light. The vessel drove before her bows two billows of liquid phosphorus and in her wake she was followed by a milky train. As far as the eye reached, the crest of every wave was bright and the sky above the horizon, from the reflected glare of these livid flames, was not so utterly obscure, as over the rest of the heavens.

Darwin also observed a luminous "jelly-fish of the genus Dianaea" and noted that "When the waves scintillate with bright green sparks, I believe it is generally owing to minute crustacea. But there can be no doubt that very many other pelagic animals, when alive, are phosphorescent." He guessed that "a disturbed electrical condition of the atmosphere" was probably responsible. Daniel Pauly comments that Darwin "was lucky with most of his guesses, but not here", noting that biochemistry was too little known and that the complex evolution of the marine animals involved "would have been too much for comfort".

Bioluminescence attracted the attention of the United States Navy in the Cold War, since submarines in some waters can create a bright enough wake to be detected; a German submarine was sunk in the First World War, having been detected in this way. The navy was interested in predicting when such detection would be possible and hence guiding their own submarines to avoid detection.

Among the anecdotes of navigation by bioluminescence, the Apollo 13 astronaut Jim Lovell recounted how as a navy pilot he had found his way back to his aircraft carrier *USS Shangri-La* when his navigation systems failed. Turning off his cabin lights, he saw the glowing wake of the ship and was able to fly to it and land safely.

The French pharmacologist Raphaël Dubois carried out work on bioluminescence in the late nineteenth century. He studied click beetles(*Pyrophorus*) and the marine bivalve mollusk *Pholas dactylus*. He refuted the old idea that bioluminescence came from phosphorus and demonstrated that the process was related to the oxidation of a specific compound, which he named luciferin, by an enzyme. He sent Harvey siphons from the mollusc preserved in sugar. Harvey had become interested in bioluminescence as a result of visiting the South Pacific and Japan and observing phosphorescent organisms there. He studied the phenomenon for many years. His research aimed to demonstrate that luciferin and the enzymes that oxidise it to produce light, were interchangeable between species, showing that all bioluminescent organisms had a common ancestor. However, he found this hypothesis to be false, with different organisms having major differences in the composition of their light-producing proteins. He spent the next thirty years purifying and studying the components, but it fell to the young Japanese chemist Osamu Shimomura to be the first to obtain crystalline luciferin. He used the sea firefly *Vargula hilgendorfii*, but it was another ten years before he discovered the chemical's structure and was able to publish his 1957 paper *Crystalline Cypridina Luciferin*. More recently, Martin Chalfie, Osamu Shimomura and Roger Y. Tsien won the 2008 Nobel Prize in Chemistry for their 1961 discovery and development of green fluorescent protein as a tool for biological research.

Chemical Mechanism

Bioluminescence is a form of chemiluminescence where light energy is released by a chemical reaction. Fireflies, anglerfish and other organisms produce the light-emitting pigment luciferin and the enzyme luciferase. Luciferin reacts with oxygen to create light:

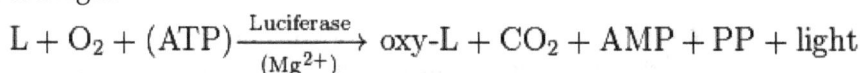

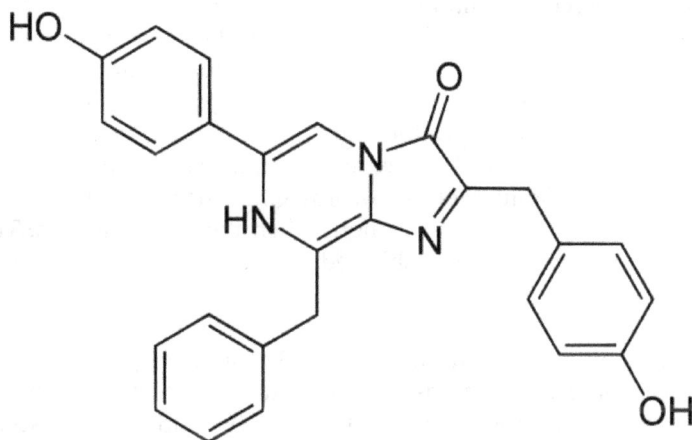

$$L + O_2 + (ATP) \xrightarrow[(Mg^{2+})]{Luciferase} oxy\text{-}L + CO_2 + AMP + PP + light$$

Coelenterazine is a luciferin found in many different marine phyla from comb jellies to vertebrates. Like all luciferins, it is oxidised to produce light.

Carbon dioxide (CO_2), adenosine monophosphate (AMP) and phosphate groups (PP) are released as waste products. Luciferasecatalyzes the reaction, which may be mediated by cofactors such as calcium (Ca^{2+}) or magnesium (Mg^{2+}) ions and for some types of luciferin (L) also the energy-carrying molecule adenosine triphosphate (ATP). The reaction can occur either inside or outside the cell. In bacteria such as *Vibrio*, the expression of genes related to bioluminescence is controlled by an operon called the lux operon. In evolution, luciferins generally vary little: one in particular, coelenterazine, is the light emitting pigment for nine ancient phyla (groups of very different organisms), including polycystine radiolaria, Cercozoa (Phaeodaria), protozoa, comb jellies, cnidaria including jellyfish andcorals, crustaceans, molluscs, arrow worms and vertebrates (ray-finned fish). Not all these organisms synthesize coelenterazine: some of them obtain it through their diet. Conversely, luciferase enzymes vary widely and tend to be different in each species. Overall, bioluminescence has arisen over forty times in evolutionary history.

Luciferin-luciferase reactions are not the only way that organisms produce light. The parchment worm *Chaetopterus* (a marine Polychaete) makes use of another photoprotein, aequorin, instead of luciferase. When calcium ions are added, the aequorin's rapid catalysis creates a brief flash quite unlike the prolonged glow produced by luciferase. In a second, much slower, step luciferin is regenerated from the oxidised (oxyluciferin) form, allowing it to recombine with aequorin, in readiness for a subsequent flash. Photoproteins are thus enzymes, but with unusual reaction kinetics. In the hydrozoan jellyfish *Aequorea victoria*, some of the blue light released by aequorin in contact with calcium ions is absorbed by green fluorescent protein; it in turn releases green light.

Uses in Nature

Bioluminescence has several functions in different taxa. Haddock *et al.* (2010) list as more or less definite functions in marine organisms the following: defensive functions of startle, counterillumination (camouflage), misdirection (smoke screen), distractive body parts, burglar alarm (making predators easier for higher predators to see) and warning to deter settlers; offensive functions of lure, stun or confuse prey, illuminate prey and mate attraction/recognition. It is much easier for researchers to detect that a species is able to produce light than to analyse the chemical mechanisms or to prove what function the light serves. In some cases the function is unknown, as with species in three families of earthworm (Oligochaeta), such as *Diplocardia longa* where the coelomic fluid produces light when the animal moves. The following functions are reasonably well established in the named organisms.

Defence

Many cephalopods, including at least 70 genera of squid, are bioluminescent. Some squid and small crustaceans use bioluminescent chemical mixtures or bacterial slurries in the same way as many squid use ink. A cloud of luminescent material is expelled, distracting or repelling a potential predator, while the animal escapes to safety. The deep sea squid *Octopoteuthis deletron* may autotomise portions of its arms which are luminous and continue to twitch and flash, thus distracting a predator while the animal flees. Dinoflagellates may use bioluminescence for defence against predators. They shine when they detect a predator, possibly making the predator itself more vulnerable by attracting the attention of predators from higher trophic levels. Grazing copepods release any phytoplankton cells that flash, unharmed; if they were eaten they would make the copepods glow, attracting predators, so the phytoplankton's bioluminescence is defensive. The problem of shining stomach contents is solved (and the explanation corroborated) in predatory deep-sea fishes: their stomachs have a black lining able to keep the light from any bioluminescent fish prey which they have swallowed from attracting larger predators. The sea-firefly is a small crustacean living in sediment. At rest it emits a dull glow but when disturbed it darts away leaving a cloud of shimmering blue light to confuse the predator. During World War II it was gathered and dried for use by the Japanese military as a source of light during clandestine operations. The larvae of railroad worms (*Phrixothrix*) have paired photic organs on each body segment, able to glow with green light; these are thought to have a defensive purpose. They also have organs on the head which produce red light; they are the only terrestrial organisms to emit light of this colour.

Communication

Communication in the form of quorum sensing plays a role in the regulation of luminesence in many species of bacteria. Small extra cellularly secreted molecules stimulate the bacteria to turn on genes for light production when cell density, measured by concentration of the secreted molecules, is high. Pyrosomes are colonial tunicates and each zooid has a pair of luminescent organs on either side of the inlet siphon. When stimulated by light, these turn on and off, causing rhythmic flashing. No neural pathway runs between the zooids, but each responds to the light produced by other individuals and even to light from other nearby colonies. Communication by light emission between the zooids enables coordination of colony effort, for example in swimming where each zooid provides part of the propulsive force. Some bioluminous bacteria infect nematodes that parasitize Lepidoptera larvae. When these caterpillars die, their luminosity may attract predators to the dead insect thus assisting in the dispersal of both bacteria and nematodes. A similar reason may account for the many species of fungi that emit light. Species in the genera *Armillaria*, *Mycena*, *Omphalotus*, *Panellus*, *Pleurotus* and others do this, emitting usually greenish light from the mycelium, cap and gills. This may attract night-flying insects and aid in spore dispersal, but other functions may also be involved. *Quantula striata* is the only known bioluminescent terrestrial mollusc. Pulses of light are emitted from a gland near the front of the foot and may have a communicative function, although the adaptive significance is not fully understood.

Biotechnology

Biology and Medicine

Bioluminescent organisms are a target for many areas of research. Luciferase systems are widely used in genetic engineering as reporter genes, each producing a different colour by fluorescence and for biomedical research using bioluminescence imaging. For example, the firefly luciferase gene was used as early as 1986 for research using transgenic tobacco plants. *Vibrio* bacteria symbiose with marine invertebrates such as the Hawaiian bobtail squid (*Euprymna scolopes*), are key experimental models for bioluminescence. Bioluminescent activated destruction is an experimental cancer treatment.

Light Production

The structures of photophores, the light producing organs in bioluminescent organisms, are being investigated by industrial designers. Engineered bioluminescence could perhaps one day be used to reduce the need for street lighting, or for decorative purposes if it becomes possible to produce light that is both bright enough and can be sustained for long periods at a workable price. The gene that makes the tails of fireflies glow has been added to mustard plants. The plants glow faintly for an hour when touched, but a sensitive camera is needed to see the glow. University of Wisconsin–Madison is researching the use of genetically engineered bioluminescent *E. coli* bacteria, for use asbioluminescent bacteria in a light bulb. In June 2013 the Glowing Plant project raised nearly $500,000 on the crowdfunding site Kickstarter to create a bioluminescent plant. An iGEM team from Cambridge (England) has started to address the problem that luciferin is consumed in the light-producing reaction by developing a genetic biotechnology part that codes for a luciferin regenerating enzyme from the North American firefly; this enzyme "helps to strengthen and sustain light output".

References

Altura, M.A., Heath-Heckman, E.A., Gillette, A., Kremer, N., Krachler, A.M., Brennan, C., Ruby, E.G., Orth, K., McFall-Ngai, M.J. (2013). "The first engagement of partners in the *Euprymna scolopes-Vibrio fischeri* symbiosis is a two-step process initiated by a few environmental symbiont cells".*Environmental Microbiology.* doi: 10.1111/1462-2920.12179.

Bioluminescence Questions and Answers. Siobiolum.ucsd.edu. Retrieved on 20 October 2011.

Bone, Quentin, Moore, Richard (2008). *Biology of Fishes.* Taylor and Francis. pp. 8: 110–111. ISBN 978-1-134-18630-3.

Bowlby, Mark R., Edith Widder, James Case (1990)."Patterns of stimulated bioluminescence in two pyrosomes (Tunicata: Pyrosomatidae)". *Biological Bulletin* 179 (3). Retrieved 2011-11-12.

Branham, Marc. "Glow-worms, railroad-worms (Insecta: Coleoptera: Phengodidae)". *Featured Creatures.* University of Florida. Retrieved 29 November 2014.

Broadley, R., Stringer, I. (2009). "Larval behaviour of the New Zealand glowworm, *Arachnocampa luminosa* (Diptera: Keroplatidae), in bush and caves". *Kerala: Research Signpost. Bioluminescence in focus,* p. 325–355.

Cha, Ariana Eunjung, "Glowing plants spark environmental debate" The Seattle Times 5 October 2013.

Copeland J., Daston, M. M. (1989). "Bioluminescence in the terrestrial snail *Quantula (Dyakia) striata".Malacologia* 30 (1–2): 317–324.

Darwin, Charles (1839). *Narrative of the surveying voyages of His Majesty's Ships Adventure and Beagle between the years 1826 and 1836, describing their examination of the southern shores of South America and the Beagle's circumnavigation of the globe. Journal and remarks. 1832-1836.* Henry Colburn. pp. 190–192.

Deheyn, Dimitri D., Wilson, Nerida G. (2010). "Bioluminescent signals spatially amplified by wavelength-specific diffusion through the shell of a marine snail". *Proceedings of the Royal Society*.doi: 10.1098/rspb.2010.2203.

Di Rocco, Giuliana, Gentile, Antonietta, Antonini, Annalisa, Truffa, Silvia, Piaggio, Giulia, Capogrossi, Maurizio C., Toietta, Gabriele (2012). "Analysis of biodistribution and engraftment into the liver of genetically modified mesenchymal stromal cells derived from adipose tissue". *Cell Transplantation* 21 (9): 1997–2008.doi: 10.3727/096368911X637452.

Douglas, R.H., Mullineaux, C.W., Partridge, J.C. (2000). "Long-wave sensitivity in deep-sea stomiid dragonfish with far-red bioluminescence: evidence for a dietary origin of the chlorophyll-derived retinal photosensitizer of Malacosteus niger". *Philosophical Transactions of the Royal Society B* 355 (1401): 1269–1272. doi: 10.1098/rstb.2000.0681.

Dr. Chris Riley, "Glowing plants reveal touch sensitivity", BBC 17 May 2000.

Encyclopedia of the Aquatic World. Marshall Cavendish. January 2004. p. 1115. ISBN 978-0-7614-7418-0.

Fordyce, William (1973). *A history of coal, coke and coal fields and the manufacture of iron in the North of England*. Graham.

Freese, Barbara (2006). *Coal: A Human History*. Arrow. p. 51. ISBN 978-0-09-947884-3.

Fulcher, Bob. "Lovely and Dangerous Lights".*Tennessee Conservationist Magazine*. Retrieved28 November 2014.

Greven, Hartmut, Zwanzig, Nadine (2013). "Courtship, Mating and Organisation of the Pronotum in the Glowspot Cockroach *Lucihormetica verrucosa* (Brunner von Wattenwyl, 1865) (Blattodea: Blaberidae)". *Entomologie heute* 25: 77–97.

Haddock, Steven H.D., Moline, Mark A., Case, James F. (2010). "Bioluminescence in the Sea".*Annual Review of Marine Science* 2: 443–493.doi: 10.1146/annurev-marine-120308-081028.

Harvey cites this as Baker, J.: 1743-1753, *The Microscope Made Easy and Employment for the Microscope*.

Harvey, E. Newton (1920). *The Nature of Animal Light*. Philadelphia and London: J. B. Lippencott. Page 1.

Hastings, J.W. (1983). "Biological diversity, chemical mechanisms and the evolutionary origins of bioluminescent systems". *J. Mol. Evol.* 19 (5): 309–21. doi: 10.1007/BF02101634. ISSN 1432-1432.

How illuminating. The Economist. 10 March 2011. Retrieved 6 December 2014.

Huth, John Edward (2013). *The Lost Art of Finding Our Way*. Harvard University Press. p. 423. ISBN 978-0-674-07282-4.

Kirkwood, Scott (2005). "Park Mysteries: Deep Blue". *National Parks Magazine* (National Parks Conservation Association). pp. 20–21. ISSN 0276-8186. Retrieved 14 June 2010.

Koo, J., Kim, Y., Kim, J., Yeom, M., Lee, I. C., Nam, H. G. (2007). "A GUS/Luciferase Fusion Reporter for Plant Gene Trapping and for Assay of Promoter Activity with Luciferin-Dependent Control of the Reporter Protein Stability". *Plant and Cell Physiology* 48 (8): 1121–31.doi: 10.1093/pcp/pcm081.

Ludwig Institute for Cancer Research (2003)."Firefly Light Helps Destroy Cancer Cells, Researchers Find That The Bioluminescence Effects Of Fireflies May Kill Cancer Cells From Within". Science Daily. Retrieved4 December 2014.

Luminescence. Encyclopaedia Britannica. Retrieved16 December 2014.

Marek, Paul, Papaj, Daniel, Yeager, Justin, Molina, Sergio, Moore, Wendy. "Bioluminescent aposematism in millipedes". *Current Biology* 21 (18): R680–R681. doi: 10.1016/j.cub.2011.08.012.

Martin, R. Aidan. "Biology of Sharks and Rays: Cookiecutter Shark". ReefQuest Centre for Shark Research. Retrieved 13 March 2013.

Merritt, David J. (2013). "Standards of evidence for bioluminescence in cockroaches". *Naturwissenschaften*100 (7): 697–698.

Meyer-Rochow, V. B., Moore, S. (1988). "Biology of *Latia neritoides* Gray 1850 (Gastropoda, Pulmonata, Basommatophora): the Only Light-producing Freshwater Snail in the World". *Internationale Revue der gesamten Hydrobiologie und Hydrographie* 73 (1): 21–42.doi: 10.1002/iroh.19880730104.

Milius, S. (1998). "Glow-in-the-dark shark has killer smudge". Science News. Retrieved 13 March 2013.

Morise, H., Shimomura, O, Johnson, F.H., Winant, J., Shimomura, Johnson, Winant (1974). "Intermolecular energy transfer in the bioluminescent system of Aequorea". *Biochemistry* 13 (12): 2656–62.doi: 10.1021/bi00709a028.

Nic Halverson (2013). "Bacteria-Powered Light Bulb Is Electricity-Free".

Nordgren, I. K., Tavassoli, A. (2014). "A bidirectional fluorescent two-hybrid system for monitoring protein-protein interactions". *Molecular BioSystems* 10 (3): 485–490.doi: 10.1039/c3mb70438f.

Ow, D.W., Wood, K.V., DeLuca, M., de Wet, J.R., Helinski, D.R., Howell, S.H. (1986). "Transient and stable expression of the firefly luciferase gene in plant cells and transgenic plants". *Science* 234 (4778) (American Association for the Advancement of Science). pp. 856—856. ISSN 0036-8075.

Pauly, Daniel (2004). *Darwin's Fishes: An Encyclopedia of Ichthyology, Ecology and Evolution*. Cambridge University Press. pp. 15–16. ISBN 978-1-139-45181-9.

Pieribone, Vincent, Gruber, David F. (2005). *Aglow in the Dark: The Revolutionary Science of Biofluorescence*. Harvard University Press. pp. 35–41. ISBN 978-0-674-01921-8.

Poisson, Jacques (2010). "Raphaël Dubois, from pharmacy to bioluminescence". *Rev Hist Pharm (Paris)* (in French) (France) 58 (365): 51–56. ISSN 0035-2349.

Reshetiloff, Kathy (2001). "Chesapeake Bay night-lights add sparkle to woods, water". *Bay Journal*. Retrieved 16 December 2014.

Ross, Alison (2005). "'Milky seas' detected from space". BBC. Retrieved 13 March 2013.

Shimomura O, Johnson FH, Saiga Y, Johnson, Saiga (1962). "Extraction, purification and properties of aequorin, a bioluminescent protein from the luminous hydromedusan, Aequorea". *J Cell Comp Physiol* 59 (3): 223–39. doi: 10.1002/jcp.1030590302.

Shimomura O. (1995). "A short story of aequorin". *The Biological bulletin* 189 (1): 1–5.doi: 10.2307/1542194.

Shimomura, O., Johnson, F.H. (1975). "Regeneration of the photoprotein aequorin". *Nature* 256 (5514): 236–238.doi: 10.1038/256236a0.

Smiles, Samuel (1862). *Lives of the Engineers*. Volume III (George and Robert Stephenson). London: John Murray. p. 107. ISBN 0-7153-4281-9

Stanger-Hall, K.F., Lloyd, J.E., Hillis, D.M. (2007). "Phylogeny of North American fireflies (Coleoptera: Lampyridae): implications for the evolution of light signals".*Molecular Phylogenetics and Evolution* 45 (1): 33–49.doi: 10.1016/j.ympev.2007.05.013.

Sullivan, Rachel. "Out of the darkness". ABC Science. Retrieved 17 December 2014.

The Nobel Prize in Chemistry 2008. 8 October 2008. Retrieved 23 November 2014.

Thomas Eisner, Michael A. Goetz, David E. Hill, Scott R. Smedley, Jarrold Meinwald (1997). "Firefly "femmes fatales" acquire defensive steroids (lucibufagins) from their firefly prey". *Proceedings of the National Academy of Sciences of the United States of America* 94 (18): 9723–9728.

Tong, D, Rozas, N.S., Oakley, T.H., Mitchell, J., Colley, N.J., McFall-Ngai, M.J. (2009). "Evidence for light perception in a bioluminescent organ". *Proceedings of the National Academy of Sciences of the United States of America* 106(24): 9836–41. Bibcode: 2009PNAS.106.9836T.doi: 10.1073/pnas.0904571106.

Viviani, Vadim (2009-02-17). "Terrestrial bioluminescence". Retrieved 2014-11-26.

Viviani, Vadim R., Bechara, Etelvino J.H. (1997)."Bioluminescence and Biological Aspects of Brazilian Railroad-Worms (Coleoptera: Phengodidae)". *Annals of the Entomological Society of America* 90 (3): 389–398.

Xiong, Yan Q., Willard, Julie, Kadurugamuwa, Jagath L., Yu, Jun, Francis, Kevin P., Bayer, Arnold S. (2004). "Real-Time *in vivo* Bioluminescent Imaging for Evaluating the Efficacy of Antibiotics in a Rat *Staphylococcus aureus* Endocarditis Model". *Antimicrobial Agents and Chemotherapy* 49 (1): 380–7. doi: 10.1128/AAC.49.1.380-387.2005.

Young, R.E., Roper, C.F. (1976). "Bioluminescent countershading in midwater animals: evidence from living squid". *Science* 191 (4231): 1046–8.Bibcode: 1976Sci.191.1046Y.doi: 10.1126/science.1251214.

Young, Richard Edward (1983). "Oceanic Bioluminescence: an Overview of General Functions". *Bulletin of Marine Science* 33 (4): 829–845.

Zhao, Dawen, Richer, Edmond, Antich, Peter P., Mason, Ralph P. (2008). "Antivascular effects of combretastatin A4 phosphate in breast cancer xenograft assessed using dynamic bioluminescence imaging and confirmed by MRI". *The FASEB Journal* 22 (7): 2445–51.doi: 10.1096/fj.07-103713.

Chapter 26

Carotenoids

Carotenoids, the colourful plant pigments some of which the body can turn into vitamin A, are powerful antioxidants that can help prevent some forms of cancer and heart disease and act to enhance your immune response to infections. These precursors to vitamin A are sometimes called provitamin A. Bright-orange beta-carotene is the most important carotenoid for adequate vitamin A intake because it yields more vitamin A than alpha- or gamma-carotene.

Some carotenoids, such as lycopene, do not convert to vitamin A at all. Lycopene, the orange-red pigment found in tomatoes and watermelon, is still of value, however, because it's an antioxidant even more potent than beta-carotene. The other carotenoids are also valuable antioxidants. Antioxidants help the body reduce the inflammatory action of singlet or free-radical oxygen. Oxygen atoms like to combine into pairs. Singlet oxygen atoms are unstable and interact with the lipids found in cell walls causing inflammation and damage. Sometimes, your own body uses these free radicals to fight infections and abnormal cells. Most of the time, these free radicals cause inflammation and damage to cells, such as those that line your arteries. Orange and yellow fruits and vegetables have high vitamin A activity because of the carotenoids they contain. Generally, the deeper the colour of the fruit or vegetable is an indication of a higher concentration of carotenoids. Carrots, for example, are especially good sources of beta-carotene. Green, leafy vegetables such as spinach, asparagus and broccoli also contain large amounts of carotenoids, but their intense green pigment, courtesy of chlorophyll, masks the tell-tale orange-yellow colour.

Most other carotenoids, such as alpha- and gamma-carotene, plus cryptoxanthin and beta-zeacarotene have less vitamin A activity than beta-carotene, but offer ample cancer prevention. Some carotenoids, such as lycopene, zeaxanthin, lutein, capsanthin and canthaxanthin are not converted into vitamin A in the body. But again, they are powerful cancer fighters, prevalent in fruits and vegetables. There is abundant evidence that lycopene in particular helps reduce the risk for prostate cancer.

Carotenes are valuable preventive medicines, too. Research shows that people who eat a lot of foods rich in beta-carotene – the carotenoid with the greatest vitamin A value – are less likely to develop lung cancer. Even among smokers, lung cancer is less likely to occur in those people who eat a diet that includes lots of vegetables and fruits containing beta-carotene. Taking a beta-carotene supplement in pill form does not always have the same effect, however. Perhaps this is because in these foods there may be other substances that offer protection as well. In three studies involving 69,000 participants, many of them smokers, beta-carotene supplements increased the rate of lung cancer. Lutein/zeaxanthin, lycopene and alpha-carotene show evidence of being significantly more protective against lung and some other cancers.

Many experts now believe that the protective effect of carotenoids depends on the timing of when you take them. If you take betacarotene before your cells have undergone any pre-cancerous changes, the antioxidant action of the carotenoids can help reduce the likelihood that any mutations will take place. The carotenoids at this point can prevent free radicals from damaging cells and the DNA inside of cells, both of which can start cancerous growth. But if you take supplemental betacarotene AFTER cells have already mutated, the beta-carotene may protect the mutated cells from being destroyed by your own immune system.

Some of the most potent cancer fighting cells in your body utilizes free radicals to fight infections and to destroy precancerous cells. So eating foods high in carotenoids or taking supplements helps with what is called primary protection from cancer - cancer never gets started. But after you have a growing colony of cancerous cells in your system, betacarotene supplements may prevent your own system from fighting the cancer, making carotenoid supplementation significantly less safe for what is called secondary prevention - stopping a recurrence of cancer.

Carotenoids are organic pigments that are found in the chloroplasts and chromoplasts of plants and some other photosynthetic organisms, including some bacteria and some fungi. Carotenoids can be produced from fats and other basic organic metabolic building blocks by all these organisms. Carotenoids can also be sourced from various animal tissues; as is often the case involving food grade additives and within the dietary supplement industry in the United States. Carniverous animal life sources Carotenoids from the consumption of animal tissue. The only animal known to produce Carotenoids is the aphid. There are over 600 known carotenoids; they are split into two classes, xanthophylls (which contain oxygen) and carotenes (which are purely hydrocarbons and contain no oxygen). All carotenoids are tetraterpenoids, meaning that they are produced from 8 isoprenemolecules and contain 40 carbon atoms. In general, carotenoids absorb wavelengths ranging from 400-550 nanometers (violet to green light). They serve two key roles in plants and algae: they absorb light energy for use in photosynthesis and they protect chlorophyll from photodamage. In humans, three carotenoids (beta-carotene, alpha-carotene and beta-cryptoxanthin) have vitamin A activity (meaning that they can be converted to retinal) and these and other carotenoids can also act as antioxidants. In the eye, certain other carotenoids (lutein, astaxanthin and zeaxanthin) apparently act directly to absorb damaging blue and near-ultraviolet light, in order to protect themacula of the retina, the part of the eye with the sharpest vision.

People consuming diets rich in carotenoids from natural foods, such as fruits and vegetables, are healthier and have lower mortality from a number of chronic illnesses.[3] Although a recent meta-analysis of 68 reliable antioxidant supplementation experiments involving a total of 232,606 individuals concluded additional β-carotene from supplements is unlikely to be beneficial and may actually be harmful, this may be due to the inclusion of studies involving smokers - β-carotene under intense oxidative stress (*e.g.* induced by heavy smoking) gives breakdown products that reduce plasma vitamin A and worsen the lung cell proliferation induced by smoke. With the notable exception of the gac fruit and crude palm oil, most carotenoid-rich fruits and vegetables are low in lipids. Since dietary lipids have been hypothesized to be an important factor for carotenoid bioavailability, a 2005 study investigated whether addition of avocado fruit or oil, as lipid sources, would enhance carotenoid absorption in humans. The study found that the addition of either avocado fruit or oil significantly enhanced the subjects' absorption of all carotenoids tested (α-carotene, β-carotene, lycopene and lutein).

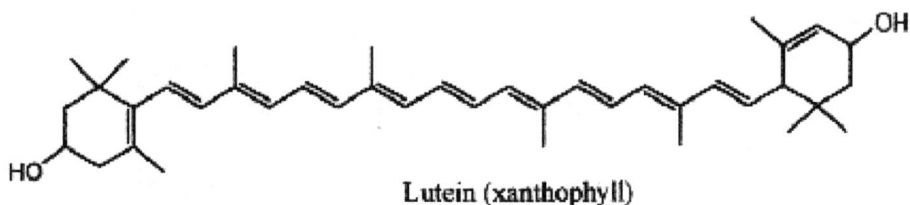

Lutein (xanthophyll)

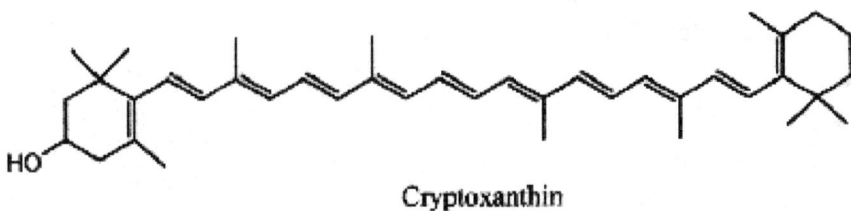

Cryptoxanthin

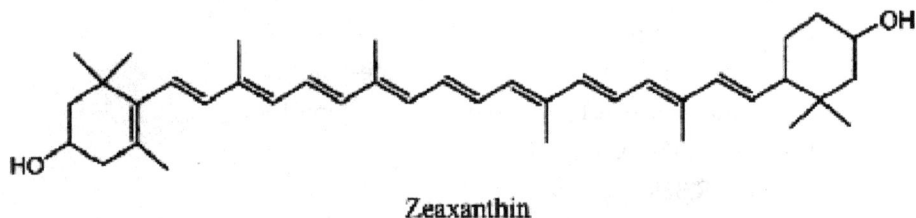

Zeaxanthin

Biosynthesis

CRT is the gene cluster responsible for the biosynthesis of carotenoids.

geranylgeranyl-PP

phytoene synthase

phyoene+ 2 pyrophosphate

crt1

zeta-carotene

crt1

lycopene

lycopene cyclase

alpha-carotene   beta-carotene

Simplified Carotenoid Synthesis Pathway

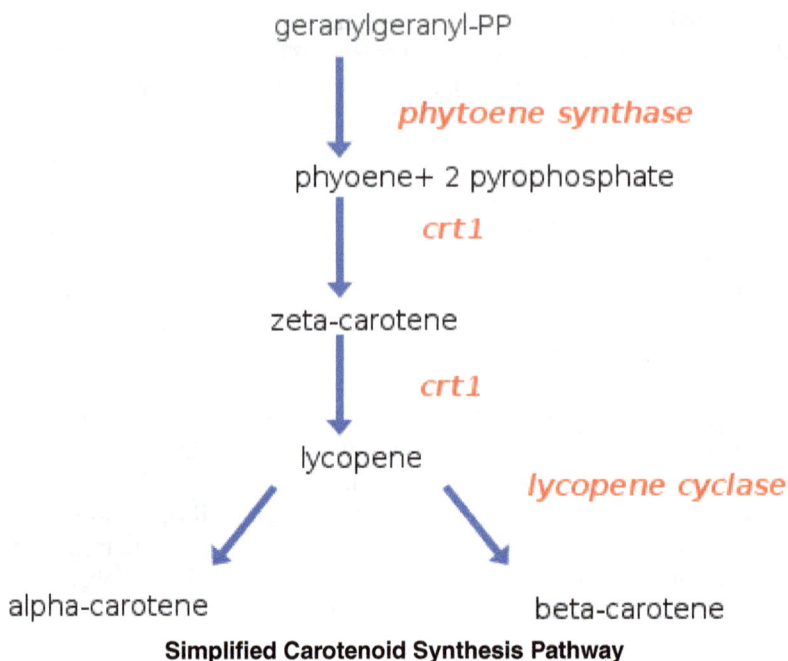

Hydrocarbons

- ☆ Lycopersene 7,8,11,12,15,7',8',11',12',15'-Decahydro-γ,γ-carotene
- ☆ Phytofluene
- ☆ Hexahydrolycopene 15-*cis*-7,8,11,12,7',8'-Hexahydro-γ,γ-carotene
- ☆ Torulene 3',4'-Didehydro-β,γ-carotene
- ☆ α-Zeacarotene 7',8'-Dihydro-ε,γ-carotene

Alcohols

- ☆ Alloxanthin
- ☆ Cynthiaxanthin
- ☆ Pectenoxanthin
- ☆ Cryptomonaxanthin (3R,3'R)-7,8,7',8'-Tetradehydro-β,β-carotene-3,3'-diol
- ☆ Crustaxanthin β,-Carotene-3,4,3',4'-tetrol
- ☆ Gazaniaxanthin (3R)-5'-cis-β,γ-Carotene-3-ol
- ☆ OH-Chlorobactene 1',2'-Dihydro-f,γ-carotene-1'-ol
- ☆ Loroxanthin β,ε-Carotene-3,19,3'-triol
- ☆ Lutein (3R,32R,62R)-β,ε-carotene-3,3-diol
- ☆ Lycoxanthin γ,γ-Carotene-16-ol
- ☆ Rhodopin 1,2-Dihydro-γ,γ-carotene-l-ol
- ☆ Rhodopinol aka Warmingol 13-*cis*-1,2-Dihydro-γ,γ-carotene-1,20-diol

☆ Saproxanthin 3',4'-Didehydro-1',2'-dihydro-β,γ-carotene-3,1'-diol

☆ Zeaxanthin

Glycosides

☆ Oscillaxanthin 2,2'-Bis(β-L-rhamnopyranosyloxy)-3,4,3',4'-tetradehydro-1,2,1',2'-tetrahydro-γ,γ-carotene-1,1'-diol

☆ Phleixanthophyll 1'-(β-D-Glucopyranosyloxy)-3',4'-didehydro-1',2'-dihydro-β,γ-carotene-2'-ol

Ethers

☆ Rhodovibrin 1'-Methoxy-3',4'-didehydro-1,2,1',2'-tetrahydro-γ,γ-carotene-1-ol

☆ Spheroidene 1-Methoxy-3,4-didehydro-1,2,7',8'-tetrahydro-γ,γ-carotene

Epoxides

☆ Diadinoxanthin 5,6-Epoxy-7',8'-didehydro-5,6-dihydro–carotene-3,3-diol

☆ Luteoxanthin 5,6: 5',8'-Diepoxy-5,6,5',8'-tetrahydro-β,β-carotene-3,3'-diol

☆ Mutatoxanthin

☆ Citroxanthin

☆ Zeaxanthin furanoxide 5,8-Epoxy-5,8-dihydro-β,β-carotene-3,3'-diol

☆ Neochrome 5',8'-Epoxy-6,7-didehydro-5,6,5',8'-tetrahydro-β,β-carotene-3,5,3'-triol

☆ Foliachrome

☆ Trollichrome

☆ Vaucheriaxanthin 5',6'-Epoxy-6,7-didehydro-5,6,5',6'-tetrahydro-β,β-carotene-3,5,19,3'-tetrol

Aldehydes

☆ Rhodopinal

☆ Warmingone 13-cis-1-Hydroxy-1,2-dihydro-γ,γ-carotene-20-al

☆ Torularhodinaldehyde 3',4'-Didehydro-β,γ-carotene-16'-al

Acids and Acid Esters

☆ Torularhodin 3',4'-Didehydro-β,γ-carotene-16'-oic acid

☆ Torularhodin methyl ester Methyl 3',4'-didehydro-β,γ-carotene-16'-oate

Ketones

☆ Astacene

☆ Astaxanthin

☆ Canthaxanthin aka Aphanicin, Chlorellaxanthin β,β-Carotene-4,4'-dione

☆ Capsanthin (3R,3'S,5'R)-3,3-Dihydroxy-β,chachuu-carotene-6'-one

☆ Capsorubin (3S,5R,3'S,5'R)-3,3'-Dihydroxy-chachuu,chachuu-carotene-6,6'-dione

☆ Cryptocapsin (3'R,5'R)-3'-Hydroxy-β,chachuu-carotene-6'-one

☆ 2,2'-Diketospirilloxanthin 1,1'-Dimethoxy-3,4,3',4'-tetradehydro-1,2,1',2'-tetrahydro-γ,γ-carotene-2,2'-dione

☆ Flexixanthin 3,1'-Dihydroxy-3',4'-didehydro-1',2'-dihydro-β,γ-carotene-4-one

☆ 3-OH-Canthaxanthin aka Adonirubin aka Phoenicoxanthin 3-Hydroxy-β,β-carotene-4,4'-dione

☆ Hydroxyspheriodenone 1'-Hydroxy-1-methoxy-3,4-didehydro-1,2,1',2',7',8'-hexahydro-γ,γ-carotene-2-one

☆ Okenone 1'-Methoxy-1',2'-dihydro-c,γ-carotene-4'-one

☆ Pectenolone 3,3'-Dihydroxy-7',8'-didehydro-β,β-carotene-4-one

☆ Phoeniconone aka Dehydroadonirubin 3-Hydroxy-2,3-didehydro-β,β-carotene-4,4'-dione

☆ Phoenicopterone β,ε-carotene-4-one

☆ Rubixanthone 3-Hydroxy-β,γ-carotene-4'-one

☆ Siphonaxanthin 3,19,3'-Trihydroxy-7,8-dihydro-β,ε-carotene-8-one

Esters of Alcohols

☆ Astacein 3,3'-Bispalmitoyloxy-2,3,2',3'-tetradehydro-β,β-carotene-4,4'-dione or 3,3'-dihydroxy-2,3,2',3'-tetradehydro-β,β-carotene-4,4'-dione dipalmitate

☆ Fucoxanthin 3'-Acetoxy-5,6-epoxy-3,5'-dihydroxy-6',7'-didehydro-5,6,7,8,5',6'-hexahydro-β,β-carotene-8-one

☆ Isofucoxanthin 3'-Acetoxy-3,5,5'-trihydroxy-6',7'-didehydro-5,8,5',6'-tetrahydro-β,β-carotene-8-one

☆ Physalien

☆ Zeaxanthin (3R,3'R)-3,3'-Bispalmitoyloxy-β,β-carotene or (3R,3'R)-β,β-carotene-3,3'-diol

☆ Siphonein 3,3'-Dihydroxy-19-lauroyloxy-7,8-dihydro-β,ε-carotene-8-one or 3,19,3'-trihydroxy-7,8-dihydro-β,å-carotene-8-one 19-laurate

Apocarotenoids

☆ β-Apo-2'-carotenal 3',4'-Didehydro-2'-apo-β-carotene-2'-al

☆ Apo-2-lycopenal

☆ Apo-6'-lycopenal 6'-Apo-y-carotene-6'-al

☆ Azafrinaldehyde 5,6-Dihydroxy-5,6-dihydro-10'-apo-β-carotene-10'-al

☆ Bixin 6'-Methyl hydrogen 9'-cis-6,6'-diapocarotene-6,6'-dioate

☆ Citranaxanthin 5',6'-Dihydro-5'-apo-β-carotene-6'-one or 5',6'-dihydro-5'-apo-18'-nor-β-carotene-6'-one or 6'-methyl-6'-apo-β-carotene-6'-one

☆ Crocetin 8,8'-Diapo-8,8'-carotenedioic acid

☆ Crocetinsemialdehyde 8'-Oxo-8,8'-diapo-8-carotenoic acid

☆ Crocin Digentiobiosyl 8,8'-diapo-8,8'-carotenedioate

☆ Hopkinsiaxanthin 3-Hydroxy-7,8-didehydro-7',8'-dihydro-7'-apo-β-carotene-4,8'-dione or 3-hydroxy-8'-methyl-7,8-didehydro-8'-apo-β-carotene-4,8'-dione

☆ Methyl apo-6'-lycopenoate Methyl 6'-apo-y-carotene-6'-oate

☆ Paracentrone 3,5-Dihydroxy-6,7-didehydro-5,6,7',8'-tetrahydro-7'-apo-b-carotene-8'-one or 3,5-dihydroxy-8'-methyl-6,7-didehydro-5,6-dihydro-8'-apo-β-carotene-8'-one

☆ Sintaxanthin 7',8'-Dihydro-7'-apo-β-carotene-8'-one or 8'-methyl-8'-apo-b-carotene-8'-one

Nor- and Seco-carotenoids

☆ Actinioerythrin 3,3'-Bisacyloxy-2,2'-dinor-b,b-carotene-4,4'-dione

☆ β-Carotenone 5,6:5',6'-Diseco-b,b-carotene-5,6,5',6'-tetrone

☆ Peridinin 3'-Acetoxy-5,6-epoxy-3,5'-dihydroxy-6',7'-didehydro-5,6,5',6'-tetrahydro-12',13',20'-trinor-β,β-carotene-19,11-olide

☆ Pyrrhoxanthininol 5,6-epoxy-3,3'-dihydroxy-7',8'-didehydro-5,6-dihydro-12',13',20'-trinor-β,β-carotene-19,11-olide

☆ Semi-α-carotenone 5,6-Seco-b,e-carotene-5,6-dione

☆ Semi-β-carotenone 5,6-seco-b,b-carotene-5,6-dione or 5',6'-seco-β,β-carotene-5',6'-dione

☆ Triphasiaxanthin 3-Hydroxysemi-b-carotenone 3'-Hydroxy-5,6-seco-β,β-carotene-5,6-dione or 3-hydroxy-5',6'-seco-b,b-carotene-5',6'-dione

Retro-Carotenoids and Retro-apo-Carotenoids

☆ Eschscholtzxanthin 4',5'-Didehydro-4,5'-retro-b,b-carotene-3,3'-diol

☆ Eschscholtzxanthone 3'-Hydroxy-4',5'-didehydro-4,5'-retro-β,β-carotene-3-one

☆ Rhodoxanthin 4',5'-Didehydro-4,5'-retro-β,β-carotene-3,3'-dione

☆ Tangeraxanthin 3-Hydroxy-5'-methyl-4,5'-retro-5'-apo-β-carotene-5'-one or 3-hydroxy-4,5'-retro-5'-apo-β-carotene-5'-one

Higher Carotenoids

☆ Nonaprenoxanthin 2-(4-Hydroxy-3-methyl-2-butenyl)-7',8',11',12'-tetrahydro-e,y-carotene

☆ Decaprenoxanthin 2,2'-Bis(4-hydroxy-3-methyl-2-butenyl)-e,e-carotene

☆ C.p. 450 2-[4-Hydroxy-3-(hydroxymethyl)-2-butenyl]-2'-(3-methyl-2-butenyl)-β,β-carotene

☆ C.p. 473 2'-(4-Hydroxy-3-methyl-2-butenyl)-2-(3-methyl-2-butenyl)-3',4'-didehydro-1',2'-dihydro-b,y-carotene-1'-ol

☆ Bacterioruberin 2,2'-Bis(3-hydroxy-3-methylbutyl)-3,4,3',4'-tetradehydro-1,2,1',2'-tetrahydro-y,y-carotene-1,1'-dio

Distinguished by their orange, yellow and red pigments, carotenoids are found in many plants, algae and bacteria. Carotenoids act as antioxidants within the body, protecting against cellular damage, the effects of aging and even some chronic diseases. These beneficial compounds cannot be synthesized by humans or animals, diet is the only way to get them. There are well over 600 known carotenoids, with beta-carotene, alpha-carotene, lutein, zeaxanthin, lycopene and astaxanthin being the most common.

Why are Carotenoids Important?

As antioxidants, carotenoids are helpful for protecting vision and combating cellular damage. Recent studies have also identified carotenoids as paramount supporters for the cardiovascular system and male reproductive health. Lycopene, a carotenoid found in tomatoes (among other fruits and vegetables), has even been linked to keeping the liver, prostate, breast, colon and lungs healthy.

Health Benefits of Consuming Carotenoids

Carotenoids have an immense value for human health and many people are seeking ways to incorporate these health-brimming compounds into their diet. Nutritionists, physicians and health organizations all recommend a diet high in antioxidants, including those that are derived from carotenoids. Here are 5 of the most pronounced benefits of carotenoids.

1. Supports Eye Health

Do you remember adults telling you as a child that carrots were good for your eyes? As it turns out, this statement is founded in scientific truth. Carrots are an excellent source of carotenoids, including retinol and pro-vitamin A, both of which have therapeutic value for degenerative diseases of the retina.

2. Cardiovascular Health

Population-based studies have demonstrated that carotenoids are effective for supporting cardiovascular health. Other nutrients that work together with carotenoids for protecting heart health include glutathione, vitamin E and vitamin C. These findings have been echoed in numerous studies.

3. Possible Anti-Tumor Properties?

It would be premature and overzealous to say that carotenoids prevent cancer, but it is known that carotenoids contain acetylenics, a group of metabolites known for combating tumor development. They also have action against harmful organisms and support the immune system. The combination of these properties have been so effective for fighting bacteria and immune-related infections that researchers are exploring their cytotoxic effects on multiple types of cancer.

4. Male Fertility

The antioxidant effect of carotenoids may protect sperm health, according a recent 2013 study. This research found that carotenoids, vitamin C, vitamin E, selenium, glutathione, N-acetylcysteine and zinc notably improved the participants' chances of becoming pregnant. These nutrients can be accessed easily with a diet rich in fresh fruits, vegetables, nuts and seeds.

5. Skin Health

Studies have reported that the carotenoids beta-carotene and beta-cryptoxanthin protect skin, tissue and cells from environmental toxins and disease. Recent research has demonstrated that non-provitamin A carotenoids, such as lutein, zeaxanthin and astaxanthin, also have protective benefits for the skin. To best support skin health, experts recommend vitamin A carotenoids from fresh vegetables and colourful fruits.

Maximize Your Carotenoid Intake

Vitamin A is considered a fat-soluble nutrient and consuming foods containing vitamin A with fat (like grass-fed butter, coconut oil, or olive oil) increases its bioavailability. Chopping and cooking the vegetables also enhances nutrient uptake, further increasing the chances your body will use the compounds efficiently.

Best Sources of Carotenoids

Carotenoids are found in orange, yellow and red fruits and vegetables, like pumpkin, carrots and tomatoes. For beta-carotene, carrots and pumpkins are forefront; however, spinach, sweet potatoes, cantaloupe and dandelion greens also represent excellent sources.

References

Armstrong GA, Hearst JE (1996). "Carotenoids 2: Genetics and molecular biology of carotenoid pigment biosynthesis". *FASEB J.* 10 (2): 228–37.

Diplock1, A. T., J.-L. Charleux, G. Crozier-Willi, F. J. Kok, C. Rice-Evans, M. Roberfroid, W. Stahl, J. Vina-Ribes. (1998). Functional food science and defence against reactive oxidative species. *British Journal of Nutrition,* 80, Suppl. 1, S77–S112

Bjelakovic G, *et al.*, Nikolova, D, Gluud, LL, Simonetti, RG, Gluud, C (2007). "Mortality in randomized trials of antioxidant supplements for primary and secondary prevention: systematic review and meta-analysis". *JAMA* 297 (8): 842–57. doi: 10.1001/jama.297.8.842.

Alija AJ, Bresgen N, Sommerburg O, Siems W, Eckl PM (2004). "Cytotoxic and genotoxic effects of β-carotene breakdown products on primary rat hepatocytes". *Carcinogenesis* 25 (5): 827–31.doi: 10.1093/carcin/bgh056.

Pauling Institute. "Micronutrient Information Center-Carotenoids". Retrieved 3 August 2013.

β-Carotene and other carotenoids as antioxidants. From U.S. National Library of Medicine. November, 2008.

Alija AJ, Bresgen N, Sommerburg O, Siems W, Eckl PM (2004). "Cytotoxic and genotoxic effects of β-carotene breakdown products on primary rat hepatocytes". *Carcinogenesis* 25 (5): 827–31.doi: 10.1093/carcin/bgh056.

Lozano, G. A. (1994). Carotenoids, parasites and sexual selection. *Oikos* 70: 309-311.

Daviesþ, Kevin M. (2004). *Plant pigments and their manipulation*. Wiley-Blackwell. p. 6. ISBN 1-4051-1737-0.

Liu GY, Essex A, Buchanan JT, *et al.* (2005). "*Staphylococcus aureus* golden pigment impairs neutrophil killing and promotes virulence through its antioxidant activity". *J. Exp. Med.* 202 (2): 209–15. doi: 10.1084/jem.20050846.

Brian H. Davies (2010). Carotenoid metabolism as a preparation for function. Pure and Applied Chemistry, Vol. 63, No. 1, pp. 131-140, 1991. available online. Accessed April 30.

Efficient Syntheses of the Keto-carotenoids Canthaxanthin, Astaxanthin and Astacene. Seyoung Choi and Sangho Koo, *J. Org. Chem.*, 2005, 70 (8), pages 3328–3331, doi: 10.1021/jo0501011

Unlu N, *et al.*, Bohn, T, Clinton, SK, Schwartz, SJ (2005). "Carotenoid Absorption from Salad and Salsa by Humans Is Enhanced by the Addition of Avocado or Avocado Oil". *Human Nutrition and Metabolism* 135 (3): 431–6.

Kidd, Parris (2011). "Astaxanthin, Cell Membrane Nutrient with Diverse Clinical Benefits and Anti-Aging Potential". *Alternative Medicine Review* 16 (4): 335–364.

Chapter 27

Chlorophyll

Chlorophyll (also chlorophyll) is a term used for several closely related green pigments found in cyanobacteria and thechloroplasts of algae and plants. Its name is derived from the Greek words *chloros* ("green") and *phyllon* ("leaf"). Chlorophyll is an extremely important biomolecule, critical in photosynthesis, which allows plants to absorb energyfrom light. Chlorophyll absorbs light most strongly in the blue portion of the electromagnetic spectrum, followed by the red portion. Conversely, it is a poor absorber of green and near-green portions of the spectrum, hence the green colour of chlorophyll-containing tissues. Chlorophyll was first isolated by Joseph Bienaimé Caventou and Pierre Joseph Pelletier in 1817.

Chlorophyll and Photosynthesis

Chlorophyll is vital for photosynthesis, which allows plants to absorb energy from light. Chlorophyll molecules are specifically arranged in and around photosystems that are embedded in the thylakoid membranes ofchloroplasts. In these complexes, chlorophyll serves two primary functions. The function of the vast majority of chlorophyll (up to several hundred molecules per photosystem) is to absorb light and transfer that light energy by resonance energy transfer to a specific chlorophyll pair in the reaction center of the photosystems. The two currently accepted photosystem units are Photosystem II and Photosystem I, which have their own distinct reaction centres, named P680 and P700, respectively. These centres are named after the wavelength (in nanometers) of their red-peak absorption maximum. The identity, function and spectral properties of the types of chlorophyll in each photosystem are distinct and determined by each other and the protein structure surrounding them. Once extracted from the protein into a solvent (such as acetone or methanol), these chlorophyll pigments can be separated in a simplepaper chromatography experiment and, based on the number of polar groups between chlorophyll a and chlorophyll b, will chemically separate out on the paper.

The function of the reaction center chlorophyll is to use the energy absorbed by and transferred to it from the other chlorophyll pigments in the photosystems to undergo a charge separation, a specific redox reaction in which the chlorophyll donates an electron into a series of molecular intermediates called an electron transport chain. The charged reaction center chlorophyll (P680[+]) is then reduced back to its ground state by accepting an electron. In Photosystem II, the electron that reduces P680[+] ultimately comes from the oxidation of water into O_2 and H^+ through several intermediates. This reaction is how photosynthetic organisms such as plants produce O_2 gas and is the source for practically all the O_2 in Earth's atmosphere. Photosystem I typically works in series with Photosystem II; thus the P700[+] of Photosystem I is usually reduced, via many intermediates in the thylakoid membrane, by electrons ultimately from Photosystem II. Electron transfer reactions in the thylakoid membranes are complex, however and the source of electrons used to reduce P700[+] can vary.

The electron flow produced by the reaction center chlorophyll pigments is used to shuttle H^+ ions across the thylakoid membrane, setting up a chemiosmotic potential used mainly to produce ATP chemical energy; and those electrons ultimately reduce NADP[+] to NADPH, a universal reductant used to reduce CO_2 into sugars as well as for other biosynthetic reductions.

Reaction center chlorophyll–protein complexes are capable of directly absorbing light and performing charge separation events without other chlorophyll pigments, but the absorption cross section (the likelihood of absorbing a photon under a given light intensity) is small. Thus, the remaining chlorophylls in the photosystem and antenna pigment protein complexes associated with the photosystems all cooperatively absorb and funnel light energy to the reaction center. Besides chlorophyll *a*, there are other pigments, calledaccessory pigments, which occur in these pigment protein antenna complexes.

Chemical Structure

Chlorophyll is a chlorin pigment, which is structurally similar to and produced through the same metabolic pathway as other porphyrinpigments such as heme. At the center of the chlorin ring is a magnesium ion. This was discovered in 1906 and was the first time that magnesium had been detected in living tissue. For the structures depicted in this article, some of the ligands attached to the Mg^{2+}center are omitted for clarity. The chlorin ring can have several different side chains, usually including a long phytol chain. There are a few different forms that occur naturally, but the most widely distributed form in terrestrial plants is chlorophyll *a*. After initial work done by German chemist Richard Willstätter spanning from 1905 to 1915, the general structure of chlorophyll *a* was elucidated by Hans Fischer in 1940. By 1960, when most of the stereochemistry of chlorophyll *a* was known, Robert Burns Woodward published a total synthesis of the molecule. In 1967, the last remaining stereochemical elucidation was completed by Ian Fleming and in 1990 Woodward and co-authors published an updated synthesis. Chlorophyll f was announced to be present in cyanobacteria and other oxygenic microorganisms that form stromatolites in 2010; a molecular formula of $C_{55}H_{70}O_6N_4Mg$ and a structure of (2-formyl)-chlorophyll *a*were deduced based on NMR, optical and mass spectra. The different structures of chlorophyll are summarized in Table 27.1.

Table 27.1: Different eStructure of Chlorophyll a

	Chlorophyll a	Chlorophyll b	Chlorophyll c_1	Chlorophyll c_2	Chlorophyll d	Chlorophyll f
Molecular formula	$C_{55}H_{72}O_5N_4Mg$	$C_{55}H_{70}O_6N_4Mg$	$C_{35}H_{30}O_5N_4Mg$	$C_{35}H_{28}O_5N_4Mg$	$C_{54}H_{70}O_6N_4Mg$	$C_{55}H_{70}O_6N_4Mg$
C2 group	$-CH_3$	$-CH_3$	$-CH_3$	$-CH_3$	$-CH_3$	$-CHO$
C3 group	$-CH=CH_2$	$-CH=CH_2$	$-CH=CH_2$	$-CH=CH_2$	$-CHO$	$-CH=CH_2$
C7 group	$-CH_3$	$-CHO$	$-CH_3$	$-CH_3$	$-CH_3$	$-CH_3$
C8 group	$-CH_2CH_3$	$-CH_2CH_3$	$-CH_2CH_3$	$-CH=CH_2$	$-CH_2CH_3$	$-CH_2CH_3$
C17 group	$-CH_2CH_2COO-Phytyl$	$-CH_2CH_2COO-Phytyl$	$-CH=CHCOOH$	$-CH=CHCOOH$	$-CH_2CH_2COO-Phytyl$	$-CH_2CH_2COO-Phytyl$
C17-C18 bond	Single(chlorin)	Single(chlorin)	Double(porphyrin)	Double(porphyrin)	Single(chlorin)	Single(chlorin)
Occurrence	Universal	Mostly plants	Various algae	Various algae	Cyanobacteria	Cyanobacteria

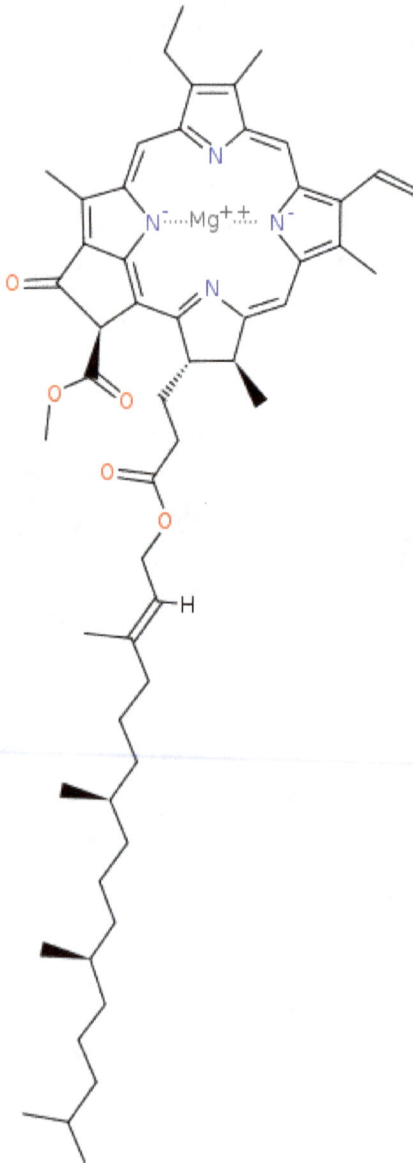

Structure of Chlorophyll *a*

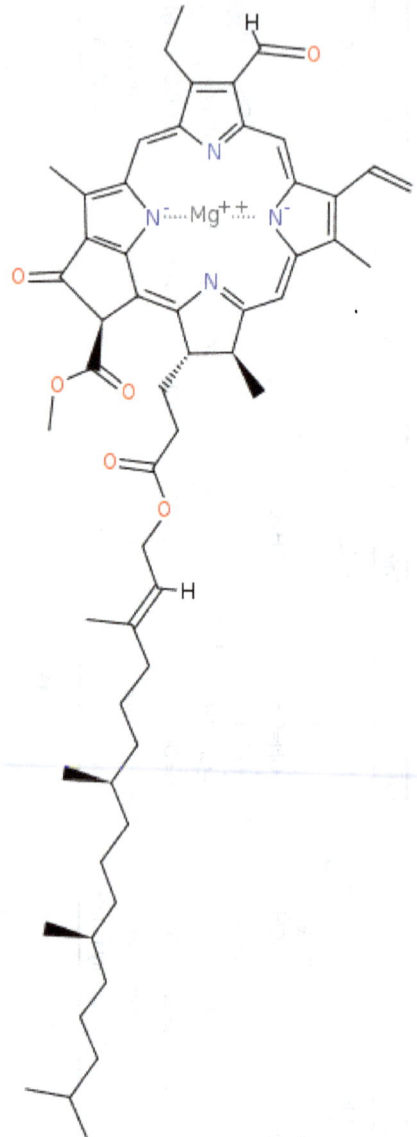

Structure of chlorophyll *b*

When leaves degreen in the process of plant senescence, chlorophyll is converted to a group of colourless tetrapyrroles known as **nonfluorescent chlorophyll catabolites** (NCC's) with the general structure:

Spectrophotometry

Measurement of the absorption of light is complicated by the solvent used to extract it from plant material, which affects the values obtained,

Structure of Chlorophyll *c₁*

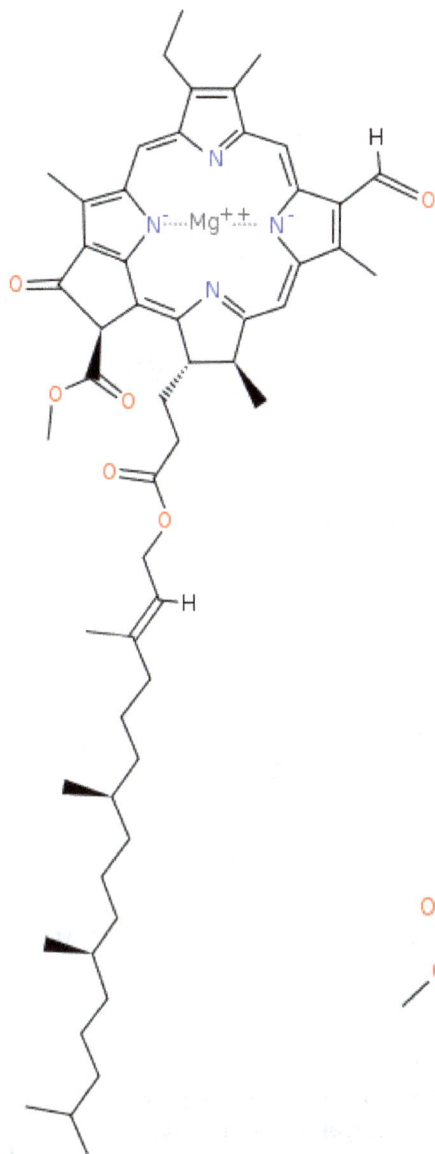

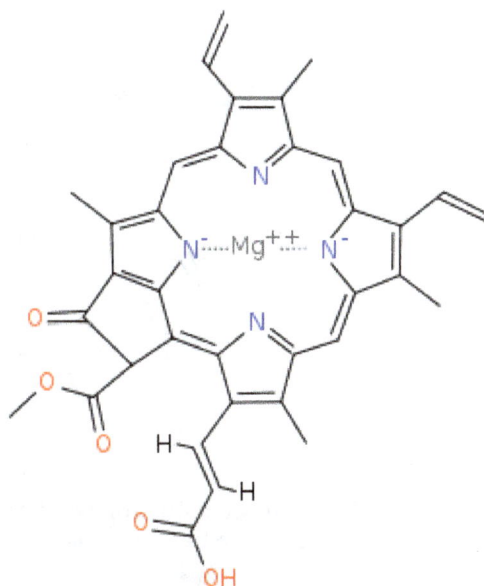

Structure of Chlorophyll *d*

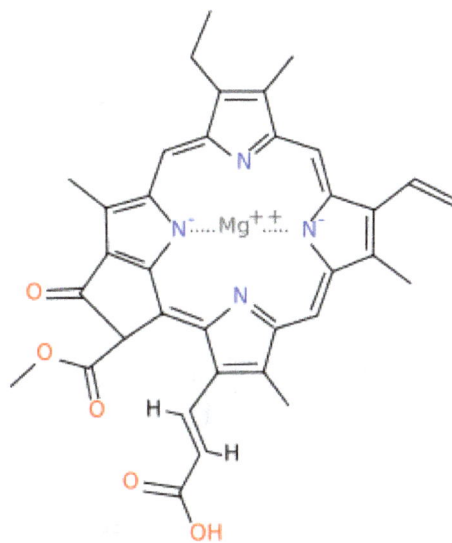

Structure of Chlorophyll *c₂*

* In diethyl ether, chlorophyll a has approximate absorbance maxima of 430 nm and 662 nm, while chlorophyll b has approximate maxima of 453 nm and 642 nm.

* The absorption peaks of chlorophyll a are at 665 nm and 465 nm. Chlorophyll a fluoresces at 673 nm (maximum) and 726 nm. The peak molar absorption coefficient of chlorophyll a exceeds 105 M–1 cm–1, which is among the highest for small-molecule organic compounds.

Structure of Chlorophyll *f*

Nonfluorescent Chlorophyll Catabolites: These compounds have also been identified in several ripening fruits.

☆ In 90 per cent acetone-water, the peak absorption wavelengths of chlorophyll a are 430 nm and 664 nm; peaks for chlorophyll b are 460 nm and 647 nm; peaks for chlorophyll c1 are 442 nm and 630 nm; peaks for chlorophyll c2 are 444 nm and 630 nm; peaks for chlorophyll dare 401 nm, 455 nm and 696 nm.

By measuring the absorption of light in the red and far red regions it is possible to estimate the concentration of chlorophyll within a leaf. In his scientific paper Gitelson (1999) states, "The ratio between chlorophyll fluorescence, at 735 nm and the wavelength range 700nm to 710 nm, F735/F700 was found to be linearly proportional to the chlorophyll content (with determination coefficient, r2, more than 0.95) and thus this ratio can be used as a precise indicator of chlorophyll content in plant leaves." The fluorescent ratio chlorophyll content meters use this technique.

Absorbance spectra of free chlorophyll *a* (blue) and *b* (red) in a solvent. The spectra of chlorophyll molecules are slightly modified *in vivo* depending on specific pigment-protein interactions.

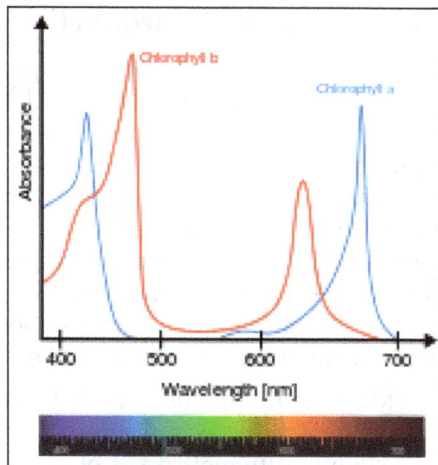

Biosynthesis

In plants, chlorophyll may be synthesized from succinyl-CoA and glycine, although the immediate precursor to chlorophyll *a* and *b* is protochlorophyllide. In Angiosperm plants, the last step, conversion of protochlorophyllide to chlorophyll, is light-dependent and such plants are pale (etiolated) if grown in the darkness. Non-vascular plants and green algae have an additional light-independent enzyme and grow green in the darkness instead.

Chlorophyll itself is bound to proteins and can transfer the absorbed energy in the required direction. Protochlorophyllide occurs mostly in the free form and, under light conditions, acts as a photosensitizer, forming highly toxic free radicals. Hence, plants need an efficient mechanism of regulating the amount of chlorophyll precursor. In angiosperms, this is done at the step of aminolevulinic acid (ALA), one of the intermediate compounds in the biosynthesis pathway. Plants that are fed by ALA accumulate high and toxic levels of protochlorophyllide; so do the mutants with the damaged regulatory system.

Chlorosis is a condition in which leaves produce insufficient chlorophyll, turning them yellow. Chlorosis can be caused by a nutrient deficiency of iron called iron chlorosis or by a shortage of magnesium or nitrogen. Soil pH sometimes plays a role in nutrient-caused chlorosis; many plants are adapted to grow in soils with specific pH levels and their ability to absorb nutrients from the soil can be dependent on this. Chlorosis can also be caused by pathogens including viruses, bacteria and fungal infections, or sap-sucking insects.

Protochlorophyllide, or monovinyl protochlorophyllide, is an immediate precursor of chlorophyll *a* that lacks the phytol side-chain of chlorophyll. Protochlorophyllide is highly fluorescent; mutants that accumulate it glow red if irradiated with blue light. In Angiosperms, the last step, conversion of protochlorophyllide to chlorophyll, is light-dependent and such plants are pale (chlorotic) if grown in the darkness. Gymnosperms, algae and photosynthetic bacteria have another, light-independent enzyme and grow green in the darkness as well.

Conversion to Chlorophyll

The enzyme that converts protochlorophyllide to chlorophyllide is protochlorophyllide reductase, EC 1.3.1.33. There are two structurally unrelated proteins with this activity: the light-dependent and the dark-operative. The light-dependent reductase needs light to operate. The dark-operative version is a completely different protein, consisting of three subunits that exhibit significant sequence similarity to the three subunits of nitrogenase, which catalyzes the formation of ammonia from dinitrogen. This enzyme might be evolutionary older but (being similar to nitrogenase) is highly sensitive to free oxygen and does not work if its concentration exceeds about 3 per cent. Hence, the alternative, light-dependent version needed to evolve.

Most of the photosynthetic bacteria have both light-dependent and light-independent reductases. Angiosperms have lost the dark-operative form and rely on 3 slightly different copies of light-dependent version, frequently abbreviated as POR A, B and C. Gymnosperms have much more copies of the similar gene (Loblolly pine has about 11). In plants, POR is encoded in the cell nucleus and only later transported to its place of work, chloroplast. Unlike with POR, in plants and algae that have the dark-operative enzyme it is at least partially encoded in the chloroplast genome.

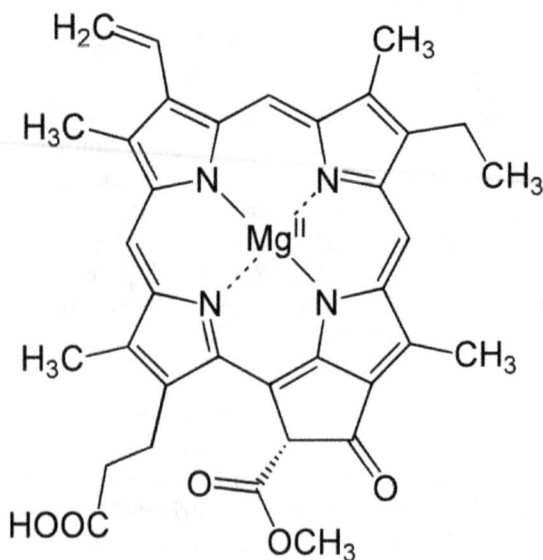

The Structure of Monovinyl Protochlorophyllide.

Complementary Light Absorbance of Anthocyanins with Chlorophylls

Anthocyanins are other plant pigments. The absorbance pattern responsible for the red colour of anthocyanins may be complementary to that of green chlorophyll in photosynthetically active tissues such as young *Quercus coccifera* leaves. It may protect the leaves from attacks by plant eaters that may be attracted by green colour.

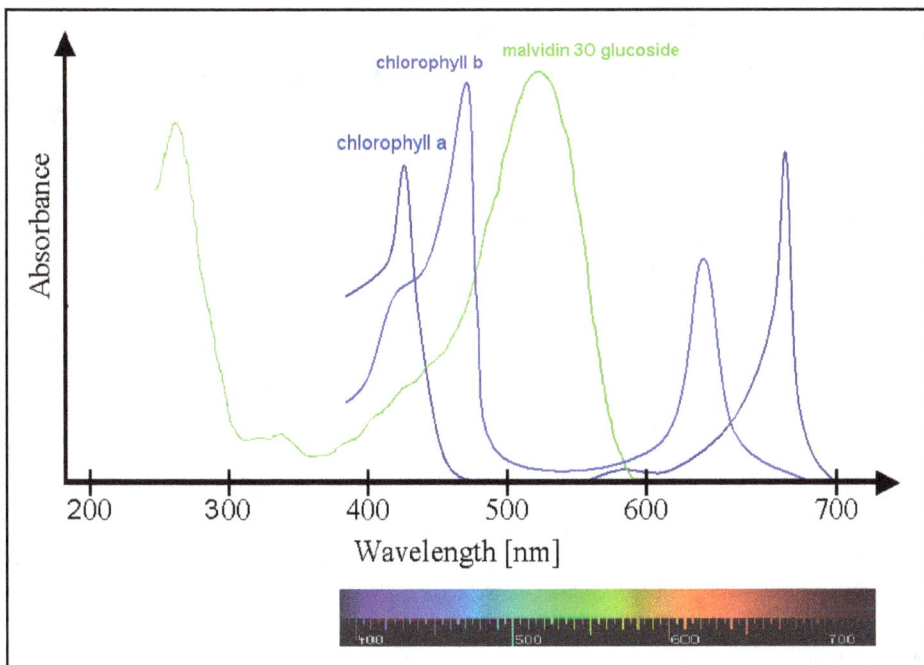

Superposition of spectra of chlorophyll a and b with oenin (malvidin 3O glucoside), a typical anthocyanidin, showing that, while chlorophylls absorb in the blue and yellow/red parts of the visible spectrum, oenin absorbs mainly in the green part of the spectrum, where chlorophylls don't absorb at all.

Culinary Use

Chlorophyll is registered as a food additive (colourant) and its E number is **E140**. Chefs use chlorophyll to colour a variety of foods and beverages green, such as pasta and absinthe. Chlorophyll is not soluble in water and it is first mixed with a small quantity of vegetable oil to obtain the desired solution. Extracted liquid chlorophyll was considered to be unstable and always denatured until 1997, when Frank S. and Lisa Sagliano used freeze-drying of liquid chlorophyll at the University of Florida and stabilized it as a powder, preserving it for future use.

Alternative Medicine

Many claims are made about the healing properties of chlorophyll, but most have been disproved or are exaggerated by the companies that are marketing them. Quackwatch, a website dedicated to debunking false medical claims, has a quote from Toledo Blade (1952) which claims "Chlorophyll Held Useless as Body Deodorant", but later has John C. Kephart pointing out "No deodorant effect can possibly occur from the quantities of chlorophyll put in products such as gum, foot powder, cough drops, etc. To be effective, large doses must be given internally".

Nutritional Value of Chlorophyll

Chlorophyll is a super food which is packed with a range of powerful nutrients. It is a good source of vitamins such as vitamin A, vitamin C, vitamin E, vitamin K and beta carotene. Chlorophyll is rich in antioxidants, vital minerals such as magnesium, iron, potassium, calcium and essential fatty acids.

Health Benefits from Chlorophyll

You may have heard a lot of buzz surrounding the healthiness of vegetables, especially leafy green vegetables with deep, rich colour tones. In many ways, they're perfect foods and part of the credit is owed to the chlorophyll they contain. Chlorophyll is a green pigment that's instrumental in photosynthesis. Research also suggests it's instrumental in promoting good health. The following 10 benefits demonstrate its amazing versatility and power.

1. Helps Control Hunger and Cravings

A study released in September of 2013 determined that compounds containing chlorophyll might help suppress hunger. Twenty moderately overweight women were given test meals on three separate occasions one week apart two meals with chlorophyll, one without. They reported reduced hunger following the meals containing the chlorophyll and blood tests showed stable blood sugar levels. These results suggest chlorophyll should be considered as an effective way to manage hunger and promote weight loss.

2. Controls Body Odor

Chlorophyll is good for your body from the inside out, literally. This includesbody and fecal odors. When tested in a nursing home with 62 patients, one study reported improvements in the odors emanating from the test subjects. In addition to odor reduction, it also alleviated constipation and gas.

3. Encourages Healing

The use of chlorophyll to promote the healing process was first reported nearly 100 years ago. One report published by Butler University (in 1950!) discussed its use for improving healing times. A more recent study found that, when used as a spray, chlorophyll significantly helped wound therapy. The research clearly backs its effectiveness in this area.

4. Promotes Cleansing

Environmental pollutants such as toxic metals can quickly destroy health. Chlorophyll binds with toxic metals to hamper absorption and research has shown it can do the same with some carcinogens. A crossover study with 4 human volunteers tested chlorophyll's ability to remove ingested aflatoxin, a known carcinogen. Each volunteer received three servings of the aflatoxin, followed with a meal, the latter two included a tablet of chlorophyll. The results showed rapid elimination with 95 per cent evacuated within 24 hours.

5. Protects DNA against Fried Foods

Fried foods aren't just bad for your waistline; they also contain chemicals known to damage colon tissue and the DNA of colon cells. In one study, participants who ate fried foods coupled with protective foods, including chlorophyll, were observed to have less DNA damage of colon cells. Don't take that as just cause to go crazy with the fried foods, the best idea is to avoid them altogether.

6. Super Potent Antioxidant Action

Chlorophyll's status as a superfood is due to its nutritional and potent antioxidant properties. It protects cells from oxidative damage by eliminating free radicals. One plant known for its high chlorophyll levels, *Conyza triloba*, showed very active superoxide scavenging behavior. An *in vitro* study found it capable of reducing free radicals while reducing cell damage.

7. Promising Potential for Cancer Therapy

Although the research doesn't currently show that chlorophyll is an all out cancer cure, researchers are excited about its potential as a therapy. One study found men with lower levels of dietary chlorophyll had higher risks of colon cancer. Another study noted it effective at stimulating liver detoxifying enzymes and may protect against other carcinogens.

8. Effective against *Candida albicans*

Candida infection is a big problem for many individuals and can lead to fatigue, depression and digestive problems; just to name a few. Research has shown that isolated chlorophyll solutions stop *Candida albicans* growth.

9. Relieves Systemic Redness and Swelling

Traditional medicine has long used green leaves for infections. With modern approaches that have limited effectiveness and a long list of side effects, natural remedies are getting another look. Animal studies have found that green leaves help to reduce swelling and redness.

10. Promotes Healthy Iron Levels

A modified form of chlorophyll known as chlorophyllin may be highly effective for anemia. By swapping iron for magnesium at the center of the molecule, it provides a bioavailable form of iron the body can use. While scientists may look for ways to develop patent-able medicines, no one needs to wait to take chlorophyll to gain its benefits.

Chlorophyll: An Excellent Dietary Supplement

Chlorophyll possesses an amazing range of benefits and is readily available to everyone. Chlorophyll rich foods are also healthy in their own right and contain essential vitamins and nutrients needed for cardiovascular, muscular and neural health. Consume foods containing chlorophyll or take a dietary supplement for best effect.

Red Blood Cells

Chlorophyll aids in restoring and replenishing the red blood cells. Chlorophyll works at molecular and cellular levels and has the ability to regenerate our body. Chlorophyll is rich in live enzymes which helps in cleansing of blood and enhances the ability of the blood to carry more oxygen. Chlorophyll is a blood builder and is also effective against anemia which is caused by the deficit of red blood cells in the body.

Cancer

Chlorophyll is effective against cancer such as human colon cancer and stimulates the induction of apoptosis. It provides protection against a wide range of carcinogens present in the air, cooked meats and grains. Studies have made it evident that chlorophyll helps in restraining the gastrointestinal absorption of harmful toxins also known as aflatoxin in the body. Chlorophyll and its derivative chlorophyll in inhibit the metabolism of these procarcinogens which may impair the DNA and also lead to liver cancer and hepatitis. Further studies conducted in this regard have advocated the chemo preventive effects of chlorophyll attributing to its antimutagenic properties. Another research has shown the efficacy of dietary chlorophyll as a phytochemical compound in the reduction of tumorigenesis.

Antioxidant Power

Chlorophyll possesses strong antioxidant capacity along with a significant amount of essential vitamins. These effective radical scavengers in chlorophyll help to neutralize harmful molecules and guard against the development of various diseases and damages caused as a consequence of oxidative stress caused by free radicals.

Arthritis

The anti-inflammatory properties of chlorophyll are beneficial for arthritis. Studies have shown that chlorophyll and its derivatives interfere with the growth of bacteria induced inflammation. This protective nature of chlorophyll makes it a potent ingredient for the preparation of phytomedicines for treating painful medical conditions such as fibromyalgia and arthritis.

Detoxification

Chlorophyllhas purifying qualities which helps in detoxification of the body. Abundance of oxygen and healthy flow of blood because of chlorophyll in the body encourages it to get rid of harmful impurities and toxins. Chlorophyll forms complexes with the mutagens and has the ability to bind and flush out the toxic chemicals and heavy metals such as mercury from the body. This helps in the detoxifying and reviving the liver. It is also effective in reducing the harmful effects of radiations and helps eliminate pesticides and drug deposits from the body.

Anti-aging

Chlorophyll helps in combating the effects of aging and supports in maintaining healthy tissues attributing to the richness of antioxidants and presence ofmagnesium.

It stimulates the anti aging enzymes and encourages healthy and youthful skin. In addition to this, vitamin K present in chlorophyll clean and rejuvenates the adrenal glands and improves the adrenal functions in the body.

Digestive Health

Chlorophyll promotes healthy digestion by maintaining intestinal flora and stimulating the bowel movements. It acts as a natural drug to the intestinal tract and helps in the renewal of wounded bowel tissues. Diets which are deficient in green vegetables and include majorly red meat pose an enhanced risk of colon disorders. As per the research, chlorophyll facilitates colon cleansing by inhibiting the cytotoxicity induced by dietary heme and preventing the proliferation of colonocytes. It is effective in constipation and alleviates the discomfort caused by gas.

Insomnia

Chlorophyll has a calming effect on the nerves and is helpful in reducing the symptoms of insomnia, nervous irritability and general body fatigue.

Anti-microbial Properties

Chlorophyll has effective antimicrobial properties. Recent studies have shown the healing effect of alkaline chlorophyll based solution in combating a medical condition called *Candida albicans* which is an infection caused by the overgrowth a kind of yeast called candida already present in small amounts in the human body.

Stronger Immunity

Chlorophyll aids in strengthening of the cell walls and overall immune system of the body attributing to its alkaline nature. Anaerobic bacteria which promote the development of diseases cannot survive in the alkaline environment offered by chlorophyll. Along with this, chlorophyll is an oxygenator which encourages body's ability to fight diseases and boosts the energy levels and accelerates healing process.

Deodorizing Properties

Chlorophyll is a valued for its deodorizing properties. It is an effective remedy to combat bad breath and is used in mouthwashes. Poor digestive health is one of the major causes of bad breath. Chlorophyll performs dual action by eliminating the stink from mouth and throat and also stimulates digestive health by cleansing the colon and the blood stream. The deodorizing effect of chlorophyll is also effectual on stinking wounds. It is administered orally to the patients suffering from colostomies and metabolic disorders such as trimethylaminuria to reduce fecal and urinary odor.

Wound Healing

Chlorophyll is a super wound healer. Studies have stated that topical usage of chlorophyll solutions is effective in the treatment of exposed wounds and burns. It helps in reducing local inflammation, strengthens the body tissues, aids in killing germs and improves the resistance of cells against infections. Chlorophyll prevents the growth of bacteria by disinfecting the environment and making it hostile for bacterial growth and speeds up healing. Chlorophyll therapy is also quite effective in the treatment of chronic varicose ulcers of the leg.

Acid-alkali Ratio

Consumption of chlorophyll rich foods helps in balancing the acid-alkali ratio in the body. Magnesium present in chlorophyll is a highly alkaline mineral. By maintaining the appropriate alkalinity and oxygen levels of the body, chlorophyll prevents development of thriving environment for the growth of pathogens. Magnesium present in chlorophyll also plays a vital role in maintaining cardiovascular health, functioning of kidney, muscles, liver and brain.

Strong Bones and Muscles

Chlorophyll helps in the formation and maintenance of strong bones. The central atom of the chlorophyll molecule *i.e.* magnesium plays an important role in bone health along with other essential nutrients such as calcium and vitamin D. Magnesium in chlorophyll also contributes towards muscle toning, contraction and relaxation.

Blood Clotting

Chlorophyll contains vitamin K which is vital for normal clotting of blood. It is used in naturopathy for the treatment of nose bleeds and for the females suffering from anemic conditions and heavy menstrual bleeding.

Kidney Stones

Chlorophyll helps in preventing the formation of kidney stones. The vitamin K present in chlorophyll form essential compounds in the urine and aids in reducing the growth of calcium oxalate crystals.

Sinusitis

Chlorophyll is effective in the treatment of various respiratory infections and other conditions such as cold, rhinitis and sinusitis.

Hormonal Balance

Chlorophyll is beneficial for maintaining sexual hormonal balance in both males and females. Vitamin E present in chlorophyll helps to stimulate the production of testosterone in males and estrogen in females.

Pancreatitis

Chlorophyll is administered intravenously by the healthcare professionals in the treatment of chronic pancreatitis. According to the research conducted in this regard, chlorophyll helps in relieving the fever and alleviates the abdominal pain and discomfort caused by pancreatitis without causing any side effects.

Oral Health

Chlorophyll helps in the treatment of dental problems such as pyorrhea. It is utilized for curing the symptoms of oral infections and calming inflamed and bleeding gums.

Culinary Usage

Along with its regular consumption, chlorophyll and its derivative chlorophyll in also serve as a food additive and are used to provide green colour to a variety of foods and beverages. Chlorophyll has a registered E number E141 and is known as natural green 3.

Summary

Chlorophyll provides energy of sun to in the concentrated form to our bodies and is the one of the most useful nutrient present around us. It enhances the energy levels and amplifies the state of general well being. Chlorophyll is also beneficial in obesity, diabetes, gastritis, hemorrhoids, asthma and skin disorders such as eczema. It helps in curing rashes and fighting skin infections. Consumption of chlorophyll prophylactically also averts the adverse effects of surgery and is advised to be administered before and after the surgeries. The magnesium content in chlorophyll aids in maintaining the blood flow in the body and sustains normal levels of blood pressure. Chlorophyll as a whole improves cellular growth and restores health and vivacity in the body.

References

Adams, Jad (2004). *Hideous absinthe : a history of the devil in a bottle*. Madison, Wisconsin: University of Wisconsin Press. p. 22. ISBN 978-0-299-20000-8.

Carter, J. Stein (1996). "Photosynthesis". University of Cincinnati.

Cate, Thomas, Perkins, T. D. (2003). "Joseph Pelletier and Joseph Caventou" *Journal of Tree Physiology* 23 (15): 1077–1079. doi: 10.1093/treephys/23.15.1077.

Chen, Min, Schliep, Martin, Willows, Robert D., Cai, Zheng-Li, Neilan, Brett A., Scheer, Hugo (2010). "A Red-Shifted Chlorophyll". *Science* 329 (5997): 1318–1319.Bibcode: 2010Sci.329.1318C. doi: 10.1126/science.1191127.

Chlorophyll molecules are specifically arranged in and around photosystems that are embedded in the thylakoid membranes of chloroplasts. Two types of chlorophyll exist in the photosystems: chlorophyll a and b. Speer, Brian R. (1997). "Photosynthetic Pigments".*UCMP Glossary (online)*. University of California Museum of Paleontology. Retrieved2010-07-17.

Chlorophyll : Global Maps. Earthobservatory.nasa.gov. Retrieved on 2014-02-02.

Delépine, Marcel (1951). "Joseph Pelletier and Joseph Caventou". *Journal of Chemical Education* 28 (9): 454. Bibcode: 1951JChEd. 28.454D. doi: 10.1021/ed028p454.

Duble, Richard L. "Iron Chlorosis in Turfgrass". Texas A and M University. Retrieved 2010-07-17.

Falciatore, A., L Merendino, F Barneche, M Ceol, R Meskauskiene, K Apel, JD Rochaix (2005). The FLP proteins act as regulators of chlorophyll synthesis in response to light and plastid signals in *Chlamydomonas*. *Genes and Dev*, 19: 176-187.

Fleming, Ian (1967). "Absolute Configuration and the Structure of Chlorophyll". *Nature* 216 (5111): 151–152. Bibcode: 1967Natur. 216.151F. doi: 10.1038/216151a0.

Gilpin, Linda (2001). "Methods for analysis of benthic photosynthetic pigment". School of Life Sciences, Napier University. Retrieved 2010-07-17.

Gitelson, Anatoly A, Buschmann, Claus, Lichtenthaler, Hartmut K (1999). "The Chlorophyll Fluorescence Ratio F735/F700 as an Accurate Measure of the Chlorophyll Content in Plants".*Remote Sensing of Environment* 69 (3): 296. doi: 10.1016/S0034-4257(99)00023-1.

Gross, Jeana (1991). *Pigments in vegetables: chlorophylls and carotenoids.* Van Nostrand Reinhold, ISBN 0442006578.

Infrared chlorophyll could boost solar cells. New Scientist. August 19, 2010. Retrieved on 2012-04-15.

Jabr, Ferris (2010) A New Form of Chlorophyll?. *Scientific American.* Retrieved on 2012-04-15.

Jeffrey, S. W., Shibata, Kazuo (1969). "Some Spectral Characteristics of Chlorophyll c from *Tridacna crocea* Zooxanthellae". *Biological Bulletin* (Marine Biological Laboratory) 136(1): 54–62. doi: 10.2307/1539668. JSTOR 1539668.

Karageorgou, P., Manetas, Y. (2006). "The importance of being red when young: Anthocyanins and the protection of young leaves of *Quercus coccifera* from insect herbivory and excess light". *Tree Physiology* 26 (5): 613–21. doi: 10.1093/treephys/26.5.613.

Kephart, John C. (1955). "Chlorophyll derivatives–Their chemistry? Commercial preparation and uses". *Journal of Ecological Botany* 9: 3. doi: 10.1007/BF02984956.

Larkum, edited by Anthony W. D. Larkum, Susan E. Douglas and John A. Raven (2003). *Photosynthesis in algae.* London: Kluwer. ISBN 0-7923-6333-7.

Li, J., M Goldschmidt-Clermont, M P Timko (1997). Chloroplast-encoded chlB is required for light-independent protochlorophyllide reductase activity in *Chlamydomonas reinhardtii. Plant Cell* 5(12): 1817–1829.

Marker, A. F. H. (1972). "The use of acetone and methanol in the estimation of chlorophyll in the presence of phaeophytin". *Freshwater Biology* 2 (4): 361. doi: 10.1111/j.1365-2427.1972.tb00377.x.

Meskauskiene R, Nater M, Goslings D, Kessler F, op den Camp R, Apel K. (2001). "FLU: A negative regulator of chlorophyll biosynthesis in *Arabidopsis thaliana*". *Proceedings of the National Academy of Sciences* 98 (22): 12826–12831.doi: 10.1073/pnas.221252798.

Meskauskiene R, Nater M, Goslings D, Kessler F, op den Camp R, Apel K. FLU: a negative regulator of chlorophyll biosynthesis in *Arabidopsis thaliana.* Proceedings of the National Academy of Sciences of the United States of America. 2001, 98(22): 12826-31.

Motilva, Maria-José (2008). "Chlorophylls – from functionality in food to health relevance".*5th Pigments in Food congress- for quality and health* (Print). University of Helsinki. ISBN 978-952-10-4846-3.

Müller, Thomas, Ulrich, Markus, Ongania, Karl-Hans, Kräutler, Bernhard (2007). "Colourless Tetrapyrrolic Chlorophyll Catabolites Found in Ripening Fruit Are Effective Antioxidants".*Angewandte Chemie* 46 (45): 8699–8702. doi: 10.1002/ anie. 200703587.

Nature (2013). "Unit 1.3. Photosynthetic Cells". *Essentials of Cell Biology*. nature.com.

Tigrina, D, required for regulating the biosynthesis of tetrapyrroles in barley, is an ortholog of the FLU gene of *Arabidopsis thaliana. FEBS Letters*, 553, 119-124.

Ubell, Earl (1952) "Chlorophyll Held Useless As Body Deodorant", Toledo Blade.

US patent 5820916, Sagliano, Frank S. and Sagliano, Elizabeth A., "Method for growing and preserving wheat grass nutrients and products thereof", issued 1998-10-13.

Woodward, R. B., Ayer, W. A., Beaton, J. M., Bickelhaupt, F., Bonnett, R., Buchschacher, P., Closs, G. L., Dutler, H., Hannah, J. (1960). "The total synthesis of chlorophyll".*Journal of the American Chemical Society* 82 (14): 3800–3802. doi: 10.1021/ja01499a093.

Woodward, R. B., Ayer, William A., Beaton, John M., Bickelhaupt, Friedrich, onnett, Raymond, Buchschacher, Paul, Closs, Gerhard L., Dutler, Hans, Hannah, John (1990)."The total synthesis of chlorophyll a" (PDF). *Tetrahedron* 46 (22): 7599–7659.doi: 10.1016/0040-4020(90)80003-Z.

Yamazaki, S., J.Nomata, Y.Fujita (2006) Differential operation of dual protochlorophyllide reductases for chlorophyll biosynthesis in response to environmental oxygen levels in the cyanobacterium *Leptolyngbya boryana. Plant Physiology*, 142, 911-922.

Yuichi FujitaDagger and Carl E. Bauer (2000). Reconstitution of Light-independent Protochlorophyllide Reductase from Purified Bchl and BchN-BchB Subunits. *J. Biol. Chem.*, Vol. 275, Issue 31, 23583-23588.

Chapter 28

Chromatophore

A pigment-containing or pigment-producing cell, especially in certain lizards, that by expansion or contraction can change the colour of the skin. Also called pigment cell. A specialized pigment-bearing organelle in certain photosynthetic bacteria and cyanobacteria. Chromatophores are pigment-containing and light-reflecting cells, or groups of cells, found in a wide range of animals including amphibians, fish, reptiles, crustaceans, cephalopods and bacteria. Mammals and birds, in contrast, have a class of cells called melanocytes for colouration. Chromatophores are largely responsible for generating skin and eye colour in cold-blooded animals and are generated in the neural crest during embryonic development. Mature chromatophores are grouped into subclasses based on their colour ("hue") under white light: xanthophores (yellow), erythrophores (red),

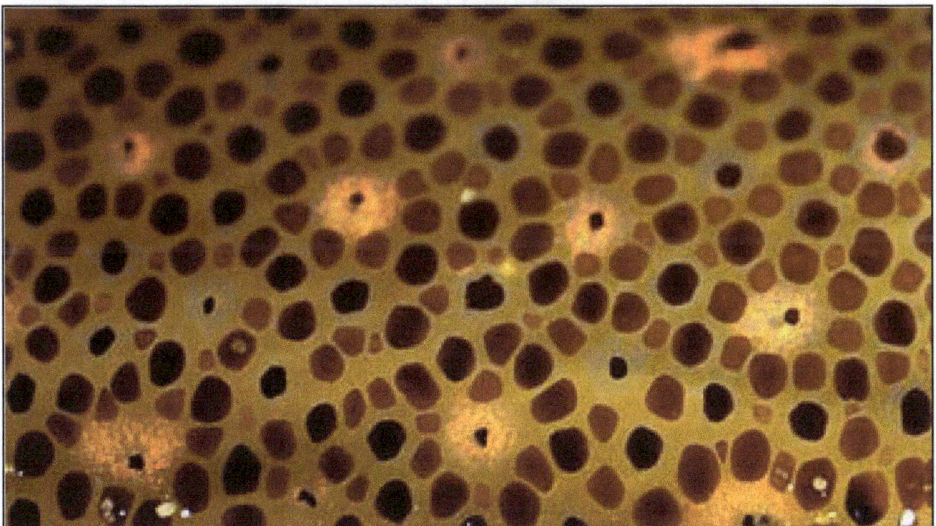

Chromatophores in the Skin of a Squid

iridophores (reflective/iridescent), leucophores (white), melanophores (black/brown) and cyanophores (blue). The term *chromatophore* can also refer to coloured, membrane-associated vesicles found in some forms of photosynthetic bacteria.

Some species can rapidly change colour through mechanisms that translocate pigment and reorient reflective plates within chromatophores. This process, often used as a type of camouflage, is called physiological colour change or metachrosis. Cephalopods such as the octopus have complex chromatophore organs controlled by muscles to achieve this, whereas vertebrates such as chameleons generate a similar effect by cell signalling. Such signals can behormones or neurotransmitters and may be initiated by changes in mood, temperature, stress or visible changes in the local environment. Chromatophores are studied by scientists to understand human disease and as a tool in drug discovery.

Classification

The term *chromatophore* was adopted (following Sangiovanni's *chromoforo*) as the name for pigment-bearing cells derived from the neural crest of cold-blooded vertebrates and cephalopods. The word itself comes from the Greek words *khrôma* (÷ñùµα) meaning "colour," and *phoros* (öïñïò) meaning "bearing". In contrast, the word *chromatocyte* (*cyte* orchachuuõôå being Greek for "cell") was adopted for the cells responsible for colour found in birds and mammals. Only one such cell type, the melanocyte, has been identified in these animals. It was only in the 1960s that chromatophores were well enough understood to enable them to be classified based on their appearance. This classification system persists to this day, even though the biochemistry of the pigments may be more useful to a scientific understanding of how the cells function.

Colour-producing molecules fall into two distinct classes: biochromes and structural colours or "schemochromes". The biochromes include true pigments, such as carotenoids and pteridines. These pigments selectively absorb parts of the visible light spectrum that makes up white light while permitting other wavelengths to reach the eye of the observer. Structural colours are produced by various combinations of diffraction, reflection or scattering of light from structures with a scale around a quarter of the wavelength of light. Many such structures interfere with some wavelengths (colours) of light and transmit others, simply because of their scale, so they often produce iridescence, creating different colours when seen from different directions. Whereas all chromatophores contain pigments or reflecting structures (except when there has been a mutation, as in albinism), not all pigment-containing cells are chromatophores. Haem, for example, is a biochrome responsible for the red appearance of blood. It is found primarily in red blood cells (erythrocytes), which are generated in bone marrow throughout the life of an organism, rather than being formed during embryological development. Therefore erythrocytes are not classified as chromatophores.

Xanthophores and Erythrophores

Chromatophores that contain large amounts of yellow pteridine pigments are named xanthophores; those with mainly red/orangecarotenoids are termed erythrophores. However, vesicles containing pteridine and carotenoids are sometimes found in the same cell, in which case the overall colour depends on the ratio of red and yellow pigments. Therefore, the distinction between these chromatophore types is not always clear. Most chromatophores can generate pteridines from guanosine triphosphate, but xanthophores appear to have supplemental biochemical pathways enabling them to accumulate yellow pigment. In contrast, carotenoids are metabolised and transported to erythrophores. This was first demonstrated by rearing normally green frogs on a diet of carotene-restricted crickets. The absence of carotene in the frogs' diet meant that the red/orange carotenoid colour 'filter' was not present in their erythrophores. This made the frogs appear blue instead of green.

Iridophores and Leucophores

Iridophores, sometimes also called guanophores, are pigment cells that reflect light using plates of crystalline chemochromes made from guanine. When illuminated they generate iridescent colours because of the diffraction of light within the stacked plates. Orientation of the schemochrome determines the nature of the colour observed. By using biochromes as coloured filters, iridophores create an optical effect known as Tyndall or Rayleigh scattering, producing bright blue or green colours. A related type of chromatophore, the leucophore, is found in some fish, in particular in the tapetum lucidum. Like iridophores, they utilize crystalline purines (often guanine) to reflect light. Unlike iridophores, however, leucophores have more organized crystals that reduce diffraction. Given a source of white light, they produce a white shine. As with xanthophores and erythrophores, in fish the distinction between iridophores and leucophores is not always obvious, but, in general, iridophores are considered to generate iridescent or metallic colours, whereas leucophores produce reflective white hues.

Melanophores

Melanophores contain eumelanin, a type of melanin, that appears black or dark-brown because of its light absorbing qualities. It is packaged in vesicles called melanosomes and distributed throughout the cell. Eumelanin is generated from tyrosine in a series of catalysed chemical reactions. It is a complex chemical containing units of dihydroxyindole and dihydroxyindole-2-carboxylic acid with some pyrrole rings. The key enzyme in melanin synthesis is tyrosinase. When this protein is defective, no melanin can be generated resulting in certain types of albinism. In some amphibian species there are other pigments packaged alongside eumelanin. For example, a novel deep (wine) red-colour pigment was identified in the melanophores of phyllomedusine frogs. This was subsequently identified aspterorhodin, a pteridine dimer that accumulates around eumelanin core and it is also present in a variety of tree frog species from Australia and Papua New Guinea. While it is likely that other lesser-studied species have complex melanophore pigments, it is nevertheless true that the majority of melanophores studied to date do contain eumelanin exclusively.

Humans have only one class of pigment cell, the mammalian equivalent of melanophores, to generate skin, hair and eye colour. For this reason and because the large number and contrasting colour of the cells usually make them very easy to visualise, melanophores are by far the most widely studied chromatophore. However, there are differences between the biology of melanophores and that of melanocytes. In addition to eumelanin, melanocytes can generate a yellow/red pigment called phaeomelanin.

Cyanophores

Nearly all the vibrant blues in animals and plants are created by structural colouration rather than by pigments. However, some types of mandarinfish do possess vesicles of a cyan biochrome of unknown chemical structure in cells named cyanophores. Although they appear unusual in their limited taxonomic range, there may be cyanophores (as well as further unusual chromatophore types) in other fish and amphibians. For example, brightly coloured chromatophores with undefined pigments are found in both poison dart frogs and glass frogs and atypical dichromatic chromatophores, named *erythro-iridophores* have been described in *Pseudochromis diadema*.

Pigment Translocation

Many species are able to translocate the pigment inside their chromatophores, resulting in an apparent change in body colour. This process, known as *physiological colour change*, is most widely studied in melanophores, since melanin is the darkest and most visible pigment. In most species with a relatively thin dermis, the dermal melanophores tend to be flat and cover a large surface area. However, in animals with thick dermal layers, such as adult reptiles, dermal melanophores often form three-dimensional units with other chromatophores. These dermal chromatophore units (DCU) consist of an uppermost xanthophore or erythrophore layer, then an iridophore layer and finally a basket-like melanophore layer with processes covering the iridophores.

Both types of melanophore are important in physiological colour change. Flat dermal melanophores often overlay other chromatophores, so when the pigment is dispersed throughout the cell the skin appears dark. When the pigment is aggregated toward the centre of the cell, the pigments in other chromatophores are exposed to light and the skin takes on their hue. Likewise, after melanin aggregation in DCUs, the skin appears green through xanthophore (yellow) filtering of scattered light from the iridophore layer. On the dispersion of melanin, the light is no longer scattered and the skin appears dark. As the other biochromatic chromatophores are also capable of pigment translocation, animals with multiple chromatophore types can generate a spectacular array of skin colours by making good use of the divisional effect.

The control and mechanics of rapid pigment translocation has been well studied in a number of different species, in particular amphibians and teleost fish. It has been demonstrated that the process can be under hormonal or neuronal control or both. Neurochemicals that are known to translocate pigment include noradrenaline, through itsreceptor on the surface on melanophores. The primary hormones involved in regulating translocation appear to be the melanocortins,melatonin and melanin-concentrating hormone (MCH), that are produced mainly in the pituitary, pineal gland and hypothalamus, respectively. These hormones may also be generated in a paracrine fashion by cells in the skin. At the surface of the melanophore, the hormones have been shown to activate specific G-protein-coupled receptors that, in turn, transduce the signal into the cell. Melanocortins result in the dispersion of pigment, while melatonin and MCH results in aggregation. Numerous melanocortin, MCH and melatonin receptors have been identified in fish and frogs, including a homologue of *MC1R*, a melanocortin receptor known to regulate skin and hair colour in humans. It has been demonstrated that *MC1R* is required in zebrafish for dispersion of melanin. Inside the cell, cyclic adenosine monophosphate (cAMP) has been shown to be an important second messenger of pigment translocation. Through a mechanism not yet fully understood, cAMP influences other proteins such as protein kinase A to drive molecular motors carrying pigment containing vesicles along both microtubules and microfilaments.

Background Adaptation

Most fish, reptiles and amphibians undergo a limited physiological colour change in response to a change in environment. This type of camouflage, known as *background adaptation*, most commonly appears as a slight darkening or lightening of skin tone to approximately mimicthe hue of the immediate environment. It has been demonstrated that the background adaptation process is vision-dependent (it appears the animal needs to be able to see the environment to adapt to it) and that melanin translocation in melanophores is the major factor in colour change. Some animals, such as chameleons and anoles, have a highly developed background adaptation response capable of generating a number of different colours very rapidly. They have adapted the capability to change colour in response to temperature, mood, stress levels and social cues, rather than to simply mimic their environment.

Development

During vertebrate embryonic development, chromatophores are one of a number of cell types generated in the neural crest, a paired strip of cells arising at the margins of the neural tube. These cells have the ability to migrate long distances, allowing chromatophores to populate many organs of the body, including the skin, eye, ear and brain. Leaving the neural crest in waves, chromatophores take either a dorsolateral route through the dermis, entering the ectoderm through small holes in the basal lamina, or a ventromedial route between the somites and the neural tube. The exception to this is the melanophores of the retinal pigmented epithelium of the eye. These are not derived from the neural crest. Instead, an outpouching of the neural tube generates the optic cup, which, in turn, forms the retina.

When and how multipotent chromatophore precursor cells (called *chromatoblasts*) develop into their daughter subtypes is an area of ongoing research. It is known in zebrafish embryos, for example, that by 3 days after fertilization each of the cell classes found in the adult fish melanophores, xanthophores and iridophores are already present. Studies using mutant fish have demonstrated that transcription factors such as *kit, sox10* and *mitf* are important in controlling chromatophore differentiation. If these proteins are defective, chromatophores may be regionally or entirely absent, resulting in a leucistic disorder.

Practical Applications

Chromatophores are sometimes used in applied research. For example, zebrafish larvae are used to study how chromatophores organise and communicate to accurately generate the regular horizontal striped pattern as seen in adult fish. This is seen as a useful model system for understanding patterning in the evolutionary developmental biology field. Chromatophore biology has also been used to model human condition or disease, including melanoma and albinism. Recently, the gene responsible for the melanophore-specific *golden* zebrafish strain, *Slc24a5*, was shown to have a human equivalent that strongly correlates with skin colour.

Chromatophores are also used as a biomarker of blindness in cold-blooded species, as animals with certain visual defects fail to background adapt to light environments. Human homologues of receptors that mediate pigment translocation in melanophores are thought to be involved in processes such as appetite suppression and tanning, making them attractive targets for drugs. Therefore, pharmaceutical companies have developed a biological assay for rapidly identifying potential bioactive compounds using melanophores from the African clawed frog. Other scientists have developed techniques for using melanophores as biosensors and for rapid disease detection (based on the discovery that pertussis toxin blocks pigment aggregation in fish melanophores). Potential military applications of chromatophore-mediated colour changes have been proposed, mainly as a type of active camouflage, which could as in cuttlefish make objects nearly invisible.

Cephalopod Chromatophores

Coleoid cephalopods (including octopuses, squids and cuttlefish) have complex multicellular organs that they use to change colour rapidly, producing a wide variety of bright colours and patterns. Each chromatophore unit is composed of a single chromatophore cell and numerous muscle, nerve, glial and sheath cells. Inside the chromatophore cell, pigment granules are enclosed in an elastic sac, called the cytoelastic sacculus. To change colour the animal distorts the sacculus form or size by muscular contraction, changing its translucency, reflectivity, or opacity. This differs from the mechanism used in fish, amphibians and reptiles in that the shape of the sacculus is changed, rather than translocating pigment vesicles within the cell. However, a similar effect is achieved.

Octopuses can operate chromatophores in complex, wavelike chromatic displays, resulting in a variety of rapidly changing colour schemes. The nerves that operate the chromatophores are thought to be positioned in the brain in a pattern similar to that of the chromatophores they each control. This means the pattern of colour change matches the pattern of neuronal activation. This may explain why, as the neurons are activated one after another, the colour change occurs in waves. Like chameleons, cephalopods use physiological colour change for social interaction. They are also among the most skilled at background adaptation, having the ability to match both the colour and the texture of their local environment with remarkable accuracy.

Bacteria

Chromatophores are found in membranes of phototrophic bacteria. Used primarily for photosynthesis, they contain bacteriochlorophyll pigments and carotenoids. In purple bacteria, such as *Rhodospirillum rubrum*, the light-harvesting proteins are intrinsic to the chromatophore membranes. However, in green sulfur bacteria, they are arranged in specialised antenna complexes called chlorosomes.

Chloroplasts are organelles, specialized subunits, in plant and algal cells. Their main role is to conduct photosynthesis, where the photosynthetic pigment chlorophyll captures the energy from sunlight and stores it in the energy storage molecules ATP and NADPH while freeing oxygen from water. They then use the ATP and NADPH to make organic molecules from carbon dioxide in a process known as the Calvin cycle. Chloroplasts carry out a number of other functions, including fatty acid synthesis, much amino acid synthesis and the immune response in plants. The number of chloroplasts per cell varies from 1 chloroplast per cell in algae and can get up to 100 chloroplasts per cell in plants like arabidopsis and wheat. A chloroplast is one of three types of plastids, characterized by its high concentration of chlorophyll. (The other two types, the leucoplast and the chromoplast, contain little chlorophyll and do not carry out photosynthesis.) Chloroplasts are highly dynamic they circulate and are moved around within plant cells and occasionally pinch in two to reproduce. Their behavior is strongly influenced by environmental factors like light colour and intensity. Chloroplasts, like mitochondria, contain their own DNA, which is thought to be inherited from their ancestor a photosyntheticcyanobacterium that was engulfed by an early eukaryotic cell. Chloroplasts cannot be made by the plant cell and must be inherited by each daughter cell during cell division. With one exception (the amoeboid *Paulinella chromatophora*), all chloroplasts can probably be traced back to a single endosymbiotic event (the cyanobacterium being engulfed by the eukaryote). Despite this, chloroplasts can be found in an extremely wide set of organisms, some not even directly related to each other a consequence of many secondary and even tertiary endosymbiotic events.

References

Andersson, TP, Filippini, D, Suska, A, Johansson, TL, Svensson, SP, Lundström, I (2005). "Frog melanophores cultured on fluorescent microbeads: biomimic-based biosensing". *Biosens Bioelectron* 21 (1): 111–20. doi: 10.1016/j.bios. 2004.08.043.

Aristotle. *Historia Animalium*. IX, 622a: 2-10. About 400 BC. Cited in Luciana Borrelli, Francesca Gherardi, Graziano Fiorito. *A catalogue of body patterning in Cephalopoda*. Firenze University Press, 2006. Abstract Google books

Aspengren, S, Sköld, HN, Quiroga, G, Mårtensson, L, Wallin, M (2003). "Noradrenaline- and melatonin-mediated regulation of pigment aggregation in fish melanophores". *Pigment Cell Res* 16 (1): 59–64. doi: 10.1034/j.1600-0749. 2003. 00003.x.

Bagnara JT. (1998). Comparative Anatomy and Physiology of Pigment Cells in Nonmammalian Tissues. In: *The Pigmentary System: Physiology and Pathophysiology*, Oxford University Press. ISBN 0-19-509861-7

Bagnara, J.T. (2003). "Enigmas of Pterorhodin, a Red Melanosomal Pigment of Tree Frogs". *Pigment Cell Research* 16 (5): 510–516. doi: 10.1034/j.1600-0749.2003.00075.x.PMID 12950730.

Bagnara, J.T., Taylor, JD, Prota, G (1973). "Colour changes, unusual melanosomes and a new pigment from leaf frogs". *Science* 182 (4116): 1034–5. doi: 10.1126/science.182.4116.1034.

Bagnara, JT (1966). "Cytology and cytophysiology of non-melanophore pigment cells". *International Review of Cytology*. International Review of Cytology 20: 173–205.doi: 10.1016/S0074-7696(08)60801-3. ISBN 978-0-12-364320-9.

Bagnara, JT, Taylor, JD, Hadley, ME (1968). "The dermal chromatophore unit". *J Cell Biol*38 (1): 67–79. doi: 10.1083/jcb.38.1.67.

Cloney, RA, Florey, E (1968). "Ultrastructure of cephalopod chromatophore organs". *Zeitschrift fur Zellforschung und mikroskopische Anatomie (Vienna, Austria: 1948)* 89 (2): 250–80. doi: 10.1007/BF00347297.

Darwin, Charles (1860). "Chapter 1. Habits of a Sea-slug and Cuttle-fish". *Journal Of Researches Into The Natural History And Geology Of The Countries Visited During The Voyage Round The World Of H.M.S. 'Beagle' Under The Command Of Captain Fitz Roy, R.N.* John Murray, London. p. 7.

Deacon, SW, Serpinskaya, AS, Vaughan, PS, Lopez Fanarraga, M, Vernos, I, Vaughan, KT, Gelfand, VI (2003). "Dynactin is required for bidirectional organelle transport". *The Journal of Cell Biology* 160 (3): 297–301. doi: 10.1083/jcb.200210066.

Demski, LS (1992). "Chromatophore systems in teleosts and cephalopods: a levels oriented analysis of convergent systems". *Brain, behavior and evolution* 40 (2–3): 141–56.doi: 10.1159/000113909.

Fox, DL. (1976). *Animal Biochromes and Structural Colours: Physical, Chemical, Distributional and Physiological Features of Coloured Bodies in the Animal World.* University of California Press, Berkeley. ISBN 0-520-02347-1

Frigaard, NU, Bryant, DA (2004). "Seeing green bacteria in a new light: genomics-enabled studies of the photosynthetic apparatus in green sulfur bacteria and filamentous anoxygenic phototrophic bacteria". *Archives of Microbiology* 182 (4): 265–76. doi: 10.1007/s00203-004-0718-9.

Fujii, R. (2000). "The regulation of motile activity in fish chromatophores". *Pigment Cell Res* 13 (5): 300–19. doi: 10.1034/j.1600-0749.2000.130502.x.

Goda M, Ohata M, Ikoma H, Fujiyoshi Y, Sugimoto M, Fujii R (2011). Integumental reddish-violet colouration owing to novel dichromatic chromatophores in the teleost fish, Pseudochromis diadema". *Pigment Cell Melanoma Res* 24 (4): 614–7. doi: 10.1111/j.1755-148X.2011.00861.x.

Hansford, Dave (2008). "Cuttlefish Change Colour, Shape-Shift to Elude Predators". *National Geographic News* (Wellington, New Zealand). [.] cuttlefish have instead relied on invisibility, a talent that may have applications for human technology. Norman said the military has shown interest in cuttlefish camouflage with a view to one day incorporating similar mechanisms in soldiers' uniforms.

Ito, S, Wakamatsu, K. (2003). "Quantitative analysis of eumelanin and pheomelanin in humans, mice and other animals: a comparative review". *Pigment Cell Res* 16 (5): 523–31.doi: 10.1034/j.1600-0749.2003.00072.x.

Jayawickreme, CK, Sauls, H, Bolio, N, Ruan, J, Moyer, M, Burkhart, W, Marron, B, Rimele, T, Shaffer, J (1999). "Use of a cell-based, lawn format assay to rapidly screen a 442,368 bead-based peptide library". *J Pharmacol Toxicol Methods* 42 (4): 189–97. doi: 10.1016/S1056-8719(00)00083-6.

Karlsson, JO andersson, RG, Askelöf, P, Elwing, H, Granström, M, Grundström, N, Lundström, I, Ohman, L (1991). "The melanophore aggregating response of isolated fish scales: a very rapid and sensitive diagnosis of whooping cough". *FEMS Microbiol Lett* 66 (2): 169–75.

Kashina, AS, Semenova, IV, Ivanov, PA, Potekhina, ES, Zaliapin, I, Rodionov, VI (2004). "Protein kinase A, which regulates intracellular transport, forms complexes with molecular motors on organelles". *Current biology : CB* 14 (20): 1877–81. doi: 10.1016/j.cub.2004.10.003.

Kelsh, RN (2004). "Genetics and evolution of pigment patterns in fish". *Pigment cell research/sponsored by the European Society for Pigment Cell Research and the International Pigment Cell Society* 17 (4): 326–36. doi: 10.1111/j.1600-0749. 2004.00174.x.

Kelsh, RN, Schmid, B, Eisen, JS (2000). "Genetic analysis of melanophore development in zebrafish embryos". *Dev Biol* 225 (2): 277–93. doi: 10.1006/dbio.2000.9840.

Lamason, RL, Mohideen, MA, Mest, JR, Wong, AC, Norton, HL, Aros, MC, Jurynec, MJ, Mao, X. (2005). "SLC24A5, a putative cation exchanger, affects pigmentation in zebrafish and humans". *Science* 310 (5755): 1782–6. doi: 10.1126/science.1116238.

Lee I. Nanotubes for noisy signal processing *PhD Thesis*. 2005, University of Southern California.

Logan, D. W., Burn, SF, Jackson, IJ (2006). "Regulation of pigmentation in zebrafish melanophores". *Pigment Cell Research* 19 (3): 206–213. doi: 10.1111/j.1600-0749.2006.00307.x.

Logan, DW, Bryson-Richardson, RJ, Pagán, KE, Taylor, MS, Currie, PD, Jackson, IJ (2003). "The structure and evolution of the melanocortin and MCH receptors in fish and mammals". *Genomics* 81 (2): 184–91. doi: 10.1016/S0888-7543(02)00037-X.

Logan, DW, Bryson-Richardson, RJ, Taylor, MS, Currie, P, Jackson, IJ (2003). "Sequence characterization of teleost fish melanocortin receptors". *Ann N Y Acad Sci* 994: 319–30.doi: 10.1111/j.1749-6632.2003.tb03196.x

Matsumoto, J (1965). "Studies on fine structure and cytochemical properties of erythrophores in swordtail, *Xiphophorus helleri*, with special reference to their pigment granules (pterinosomes)". *J Cell Biol* 27 (3): 493–504. doi: 10.1083/jcb.27.3.493.

Morrison, RL. (1995). "A transmission electron microscopic (TEM) method for determining structural colours reflected by lizard iridophores". *Pigment Cell Res* 8 (1): 28–36.doi: 10.1111/j.1600-0749.1995.tb00771.x.

Neuhauss, SC. (2003). "Behavioral genetic approaches to visual system development and function in zebrafish" (PDF). *J Neurobiol* 54 (1): 148–60. doi: 10.1002/neu.10165.

Palazzo, RE, Lynch, TJ, Lo, SJ, Taylor, JD, Tchen, TT (1989). "Rearrangements of pterinosomes and cytoskeleton accompanying pigment dispersion in goldfish xanthophores". *Cell Motil Cytoskeleton* 13 (1): 9–20. doi: 10.1002/cm.970130103.

Porras, MG, De Loof, A, Breuer, M, Aréchiga, H (2003). "Procambarus clarkii". *Peptides* 24(10): 1581–9. doi: 10.1016/j.peptides.2003.08.016.

Richardson, J, Lundegaard, PR, Reynolds, NL, Dorin, JR, Porteous, DJ, Jackson, IJ, Patton, EE (2008). "mc1r Pathway regulation of zebrafish melanosome dispersion". *Zebrafish* 5 (4): 289–95. doi: 10.1089/zeb.2008.0541.

Rodionov, VI, Hope, AJ, Svitkina, TM, Borisy, GG (1998). "Functional coordination of microtubule-based and actin-based motility in melanophores". *Current biology: CB* 8 (3): 165–8. doi: 10.1016/S0960-9822(98)70064-8.

Salton, MR (1987). "Bacterial membrane proteins". *Microbiological sciences* 4 (4): 100–5.

Sangiovanni G. (1819). Descrizione di un particolare sistema di organi cromoforo espansivo-dermoideo e dei fenomeni che esso produce, scoperto nei molluschi cefaloso. *G. Enciclopedico Napoli*. 9: 1–13.

Schwalm, PA, Starrett, PH, McDiarmid, RW (1977). "Infrared reflectance in leaf-sitting neotropical frogs". *Science* 196 (4295): 1225–7. doi: 10.1126/science.860137.

Scott M. Boback and Lynn M. Siefferman (2010). "Variation in Colour and Colour Change in Island and Mainland Boas (*Boa constrictor*)". *Journal of Herpetology* 44 (4): 506–515. doi: 10.1670/09-026.1.

Snider, J, Lin, F, Zahedi, N, Rodionov, V, Yu, CC, Gross, SP (2004). "Intracellular actin-based transport: How far you go depends on how often you switch". *Proceedings of the National Academy of Sciences of the United States of America* 101 (36): 13204–9.doi: 10.1073/pnas.0403092101.

Sugden, D, Davidson, K, Hough, KA, Teh, MT (2004). "Melatonin, melatonin receptors and melanophores: a moving story". *Pigment Cell Res* 17 (5): 454–60. doi: 10.1111/j.1600-0749.2004.00185.x.

Taylor, JD. (1969). "The effects of intermedin on the ultrastructure of amphibian iridophores".*Gen Comp Endocrinol* 12 (3): 405–16. doi: 10.1016/0016-6480(69) 90157-9.

Valverde, P, Healy, E, Jackson, I, Rees, JL, Thody, AJ (1995). "Variants of the melanocyte-stimulating hormone receptor gene are associated with red hair and fair skin in humans". *Nat Genet* 11 (3): 328–30. doi: 10.1038/ng1195-328.

Chapter 29

Melanin

Melanin is a complex polymer derived from the amino acid tyrosine. Melanin is responsible for determining skin and hair colour and is present in the skin to varying degrees, depending on how much a population has been exposed to the sun historically.

Melanin Synthesis

Numerous steps are involved in the biosynthesis of melanin. The first step is catalysis of the chemical L-3,4-dihydroxyphenylalanine by tyrosinase. A lack of tyrosine can lead to albinism. Tyrosine is only found in specialized cells called melanocytes, inside which tiny granules of melanin pigment are contained in vesicles called melanosomes. These melanosomes leave the melanocytes and move into other cells in the epidermis. Mostly brown or black in colour, melanin deposits determine the skin pigment which varies depending on the number and distribution of the melanosomes. Aside from determining skin colour, the light absorbent melanin protects the DNA against UV radiation from the sun and it is considered a potential candidate in melanoma treatment.

Melanin is found in several areas of the human body including:

☆ Skin where it provides skin colour

☆ Hair

☆ Pupils or irises of the eyes

☆ Stria vascularis of the inner ear

☆ Areas of the brain, the substantia nigra and locus coeruleus

☆ The medulla and zona reticularis of the adrenal gland

Types of Melanin

Some of the different types of melanin include eumelanin, pheomelanin and neuromelanin. Eumelanin is found in the hair, skin and dark areas around the nipples. It is particularly abundant among black populations and provides black and brown pigment to the hair, skin and eyes. When eumelanin is present only in small amounts, hair may be blonde.

Pheomelanin is also found in the hair and skin. This type of melanin provides pink and red colours and is the main pigment found among red-haired individuals. This type of melanin is not as protective against UV- radiation induced cancer as eumelanin. Neuromelanin is a form of melanin found in different areas of the brain and loss of this melanin may cause several neurological disorders.

Melanin is a broad term for a group of natural pigments found in most organisms (arachnids are one of the few groups in which it has not been detected). Melanin is produced by the oxidation of the amino acid tyrosine, followed by polymerization. The pigment is produced in a specialized group of cells known as melanocytes. There are three basic types of melanin: eumelanin, pheomelanin and neuromelanin. The most common type of melanin is eumelanin. There are two types of eumelanin- brown eumelanin and black eumelanin. Pheomelanin is a cysteine-containing red polymer of benzothiazine units largely responsible for red hair, among other pigmentation. Neuromelanin is found in the brain, though its function remains obscure. In the skin, melanogenesis occurs after exposure to UV radiation, causing the skin to visibly tan. Melanin is an effective absorber of light; the pigment is able to dissipate over 99.9 per cent of absorbed UV radiation. Because of this property, melanin is thought to protect skin cells from UVB radiation damage, reducing the risk of cancer. Furthermore, though exposure to UV radiation is associated with increased risk of malignant melanoma, a cancer of the melanocytes, studies have shown a lower incidence for skin cancer in individuals with more concentrated melanin, *i.e.* darker skin tone. Nonetheless, the relationship between skin pigmentation and photoprotection is still being clarified.

Humans

In humans, melanin is the primary determinant of skin colour. It is also found in hair, the pigmented tissue underlying the iris of the eye and the stria vascularis of the inner ear. In the brain, tissues with melanin include the medulla and pigment-bearing neurons within areas of thebrainstem, such as the locus coeruleus and the substantia nigra. It also occurs in the zona reticularis of the adrenal gland. The melanin in the skin is produced by melanocytes, which are found in the basal layer of the epidermis. Although, in general, human beings possess a similar concentration of melanocytes in their skin, the melanocytes in some individuals and ethnic groups produce variable amounts of melanin. Some humans have very little or no melanin synthesis in their bodies, a condition known as albinism. Because melanin is an aggregate of smaller component molecules, there are many different types of melanin with differing proportions and bonding patterns of these component molecules. Both pheomelanin and eumelanin are found in human skin and hair, but eumelanin is the most abundant melanin in humans, as well as the form most likely to be deficient in albinism.

Eumelanin

Eumelanin polymers have long been thought to comprise numerous cross-linked 5,6-dihydroxyindole (DHI) and 5,6-dihydroxyindole-2-carboxylic acid (DHICA) polymers.

There are two types of eumelanin- brown eumelanin and black eumelanin, which chemically differ from each other in their pattern of polymeric bonds. A small amount of black eumelanin in the absence of other pigments causes grey hair. A small amount of brown eumelanin in the absence of other pigments causes yellow (blond) colour hair.

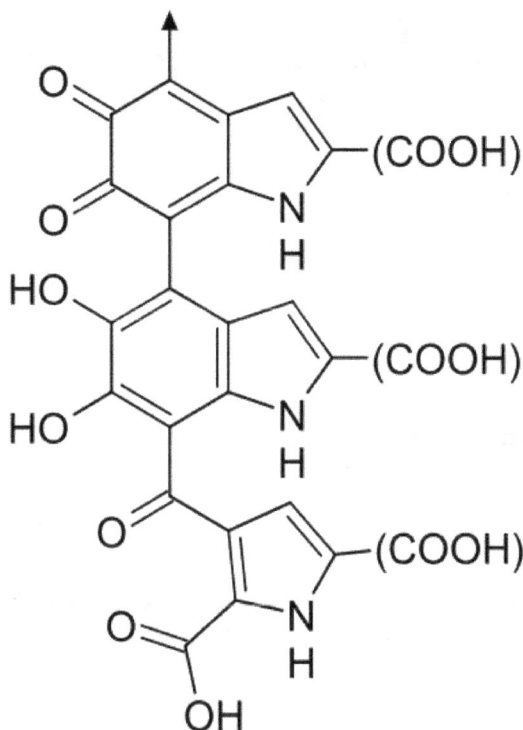

Part of the structural formula of eumelanin. "(COOH)" can be COOH or H, or (more rarely) other substituents. The arrow denotes where the polymer continues.

Pheomelanin

Pheomelanin imparts a pink to red hue, depending upon its concentration. Pheomelanin is particularly concentrated in the lips, nipples, glans of the penis and vagina. When a small amount of brown eumelanin in hair, which would otherwise cause blond hair, is mixed with red pheomelanin, the result is red hair. In chemical terms, pheomelanin differs from eumelanin in that its oligomer structure incorporatesbenzothiazine and benzothiazole units that are produced, instead of DHI and DHICA, when the amino acid L-cysteine is present.

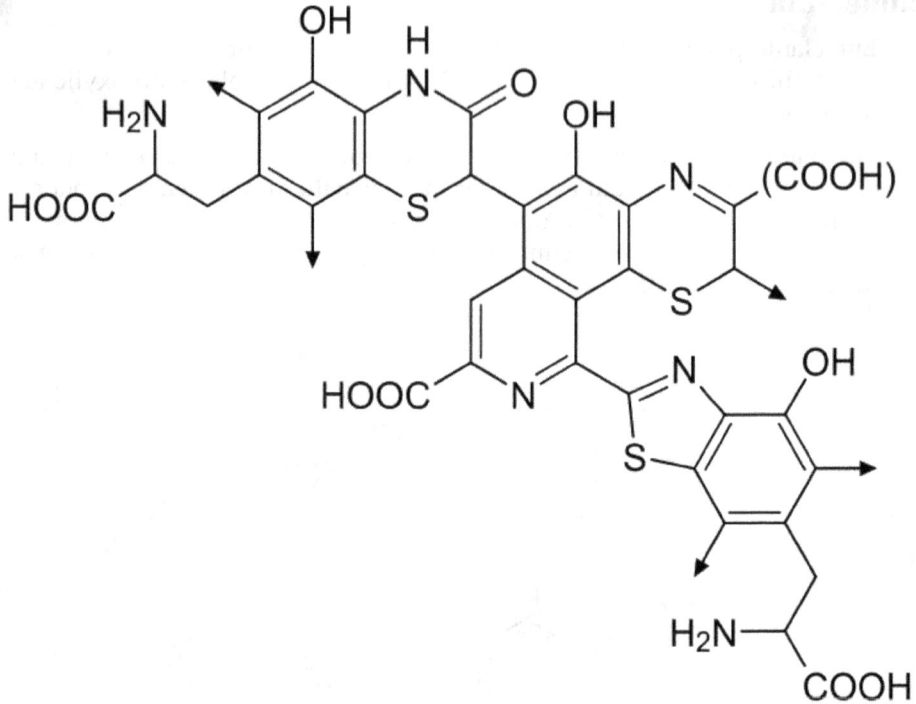

Part of the structural formula of pheomelanin. "(COOH)" can be COOH or H, or (more rarely) other substituents. The arrows denote where the polymer continues.

Neuromelanin

Neuromelanin (NM) is a dark polymer pigment produced in specific populations ofcatecholaminergic neurons in the brain, or in the upper vocal cords. Humans have the largest amount of NM, while it is in lesser amount in other non-human primates and totally absent in other species. However, the biological function remains unknown, although human NM has been shown to efficiently bind transition metals such as iron, as well as other potentially toxic molecules. Therefore, it may play crucial roles in apoptosis and the related Parkinson's disease.

Other Organisms

Melanins have very diverse roles and functions in various organisms. A form of melanin makes up the ink used by many cephalopods (see cephalopod ink) as a defense mechanism against predators. Melanins also protect microorganisms, such as bacteria and fungi, against stresses that involve cell damage such as UV radiation from the sun andreactive oxygen species. Melanin also protects against damage from high temperatures, chemical stresses (such as heavy metals and oxidizing agents) and biochemical threats (such as host defenses against invading microbes). Therefore, in many pathogenic microbes (for example, in *Cryptococcus neoformans*, a fungus) melanins appear to play important roles in virulence and pathogenicity by protecting the microbe against immune responses of its host. In invertebrates, a major aspect of the innate immune defense system against invading pathogens involves melanin. Within minutes after infection, the microbe is encapsulated within melanin (melanization) and the generation of free radical byproducts during the formation of this capsule is thought to aid in killing them. Some types of fungi, called radiotrophic fungi, appear to be able to use melanin as a photosynthetic pigment that enables them to capture gamma rays and harness this energy for growth.

Melanin is important in mammalian pigmentation. The black feathers of birds owe their colour to melanin; they are less readily degraded by bacteria than white feathers, or those containing other pigments such as carotenes. In a bird's eye, a specialized organ, rich in blood vessels, the pecten oculi, is also extremely rich in melanin, which has been considered to have role in absorption of light falling on optic disc and using it to warm up the eye. This, in turn may stimulate release of nutrients from pecten oculi to retina, via vitreous humor; it is plausible as bird retina is devoid of its own blood vessels. In pigment epithelium of retina, presence of high amounts of melanin granules may also minimize back-scatter of image light on the retina.

In some mice, melanin is used slightly differently. For instance, in Agouti mice, the hair appears brown because of alternation between black eumelanin production and a yellow variety of pheomelanin. The hairs are actually banded black and yellow and the net effect is the brown colour of most mice. Some genetic irregularities can produce either fully black or fully yellow mice.

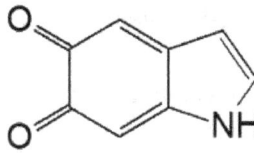

Chemical Structure of indole-5,6-quinone

Melanins produced by plants are sometimes referred to as 'Catechol Melanins' as they can yield catechol on alkali fusion. It is commonly seen in the enzymatic browning of fruits such as bananas. Biosynthesis involves the oxidation of indole-5,6-quinone by the tyrosinase typepolyphenol oxidase from tyrosine and catecholamines leading to the formation of catechol melanin. Despite this many plants contain compounds which inhibit the production of melanins.

Biosynthetic Pathways

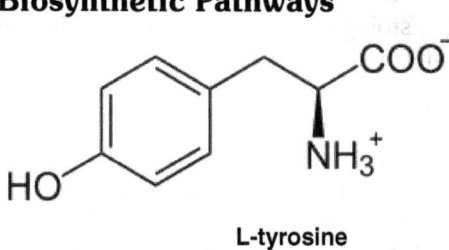

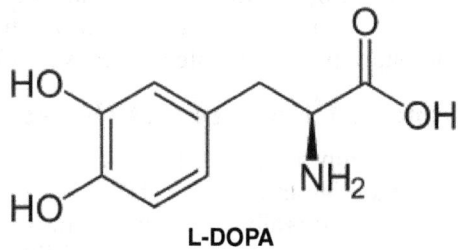

L-tyrosine   L-DOPA

L-dopaquinone   **L-leucodopachrome**

L-dopachrome

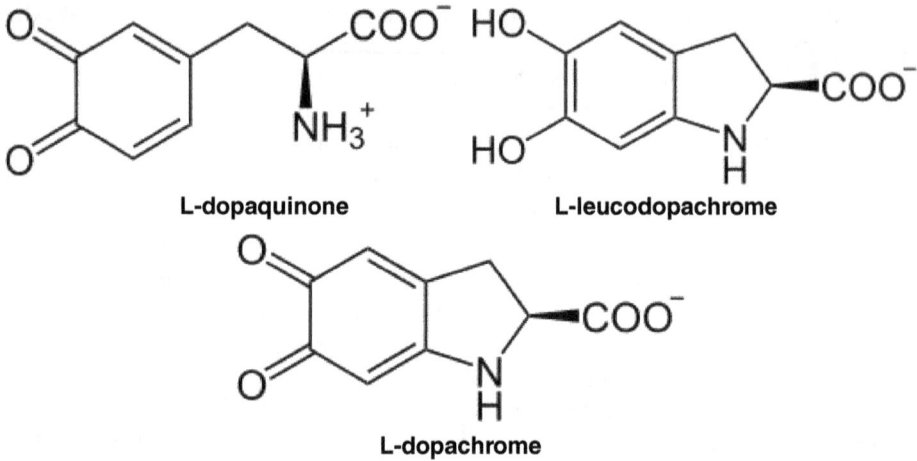

The first step of the biosynthetic pathway for both eumelanins and pheomelanins is catalysed by tyrosinase:

Tyrosine → DOPA → dopaquinone

Dopaquinone can combine with cysteine by two pathways to benzothiazines and pheomelanins

Dopaquinone + cysteine → 5-S-cysteinyldopa → benzothiazine intermediate → pheomelanin

Dopaquinone + cysteine → 2-S-cysteinyldopa → benzothiazine intermediate → pheomelanin

Also, dopaquinone can be converted to leucodopachrome and follow two more pathways to the eumelanins

Dopaquinone → leucodopachrome → dopachrome → 5,6-dihydroxyindole-2-carboxylic acid → quinone → eumelanin

Dopaquinone → leucodopachrome → dopachrome → 5,6-dihydroxyindole → quinone → eumelanin

Microscopic Appearance

Melanin is brown, non-refractile and finely granular with individual granules having a diameter of less than 800 nanometers. This differentiates melanin from common blood breakdown pigments, which are larger, chunky and refractile and range in colour from green to yellow or red-brown. In heavily pigmented lesions, dense aggregates of melanin can obscure histologic detail. A dilute solution of potassium permanganate is effective melanin bleach.

Genetic Disorders and Disease States

Melanin deficiency has been connected for some time with various genetic abnormalities and disease states. There are approximately nine different types of oculocutaneous albinism, which is mostly an autosomal recessive disorder. Certain ethnicities have higher incidences of different forms. For example, the most common type, called oculocutaneous albinism type 2 (OCA2), is especially frequent among people of black African descent. It is an autosomal recessive disorder characterized by a congenital reduction or absence of melanin pigment in the skin, hair and eyes. The estimated frequency of OCA2 among African-Americans is 1 in 10,000, which contrasts with a frequency of 1 in 36,000 in white Americans. In some African nations, the frequency of the disorder is even higher, ranging from 1 in 2,000 to 1 in 5,000. Another form of Albinism, the "yellow oculocutaneous albinism", appears to be more prevalent among the Amish, who are of primarily Swiss and German ancestry. People with this IB variant of the disorder commonly have white hair and skin at birth, but rapidly develop normal skin pigmentation in infancy.

Ocular albinism affects not only eye pigmentation but visual acuity, as well. People with albinism typically test poorly, within the 20/60 to 20/400 range. In addition, two forms of albinism, with approximately 1 in 2700 most prevalent among people of Puerto Rican origin, are associated with mortality beyond melanoma-related deaths. Mortality also is increased in patients with Hermansky-Pudlak syndrome and Chediak-Higashi syndrome. Patients with Hermansky-Pudlak syndrome have a bleeding diathesis secondary to platelet dysfunction and also experience restrictive lung disease (pulmonary fibrosis), inflammatory bowel disease, cardiomyopathy and renal disease. Patients with Chediak-Higashi syndrome are susceptible to infection and also can develop lymphofollicular malignancy.

The role that melanin deficiency plays in such disorders remains under study.

The connection between albinism and deafness is well known, though poorly understood. In his 1859 treatise *On the Origin of Species*, Charles Darwin observed that "cats which are entirely white and have blue eyes are generally deaf". In humans, hypopigmentation and deafness occur together in the rare Waardenburg's syndrome, predominantly observed among the Hopi in North America. The incidence of albinism in Hopi Indians has been estimated at approximately 1 in 200 individuals. It is interesting to note that similar patterns of albinism and deafness have been found in other mammals, including dogs and rodents. However, a lack of melanin *per se* does not appear to be directly responsible for deafness associated with hypopigmentation, as most individuals lacking the enzymes required to synthesize melanin have normal auditory function. Instead the absence of melanocytes in the stria vascularis of the inner ear results in cochlear impairment, though why this is, is not fully understood.

In Parkinson's disease, a disorder that affects neuromotor functioning, there is decreased neuromelanin in the substantia nigra and locus coeruleus as consequence of specific dropping out of dopaminergic and noradrenergic pigmented neurons. This results in diminished dopamine and norepinephrine synthesis. While no correlation between race and the level of neuromelanin in the substantia nigra has been reported, the significantly lower incidence of Parkinson's in blacks than in whites has "prompt[ed] some to suggest that cutaneous melanin might somehow serve to protect the neuromelanin in substantia nigra from external toxins." Also see Nicolaus review article on the function of neuromelanins.

In addition to melanin deficiency, the molecular weight of the melanin polymer may be decreased by various factors such as oxidative stress, exposure to light, perturbation in its association with melanosomal matrix proteins, changes in pH, or in local concentrations of metal ions. A decreased molecular weight or a decrease in the degree of polymerization of ocular melanin has been proposed to turn the normally anti-oxidant polymer into a pro-oxidant. In its pro-oxidant state, melanin has been suggested to be involved in the causation and progression of macular degeneration and melanoma. Rasagiline, an important monotherapy drug in Parkinson's disease, has melanin binding properties and melanoma tumor reducing properties.

Higher eumelanin levels also can be a disadvantage, however, beyond a higher disposition toward vitamin D deficiency. Dark skin is a complicating factor in the laser removal of port-wine stains. Effective in treating white skin, in general, lasers are less successful in removing port-wine stains in people of Asian or African descent. Higher concentrations of melanin in darker-skinned individuals simply diffuse and absorb the laser radiation, inhibiting light absorption by the targeted tissue. In similar manner, melanin can complicate laser treatment of other dermatological conditions in people with darker skin. Freckles and moles are formed where there is a localized concentration of melanin in the skin. They are highly associated with pale skin. Nicotine has an affinity for melanin-containing tissues because of its precursor function in melanin synthesis or its irreversible binding of melanin. This has been suggested to underlie the increased nicotine dependence and lower smoking cessation rates in darker pigmented individuals.

Human Adaptation

Melanocytes insert granules of melanin into specialized cellular vesicles called melanosomes. These are then transferred into the other skin cells of the human epidermis. The melanosomes in each recipient cell accumulate atop the cell nucleus, where they protect the nuclear DNA from mutations caused by the ionizing radiation of the sun's ultraviolet rays. In general, people whose ancestors lived for long periods in the regions of the globe near the equator have larger quantities of eumelanin in their skins. This makes their skins brown or black and protects them against high levels of exposure to the sun, which more frequently result in melanomas in lighter-skinned people. With humans, exposure to sunlight stimulates the skin to produce vitamin D. Because high levels of cutaneous melanin act as a natural sun screen, dark skin can be a risk factor for vitamin D deficiency in regions of the Earth known as cool temperate zones, *i.e.*, above 36 degrees latitude in the Northern hemisphere and below 36 degrees in the Southern hemisphere. As a result of this, health authorities in Canada and the USA have issued recommendations for people with darker complexions (including people of southern European descent) to consume between 1000-2000 IU (International Units) of vitamin D, daily, autumn through spring. The most recent scientific evidence indicates that all humans originated in Africa, then populated the rest of the world through successive radiations. It seems likely that the first modern humans had relatively large numbers of eumelanin-producing melanocytes. In accordance, they had darker skin as with the indigenous people of Africa today. As some of these original peoples migrated and settled in areas of Asia and Europe, the selective pressure for eumelanin production decreased in climates where radiation from the sun was less intense. Of the two common gene variants known to be associated with pale human skin, *Mc1* does not appear to have undergone positive selection, while *SLC24A5* has. As with peoples having migrated northward, those with light skin migrating toward the equator acclimatizes to the much stronger solar radiation. Most people's skin darkens when exposed to UV light, giving them more protection when it is needed. This is the physiological purpose of sun tanning. Dark-skinned people, who produce more skin-protecting eumelanin, have a greater protection against sunburn and the development of melanoma, a potentially deadly form of skin cancer, as well as other health problems related to exposure to strong solar radiation, including the photodegradation of certain vitamins such as riboflavins, carotenoids, tocopherol and folate.

Melanin in the eyes, in the iris and choroid, helps protect them from ultraviolet and high-frequency visible light; people with gray, blue and green eyes are more at risk for sun-related eye problems. Further, the ocular lens yellows with age, providing added protection. However, the lens also becomes more rigid with age, losing most of its accommodation the ability to change shape to focus from far to near a detriment due probably to protein cross linking caused by UV exposure.

Recent research suggests that melanin may serve a protective role other than photoprotection. Melanin is able to effectively ligate metal ions through its carboxylate and phenolic hydroxyl groups, in many cases much more efficiently than the powerful chelating ligand ethylenediaminetetraacetate (EDTA). Thus, it may serve to sequester potentially toxic metal ions, protecting the rest of the cell. This hypothesis is supported by the fact that the loss of neuromelanin observed in Parkinson's disease is accompanied by an increase in iron levels in the brain.

Physical Properties and Technological Applications

Evidence exists in support of a highly cross-linked heteropolymer bound covalently to matrix scaffolding melanoproteins. It has been proposed that the ability of melanin to act as an antioxidant is directly proportional to its degree of polymerization or molecular weight. Suboptimal conditions for the effective polymerization of melanin monomers may lead to formation of lower-molecular-weight, pro-oxidant melanin that has been implicated in the causation and progression of macular degeneration and melanoma. Signaling pathways that upregulate melanization in the retinal pigment epithelium (RPE) also may be implicated in the down regulation of rod outer segment phagocytosis by the RPE. This phenomenon has been attributed in part to foveal sparing in macular degeneration.

Known Benefits of having Melanin

Essential to Brain, Nerve and Organ Function

In humans, melanin is the primary determinant of skin and hair colour. However, few people know that melanin is found in almost every organ of the body and is necessary for the brain and nerves to operate, the eyes to see and the cells to reproduce.

It is also found in the strain vascularis of the inner ear. In the brain, tissues with melanin include the medulla and pigment-bearing neurons within areas of the brainstem, such as the locus coeruleus and the substantia nigra. It also occurs in the zona reticularis of the adrenal gland.

Targets Free Radicals/Destroys Free Radicals

Free radicals have been implicated as the cause of widespread damage to human cells. In an article written for *The Sun and Your Skin* website, Diana Clarke, the website's founder, writes about melanin's role in scavenging free radicals, preventing the skin damage they can cause.

Superior Protection against the Effects of Ultraviolet Radiation

Exposure to the sun has the potential to cause premature aging of the skin, as well as various skin cancers. Your ability to withstand the potentially damaging effects of the sun's ultraviolet radiation depends on the amount of melanin in your skin, which is determined by the number of melanocytes that are active beneath the surface of your skin. Melanin is an effective absorber of light; the pigment is able to dissipate more than 99.9 per cent of absorbed UV radiation. In even the most light-skinned people, the body's melanocytes respond to sun exposure by producing more melanin, which creates the effect known as tanning. However, there is a limit to the degree of protection that melanin can provide and it's significantly higher in people with naturally darker skin.

Neutralizes Harmful Effects of Other Dangerous Radiation other than Ultraviolet

Melanin can absorb a great amount of energy and yet not produce a tremendous amount of heat when it absorbs this energy, because it can transform harmful energy into useful energy. According to dermatologist and derma-pathologist Dr. Leon Edelstein, director of the National American West Skin Pathology Consultation Service, melanin can absorb tremendous quantities of energy of all kinds, including energy from sunlight, x-ray machines and energy that is formed within cells during the metabolism of cells. His theory is that melanin has the ability to neutralize the potentially harmful effects of these energies.

Causes Younger Looking Skin

Exposure to the sun has the potential to cause premature aging of the skin. Darkly pigmented people tend to exhibit less signs of aging. Dermatologist Susan C. Taylor, author of "Brown Skin," points out that Blacks and other people of colour generally look younger than their lighter-skinned peers because of the higher levels of melanin in their skin. The increased melanin protects those who have it from short-term damage from the sun, as well as the long-term signs of aging, such as age spots, deep wrinkles and rough texture, according to Taylor.

Melanin Aids in Human Reproduction

The dark pigmentation protects from DNA damage and absorbs the right amounts of UV radiation needed by the body, as well as protects against folate depletion. Folate is water soluble vitamin B complex which naturally occurs in green, leafy vegetables, whole grains and citrus fruits. Women need folate to maintain healthy eggs, for proper implantation of eggs and for the normal development of placenta after fertilization. Folate is needed for normal sperm production in men. Furthermore, folate is essential for fetal growth, organ development and neural tube development. Folate breaks down in high intense UVR. Dark-skinned women suffer the lowest level of neural tube defects.

Melanin: The Organizing Molecule

Dr. Frank Barr, pioneering discoverer of melanin's organizing ability and other properties, theorizes in his technical work, *Melanin: The Organizing Molecule*: The hypothesis is advanced that (neuro) melanin (in conjunction with other pigment molecules such as the isopentenoids) functions as the major organizational molecule in living systems. Melanin is depicted as an organizational "trigger" capable of using established properties such as photon- (electron)- photon conversions, free radical redox mechanism, ion exchange mechanisms, ion exchange mechanisms and semi-conductive switching capabilities to direct energy to strategic molecular systems and sensitive hierarchies of protein enzyme cascades. Melanin is held capable of regulating a wide range of molecular interactions and metabolic processes".

References

Brenner M, Hearing VJ (2008). "The protective role of melanin against UV damage in human skin". *Photochemistry and Photobiology* 84 (3): 539–49. doi: 10.1111/j.1751-1097.2007.00226.x.

Cable J, Huszar D, Jaenisch R, Steel KP (1994). "Effects of mutations at the W locus (c-kit) on inner ear pigmentation and function in the mouse". *Pigment Cell Research* 7(1): 17–32. doi: 10.1111/j.1600-0749.1994.tb00015.x.

Castelvecchi, Davide (2007). "Dark Power: Pigment seems to put radiation to good use". *Science News* 171 (21): 325. doi: 10.1002/scin.2007.5591712106.

Cerenius L, Söderhäll K (2004). "The prophenoloxidase-activating system in invertebrates". *Immunological Reviews* 198: 116–26. doi: 10.1111/j.0105-2896. 2004.00116.x.

Dadachova E, Bryan RA, Huang X. (2007). "Ionizing radiation changes the electronic properties of melanin and enhances the growth of melanized fungi". *Plos One* 2 (5): e457.doi: 10.1371/journal.pone.0000457.

Donatien PD, Orlow SJ (1995). "Interaction of melanosomal proteins with melanin". *European Journal of Biochemistry* 232 (1): 159–64. doi: 10.1111/j.1432-1033.1995. tb20794.x.

Double KL (2006). "Functional effects of neuromelanin and synthetic melanin in model systems". *J Neural Transm* 113 (6): 751–756. doi: 10.1007/s00702-006-0450-5.

Fedorow H, Tribl F, Halliday G, Gerlach M, Riederer P, Double KL (2005). "Neuromelanin in human dopamine neurons: comparison with peripheral melanins and relevance to Parkinson's disease". *Prog Neurobiol* 75 (2): 109–124. doi: 10.1016/j.pneurobio. 2005.02.001.

Grande, Juan Manuel, Negro, Juan José, María Torres, José (2004). "The evolution of bird plumage colouration, a role for feather-degrading bacteria?". *Ardeola* 51 (2): 375–83.doi: 10.1007/s00114-008-0462-0.

GrecoG,PanzellaL,VerottaL,d'IschiaM,NapolitanoA(2011)."Uncoveringthestructure of human red hair pheomelanin: benzothiazolylthiazinodihydroisoquinolines as key building blocks". *Journal of Natural Products* 74 (4): 675–82. doi: 10.1021/np100740n.

Hamilton AJ, Gomez BL (2002). "Melanins in fungal pathogens". *Journal of Medical Microbiology* 51 (3): 189–91.

Harding RM, Healy E, Ray AJ et al. (2000). "Evidence for variable selective pressures at MC1R". *American Journal of Human Genetics* 66 (4): 1351–61. doi: 10.1086/302863.

Jimbow, K, Quevedo WC, Jr, Fitzpatrick, TB, Szabo, G (1976). "Some aspects of melanin biology: 1950-1975.". *The Journal of investigative dermatology* 67 (1): 72–89.doi: 10.1111/1523-1747.ep12512500.

Kim, Y.-J., Uyama, H. (2005). "Tyrosinase inhibitors from natural and synthetic sources: structure, inhibition mechanism and perspective for the future". *Cellular and Molecular Life Sciences* 62 (15): 1707–1723. doi: 10.1007/s00018-005-5054-y.

King G, Yerger VB, Whembolua GL, Bendel RB, Kittles R, Moolchan ET (2009). "Link between facultative melanin and tobacco use among African Americans". *Pharmacology, Biochemistry and Behavior* 92 (4): 589–96. doi: 10.1016/j. pbb.2009.02.011.

Lamason RL, Mohideen MA, Mest JR et al. (2005). "SLC24A5, a putative cation exchanger, affects pigmentation in zebrafish and humans". *Science* 310 (5755): 1782–6.doi: 10.1126/science.1116238.

Liu Y, Hong L, Kempf VR, Wakamatsu K, Ito S, Simon JD (2004). "Ion-exchange and adsorption of Fe(III) by Sepia melanin". *Pigment Cell Research* 17 (3): 262–9. doi: 10.1111/j.1600-0749.2004.00140.x.

Meier-Davis SR, Dines K, Arjmand FM et al. (2012). "Comparison of oral and transdermal administration of rasagiline mesylate on human melanoma tumor growth *in vivo*". *Cutaneous and Ocular Toxicology* 31 (4): 312–7. doi: 10.3109/15569527. 2012.676119.

Meredith P, Riesz J (2004). "Radiative relaxation quantum yields for synthetic eumelanin". *Photochemistry and Photobiology* 79 (2): 211–6. doi: 10.1111/j.1751-1097.2004.tb00012.x.

Meyskens FL, Farmer P, Fruehauf JP (2001). "Redox regulation in human melanocytes and melanoma". *Pigment Cell Research* 14 (3): 148–54. doi: 10.1034/j.1600-0749.2001.140303.x.

Meyskens FL, Farmer PJ, Anton-Culver H (2004). "Etiologic pathogenesis of melanoma: a unifying hypothesis for the missing attributable risk". *Clinical Cancer Research* 10 (8): 2581–3. doi: 10.1158/1078-0432.ccr-03-0638.

Nicolaus BJ (2005). "A critical review of the function of neuromelanin and an attempt to provide a unified theory". *Med. Hypotheses* 65 (4): 791–6. doi: 10.1016/j. mehy.2005.04.011.

Sarangarajan R, Apte SP (2005). "Melanin aggregation and polymerization: possible implications in age-related macular degeneration". *Ophthalmic Research* 37 (3): 136–41.doi: 10.1159/000085533.

Sarangarajan R, Apte SP (2005). "Melanization and phagocytosis: implications for age related macular degeneration". *Molecular vision* 11: 482–90.

Tishkoff SA, Reed FA, Friedlaender FR (2009). "The genetic structure and history of Africans and African Americans". *Science* 324 (5930): 1035–44.doi: 10.1126/ science.1172257.

Chapter 30

Photo-Protective Pigments

Some carotenoids play an important role in preventing photo-oxidative damage to the photosynthetic apparatus or to the intracellular materials of plants under high-irradiance conditions (Raw, 1988; Young and Britton, 1990). In chromophyte algae such as diatoms, dinoflagellates and prymnesiophytes, xanthophylls, diadinoxanthin (DD) and diatoxanthin (DT), function as photoprotective pigments (Demers et al., 1991; Arsalaneet al., 1994; Moisan et al., 1998). When algae are exposed to high irradiance, some excess energy is dissipated by the interconversion of DD and DT, referred to as the xanthophyll cycle. Several culture experiments with chromophyte algae have shown that the chlorophyll (Chl) a-specific xanthophyll pool [(DD+DT)/Chl a] was higher in cells grown under high light irradiances compared with those grown under lower irradiances (Demers et al., 1991; Willemoës and Monas, 1991; Moisan et al., 1998). Thus, the increase in the xanthophyll pool constitutes an important strategy for quenching photo-oxidative damage. Moreover, the xanthophyll pool has been suggested as an index to assess phytoplankton light histories and the degree of water column stability (Moline, 1998; Sigleo et al., 2000).

The area of the present study, Sagami Bay, is strongly influenced by the Kuroshio Current, which runs along the eastern coast of Japan. Oligotrophic oceanic waters flowing into the bay are mixed with eutrophic riverine waters. In the bay, rich blooms of diatoms or dinoflagellates are often observed during the period from May to July (Nakata, 1985; Satoh et al., 2000). During the period of study, the incident irradiance on sunny days was >1000 μmol m^{-2} s^{-1}, although cloud cover due to monsoon altered the light environment on a time scale of days (F. Satoh, personal communication). We hypothesize that the xanthophyll pigments serve as a photoprotection against excess light energy during the phytoplankton blooms in the bay. In the present study, we report the variability of xanthophyll pigments and their relationship to the changing light environment on a time scale of days. Assessing the ability of phytoplankton to adapt to high-irradiance conditions may lead to an improvement in our understanding of the ecological strategies of phytoplankton assemblages in the ocean.

Studies were conducted at Station G (35°092303N, 139°092253E, depth 20 m) (Figure 30.1). Surface seawater samples were collected at about 9 a.m. every other day between 19 April and 30 July 2000. Seawater temperature was measured with a mercury thermometer immediately on collection of the samples, while salinity was measured using an inductive salinometer at the Manazuru Marine Laboratory, Yokohama National University. Photosynthetically active radiation (PAR) was measured continuously during the period from 17 April to 31 July 2000 with a LI-190 SA surface radiometer (LI-COR) placed on the roof of the laboratory. Water samples were pre-screened through a 333 μm mesh net to eliminate large zooplankton and debris. For pigment analysis, duplicate subsamples of >200 ml were filtered onto Whatman GF/F glass fiber filters, which were immediately soaked in *N,N*-dimethyl-formamide (DMF) (Suzuki and Ishimaru, 1990). Cells were sonicated and left in the dark at 4°C for 24 h to complete extraction of pigments. Pigments were separated and identified in a HPLC system equipped with a Beckman C18 reversed-phase 3 mm Ultrasphere column, using a solvent gradient system similar to that described by Head and Horne (Head and Horne, 1993). For size fractionation of phytoplankton assemblages, duplicate subsamples of >100 ml were filtered sequentially through 10, 2.0 and 0.2 μm Millipore polycarbonate filters to enable size fractionation. Each filter was extracted in DMF and Chl *a* concentration was determined fluorometrically as described in Holm-Hansen *et al.* (Holm-Hansen *et al.*, 1965). The relative contribution of each size fraction was expressed as the percentage Chl *a* of the total Chl *a* in all three fractions.

As the season progressed from spring to summer, seawater temperature rose from 15°C in the latter part of April to 26°C at the end of July (Figure 30.2a). Salinity

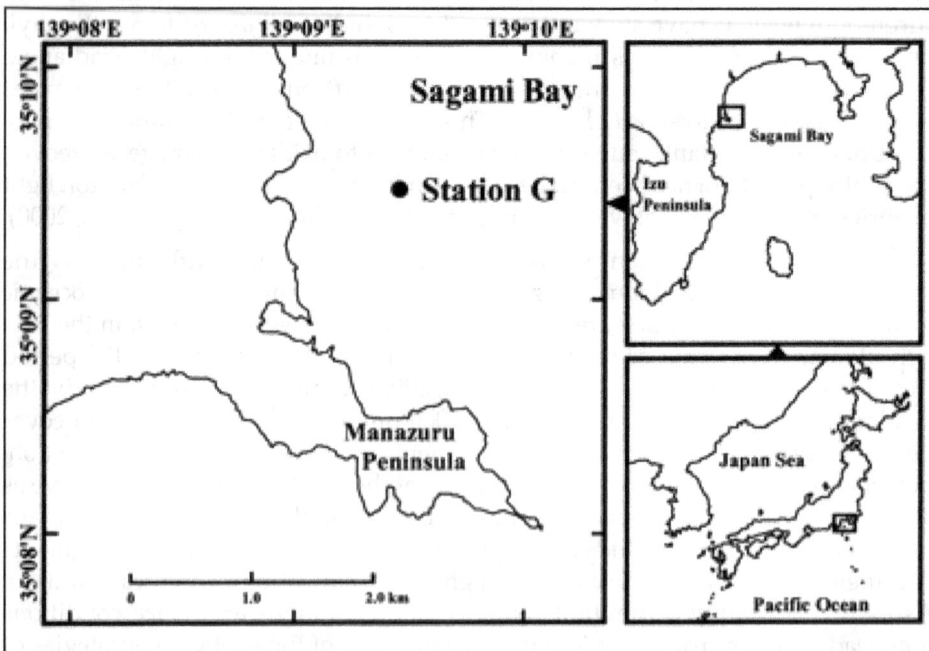

Figure 30.1: Map of Sampling Station G in Sagami Bay, Japan.

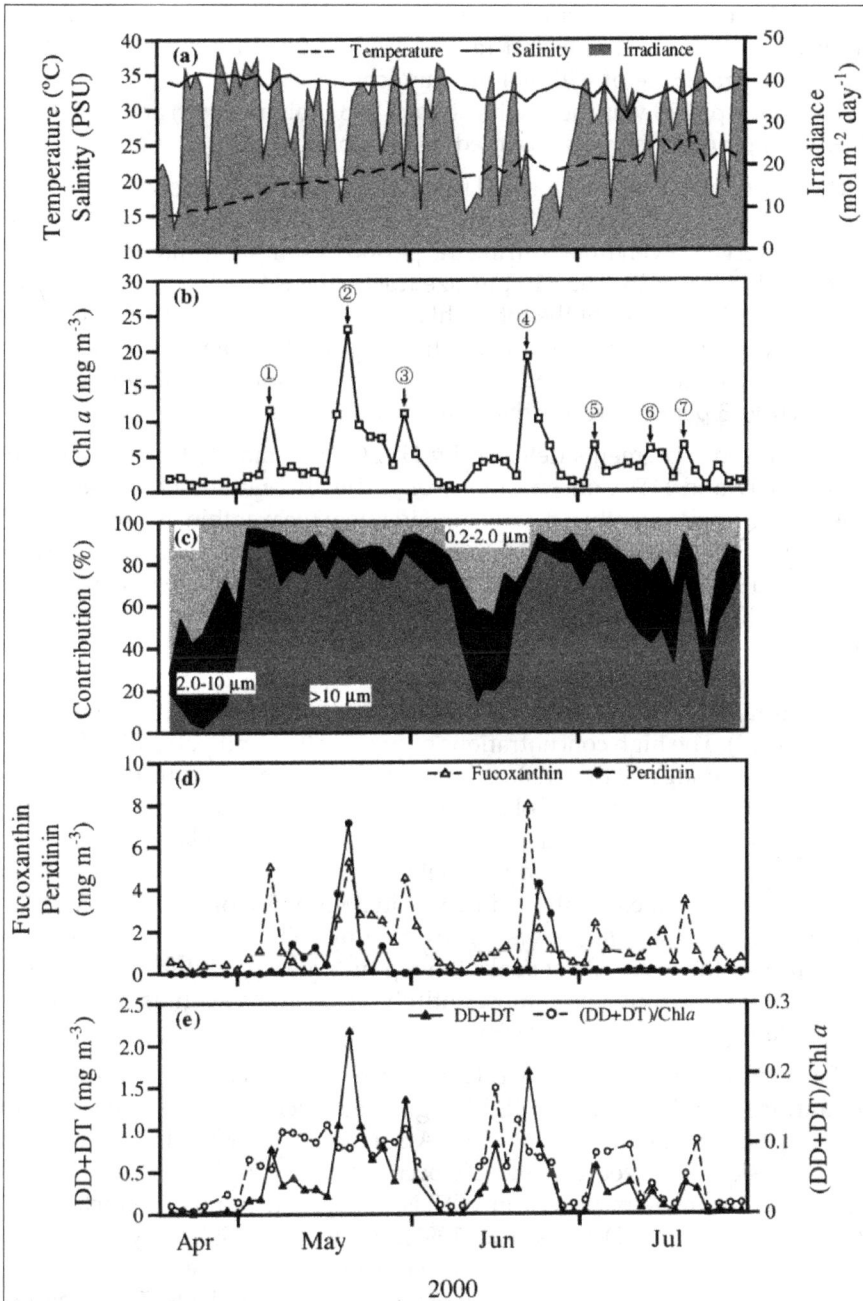

Figure 30.2: Temporal Variations of (a) Temperature, Salinity and Daily Irradiance, (b) Total Chl *a* Concentration, (c) The Percentage Contribution of each Size Fraction (0.2–2.0, 2.0–10 and >10 μm) to Total Chl *a* Concentration, (d) Phytoplankton Taxa-specific Pigments, Fucoxanthin and Peridinin and (e) DD Plus DT Concentrations [DD+DT] and Chl *a*-specific DD+DT [(DD+DT)/Chl *a*] at Station G from 17 April to 31 July 2000. Numbers in circles indicate Chl *a* peaks >5 mg m^{-3} (b).

decreased during the rainy season from early June through early July (Figure 30.2a). Daily irradiance was variable and ranged from ~20 to 50 mol m^{-2} day^{-1} (Figure 30.2a). Daily irradiance of <20 mol m^{-2} day^{-1}continued for a few days during the rainy season. April 28 was the brightest day (47.3 mol m^{-2} day^{-1}) and irradiance as high as 1730 µmol m^{-2} s^{-1} was observed before noon.

The Chl a concentration, obtained from HPLC measurements, ranged from 0.36 mg m^{-3} on 10 June to 23.2 mg m^{-3} on 21 May (Figure 30.2b). Peaks with Chl a >5 mg m^{-3} were observed seven times during the period of sampling and are indicated by numbers in Figure 30.2b. The >10 µm size fraction was the dominant fraction and averaged 59 ± 26 per cent of the total Chl a (Figure 30.2c). The other size fractions (0.2–2.0 and 2.0–10 µm) predominated during periods when total Chl a was low (roughly <2.0 mg m^{-3}). The 0.2–2.0 and 2.0–10 µm fractions contributed 21 ± 17 per cent and 20 ± 13 per cent, respectively, to the total Chl a.

The accessory pigments detected by HPLC were mainly Chl c, fucoxanthin, peridinin, DD and DT. Concentrations of Chl b, 192-butanoyloxy-fucoxanthin, 192-hexanoyloxy-fucoxanthin, β-carotene and lutein/zeaxanthin (not resolved) were low or below the detection limit. Fucoxanthin was the most abundant accessory pigment and ranged from 0.033 to 8.0 mg m^{-3}, with the highest concentration recorded on 22 June (Figure 30.2d). The peridinin concentration increased to 7.1 mg m^{-3} on 21 May and to 4.2 mg m^{-3} on 24 June, but quickly declined thereafter (Figure 30.2d). Accessory pigments are frequently used as indicators of phytoplankton taxa in natural waters (Everitt *et al.*, 1990; Letelier *et al.*, 1993; Bidigare *et al.*, 1996; Goericke, 1998). The high concentrations of fucoxanthin and peridinin observed in the present study suggest that diatoms and dinoflagellates were the primary classes of phytoplankton in the area of observation. Microscopic examination of samples (fixed with glutaraldehyde, final concentration 2.5 per cent) (Hasle, 1978) collected in conjunction with this study showed that the increase in peridinin on 21 May and on 24 June was consistent with the increase in abundance of large dinoflagellates, *Ceratium furca* and *Ceratium fusus* (H. Aono, unpublished data). Phytoplankton assemblages of the seven Chl a peaks (>5 mg m^{-3}) measured during the period of observation were dominated almost entirely by diatoms such as *Nitzschia* spp., *Thalassiosira* spp. and *Chaetoceros* spp.

The xanthophylls, DD and DT, act as photoprotectors, undergoing de-epoxidation/epoxidation in a light-regulated cycle which is known as the xanthophyll cycle (Demers *et al.*, 1991; Arsalane *et al.*, 1994). In phytoplankton cells, conversions between DD and DT occur at a very rapid time scale (seconds to minutes), but the sum of DD and DT is recognized to remain unchanged on such a short time scale (Demers *et al.*, 1991;Moisan *et al.*, 1998; Meyer *et al.*, 2000). Here, in order to avoid possible influence of short-term conversions, DD and DT concentrations are expressed as the sum of the two pigments. DD+DT concentration ranged from 0.0038 to 2.2 mg m^{-3}, with the highest on 21 May (Figure 30.2e). The xanthophyll pool expressed as Chl a-specific DD+DT [(DD+DT)/Chl a] is generally used as an index of photoadaptation by the xanthophyll pigments (Willemoës and Monas, 1991; Brunet *et al.*, 1993; Moline, 1998). The (DD+DT)/Chl a ratios ranged from 0.0053 to 0.18 (Figure 30.2e). Almost all (DD+DT)/Chl a ratios were >0.1 in May, when a long sustained period of high daily irradiance (monthly average of 33.4 mol m^{-2} day^{-1}) was observed. The highest (DD+DT)/Chl a of 0.18 was observed on 16 June (Figure 30.2e), 1 day after high irradiance (36.2 mol m^{-2} day^{-1}). Prior to 15 June, low irradiance (<15 mol m^{-2} day^{-1}) continued for 4 days and (DD+DT)/Chl a on 13 and 14 June was 0.066 and 0.075, respectively. Since different species have different values of (DD+DT)/Chl a (Demers *et al.*, 1991), changes in species composition might affect the (DD+DT)/Chl $a$response of the phytoplankton assemblage. Earlier work conducted using cultures of diatoms *Phaeodactylum tricornutum*, *Chaetoceros gracilis* and *Thalassiosira weissflogii* by Fujiki and Taguchi (Fujiki and Taguchi, 2001) showed that the (DD+DT)/Chl a of these three diatoms had a 2-fold range when all three cultures were exposed to the same light conditions. Although the increase in (DD+DT)/Chl a observed on 16 June could result from changing phytoplankton species, no marked change in phytoplankton composition was observed during the period between 13 and 16 June, as demonstrated by the size fraction (>10 μm, 18 ± 3 per cent; 2.0–10 μm, 38 ± 4 per cent; 0.2–2.0 μm, 43 ± 2 per cent) and pigment composition (Chl c, 0.15 ± 0.01; fucoxanthin, 0.19 ± 0.01; peridinin, 0.012 ± 0.000; all normalized to Chl a), which remained almost constant on 13, 14 and 16 June. Moreover, microscopic examination showed that the large size cells (>2.0 μm) were mainly composed of *Nitzschia* spp. and *Thalassiosira* spp. (H. Aono, unpublished data). Therefore, the increase in (DD+DT)/Chl $a$observed on 16 June was due to xanthophyll pigments that served as photoprotection against high irradiances.

For Chl a concentrations >2 mg m^{-3}, a linear correlation was found between (DD+DT)/Chl a and the irradiance of the previous day (Figure 30.3a). However, when integrated irradiance from 2 days ago was included, the correlation coefficient of the regression (r^2) decreased from 0.25 to 0.11 ($n = 34$), indicating that phytoplankton assemblages regulated the xanthophyll pool within a day, depending

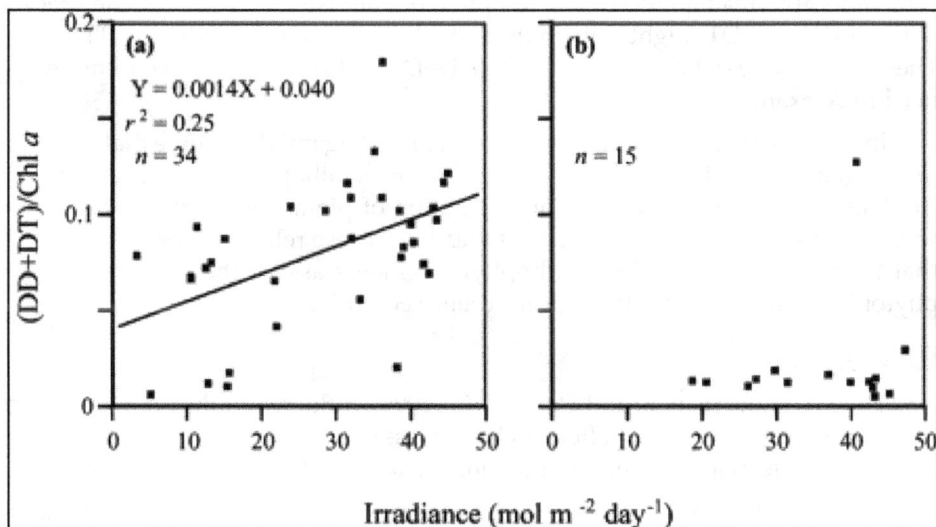

Figure 30.3: Relationship between (DD+DT)/Chl a and Irradiance for Chl a Concentrations >2 mg m^{-3} (a) and <2 mg m^{-3} (b). Irradiance values are for the day before sampling.

on the irradiance of the previous day. In the eastern English Channel, Brunet *et al.*, have shown that (DD+DT)/Chl *a* increased with monthly average surface irradiance (Brunet*et al.*, 1993). Our study, conducted on a short temporal interval, suggests that phytoplankton assemblages adjust the xanthophyll pool rapidly to variations in ambient irradiance on a time scale of days. Our view is supported by several culture experiments conducted using chromophyte algae, which showed that the (DD+DT)/Chl *a* increased in response to higher growth irradiances over time scales of hours to days (Demers *et al.*, 1991; Johnsen *et al.*, 1992; Casper-Lindley and Björkman, 1998; Moisan *et al.*, 1998). An increase in UV-B radiation with increasing solar irradiance has also been observed at this location during the spring–summer period (Kuwahara *et al.*, 2000). Goss *et al.*, found that in the diatom *P. tricornutum*, the xanthophyll cycle was stimulated by UV-B radiation (Goss*et al.*, 1999). Protection against the potentially harmful effects of increased UV-B radiation might reinforce the increase in (DD+DT)/Chl *a* in the present study.

In the case of Chl *a* concentrations <2 mg m^{-3}, (DD+DT)/Chl *a* remained low and independent of irradiance of the previous day (Figure 30.3b). For example, (DD+DT)/Chl *a* as low as 0.029 was measured on 29 April when Chl *a* was 1.4 mg m^{-3}, 1 day after the brightest day (47.3 mol m^{-2} day^{-1}). The inconsistency in (DD+DT)/Chl *a* with irradiance for Chl *a*concentrations <2 mg m^{-3} may be due to the following. (i) As the xanthophyll cycle is an enzymatic reaction (Demmig-Adams and Adams, 1996), temperature may affect the turnover rates of the xanthophyll pool. Thus, lack of xanthophyll photoadaptation at the end of April may be the result of low temperatures (15–17°C; Figure 30.2). (ii) Chl *a* concentrations <2 mg m^{-3} measured in the present study were 50 per cent below the average Chl *a* concentration of 4.0 mg m^{-3} measured from 1995 to 1999 at the same sampling site (Toda *et al.*, 2000). Therefore, the low physiological state of phytoplankton cells could affect photoadaptation by xanthophyll pigments. (iii) Phytoplankton species without DD and DT might contribute to the lack of variation in (DD+DT)/Chl *a*. These arguments on the inconsistency of (DD+DT)/Chl *a* at low Chl *a* concentrations need to be examined in greater detail.

In conclusion, phytoplankton assemblages in Sagami Bay where diatoms and dinoflagellates dominated appear to regulate the xanthophyll pool depending on irradiance of the previous day. This regulation of pigments for photoadaptation was not observed when Chl *a* biomass was low. These relationships may suggest that the photoprotection by xanthophyll pigments assists the development of phytoplankton blooms under high-irradiance conditions.

References

Arsalane, W., Rousseau, B. and Duval, J.-C. (1994). Influence of the pool size of the xanthophyll cycle on the effects of light stress in a diatom: competition between photoprotection and photoinhibition. *Photochem. Photobiol.*, 60, 237–243.

Bidigare, R. R., Iriarte, J. L., Kang, S.-H., Ondrusek, M. E., Karentz, D. and Fryxell, G. A. (1996). Phytoplankton: quantitative and qualitative assessments. In Ross, R., Hofmann, E. and Quetin, L. (eds), *Foundations for Ecosystem Research in the Western Antarctic Peninsula Region. Antarctic Research Series 70.* American Geophysical Union, Washington, DC, pp. 173–197.

Brunet, C., Brylinski, J. M. and Lemoine, Y. (1993). *In situ* variations of the xanthophylls diatoxanthin and diadinoxanthin: photoadaptation and relationships with a hydrodynamical system in the eastern English Channel. *Mar. Ecol. Prog. Ser.*, 102, 69–77.

Casper-Lindley, C. and Björkman, O. (1998). Fluorescence quenching in four unicellular algae with different light-harvesting and xanthophyll-cycle pigments. *Photosynth. Res.*, 56, 277–289.

Demers, S., Roy, S., Gagnon, R. and Vignault, C. (1991). Rapid light-induced changes in cell fluorescence and in xanthophyll-cycle pigments of *Alexandrium excavatum* (Dinophyceae) and *Thalassiosira pseudonana* (Bacillariophyceae): a photo-protection mechanism. *Mar. Ecol. Prog. Ser.*, 76, 185–193.

Demmig-Adams, B. and Adams, W. W., III (1996). The role of xanthophyll cycle carotenoids in the protection of photosynthesis. *Trends Plant Sci.*, 1, 21–26.

Everitt, D. A., Wright, S. W., Volkman, J. K., Thomas, D. P. and Lindstrom, E. J. (1990). Phytoplankton community compositions in the western equatorial Pacific determined from chlorophyll and carotenoid pigment distributions. *Deep-Sea Res.*, 37,975–997.

Fujiki, T. and Taguchi, S. (2001). Relationship between light absorption and the xanthophyll-cycle pigments in marine diatoms. *Plankton Biol. Ecol.*, 48, 96–103.

Goericke, R. (1998). Response of phytoplankton community structure and taxon-specific growth rates to seasonally varying physical forcing in the Sargasso Sea off Bermuda. *Limnol. Oceanogr.*, 43, 921–935.

Goss, R., Mewes, H. and Wilhelm, C. (1999). Stimulation of the diadinoxanthin cycle by UV-B radiation in the diatom *Phaeodactylum tricornutum. Photosynth. Res.*, 59, 73–80.

Hasle, G. R. (1978). Using the inverted microscope. In Sournia, A. (ed.),*Phytoplankton Manual. Monographs on Oceanographic Methodology 6.* UNESCO, Paris, pp. 191–196.

Head, E. J. H. and Horne, E. P. W. (1993). Pigment transformation and vertical flux in an area of convergence in the North Atlantic. *Deep-Sea Res.*, 40, 329–346.

Holm-Hansen, O., Lorenzen, C. J., Holmes, R. N. and Strickland, J. D. H. (1965). Fluorometric determination of chlorophyll. *J. Cons. Perm. Int. Explor. Mer*, 30, 3–15.

Johnsen, G., Sakshaug, E. and Vernet, M. (1992). Pigment composition, spectral characterization and photosynthetic parameters in *Chrysochromulina polylepis. Mar. Ecol. Prog. Ser.*, 83, 241–249.

Kuwahara, V. S., Toda, T., Hamasaki, K., Kikuchi, T. and Taguchi, S. (2000). Variability in the relative penetration of ultraviolet radiation to photosynthetically available radiation in temperate coastal waters, *Japan. J. Oceanogr.*, 56, 399–408.

Letelier, R. M., Bidigare, R. R., Hebel, D. V., Ondrusek, M. E., Winn, C. D. and Karl, D. M. (1993). Temporal variability of phytoplankton community structure based on pigment analysis. *Limnol. Oceanogr.*, 38, 1420–1437.

Meyer, A. A., Tackx, M. and Daro, N. (2000). Xanthophyll cycling in *Phaeocystis globosa* and *Thalassiosira* sp.: a possible mechanism for species succession. *J. Sea Res.*, 43, 373–384.

Moisan, T. A., Olaizola, M. and Mitchell, B. G. (1998). Xanthophyll cycling in*Phaeocystis antarctica*: changes in cellular fluorescence. *Mar. Ecol. Prog. Ser.*, 16 9, 113–121.

Moline, M. A. (1998). Photoadaptive response during the development of a coastal Antarctic diatom bloom and relationship to water column stability. *Limnol. Oceanogr.*, 43, 146–153.

Nakata, N. (1985). Sagami Bay: biology. In Oceanographical Society of Japan (ed.), *Coastal Oceanography of Japanese Islands*. Tokai University Press, Tokyo, pp. 417–427 (in Japanese).

Raw, W. (1988). Functions of carotenoids other than in photosynthesis. In Goodwin, T. W. (ed.), *Plant Pigments*. Academic Press, London, pp. 231–255.

Satoh, F., Hamasaki, K., Toda, T. and Taguchi, S. (2000). Summer phytoplankton bloom in Manazuru Harbor, Sagami Bay, central Japan. *Plankton Biol. Ecol.*, 47, 73–79.

Sigleo, A. C., Neale, P. J. and Spector, A. M. (2000). Phytoplankton pigments at the Weddell–Scotia confluence during the 1993 austral spring. J. *Plankton Res.*, 22, 1989–2006.

Suzuki, R. and Ishimaru, T. (1990). An improved method for the determination of phytoplankton chlorophyll using *N,N*-dimethylformamide. *J. Oceanogr. So. Jpn.*, 46,190–194.

Toda, T., Kikuchi, T., Hamasaki, K., Takahashi, K., Fujiki, T., Kuwahara, V. S., Yoshida, T. and Taguchi, S. (2000). Seasonal and annual variations of environmental factors at Manazuru Port, Sagami Bay, Japan from 1995 to 1999. *Actinia*, 13, 31–41 (in Japanese).

Willemoës, M. and Monas, E. (1991). Relationship between growth irradiance and the xanthophyll cycle pool in the diatom *Nitzschia palea*. *Physiol. Plant.*, 83, 449–456.

Young, A. and Britton, G. (1990). Carotenoids and stress. In Alscher, R. G. and Cumming, J. R. (eds), *Stress Responses in Plants: Adaptation and Acclimation Mechanisms*. Wiley-Liss, New York, pp. 87–112.

Chapter 31

Lighting and Physiology

This study is one of a series outlining the major design elements of the new Council House 2 (CH_2) building in Melbourne, Australia. CH_2 is an environmentally significant project that involves a design approach inspired by the mimicry of natural systems to produce comfortable, healthy and productive indoor conditions, an approach termed biomimicry: to mimic nature or living systems. This study focuses on lighting and physiology and examines the artificial and natural lighting options used in CH_2 and the likely effects these will have on building occupants. The purpose of the study is to critically comment on the adopted strategy and mindful of contemporary thinking in lighting design, to judge the effectiveness of this aspect of the project with a view to later verification and post-occupancy review. The study concludes that CH_2 is a prime example of lighting innovation that provides valuable lessons to designers of office buildings, particularly in a CBD environment.

Introduction - Scope of this Study

The lighting of a workplace can positively influence the health of office personnel, improve efficiency, reduce unnecessary sick leave and result in greater productivity. In particular, natural light, with its variations and spectral composition, together with the provision for external views, is of great importance for personal well-being and mental health, reducing suppressed feelings of panic, anxiety, disorientation and melancholy. The careful management of natural and artificial lighting, including the use of shading devices, can also bring tangible energy savings, preserving the natural colours of the outside environment, while preventing glare and minimizing heat gains.

A transparent façade should always be designed to fulfil the needs of the users and the requirements of the building and find a balance between needs of transmission and requirements of protection. For buildings with high percentage of glass, it is even more important to find the right balance between opening to the environment and protecting from its extremes.

Selecting façade and lighting solutions for comfort and energy efficiency can be a very complex problem. There are many design and context variables that interact with each other, making selection and optimisation more difficult.

As Guzowski explains, a good lighting strategy should maximise the potential of architectural form while taking advantage of technologies to further refine solutions. The goals of a lighting strategy can be defined from a wide variety of perspectives such as *ecological issues* (energetic and natural resource depletion, environmental impact), *tasks* and *activities* (lighting needs in both qualitative and quantitative terms), *systems integration* (lighting, HVAC), *human experience* (visual and thermal comfort, health, orientation in space and time, connection to the beat of outside life), *aesthetic considerations* (form, dimension and articulation of spaces, materials), as well as other concerns. For this reason it is not always possible or even necessary to address these objectives simultaneously; yet, analysing their potential can clarify design intentions, determine priorities and reveal possible contradictions. For example, lighting is often designed to remain fairly constant during day and night in working environments, not considering that occupant needs can vary in terms of preferences and associated differences in clothing, metabolic levels and the nature of visual tasks occurring in a given working environment.

This study aims to show how the paradox of opening to natural forces and protecting from its extremes can be resolved and to promote a new design attitude that modulates the relationship between users' needs and sustainability. As a reference case of international best practice, the lighting strategies of the newly developed Council House 2 (CH_2) building in Melbourne will be analysed and its contribution to sustainable design discussed.

Light and Physiology

The use of daylight as a light source in buildings is important to achieve ecological sustainable development (ESD) objectives, being assumed to minimise resource consumption, waste generation and improve human well-being. However, the widespread use of shared spaces that inhibit direct lighting control by every user, the inherent limits of automatic lighting control systems and the reductions in terms of energy consumption of modern electric lighting have made it difficult to justify the cost of extensive natural day lighting design solutions on the basis of economic paybacks from potential energy savings. To substantiate the use of daylight in buildings it is necessary to demonstrate beneficial effects in other areas that have, potentially, a more significant impact on occupants, tenants and owners of buildings.

Light (in all its forms) is not only a resource and a vital sustenance, but can also create meaningful architectural experiences. The mood and quality of an architectural space can vary greatly depending on its lighting and colour conditions, transforming a sometimes dark, sober and oppressive place into a captivating, enthralling and stimulating one. In addition, scientific research has recently proven that a close relationship exists between lighting conditions, health, well-being and our perception of the environment. Daylight, for example, represents one of the most important means of maintaining our biological rhythm and connection to rhythms of nature and is a key way of marking important daily events (dawn, morning, noon, afternoon, sunset and evening).

When light passes through the eye, the signals are carried not only to the visual areas of the brain but also to areas responsible for emotion and hormonal regulation. Ocular light stimuli from the retina result in signals being sent to various glands, involving the whole of the *physical* (energetic exchanges), *physiological* (transformation of energetic fluxes into nervous stimuli) and *psychological* (brain interpretations of those stimuli). The combination of these activities create the 'process of perception' informing us about the characteristics of the surrounding environment.

Regardless of this awareness, a great part of our social interaction is temporally organised in relation to a rather 'mechanical time', which is largely independent of the rhythms of our body's impulses and needs. In other words, we are increasingly deviating from the organic and functional recurrence dictated by the natural colour, angle and intensity of daylight and replacing it with an artificial timetable which is imposed by work schedules, the calendar and the clock. As Van den Beld suggests, the species *Homo sapiens* appeared on Earth around 250,000 years ago and evolved under the daily 24-hour light-dark cycle. To a large extent life has been regulated by a natural wake/sleep rhythm: active, mostly outside during the day and resting at night. During the last couple of centuries, this natural pattern has changed rapidly, initially due to the industrial revolution and then to some technological innovations (such as electric light) that are now moving us towards a global 24-hour society. Most people nowadays spend more than 90 per cent of their time indoors, often in offices and in all cases the lighting is based on the requirement that, whatever the time of day or night and regardless of the physiological needs of the human body, the task should be accomplished efficiently, safely and with a degree of visual comfort.

Medical research has recently discovered that almost all human physiological and psychological processes are based on rhythms directly linked to the natural daily (*circadian*) and seasonal (*annual*) cycles of light. In particular, the human brain has been discovered to contain an internal 'biological' clock, daily synchronised to the periodicity of nature through the medium of ocular light received by the eye. Day/night light patterns regulate many body processes such as body temperature, heart rate, mood, fatigue and thus alertness, performance, productivity, etc. Sufficient light received during the natural light period (daytime) synchronises the 'biological clock' contained in the human brain, stimulating circulation, increasing the production of vitamin D, enhancing the uptake of calcium in the intestine, regulating protein metabolism, controlling the levels of serotonin, dopamine (pleasure hormones), melatonin (sleep hormone) and cortisol. In other words, light provides the direct stimuli needed for the human body to function and feel well and healthy.

Exposure to daylight is usually the major factor for setting the human circadian rhythm, since it usually produces a high illuminance at the eye with a spectrum that perfectly matches the specific sensitivity of the circadian system and peaks at about 465 nm. Sufficient retinal illumination to entrain the circadian system can be provided by artificial lighting alone, even though this solution is less likely to obtain the same results as natural daylight. If we consider a daytime office worker, daylight deficiency may result in a de-synchronisation of his or her biological clock. As such the body and mind may prefer to rest but are required to remain active. The effects of this de-synchronisation are lower performance, a decrease in alertness, diminished sleep quality and, in the longer term, impacts on well-being and health.

Research shows that lack of exposure to sufficient light during the day may foster negative effects on various physiological aspects of the human body; this is more evident in particular during the 'dark' winter season or in regions characterised by cold and sombre climate, where there is less light and days are short. About three per cent of the population in those regions suffer from winter depression (SAD, *Seasonal Affective Disorder*) and the so-called 'winter blues' are common. Intensive bright light through the eye can mitigate those feelings and is the first line of treatment for SAD.

The combination of medical and scientific research leads to the hypothesis that "healthy" lighting for daytime indoor activity is influenced by many more factors than what is suggested in most lighting standards and regulations. This should preferably be a combination of natural and artificial sources, the electric light alone serving to take over when natural daylight falls in the winter period or in the later part of the working day.

Daylight Quality vs Indoor Light

It is now important to establish how serious the consequences of working and living indoors at 'unnatural' times are and whether a 'healthy' lighting system can be designed to compensate for this. A number of interesting facts and figures are relevant to this issue and are discussed below.

For example, although human beings are accustomed to significant variations in the level and duration of daylight, office lighting practice seems to ignore this fact. Natural outdoor illumination varies from over 100,000 lux on a sunny day to a few thousand lux on a dark, overcast winter day and for periods of between two and almost 16 hours per day. External lighting levels are therefore, on average, at least 800 lux higher than the accepted horizontal illumination in working spaces (300-600 lux).

Secondly, just as the spectral composition of daylight shows large variations during the day ('cold' light in the morning and 'warm' light at sunset), people prefer variations in the correlated colour temperature (CCT) of artificial light. In particular, according to the Curve of Amenity (*Kruithof Diagram*), the higher the overall lighting level, the higher its colour temperature should be. Most current lighting systems are only adjustable in output levels and not in terms of colour temperature, as a result they can rarely add significant meaning to the variability of a workplace, often creating simple, repetitive and arbitrary indoor lighting environment configuration.

Thirdly, daylight is highly dynamic in its intensity and direction and research shows people would prefer to be aware of these changes and desire continuous contact with the world outside.

Another important issue is the influence of colour on physiology, an issue that may involve subjective as well as objective responses. There are reliable and measurable physiological reactions to colour in addition to those generally associated with vision. These reactions may be revealed by objective measurements such as *galvanic skin response, electroencephalograms, heart rate, respiration rate, oxiometry, eye blink frequency* and *blood pressure*. Whether the association between colour and one or more of the above physiological index is direct (*i.e.* colour causes the physiological response to be elicited without mediation by a cognitive intermediary response) or indirect (*i.e.* exposed to a colour the observer makes certain associations), is yet to be clearly defined.

Comfort, Satisfaction and Productivity Implications

The above considerations lead to the conclusion that the way in which lighting conditions influence the comfort and performance of individuals involves not only the *visual* (amount, spectrum and distribution of the light) and the *circadian* system, but also the *perceptual* system, which takes over once the retinal image has been processed. Poor lighting conditions are generally considered uncomfortable and can lead to distraction from the task or fatigue due to the presence of glare or flicker. However, perception is much more sophisticated than simply producing a sense of visual discomfort because of the influence of these conditions on an observer's mood and behaviour.

The knowledge gained from these recent discoveries on the effects of light on well-being and human health lead to new demands for lighting design solutions. In addition to the well-known visual comfort criteria and direct stimulation of the brain, additional non-visual issues for health and well-being are being formulated in the scientific literature.

The combination of medical and scientific research leads to the hypothesis that 'healthy' lighting for daytime indoor activity is influenced by many more factors than those in most lighting standards and regulations. This research indicates that there should preferably be a combination of natural and artificial lighting sources in office environments, with the electric light taking over when natural daylight is reduced in winter or later in the working day.

Dark or windowless spaces are generally disliked by occupants, particularly when rooms are small and there is a lack of external stimulation. However, given the general preference for daylight and considering the number of factors involved and the fact that people will give up daylight if it is associated with glare, contrast, reflection, solar heat gain or a perceived loss of privacy, it is hard to demonstrate that the presence of windows alone can improve an occupant's productivity.

Meeting the Lighting Requirements

The Daylight Design Challenge

Many people perform daily activities that are best described as office tasks such as files processing, communication with other people, thinking, organising and other related tasks. Each activity involves a different relationship with the spaces that surrounding a specific workstation and has to meet very complex requirements, including a number of basic human needs that necessarily have to be taken into account in the design of office spaces. Those needs reflect people's desire for a specific orientation in space and time (*genius loci*) - including aspects related to physiological biorhythm, but also to society and culture, privacy and communication, information and familiarity, variation and surprise. Lighting, both natural and artificial, through the choice of form, colour, material and details, plays a key role in creating an atmosphere that meets occupant's expectations (functionality, aesthetics, ergonomics, of the rooms and their furnishings) and demands (privacy, concentration, appreciation of details, etc.). Lighting can also facilitate perception and create a mood or ambiance of its own.

During the day, the presence of daylight should render spaces lively, activating and motivating in line with the human biorhythm. In addition, daylight is often associated with a view, which provides information about the time of day, the season and the weather. Views and variations in intensity and colour are extremely stimulating for the brain and the visual apparatus, contributing to a person's well-being and improving their sense of orientation and feeling of spaciousness. In addition, various screen-based tasks require a limited eye movement or change of focus that can be very fatiguing. Views can reduce muscle strain by allowing the eyes to shift focus from the near field surrounding the work area towards distant objects.

There is absolutely no doubt that occupants, given a choice, would prefer to live and work by daylight and to enjoy a view to the outside. Small and artificially lit spaces are usually disliked, even though they are sometimes accepted due to external factors (working groups, stringent visual tasks, etc.) Nevertheless, daylight can have major drawbacks on visual comfort such as direct sunlight, bright clouds and reflective surfaces that create glare and contrast. Luminance ratios in the field of vision should always be contained within certain limits: too large and it is difficult for the eyes to adapt; too small and there are difficulties in estimating depth and distance.

Since people are also phototropic (attracted to light) and areas of high luminance in the background of the visual task should be avoided. As the eye attempts to even out the contrast between the two differently illuminated surfaces, the muscles of the eye have to work harder and more frequently resulting in tired eyes and an increased level of stress. As such, the task should always be the area of major visual attention and brighter than its surroundings, to enhance comfort and minimise glare.

Glare in particular is a potential source of visual discomfort due to the illuminance from a bright source (direct or reflected) relative to the average illuminance in the field of view of the observer. The ratio at which this contrast becomes a source of discomfort depends on the specific function being performed. Sources of glare in a workplace can be the sky vault and the sun and/or its reflections on external surfaces or lighting installations. Glare can be categorised in two different ways: 'disability' glare and 'discomfort' glare, the former preventing the viewer from performing the visual task and the latter causing a decrease in visual comfort (and hence productivity). In addition to these two categories, there is also 'direct' and 'indirect' glare; direct glare is caused when a person views a source of illumination, indirect glare results from light being reflected off surfaces. It is important to note that some studies show that occupants are more tolerant of glare if the light source is accompanied by a view and that lighting fluctuations coming from a natural source are generally quite well accepted, while people tend to find changes in the artificial lighting environment rather disturbing.

In relation to glare and reflections on computer screens, direct illumination can decrease visibility by reducing contrast or washing out the screen image. In general, old cathode screens are more susceptible to those problems, while newer display technologies, such as liquid-crystal flat screens with anti-reflection coatings, can be viewed under some direct sun conditions.

Predicting Glare and Controlling Natural Light

The degree of glare in an interior space can be predicted by the determination of a *Daylight Glare Index* (*DGI*) at a specified location and orientation within the space. This index is based on a subjective response to brightness within a person's field of view, with higher values indicating a greater probability of discomfort glare and vice-versa. The least perceptible difference of glare index which can be visually appreciated is one unit, while the least difference which makes a significant change in the perception of discomfort glare is three units. Each step in the scale of the DGI (13, 16, 19, 22, 25 and 28) represents one significant change in glare effect. A DGI of 10 is the threshold for *just perceptible* glare and a glare index of 16 is the threshold where glare is *just acceptable*. For a normal range of tasks in an office environment, the typical maximum glare index is assumed to be 19.

To control the intensity of light sources entering a work space and to guarantee a comfortable luminous environment, a good daylight solution should generally be composed of more than a simple window or skylight. Depending on climate, building orientation and environment, additional elements or adaptations may be needed to increase visual comfort.

Daylight systems range from simple static elements (such as louvers or fixed overhangs) to adaptable dynamic elements (such as blinds or movable lamellae) and/or combinations of these. Good solutions start from exploring simple techniques and adding advanced elements, if required. The performance of complex systems such as highly reflective surfaces is very dependent on maintenance and durability of components. Dust, condensation or surface deterioration can also reduce the optical efficiency of these systems, sometimes by more than 50 per cent. Simple daylight systems to reduce glare include a light shelf that can be used to deliver daylight at greater depths into the room by reflecting light on the ceiling and reducing glare in areas close to the perimeter without significantly reducing light levels near the window.

Additional elements comprising a daylight system include indoor or outdoor solar blinds, which are used to help control the intrusiveness of solar energy, in terms of its light and heat components. In office environments, blinds are mostly either horizontal or vertical and should preferably be composed of light, diffusive materials. The individual strips making up the blinds should be as narrow as possible, for the wider the strips, the larger the undesirable light-dark patterns. Slim, light-coloured, horizontal blinds offer the best control of brightness and light distribution. It is important that individual employees are able to close or open blinds to suit their preference. Since no two people are the same, nor do they perform the same task all day long, daylight and lighting controls should be as versatile and flexible as possible. If manually-controlled interior shading is the only option, many occupants will keep the device closed meaning the window is no longer transparent. For efficient use of daylight and to allow continuous adjustments, automatically-operated movable devices are preferable, although the initial cost and maintenance could be slightly higher than with fixed devices.

Glazing and Colour Affects

Glazing type should be selected according to daylight effectiveness, occupant comfort and energy efficiency, while still meeting architectural objectives. Glass for windows should be evaluated according to the specific optical and thermal characteristics of the glazing. In particular, glazing colour affects colour appearance and colour rendering of interior finishes and tasks in day lit areas.

In relation to interior finishes, light-coloured surfaces reflect more daylight than dark hues. Specular surfaces, such as glazed tiles or mirrored glazing, can create glare if viewed directly from a task position, while diffuse ground-reflected daylight can increase daylight availability. Deep reveals, ceiling baffles, exterior fins and shelves - if they are light in colour - may help keep daylight more even.

As a person's mood can be significantly affected by his or surroundings, colour monotony should generally be avoided as much as colour fatigue. Whilst the colour palette on interior surfaces should preferably be simple, attractive visual centres (such as colourful hangings, paintings, posters) are desirable in order to produce a visually pleasant environment. Plants and individually coloured screens play the same role, while a view outside always enhances visual rest and connection with the outdoors. Design for natural and artificial light

In a work environment, a combination of daylight and artificial light is preferable, combined to produce sufficient and suitable lighting for tasks throughout the room, day and night. Good integration between these two sources of light makes it possible to gradually dim the amount of electric light when available daylight is sufficient for the task.

The design of an office lighting system should also allow for the various requirements of its occupants, allowing users flexibility and personal over-ride to adjust (at least partially) the luminous environment according to their individual needs. Privacy and personal needs, in particular, require that each area be separate from other workstations in terms of the luminous environment and can be fitted out in a personalised way. Depending on the different tasks and activities performed in a space, several adjustable lighting systems are preferable to evenly distributed ceiling lights. In terms of light distribution, a combination of diffuse and direct light - with directional lighting and some diffuse light needed to avoid dark areas with dense shadows - can assist in the perception of three dimensional objects and give 'life' to an environment.

In an office, daylight can provide adequate ambient light for most working hours and when supplementary light is needed, user-controlled task lights can ensure work requirements are met. As mentioned previously, ambient illumination should be significantly lower than task requirements. Likewise, different artificial light is required during the day than at night, when, as a general rule, light should be calming and restful, in accordance with the human biorhythm.

In a day lit space, it is obvious that people close to windows will often use natural light as their primary illumination source. For other locations, direct/indirect lighting should be designed to pair with daylight distribution. To ensure adequate illumination, fixtures and lighting circuits should preferably be grouped by areas of similar daylight availability (*e.g.* in rows parallel to window wall), in order to allow the possibility for control to be added as retrofit.

If computers are present, ambient lighting should not exceed 300 lux, in which case user-controlled task lighting is available as a supplement. A rule of thumb for spaces that host Video Display Terminals (VDTs) provides as little light as possible on computer screens (150-300 lux) for surround lighting and up to 500 lux on adjacent task space. However, if glare from windows is expected, interior luminance should be kept high to balance window brightness and decrease the risk of visual contrast. High-reflectance, light-coloured walls and partitions are preferable.

The choice of the most appropriate 'colour temperature' for a light source is largely determined by the function of the room, involving psychological aspects such as the impression of warmth, relation, clarity and other considerations. For best colour temperature pairing with daylight, a generally well-accepted choice is to install fluorescent lamps with a minimum colour temperature of 4000 K. However, when there is significant night-time use of the building, lamps with a CCT lower than 4000 K may be required.

In order to save energy and ensure optimum light distribution at all times, a control system that can adjust the lights and/or turn them off when there is adequate daylight may reduce consumption and result in minimal complaints. Typical artificial lighting control systems for commercial buildings include a photosensor strategically located either under the luminaire close to the external opening or on the ceiling. Their sensitivity to light may vary sensibly within the cone of view according to the specific need. Sensors 'measure' light by looking at a wide area of the room floor and work surfaces and send signals to the control system to dim or switch the lights according to daylight availability and link to a host of other inputs (*e.g.* lumen maintenance, tuning, occupancy sensors, weekend/holiday/night time schedules, etc.).

Dimming control is generally twice as expensive as switching because it is the best strategy for implementing energy savings and the most acceptable to occupants since observable changes in the artificial light levels can be made less disturbing. Dimming is not a cost-effective strategy in non day lit areas unless coupled with scheduling controls. In relation to lighting control systems, dimming electronic ballast are fast becoming cost-effective to operate fluorescent lamps in rapid-start mode (*i.e.* the fluorescent lamp cathodes are supplied with power at all times during operation).

International Best Practice - BRE Fire Research Station

The principles of best practice discussed so far have been applied on a number of buildings around the world. Some of these include the new headquarters of the British Research Establishment (BRE) Fire Research Station, built in Garston (UK) by Feilden Clegg Architects in 1996; a landmark-building intended to be a replicable example of cutting-edge environmental design that shares a number of elements in common with the Council House 2 building in Melbourne.

The design of the BRE Fire Station was based on a new performance specification, *Energy Efficient Office of the Future*, founded on environmentally friendly principles with energy performances corresponding to an 'Excellent' Building Research Establishment's Environmental Assessment Method (BREEAM) rating. In the case of the CH_2 building, all consultants were involved from the earliest stage of design and were joined by the main contractor during the documentation and specification stage.

The BRE Fire Station is an L-shaped block with a three storey office wing (30m by 13.5m), fronting a landscaped area. The relatively shallow office plan, with highly glazed façades, exploits natural daylight and is well suited to cross ventilation via Building Management System controlled windows at high level and manually operable windows at lower levels. Daylight is maximised with large areas of glazing on the north and south façades, providing daylight factors of over two per cent across the floorplate. Penetration is assisted by 3.7m ceiling heights, which are also painted white.

As in CH_2, a wave form floor slab design incorporates ventilation routes that pass over the ceiling of cellular spaces; the high points of the wave have corresponding high level windows allowing daylight to penetrate deep into the plan. Fully glazed façades, in combination with high ceilings and a relatively shallow plan depth, minimise the need for artificial lighting and the consequent electrical energy load is significantly reduced when compared to a conventional office building. However, the need to control glare and solar gain is more important.

At the BRE Fire Station, these factors are controlled by BMS-controlled external motorised glass louvers. The louvers are extremely slim when rotated to their horizontal position (10mm) but, being wide (400mm), are set well apart (1.2m) so that an excellent view is maintained when they are not angled to provide shading. It is also possible to rotate the blades beyond the horizontal so they can act as adjustable light shelves to reflect direct sunlight onto the ceiling and deeper into the plan. The louvers are controlled by the BMS with the specific purpose of managing solar gain. Occupants can over-ride the automatic setting to reduce glare if they wish, however the system is reset to an optimum position at the end of the day. In terms of artificial lighting, internal sensors measure light levels and movement, dimming high-efficiency T5 fluorescent lamps (from 100 to 0 per cent) when there is sufficient daylight, or switching them off if a room is unoccupied. The sensors have an infra-red receiver, which allows users to over-ride the automatic control by means of a remote control unit. The lamps provide general lighting at around 300 lux, which is supplemented by task lighting when required and have an up lighting component to guarantee a balanced visual environment. Each of the lamps can be controlled separately by the BMS to allow different light output levels across the floor plan and to make maximum advantage of daylight for the buildings lighting requirements.

International Best Practice – Other Notable Daylight Examples

In addition to the BRE edifice, there are a number of other buildings where natural light is carefully exploited via the use of advanced solutions and devices. One of these buildings is the Parliamentary Office (Portcullis House) in Westminster, London, designed in 1998 by Michael Hopkins and Partners. In this building solar shading is controlled by adjustable louver blinds built into the cavity between a double skin façade and by light shelves positioned above the bays. These shelves shade the area immediately next to the window and reflect incoming light deeper into the rooms.

Light shelves to enhance solar penetration and provide sun shading are also used in the EOS Building in Lausanne, designed by Richeret and Rocha Architects, in the form of 0.8m outside-tilted anodised aluminium blades located at the height of 2m above the floor surface, a choice which offers a reasonable compromise between reflective properties and durability. Deep 0.9m aluminium light shelves, mounted on the inside of two south-facing window strips (northern hemisphere), are also used in the Tax Office Extension building in Enschede (The Netherlands), a project partly founded by the European Commission THERMIE Program and designed by a team from the Dutch Government Building Agency. Rather than using a reflective device to direct natural light towards the interior, the newly refurbished SUVA Company building in Basel, by Herzog and De Meuron architects, exploits a glazed façade divided at each floor into three horizontal bands of motorised top-hinged windows, designed to perform different functions. In particular, the upper 'day lighting' band consists of insulated prismatic glass automatically manoeuvred to follow the solar path. From the inside, they are essentially translucent, thereby inhibiting glare from the sky vault. In response to lighting levels detected by an internal photosensor, they can be adjusted perpendicular to the solar altitude to refract light directly inside.

Energy saving targets aside, in relation to the application of advanced lighting control systems, the design of the ABN AMRO Head Offices in Amsterdam (NL), is able to satisfy two other very important requirements: firstly, that the occupants are continuously able to control their thermal and luminous environment; and secondly, that the technical installations can accommodate any changes to the layout of office space. To satisfy these requirements, the extremely flexible LONWorks platform a widely adoptedopen-bus system used to build automated control applications - has been employed in conjunction with an HELIO lighting control system. With this equipment, the occupant can simply use a remote-control unit to select his or her personal preferences for lighting, temperature, setting of the sun-blinds and even the do-not-disturb sign over the office door. The HELIO sensor passes this information on to the LON that accordingly instructs all the individual building systems. Another important aspect critical to the performance of the ABN AMRO building and the productivity of the organisation is related to the fact that the company experiences frequent organisational changes, requiring constant adaptation of the office layout. With traditional vertical cabling methods, this would involve considerable time and money because internal walls would need to be moved and cabling re-routed. Thanks to the LONWork system, all modifications to the technical control network can be done using computer software rather than having to make changes to the physical layout because all comfort system connections are soft wired. In addition, the lighting installation at ABN AMRO is particularly energy efficient and offers optimal lighting control. The luminaries employ T5 tubular fluorescent lamps, specially designed with mirror optics to produce more uniform distribution of direct and the indirect light and therefore increase user comfort when working on computers. The luminaries also ensure constant lighting levels on desktops due to a built-in light sensor, achieving a significant saving in energy.

CH_2 Design Features, Expected Performance and Operation

The Daylight Strategy

The CH_2 building has been designed as a world leader in ecologically sustainable design and commercial green building technologies. The initial Melbourne City Council brief called for 'a landmark building that (would) provide a healthy, stimulating workplace'. Most of the principles followed in CH_2 are not completely new, yet never before in Australia have they been integrated and pursued in such a comprehensive and inter-related way in a multi-storey office building.

The building was designed foremost to be comfortable and healthy for its occupants, whose physiology and experiential feelings have been regarded as key-factors in every single decision as it is recognised that an occupant's positive response to the indoor environment can contribute noticeably to their performance and hence the overall productivity of the organisation. The challenge of getting natural light into the building given its location and overshadowing by surrounding buildings was an issue. The simple rectilinear form has been dictated by the boundaries of the site, with the largest length oriented along the east and west axis to maximises northern solar access and daylight while minimising unwanted solar heat gain being absorbed by the buildings east and west façades. Nevertheless, the major drawbacks of exposing the façades to excessive sunlight have been pointed out, since the beginning, by the project team and their implications have been discussed earlier in this study.

One strategy to reduce these unwanted consequences, for the northern and southern façade, was to progressively widen the lower windows of the building, as more daylight is needed at lower levels due to reduced natural light availability in surrounding narrow city lanes. While optimising access to available natural light at different levels this strategy has other advantages, such as reducing the total amount of glass used and minimising energy losses, heat gain and glare risks. The narrowing window design strategy also combined synergistically with the variable size requirements of the ventilation air supply and exhaust ducts that are integrated into the north and south façade between the window panels. Progressively increasing the size of the windows at lower levels matched a reduction in the width (volume) of the ventilation ducts sizing requirements at lower levels. This nice match of design requirements occurs because the volume of air to be transported via the façade ducts through the roof top intakes and outlets is reduced with every air take off at each lower floor, decreasing progressively from the tenth level down to the first.

The glazing has been selected to achieve a visible light transmittance greater than 50 per cent, combined with a solar transmittance smaller than 35 per cent. This choice allows for relatively high daylight levels and, above all, reduces solar heat gains, an issue of significant importance in a climate such as Melbourne both during mid-seasons and summer. Internal and external visible light reflectance are respectively <15 and <20 per cent, while the colour of the glass is absolutely neutral for better internal and external colour rendering. To neutralise harmful low afternoon sun on the western façade, recycled timber louvre screens run across the entire elevation. Movement of the louvres is powered by photovoltaic panels located on the roof and controlled by a computer that dictates the tilt angle and position to ensure optimum shading while still allowing filtered daylight and views.

Working within the constrained nature of the site and the strict requirements of an open plan office setting has presented many challenges to the design team in terms of daylight accessibility and distribution, especially in order to achieve high indoor environmental quality and energy savings. The solution is a unique system of day light distribution that also synergise with the cooling and ventilation strategy. A barrel vault concrete ceiling, running like waves in north and south directions, enables light to penetrate deep into the space, while light shelves (made of 50 per cent perforated steel internally and movable fabric externally), situated 2.2m above the floor level on the northern elevation, enhance daylight penetration and increase reflection onto and off the vaulted roof. As Fontoynont suggests, this strategy results in a more uniform and indirect daylight distribution, while providing significant artificial light reductions.

Despite this extremely effective strategy, it must be pointed out that no more than a quarter of the total floor area will achieve a daylight factor greater than 2 per cent due to the overshadowing from the surrounding buildings. Conversely, a detailed study has been simulated on each façade at different times of the day and the year (solstices and equinox) in order to define the areas of the façades which may require additional shading or may not receive enough natural light.

Based on the result of those simulations, some assumptions can be made. The eastern façade will be characterised in winter by a direct illumination occurring mainly on the upper floors during the last hours of the morning and the first few hours of the afternoon. In mid seasons (spring and fall), the façade will be directly lit on the top two floors between 11 and 12am, before the sun moves northerly. During summer the balconies at all levels will increasingly receive direct light from 6.45am to 9am, while after 9am, the façade will be totally lit until 11.45am. In the afternoon, the façade will be entirely in shadow.

Direct solar illumination will not be a concern for visual comfort since during the early hours of summer the building is expected to be only sparsely occupied, while the risk of solar overheating from early morning sun will be significantly less than for western orientation due to lower air temperatures. In addition, it is highly probable that direct light on the balconies during the morning could present opportunities for workers to relax and take a short break, which is important physiologically and psychologically.

With regard to the north façade, in winter, the eastern part of the elevation will be almost totally directly lit from 7.15am to 9am. However, even in this case, glare should not present a major risk of discomfort since during those hours the building is still unlikely to be heavily occupied, while direct solar penetration could provide some passive solar heating, particularly appreciated during clear, cold winter mornings. Top floors of the western part of the façade will be directly lit from 11am till 12.45pm and this direct illumination will cause glare due to a relatively low-angle winter sun unless shaded accordingly. The major concern for the north façade during winter seems to be overshadowing of a neighbouring building during morning hours, which could generate low levels of natural illumination in the office spaces (particularly at lower floors) and consequently greater energy consumption for artificial lighting. The top three floors and western part of the north façade will be increasingly directly lit during the afternoon. At this time of the day the relatively low north-western sun will probably increase daylight penetration in the office space and a light shelf mounted on top of the windows will act as a shading device rather than a tool to redistribute light. However, on a clear and sunny winter day, the high luminance off the glazed surfaces and the light shelf may require a thorough masking by blinds mounted in the clerestory area and in the later hours of the afternoon, by proposed vertically-shading timber louvers or evergreen vines.

In mid-seasons, the eastern part of the north façade receives direct sunlight from 6.30am to 9am. Again, this is not a concern in terms of quality of daylight since the building will not be heavily occupied at that time of the day. At 9am, almost the entire façade will be in shadow except for a small portion of the western part on the top two floors. Throughout the morning almost half of the façade will be lit by a comfortable high sun. At 3pm, three quarters of the façade will be day lit and this portion will increase gradually until 5.30pm. During the last hours of the afternoon, there may be the risk of glare from a low western sun, although the presence of evergreen vines, together with blinds that have been designed to rise up from the floor to counter act NW sun in late afternoon, will reduce the risk of visual discomfort.

During summer, the northern façade will be increasingly lit by direct sunlight from 6.45am. From 9am to 10.45am there may be some glare due to the relatively low-angle eastern sun, even though this discomfort could be alleviated by the vines and may not be disturbing due to the positive physiological and psychological effects of morning light. After 10.15am, some shadowing off the facing building may occur on the lower and central area of the façade and this pattern of shadow will cover almost all of the eastern part of the northern façade until 2.30pm. From 2.30pm to 5pm, the façade will be entirely directly lit and good daylight penetration will be assured by the light shelf that maximizes its effectiveness when the sun is high on the horizon. In addition, the presence of the balconies and the light shelf may provide sufficient means of masking direct glare along the perimeter. Some risks of glare may occur in the later part of the afternoon, from 5.15pm to 6.45pm, when the building will not be heavily occupied.

In terms of the west façade, the patterns of shadow are more straightforward due to the absence of major obstructions projecting shade over the building. In winter, the western façade is directly lit from 2pm until sunset (5.30pm), while, in summer, the façade receives direct illumination from 1pm. The risk of glare and overheating are significant but the presence of the intelligently-controlled timber louvers (whose control is based upon a six-hour open and close cycle) can minimise those drawbacks while simultaneously providing a dynamically changing pattern to the façade.

Finally, the southern façade during winter will receive diffuse light during the morning, with some direct illumination on the upper eastern part of the elevation between 8am and 10am (due to the slight easterly inclination of the site). This low-angle sun could potentially cause glare, however low winter temperatures justify the absence of devices or strategies to block this sun penetration since some passive solar heating may be provided. Again, the presence of warm sunlight during the morning, even when it comes with slight visual discomfort, is likely to be tolerated from the occupants because of its physical, physiological and psychological benefits. During the rest of the day, the façade will be in shadow, although the upper floors will receive some indirect glare reflected off the adjacent Victoria Hotel.

During summer months, the façade will receive direct light in the first hours of the day but glare will not be an issue since direct illumination will occur when the building is largely unoccupied. However, there is the risk of excessive solar heating in the early morning that could counteract the night-time ventilation strategy adopted during summer nights. During the rest of the day, the façade will receive reflected diffuse light off the fronting building and will receive direct illumination on the upper levels after 5.15pm, when the building will not be extensively occupied.

In mid-seasons, the patterns of light and shadow will follow the same schedule, with less risk of glare and solar heating in the early mornings and late afternoons because of the (respectively) more easterly and westerly position of the sun at sunrise and sunset.

Glare Control and Vertical Gardens

Shading devices to control sun intrusiveness and reduce visual discomfort will be used on north, east and west façades of the CH_2 building. These devices will consist of vertical gardens, perforated metal light shelves and vertically-shading/pivoting timber louvres.

In order to assess the potential for glare at CH_2, a series of studies and simulations have been conducted using the *Daylight Glare Index* to measure visual comfort. In terms of glare control, the adopted lighting strategy and design aims to avoid instances of *intolerable glare* (25<DGI<28) and minimise the potential occurrence of *uncomfortable glare* (DGI<25) in office areas.

The preliminary studies, the results of which are described in a series of reports by Advanced Environmental Concepts (AEC), indicate that reflections off the Victoria Hotel may represent a significant *uncomfortable* visual source of glare, affecting the narrower field of vision (90°) as well as the peripheral one, particularly during the middle of the day in mid-seasons and summer. In addition, some *intolerable* glare at certain view angles may also occur. With these issues in mind, it must be pointed out that the south façade has been designed with no fixed or movable shading devices to mask the glazing, however, columns of plant leafs in the window mullions have been experimented with by the design team. This practical experiment seems to show that the presence of the leaves helps break up the sharp edges of high contrast, which helps the eyes to adjust. Specifically, glare is expected to be worse at the upper levels than at the lower ones due to a greater penetration of reflected daylight. However, the choice of decreasing the size of the glazed surface with the height of the building represents a favourable strategy in relation to glare problems. Based on these simulations, glare off the south façade is supposed to be *acceptable* (DGI<19) during the early morning and late afternoon in all seasons.On the north façade, glare is likely to be relatively low and generally in the *acceptable* range although the occurrence of discomfort glare may be more severe at the periphery compared with the focus of the viewer's field of vision and more significant in the winter and mid-season afternoons because of low western sun. However, it is probable that once partitions, task lighting, accent lighting and workstations are in place their presence will limit the amount of potential periphery glare, as suggested in the Australian Standard 1680.2.2-1994.To decrease the risk of *intolerable glare* on the north façade, steel trellises and balconies, supporting a series of vertical gardens that run the full height of the building alongside the windows, will filter light entering office spaces and form a 'green' microclimate. The 3-4m vines will be grown in specially designed boxes situated to the east and west of each balcony on every storey via stainless steel mesh stretching from the ground to the roof. As already pointed out, the presence of the vines will clearly increase visual comfort for users, both in terms of glare reduction and the positive physiological effect of having a 'green' visual outlook. Further shading on the north façade will be provided by balconies and the light shelves that will block high-angle sun penetration during summer, while redistributing incoming radiation deeper in the space. At the same time, the internal upward rolling retractable blinds located at the level of the light shelf and the manually-adjustable vertically sliding timber screen at the window line will guarantee the required protection from horizontal light intrusiveness during winter months and in the late hours of the afternoon, while maintaining an unrestricted view at eye-level.

Finally, since curved surfaces typical of CRT monitors and reflective glass screens can increase the level of glare and veiling reflection, brighter flat TFT screens have been chosen for CH_2 to enhance visual comfort, maximise the use of space and reduce internal heat loads.

Artificial Lighting Strategy

The CH_2 building is characterised by a deep open plan. A consequence of this building type, forced by the limitations of the small CBD site is that natural lighting is not an option for a significant part of the internal floor plate and thus has been complemented by an artificial system. This system is designed as a two-component scheme: a low-energy background lighting component (provided as part of the base building design) and a separate individual task lighting component (part of the fit-out), giving users more control over their luminous environment. Ecological, economical and psychological issues (as well as culture considerations) have been factored into the light system design to achieve the proposed objectives and to provide an optimum level of lighting for movement, security and occupant activities. A number of different artificial lighting solutions have been simulated to obtain the minimum required ambient lighting level on the floor (160 lux) as uniformly as possible, while limiting any potential glare impacts from the light fittings.

The background ambient lighting supplied by T5 fluorescent lamps provides a low illuminance (160 lux ambient lighting; 70 per cent wattage emitted to the office space, 30 per cent emitted to the ceiling), while the individually controlled lamps at workstations provide up to 320 lux on each desk. An illuminance no greater than 400 lux will be provided anywhere on the office floor, with a colour rendering index (CRI)>85 per cent. To achieve an optimal distribution of light, materials and finishes have been chosen with an overall reflectance of 30 per cent for the carpet, 50 per cent for the walls and 70-80 per cent for the ceiling and desktops. All of these choices are in accordance with the parameters suggested in Australian Standard 1680.1-1990.

The adopted artificial lighting system includes sensors that continuously monitor the amount of direct and diffuse daylight entering the building and reflected off the light shelves and accordingly dim the artificial light levels supplied, thus creating a mix of filtered natural and artificial illumination. The artificial lighting system flexibly provides a number of separated switched zones per office floor that are no greater than 100 m², which means that every single luminaire can be programmed to separately address the specific needs of a zone and to suit future fit-out requirements of the space. The deliberate use of workstation task lighting will create the illusion of "campfires" of activity that is both warm and inviting.

From an energy savings point of view, the chosen fluorescent T5 fittings incorporating high frequency dimmable electronic ballasts and individual task lighting (10W compact fluorescent) for workstations will offer a significant reduction in energy consumption with T5 lamps widely accepted as far more efficient and longer lasting than conventional T8 lamps. T5 lamps actually benefit is its optical efficiency when compared to conventional T8s due to their smaller diameter, even though they have the disadvantage of higher surface brightness, which can be a potential source of glare, however this has been accounted for in the design of CH$_2$ and has been discussed earlier. The savings for this configuration has been calculated to be equivalent to a reduction of around 65 per cent when compared to the energy consumed by the current Council house building lighting system. Overall the lighting energy load, as a proportion of the buildings total base load, is estimated to be 1.37 per cent

In terms of control strategies, one security light will be located at each lift/ lobby area and switched on from 12pm to 6am. In all the other areas light fittings will be switched off during that period of time. One push button control panel will be located at each end of the office area in front of the lifts with three different operational modes: *normal* (8am to 6pm, providing an ambient lighting level of

160 lux), *security* (6pm to 8am, two security light fittings will be switched on) and *cleaner* (6am to 8am, providing an ambient lighting of 160 lux for two hours). The presence of different lighting strategies according to the functions taking place in the building will assure optimum energy management.

From 6am to 8am, all the office lighting will be switched on, but dimmed to achieve an internal lighting level of 80 lux. From 8am to 6pm, during office hours, lighting levels will be automatically set to achieve an ambient level of 160 lux, with the photosensors detecting the availability and distribution of daylight and accordingly controlling the dimmable electronic ballasts. When daylight provides more than 160 lux, all the lights will be switched off. After office hours, the daylight sensors will be inactive.

To optimise the use of electric artificial lighting, each workstation will be provided with a local dimmer switch integrated as an icon on the PC screen, which will provide three different lighting control options: *high*, *medium* and *light*. The latter option will cause a slider to appear with a 'save' button enabling the user to set the preferred task light level in the area of their workstation. Task lighting will be provided in locations where a total lighting of 320 lux is not achievable. Local light fittings will automatically be set at 160 lux of ambient lighting once the computer screen has been turned off. An employee who works after 6pm can use his PC control (or the push button control panel) to indicate longer working hours and adjust the lighting level as required.

There were a number of opportunities, requiring more complex technologies than windows and light-shelves, available for CH$_2$ to harvest additional natural light and direct it for use within the building. Meticulous investigation was undertaken by the design team in an attempt to improve the daylight factor of the building, especially at lower floor levels, where daylight distribution systems such as a *light pipe*, *fibre optics*, *prismatic shafts*, or a *heliobus* system were investigated for their feasibility for channelling sunlight from the roof and redistributing it to internal spaces. However, these alternative strategies were found to have major drawbacks especially in terms of practical issues (*i.e.* the openness of the space arrangement and costs).

One of the main problems in implementing the adopted design, that emerged from the lighting strategy discussed earlier and the analysis of the lighting patterns down the façades of the building, is the poor availability of daylight especially at the lower floors during winter. In this case, the low sun together with the overshadowing from adjacent buildings dramatically reduces the availability of natural light, a concern that is relevant both in terms of energy savings but also from a physiological and psychological point of view.

As a daylight device, the light shelf is actually proven to be particularly useful under direct high-sun conditions, while its effectiveness in terms of daylight distribution is dramatically reduced when the available radiation diffusely comes from the sky vault. Conversely, light shelves are far more efficient in rejecting sunlight than displacing diffuse light deep into the interior, especially when their reflectivity indices may be influenced by inconsistent maintenance.

Potential improvements, although requiring additional costs, to the proposed design could have occurred by putting in place light re-directing devices on the northern façade (mounted in the clerestory area) that could have complemented the effectiveness of the light shelf; examples of those devices can be found in *louvers, highly reflective movable blinds, prismatic panels, laser-cut panels* and *anidolic ceiling.* However, after careful consideration, none of these improvements were thought to provide significant benefits to the adopted design.

Louvres and *venetian blinds* are 'classic' daylighting systems that can be applied to redirect external daylight deeper into the rooms. These systems are generally composed of evenly-spaced, horizontal, vertical or sloping slats and highly sophisticated shapes and surface finishes. Fixed systems are usually designed for solar shading, while movable devices can operate optimally according to outdoor conditions. Louvers are generally made of galvanized steel, anodized or painted aluminium or plastic (PVC) for improved durability and reduced maintenance, while venetian blinds are usually composed of small or medium-sized PVC or painted aluminium slats that are either flat or curved. Slat size varies with the location of the blind (exterior, interior or between the glass panes in a DGU unit). Some light-directing devices include perforations and concave curvature in order to reflect the maximum possible amount of light to the ceiling while having a very low brightness at angles below the horizontal.

Movable blinds can be operated either manually or automatically, the latter option are preferable for energy efficiency and to optimize the penetration of visible light according to daily and seasonal variations in solar positions. However, with automated systems, an over-ride capability via a remote-control device would almost always be required by tenants to increase satisfaction. Manually-operated systems, on the other hand, are usually less-energy efficient because occupants do not generally continuously optimise their position.

As a range of studies suggest, if adopted in clerestory areas that do not influence the perception of the external environment, highly-reflective, automatically adjusted Venetian blinds can improve daylight availability and distribution, enhancing the satisfaction and productivity of workers and result in significant cost and energy savings. Another element that could have been installed in the clerestory area of CH_2 is a device called a *prismatic* or *laser-cut panel* to enhance the penetration of diffuse skylight deeper in the floor plate.

A *prismatic panel* is generally composed of a thin, planar, saw tooth device consisting of an array of clear acrylic (PMMA) prisms with one surface of each prism forming a plane; some panels may be partially coated with an aluminium film with high specular reflectance. The system can be designed to reflect light coming from a certain angle while transmitting light coming from others. Refraction and total internal reflection (according to the critical angle of the panel) can be used to change the direction of the light beams, depending on the angle of incidence and the index of refraction of the material used.

A *laser-cut panel* is a daylight-redirecting system composed of a thin clear acrylic material divided by laser cut into an array of rectangular elements; the surface of each laser cut becomes a small internal mirror that deflects light passing through the panel. A laser-cut device strongly deflects light coming from higher elevations (>30°), while transmitting light at a near-normal incidence with little disturbance, thus maintaining external view. Light is deflected in each element of the panel by refraction, then by total internal reflection and then by refraction again. In general, the panels are cut at an angle perpendicular to the external surfaces, but it is possible to make the cuts at a different angle for added control over the direction of the deflected light.

Although these devices can provide some benefits to a lighting strategy, it should be pointed out that light shelves, louvers, reflective blinds, prismatic and laser-cut panels perform at their optimum under direct light conditions, a situation unlikely to occur at CH_2, especially during winter months. The CH_2 design team have considered this in their simulations of direct sunlight over the whole year and found that the installation of light shelves are justified on the basis that they also perform the function of shades to reduce heat gain and glare caused by direct sunlight penetration. The light shelves duel function, shading and light reflection, which may only address problems of heat gain or glare for one hour a day by operating as a shade, it will significantly reduce the likelihood of occupants drawing a blind, which is then likely to remain closed for the remainder of the day and lead to a large proportion of the days natural light gain potential being lost.

Other technical solutions for exploiting the diffuse light coming from the sky vault include *light-guiding shade* or an *anidolic system* located in the clerestory area. A *light-guiding shade* is composed of a diffusing glass aperture and two reflectors designed to redirect intercepted diffuse light within a specified angular range. In general, highly reflective material such as bright-finish aluminium can be used for the device's inner surfaces. All daylight that enters the light-guiding shade is directed over the ceiling, which means the shade becomes a source of diffuse light completely free of glare when viewed by occupants. As a consequence, the light-guiding shade could have been considered as an alternative solution to the light-shelf, since it is a fixed device installed over the upper third of the window, totally shading the glazed surface from direct sun penetration.

Finally, an *anidolic ceiling* consists of daylight-collecting, non-imaging optics coupled to a specular light duct which transports the light to the back of a room. The system exploits the optical properties of compound parabolic concentrators made of anodized aluminium surfaces and placed on both ends of a light duct to collect diffuse light from the upper area of the sky vault. In particular, on the outside of the building, an anidolic optical concentrator captures and concentrates diffuse light and efficiently introduces those rays into the light duct, where, at its aperture in the back of the room, a parabolic reflector distributes the light, avoiding reflections. In CH_2, the use of such a system would have required adequate suspended ceiling space to exploit its potential to the full. It would also have been inappropriate if coupled with the vaulted roof, which plays an important role in terms of ventilation. However, a potential solution using similar technology may exist, such as *anidolic solar blinds* placed in the clerestory areas on top of the light shelves. These blinds (at present in the prototype stage) consist of a grid of hollow reflective elements, each of which is composed of two three-dimensional compound parabolic concentrators. The portions of the blind that emit light are designed to direct daylight into the upper quadrant of the room towards the ceiling.

Considering all of the above, since the poor availability of daylight at lower levels in winter may not be counteracted without major practical (or visual) drawbacks, it may be advisable to locate employees whose work is generally done in a team in those areas. Through interaction with their colleagues these employees should get the necessary environmental stimulation they require that cannot be provided by the dynamic pattern of natural illumination.

Concerning the south façade, as reported earlier, one of the major issues with regard to visual comfort is due to the presence of glare reflected off the Victoria Hotel Building to the south, which creates high visual contrast in the field of view of the occupant and thus without accent lighting has the potential to significantly decreases the quality of the luminous environment. Of the various implementations investigated by the consultants Advanced Environmental Concepts (AEC), the best option seems to be the use of artificial lighting to illuminate the internal wall adjacent the windows in order to decrease the contrast effect between the high luminance of the windows and the comparatively dark internal environment and internal wall finishes. Even though the additional economic and environmental cost of providing such additional lighting may not be compensated by a relatively marginal improvement in luminous comfort it is well worth countering glare in this way as it can often save energy overall by reducing the tendency of people to draw blinds and thus losing all the benefits of available natural light.

An alternative solution could be the installation of vertical gardens on the south façade, possibly using plants that do not need continuous direct illumination, or adjustable downward roller blinds that could block discomfort glare but still admit a reasonable quantity of diffuse light. In this case, the use of a roller blind rather than a Venetian device would most likely be accepted by occupants since glare problems would probably be fairly constant at certain times of the day. Likewise the continuous availability of a view would not be a major concern due to the proximity of adjacent buildings and a roller blind would also increase the sense of privacy. A further, simple (but not necessarily less effective), potential response could be the installation of flexible visual tasks, as for example by mounting the flat TFT computer screens on movable arms that could be adjusted directly by the users according to the characteristics of their luminous environment and their individual preferences.

It is worth noting that this southern façade glare issue has been exceptionally difficult to respond to within the limitations of the building's design. Also, similar glare conditions are currently experienced by staff occupying the City of Melbourne's existing adjacent office building. Proper resolution of this issue indicates that a response at the urban design level is required, where urban design requirements have the potential to make consideration for the reflective properties of the facades of buildings and their potential negative impact in the urban environment, such as creating glare issues for adjacent buildings. For example, one design response would be to change the existing yellow colour of the adjacent Victoria Hotel building to a darker less highly light reflective finish or to suspend planter boxes on this light rich northern façade to support the growth of plants that naturally have low reflective properties.

In terms of future opportunities, the use of advanced chromogenic glazing technologies such as *gasochromic* or *electrochromic* devices, if thoroughly implemented with lighting control strategy, may be able to reduce indirect glare off the adjacent reflective façade, while keeping visual link with the external environment, a key-issue that is fundamental both ergonomically and physio-psychologically. An important feature of such devices as part of a lighting strategy is the possibility of being linked to an automated control system so as to provide the desired interior illuminance over a wide range of exterior lighting conditions. Those technologies, while already available on the market, have not yet been applied to large multi-storey buildings, even though they show enormous potential in terms of energy saving and occupant comfort.

Conclusion

The lighting strategy developed for the CH_2 building represents an outstanding example of international best practice, particularly, in how it manages to integrate innovative solutions with a number of functions and requirements of the building and importantly how it addresses the physiological and psychological needs of workers. The adopted design exceeds all of the commonly-accepted standards for a so-called sustainable building. There are many lessons to be learnt from CH_2. Some of these include the need to foresee and control future development of neighbouring buildings in order to develop strategies to counteract the effects on solar access and its consequences on luminous environments (overshadowing, reflections, indirect glare, etc.). This is essentially a problem beyond the grasp of the design team. As the CH_2 building demonstrates, to be truly sustainable, buildings have to be designed and operated as 'living' and complex systems rather than as a passive collection of distinct parts. This is the only way to guarantee optimum comfort for all occupants, both in perceptive and energetic terms, while also creating a pleasant place to live and work.

References

Advanced Environmental Concepts, "Artificial Lighting Study", June 2003.

Advanced Environmental Concepts, "Glare Study", Melbourne City Council, August 2003.

Advanced Environmental Concepts, "Lighting Simulation Report", Melbourne City Council, July 2003.

Advanced Environmental Concepts, "Natural Light Opportunities", Melbourne City Council, June 2003.

Advanced Environmental Concepts, "Overshadowing and Shading Images of MCC", Melbourne City Council, March 2003.

Altomonte S. (2003). "The Architectural Integration of Switchable Devices in Daylight Control Strategies", in *Proceedings of the 2003 CISBAT Conference*, Lausanne.

Altomonte S. (2003). *Switchable Façade Technology Architectural Guidelines*, European SWIFT Research Project Report.

Australian Standard AS 1680.1-1990, *Interior Lighting-Part 1: General principles and recommendations*.

Australian Standard AS1680.2.2-1994 *Interior Lighting-Part. 2.2: Office and screen-based tasks*.

Beltran L.O., Lee E.S., Selkowitz S.E. (1996). "Advanced Optical Daylighting Systems: Light Shelves and Light Pipes", in *Proceedings of the 1996 IESNA Annual Conference*, Cleveland.

Beltran L.O., Lee E.S., Selkowitz S.E. (1994). "The Design and Evaluation of Three Advanced Daylighting Systems: Light Shelves, Light pipes and Skylights", in *Proceedings of the Solar '94 Golden Opportunities for Solar Prosperity Conference*, San Jose'.

Birren F. (1982). "Light, colour and environment – Revised edition", Van Nostrand Reinhold Co., New York.

Boyce P., Hunter C., Howlett O. (2003). "The benefits of daylight through windows", Lighting Research Center.

Boyce P.R., Cuttle C. (1990). "Effect of Correlated Colour Temperature on the Perception of Interiors and Colour Discrimination Performance", in *Lighting Research and Technology*.

Brainard G., Glickman G. (2003). "The biological potency of light in humans: significance to health and behaviour", in *Proceedings of the 25th CIE Session*, San Diego

Caminada J.F. (2001). "From surfaces to spaces", in *International Lighting Review*, Philips Lighting, n. 01/2001.

Carmody J., Selkowitz S., Lee E., Arasteh D., Willmert T. (2004). *Window Systems for High-Performance Buildings*, W.W. Norton and Company, New York.

Compagno A. (1995). *Intelligent Glass Façades - Material, Practice, Design*, Artemis Verlags - AG, Basel.

Ehrlich C., Papamichael K., Lai J., Revzan K. (2001). "Simulating the Operation of Photosensor-based Lighting Controls", in *Proceedings of the 7th Building Simulation Conference*, Rio de Janeiro

Fonseca I.C.L., *et al.* (2002). "Quality of light and its impact on man's health, mood and behaviour", in *Proceedings of the PLEA 2002 Conference*, Toulouse, July.

Fontoynont M. (edited by), (1999). *Daylight performance of buildings*, James and James.

Guzowski M. (2000). *Daylighting for sustainable design*, Mc Graw-Hill.

IEA SHC Task 21, *Daylight in Buildings – A source book on daylighting systems and components*, LBNL, 2000.

Kaiser P.K. (1994). "Physiological Response to Colour: A Critical Review", in *Colour research and application*, Volume 9, Number 1, Spring.

Kramer H. (2001). "Mastering Office Lighting", in *International Lighting Review*, Philips Lighting, n. 01/2001.

Lee E.S., Di Bartolomeo D.L., Selkowitz S.E. (2000). "Electrochromic windows for commercial buildings: monitored results from a full-scale testbed", in *Proceedings of the ACEEE 2000 Summer Study on Energy Efficiency in Buildings*, Asilomar, April.

Lee E.S., Di Bartolomeo D.L., Selkowitz S.E. (1998). "The Effect of Venetian Blinds on Daylight Photoelectric Control Performance", in *Journal of the Illuminating Engineering Society*, 28, no. 1.

Lee E.S., Di Bartolomeo D.L., Vine E.L, Selkowitz S.E. (1998). "Integrated Performance of an Automated Venetian Blind/Electric Lighting System in a Full-Scale Private Office", in *Proceedings of the Thermal Performance of the Exterior Envelopes of Buildings VII*, Florida, December.

Lee E.S., *et al.* (2002). "Active Load Management with Advanced Window Wall Systems: Research and Industry Perspectives", in *Proceedings from the ACEEE 2002 Summer Study on Energy Efficiency in Buildings*, Asilomar, August.

Lee E.S., Selkowitz S.E. (1998). "Integrated Envelope and Lighting Systems for Commercial Buildings: a Retrospective", in *Proceedings of the ACEEE 1998 Summer Study on Energy Efficiency in Buildings*, Asilomar, June.

O'Connor J., Lee E., Rubinstein F., Selkowitz S. (1997). *Tips for Daylighting with Windows – The Integrated Approach*, LBNL, California.

Philips Lighting. (2001). "Fingertip comfort" – Automated control systems for ABN Ambro", in *International Lighting Review*, n. 01/2001.

Selkowitz S.E. (1999). "High Performance Glazing Systems: Architectural Opportunities for the 21st Century", in *Proceedings of the 7th International Glass Processing Days*, Tampere, Finland, June.

Selkowitz S.E. (2001). "Integrating Advanced Façade into High Performance Buildings", in *Proceedings of the 7th International Glass Processing Days*, Tampere.

Selkowitz S.E., Lee E.S. (1998). "Advanced Fenestration Systems for Improved Daylight Performance", in *Daylighting '98 Conference Proceedings*, Ottawa.

Van den Beld G. (2001). "Light and Health", in *International Lighting Review*, Philips Lighting, n. 1/2001.

Wigginton M. (1996). *Glass in Architecture*, Phaidon Press, London.

Wigginton M. (2002). Harris J., *Intelligent Skins*, Architectural Press, Oxford.

Wu Kwok-tin M. (2000). "T5 lamps and luminaries, The 3rd generation in office lighting", in *Hong Kong and Kowloon Electrical Engineering and Appliances Trade Workers Union Meeting*, August.

Zonneveldt L. (2001). "The Daylight Challenge", in *International Lighting Review*, Philips Lighting, n. 01/2001.

Chapter 32

Tetrapyrroles

Tetrapyrroles are a class of chemical compounds whose molecules contain four pyrrole rings held together by direct covalent bonds or by one-carbon bridges (= (CH)- or -CH_2-units), in either a linear or a cyclic fashion. A pyrrole ring in a molecule is a five-atom ring where four of the ring atoms are carbon and one is nitrogen. In cyclic tetrapyrroles, lone electron pairs on nitrogen atoms facing the center of the macrocycle ring can bond or chelate with a metal ion such as iron,cobalt, or magnesium.

Some tetrapyrroles are the active cores of some compounds with crucial biochemical roles in living systems, such as hemoglobin and chlorophyll). In these two molecules, in particular, the pyrrole macrocycle ring frames a metal atom, that forms a coordination compound with the pyrroles and plays a central role in the biochemical function of those molecules.

Structure

Linear tetrapyrroles (called bilanes) include:

- ☆ Heme breakdown products (*e.g.*, bilirubin, biliverdin)
- ☆ Phycobilins (found in cyanobacteria)

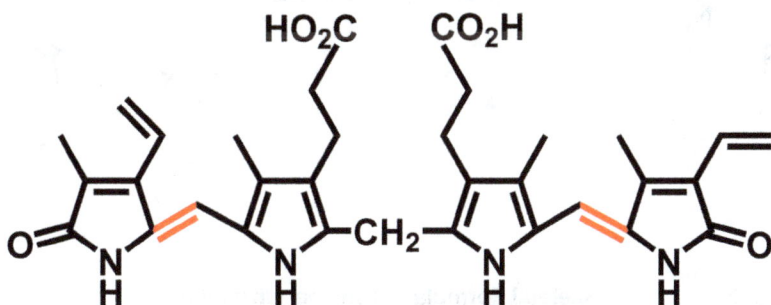

Correct Formula of Natural (Z, Z)-bilirubin

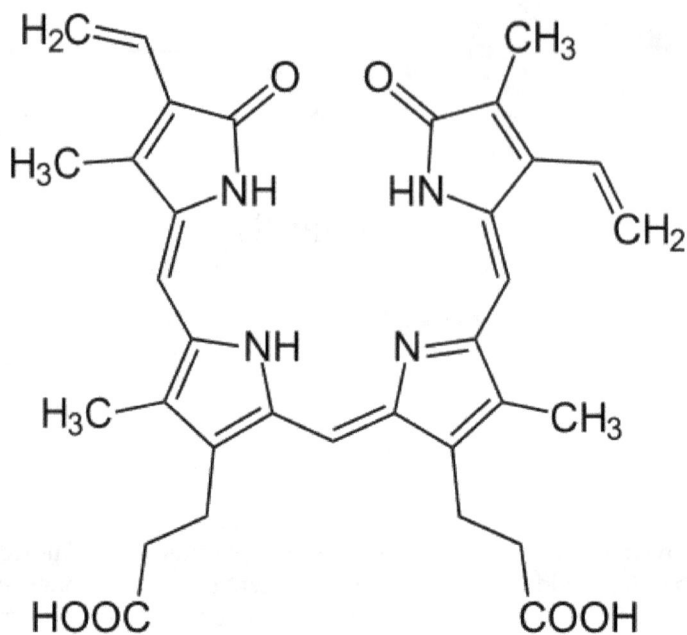

Structure of Biliverdin

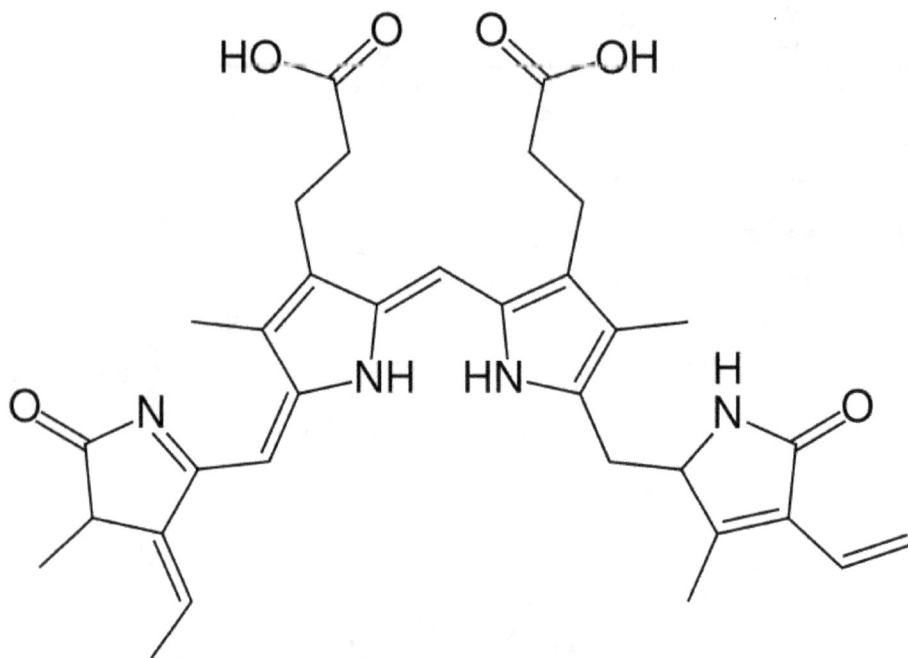

Skeletal Formula of Phycoerythrobilin

Cyclic tetrapyrroles having four one-carbon bridges include:

☆ Porphyrins, including heme, the core of hemoglobin

☆ Chlorins, including those at the core of chlorophyll.

Cyclic tetrapyrroles having three one-carbon bridges and one direct bond between the pyrroles include:

☆ Corrins, including the cores of cobalamins, when complexed with a cobalt ion.

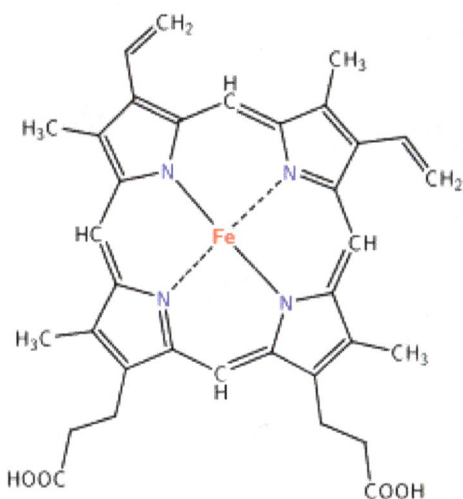

Heme Group of Hemoglobin

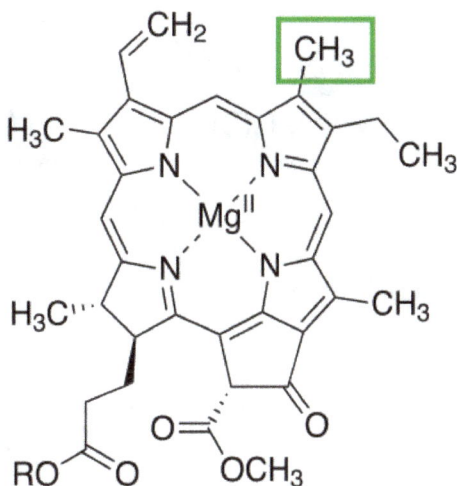

The Chlorin Section of the Chlorophyll *a* Molecule

The tetrapyrrole portions of the molecules typically act as chromophores because of a high degree of conjugation in them. Therefore, these compounds are commonly coloured.

Bilirubin (formerly referred to as haematoidin) is the yellow breakdown product of normal heme catabolism, caused by the body's clearance of aged red blood cells which contain hemoglobin. Bilirubin is excreted in bile and urine and elevated levels may indicate certain diseases. It is responsible for the yellow colour of bruises and the yellow discolouration in jaundice. It is also responsible for the brown colour of feces (via its conversion to stercobilin) and the background straw-yellow colour of urine via its breakdown product, urobilin, contributing to urine colour along with thiochrome, a breakdown product of thiamine which produces the more obvious but variable bright yellow colour of urine. It has also been found in plants.

Chemistry

Bilirubin consists of an open chain of four pyrrole-like rings (tetrapyrrole). In heme, these four rings are connected into a larger ring, called a porphyrin ring.

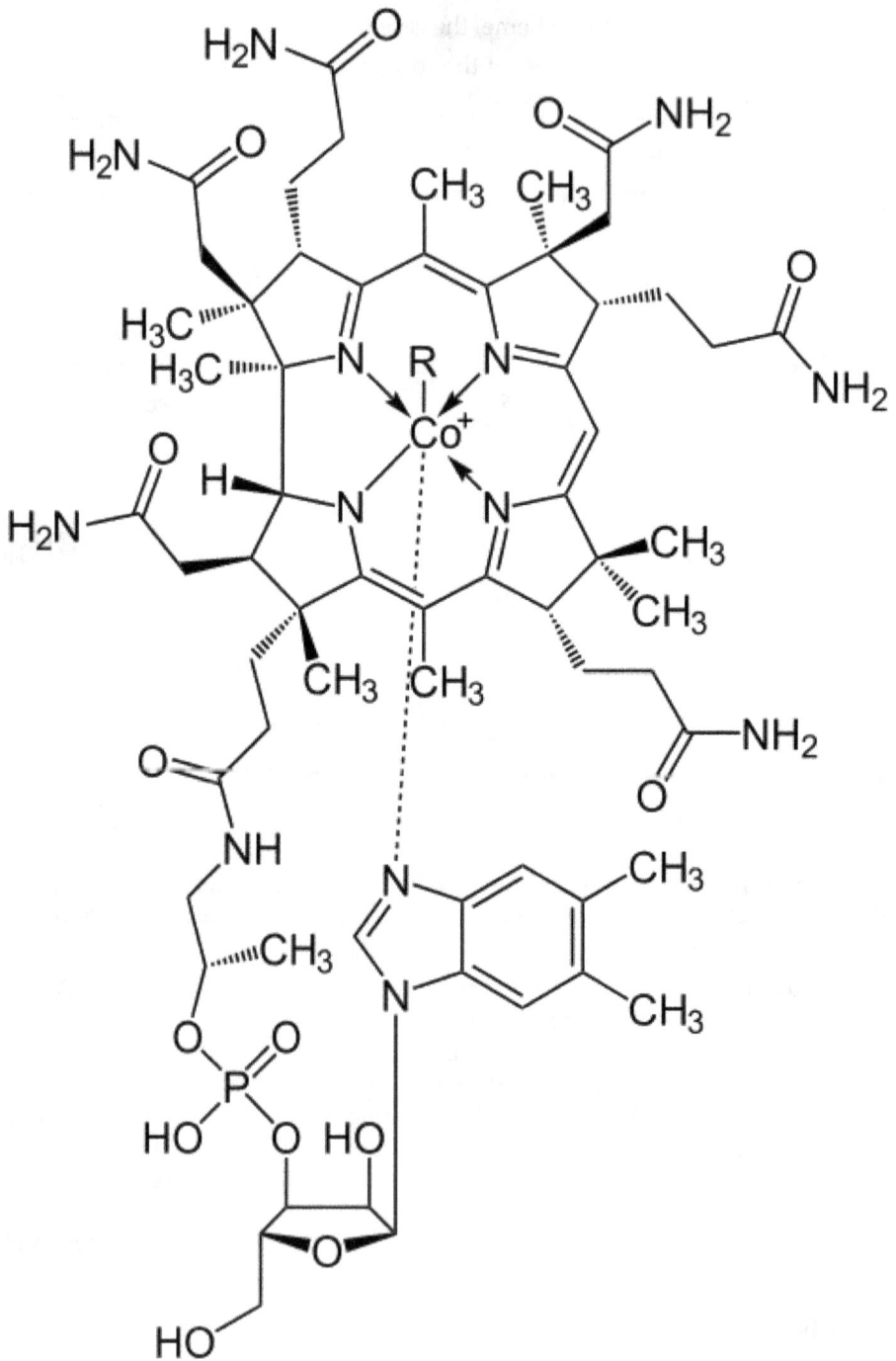

Structure of Cobalamin (Vitamin B₁₂)

Bilirubin can be "conjugated" with a molecule of glucuronic acid which makes it soluble in water (see below). This is an example of glucuronidation. Bilirubin is very similar to the pigment phycobilin used by certain algae to capture light energy and to the pigment phytochrome used by plants to sense light. All of these contain an open chain of four pyrrolic rings. Like these other pigments, some of the double-bonds in bilirubin isomerize when exposed to light. This is used in the phototherapy of jaundiced newborns: the E, Z-isomers of bilirubin formed upon light exposure are more soluble than the unilluminated Z, Z-isomer, as the possibility of intramolecular hydrogen bonding is removed. This allows the excretion of unconjugated bilirubin in bile. Some textbooks and research articles show the incorrect geometric isomer of bilirubin. The naturally occurring isomer is the Z, Z-isomer.

Function

Bilirubin is created by the activity of biliverdin reductase on biliverdin, a green tetrapyrrolic bile pigment that is also a product of heme catabolism. Bilirubin, when oxidized, reverts to become biliverdin once again. This cycle, in addition to the demonstration of the potent antioxidant activity of bilirubin, has led to the hypothesis that bilirubin's main physiologic role is as a cellular antioxidant.

Metabolism

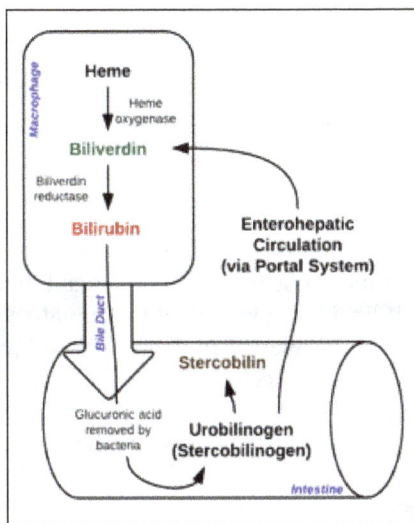

Heme Metabolism

Unconjugated ("Indirect")

Erythrocytes (red blood cells) generated in the bone marrow are disposed of in the spleen when they get old or damaged. This releases hemoglobin, which is broken down to heme as the globin parts are turned into amino acids. The heme is then turned into unconjugated bilirubin in the monocyte macrophages system of the spleen. This unconjugated bilirubin is not soluble in water, due to intramolecular hydrogen bonding. It is then bound to albumin and sent to the liver. The measurement of direct bilirubin depends on its reaction with diazosulfanilic acid to create azobilirubin. However, unconjugated bilirubin also reacts slowly with diazosulfanilic acid, so that the measured indirect bilirubin is an underestimate of the true unconjugated concentration.

Conjugated ("Direct")

In the liver, bilirubin is conjugated with glucuronic acid by the enzyme glucuronyltransferase, making it soluble in water: the conjugated version is also often called "direct" bilirubin. Much of it goes into the bile and thus out into the small intestine. Though most bile acid is resorbed in the terminal ileum to participate in enterohepatic circulation, conjugated bilirubin is not absorbed and instead passes into the colon. There, colonic bacteria deconjugate and metabolize the bilirubin into colourless urobilinogen, which can be oxidized to form urobilin and stercobilin: these give stool its characteristic brown colour. A trace (~1 per cent) of the urobilinogen is resorbed into the enterohepatic circulation to be re-excreted in the bile: some of this is instead processed by the kidneys, colouring the urine yellow. Although the terms direct and indirect bilirubin are used equivalently with conjugated and unconjugated bilirubin, this is not quantitatively correct, because the direct fraction includes both conjugated bilirubin and Δ bilirubin (bilirubin covalently bound to albumin, which appears in serum when hepatic excretion of conjugated bilirubin is impaired in patients with hepatobiliary disease). Furthermore, direct bilirubin tends to overestimate conjugated bilirubin levels due to unconjugated bilirubin that has reacted with diazosulfanilic acid, leading to increased azobilirubin levels (and increased direct bilirubin).

Urine

Under normal circumstances, a tiny amount of urobilinogen, if any, is excreted in the urine. If the liver's function is impaired or when biliary drainage is blocked, some of the conjugated bilirubin leaks out of the hepatocytes and appears in the urine, turning it dark amber. However, in disorders involving hemolytic anemia, an increased number of red blood cells are broken down, causing an increase in the amount of unconjugated bilirubin in the blood. Because the unconjugated bilirubin is not water-soluble, one will not see an increase in bilirubin in the urine. Because there is no problem with the liver or bile systems, this excess unconjugated bilirubin will go through all of the normal processing mechanisms that occur (*e.g.*, conjugation, excretion in bile, metabolism to urobilinogen, reabsorption) and will show up as an increase in urine urobilinogen. This difference between increased urine bilirubin and increased urine urobilinogen helps to distinguish between various disorders in those systems.

Toxicity

Unconjugated hyper bilirubinaemia in a newborn can lead to accumulation of bilirubin in certain brain regions (particularly the basal nuclei) with consequent irreversible damage to these areas manifesting as various neurological deficits, seizures, abnormal reflexes and eye movements. This type of neurological injury is known as kernicterus. The spectrum of clinical effect is called Bilirubin Encephalopathy. The neurotoxicity of neonatal hyper bilirubinemia manifests because the blood–brain barrier has yet to develop fully and bilirubin can freely pass into the brain interstitium, whereas more developed individuals with increased bilirubin in the blood are protected. Aside from specific chronic medical conditions that may lead to hyperbilirubinaemia, neonates in general are at increased risk since they lack the intestinal bacteria that facilitate the breakdown and excretion of conjugated bilirubin in the feces (this is largely why the feces of a neonate are paler than those of an adult). Instead the conjugated bilirubin is converted back into the unconjugated form by the enzyme β-glucuronidase (in the gut, this enzyme is located in the brush border of the lining intestinal cells) and a large proportion is reabsorbed through the enterohepatic circulation.

Associated Health Benefits

Research has indicated that in the absence of liver disease, individuals with high levels of total bilirubin may experience various health benefits exceeding those with lower levels of bilirubin.[11] Studies have found higher levels of bilirubin in elderly individuals are associated with higher functional independence. Studies have also revealed that levels of serum bilirubin are inversely related to risk of certain heart diseases.

Blood Tests

Bilirubin is degraded by light. Blood collection tubes containing blood or (especially) serum to be used in bilirubin assays should be protected from illumination. For adults, blood is typically collected by needle from a vein in the arm. In newborns, blood is often collected from a heelstick, a technique that uses a small, sharp blade to cut the skin on the infant's heel and collect a few drops of blood into a small tube. Non-invasive technology is available in some health care facilities that will measure bilirubin by using an instrument placed on the skin (transcutaneous bilirubin meter)

Bilirubin (in blood) is in one of two forms:

Abb.	Name(s)	Water-soluble	Reaction
"BC"	"Conjugated" or "direct bilirubin"	Yes (bound to glucuronic acid)	Reacts quickly when dyes (diazo reagent) are added to the blood specimen to produce azobilirubin "Direct bilirubin"
"BU"	"Unconjugated" or "indirect bilirubin"	No	Reacts more slowly, still produces azobilirubin, Ethanol makes all bilirubin react promptly, then: indirect bilirubin = total bilirubin − direct bilirubin

Total bilirubin (TBIL) measures both BU and BC. Total and direct bilirubin levels can be measured from the blood, but indirect bilirubin is calculated from the total and direct bilirubin. Indirect bilirubin is fat-soluble and direct bilirubin is water-soluble.

Measurement Methods

Originally, the Van den Bergh reaction was used for a qualitative estimate of bilirubin. This test is performed routinely in most medical laboratories and can be measured by a variety of methods. Total bilirubin is now often measured by the 2,5-dichlorophenyldiazonium (DPD) method and direct bilirubin is often measured by the method of Jendrassik and Grof.

Blood Levels

The bilirubin level found in the body reflects the balance between production and excretion. Blood test results should always be interpreted using the reference range provided by the laboratory that performed the test, but typical [0.3 to 1.9 mg/dL]s for adults and [340 µmol/L] for new borns:

	µmol/l = micromole/liter	mg/dl = milligram/deciliter
total bilirubin	<21	<1.23
direct bilirubin	1.0–5.1	0–0.3
		0.1–0.3
		0.1–0.4

Urine Tests

Urine bilirubin may also be clinically significant. Bilirubin is not normally detectable in the urine of healthy people. If the blood level of conjugated bilirubin becomes elevated, *e.g.* due to liver disease, excess conjugated bilirubin is excreted in the urine, indicating a pathological process. Unconjugated bilirubin is not water-soluble and so is not excreted in the urine. Testing urine for both bilirubin and urobilinogen can help differentiate obstructive liver disease from other causes of jaundice.

Biliverdin is a green tetrapyrrolic bile pigment and is a product of heme catabolism. It is the pigment responsible for a greenish colour sometimes seen in bruises.

Metabolism

Biliverdin results from the breakdown of the heme moiety of hemoglobin in erythrocytes. Macrophages break down senescent erythrocytes and break the heme down into biliverdin, which normally rapidly reduces to free bilirubin. Biliverdin is seen briefly in some bruises as a green colour. In bruises, its breakdown into bilirubin leads to a yellowish colour.

Role in disease

Biliverdin has been found in excess in the blood of humans suffering from hepatic diseases. Jaundice is caused by the accumulation of biliverdin or bilirubin (or both) in the circulatory system and tissues. Jaundiced skin and sclera (whites of the eyes) are characteristic of liver failure.

Role in Treatment of Disease

While typically regarded as a mere waste product of heme breakdown, evidence that suggests that biliverdin and other bile pigments have a physiological role in humans has been mounting. Bile pigments such as biliverdin naturally possess significant anti-mutagenic and antioxidant properties and therefore fulfill a useful physiological function. Biliverdin and bilirubin have been shown to be potent scavengers of peroxyl radicals. They have also been shown to inhibit the effects of polycyclic aromatic hydrocarbons, heterocyclic amines and oxidants all of which are mutagens. Studies have even found that people with higher concentration levels of bilirubin and biliverdin in their bodies have a lower frequency of cancer and cardiovascular disease.

A 1996 study by McPhee *et al.*, suggested that biliverdin as well as many other tetrapyrrolic pigments may function as an HIV-1 protease inhibitor. Of the fifteen compounds tested, biliverdin was one of the most active. *In vitro* experiments showed that biliverdin and bilirubin competitively inhibited HIV-1 proteases at low micromolar concentrations, reducing viral infectivity. However, when tested in cell culture with micromolar concentrations, it was found that biliverdin and bilirubin reduced infectivity by blocking viral entry into cells. Results were found to be similar for HIV-2 and SIV. Further research is needed to confirm these results and to determine whether unconjugated hyperbilirubinemia has any effect on the progression of HIV infection. Current research has suggested that the anti-oxidant properties of biliverdin and other bile pigments may also have a beneficial effect on asthma. This is because oxidative stress may play a vital role in the pathogenesis of asthma. A 2003 study found that asthma patients suffering from jaundice brought on by acute hepatitis B exhibited temporary relief of asthma symptoms. However, there could also have been confounding factors such as elevated levels of cortisol and epinephrine, so more research into this possibility is required.

In Fluorescence Imaging

In a complex with reengineered bacterial phytochrome, biliverdin has been employed as IR-emitting chromospheres for *in vivo* imaging. In contrast to fluorescent proteins which form their chromophore through posttranslational modifications of the polypeptide chain, phytochromes bind an external ligand (in this case, biliverdin) and successful imaging of the first bacteriophytochrome-based probe required addition of the exogenous biliverdin. Recent studies demonstrated that bacteriophytochrome-based fluorescent proteins with high affinity to biliverdin can be imaged *in vivo* utilizing endogenous ligand only and, thus, with the same ease as the conventional fluorescent proteins. Advent of the second and further generations of the biliverdin-binding bacteriophytochrome-based probes should broaden the possibilities for the non-invasive *in vivo* imaging.

Phycobilins meaning "alga" and from Latin: *bilis* meaning "bile") are light-capturing bilanes found in cyanobacteria and in the chloroplasts of red algae, glaucophytes and some cryptomonads (though not in green algae and plants). Most of their molecules consist of a chromophore which makes them coloured. They are unique among the photosynthetic pigments in that they are bonded to certain water-soluble proteins, known as phycobiliproteins. Phycobiliproteins then pass the light energy tochlorophylls for photosynthesis. The phycobilins are especially efficient at absorbing red, orange, yellow and green light, wavelengths that are not well absorbed by chlorophyll a. Organisms growing in shallow waters tend to contain phycobilins that can capture yellow/red light, while those at greater depth often contain more of the phycobilins that can capture green light, which is relatively more abundant there. The phycobilins fluoresce at a particular wavelength and are, therefore, often used in research as chemical tags, e.g., by binding phycobiliproteins to antibodies in a technique known as immunofluorescence.

Types

There are four types of phycobilins:

1. Phycoerythrobilin, which is red
2. Phycourobilin, which is orange
3. Phycoviolobilin (also known as phycobiliviolin) found in phycoerythrocyanin
4. Phycocyanobilin (also known as phycobiliverdin), which is blue.

They can be found in different combinations attached to phycobiliproteins to confer specific spectroscopic properties.

Structural Relation to other Molecules

In chemical terms, phycobilins consist of an open chain of four pyrrole rings (*tetrapyrrole*) and are structurally similar to the bile pigment bilirubin, which explains the name. (Bilirubin's conformation is also affected by light, a fact used for the phototherapy of jaundiced newborns.) Phycobilins are also closely related to the chromophores of the light-detecting plant pigment phytochrome, which also consist of an open chain of four pyrroles. Chlorophylls are composed of four pyrroles as well, but there the pyrroles are arranged in a ring and contain a metal atom in the center.

References

Austin C, Perkins S (2006). "Parasites in a biodiversity hotspot: a survey of hematozoa and a molecular phyolgenetic analysis of plasmodium in New Guinea skinks". *Journal of Parasitology* 92(4): 770-777. doi: 10.1645/GE-693R.1

Baranano, D. E., Rao, M., Ferris, C. D., Snyder, S. H. (2002). "Biliverdin reductase: A major physiologic cytoprotectant". *Proceedings of the National Academy of Sciences* 99 (25): 16093–8. Bibcode: 2002PNAS. 9916093B. doi: 10.1073/pnas. 252626999. *JSTOR* 3073913.

Bilirubin's Chemical Formula. Retrieved 2007-08-14.

Boris Rolinski, Küster, H, Ugele, B, Gruber, R, Horn, K (2001). "Total Bilirubin Measurement by Photometry on a Blood Gas Analyzer: Potential for Use in Neonatal Testing at the Point of Care". *Clinical Chemistry* 47 (10): 1845–7.

Boron W, Boulpaep E. (2005). Medical Physiology: a cellular and molecular approach. p. 984-986. Elsevier Saunders, United States. ISBN 1-4160-2328-3

Bulmer, AC, Ried, K, Blanchfield, JT, Wagner, KH (2008). "The anti-mutagenic properties of bile pigments". *Mutation research* 658 (1–2): 28–41. doi: 10.1016/j. mrrev. 2007.05.001.

Cheifetz, Adam S. (2010). *Oxford American Handbook of Gastroenterology and Hepatology.* Oxford: Oxford University Press, USA. p. 165. ISBN 0199830126.

Fang, LS, Bada, JL (1990). "The blue-green blood plasma of marine fish". *Comparative biochemistry and physiology. B, Comparative biochemistry* 97 (1): 37–45. doi: 10.1016/0305-0491(90)90174-R.

Filonov, G. S., Piatkevich, Kiryl D, Ting, Li-Min, Zhang, Jinghang, Kim, Kami, Verkhusha, Vladislav V (2011). "Bright and stable near infra-red fluorescent protein for *in vivo* imaging". *Nat Biotechnol* 29 (8): 757–761. doi: 10.1038/nbt.1918.

Golonka, Debby (2010). "Digestive Disorders Health Center: Bilirubin". WebMD. p. 3. Archived from the original on 1 January. Retrieved 2010-01-14.

Harmonisation of Reference Intervals. Pathology Harmony. Retrieved 23 Sep 2014.

Kao, T. W., Chou, C. H., Wang, C. C., Chou, C. C., Hu, J., Chen, W. L. (2012). "Associations between serum total bilirubin levels and functional dependence in the elderly". *Internal Medicine Journal* 42 (11): 1199–207. doi: 10.1111/j.1445-5994.2011.02620.x.

Kuntz, Erwin (2008). *Hepatology: Textbook and Atlas.* Germany: Springer. p. 38. ISBN 978-3-540-76838-8.

Laboratory tests. Archived from the original on 13 August 2007. Retrieved 2007-08-14.

Lee Grismer, L., Thy, Neang, Chav, Thou, Holden, Jeremy (2007). "A New Species of Chiromantis Peters 1854 (Anura: Rhacophoridae) from Phnom Samkos in the Northwestern Cardamom Mountains, Cambodia". *Herpetologica* 63 (3): 392. doi: 10.1655/0018-0831.

Liu, Y, Li, P, Lu, J, Xiong, W, Oger, J, Tetzlaff, W, Cynader, M (2008). "Bilirubin possesses powerful immunomodulatory activity and suppresses experimental autoimmune encephalomyelitis". *Journal of Immunology* 181 (3): 1887–97. doi: 10.4049/jimmunol.181.3.1887.

Lote, CJ, Saunders, H (1991). "Aluminium: gastrointestinal absorption and renal excretion". *Clinical science (London, England : 1979)* 81 (3): 289–95.

McDonagh, Antony F, Palma, Lightner, Palma, LA, Lightner DA (1980). "Blue Light and Bilirubin Excretion". *Science* 208 (4440): 145–151. Bibcode: 1980Sci. 208.145M. doi: 10.1126/science.7361112.

McPhee, F, Caldera, PS, Bemis, GW, McDonagh, AF, Kuntz, ID, Craik, CS (1996). "Bile pigments as HIV-1 protease inhibitors and their effects on HIV-1 viral maturation and infectivity *in vitro*". *The Biochemical journal.* 320 (Pt 2) (Pt 2): 681–6.

Merck Manual Jaundice Last full review/revision July 2009 by Steven K. Herrine

Mosqueda L, Burnight K, Liao S (2005). "The Life Cycle of Bruises in Older Adults". *Journal of the American Geriatrics Society.* 53(8): 1339-1343. doi: 10.1111/j.1532-5415.2005.53406.x

Moyer, Kathryn D., Balistreri, William F. (2011). "Liver Disease Associated with Systemic Disorders". In Kliegman, Robert M., Stanton, Bonita F., St. Geme, Joseph W., Schor, Nina F., Behrman, Richard E. *Nelson Textbook of Pediatrics.* Saunders. p. 1405. ISBN 978-1-4377-0755-7.

Novotný, L, Vítek, L (2003). "Inverse relationship between serum bilirubin and atherosclerosis in men: A meta-analysis of published studies". *Experimental Biology and Medicine* 228 (5): 568–71.

O'Carra P, Murphy RF, Killilea SD (May 1980). "The native forms of the phycobilin chromophores of algal biliproteins. A clarification". *Biochem. J.* 187 (2): 303–9.

Ohrui, T, Yasuda, H, Yamaya, M, Matsui, T, Sasaki, H (2003). "Transient relief of asthma symptoms during jaundice: a possible beneficial role of bilirubin". *The Tohoku journal of experimental medicine* 199 (3): 193–6. doi: 10.1620/tjem.199.193.

Pirone, Cary, Quirke, J. Martin E., Priestap, Horacio A., Lee, David W. (2009). "Animal Pigment Bilirubin Discovered in Plants". *Journal of the American Chemical Society* 131 (8): 2830.doi: 10.1021/ja809065g.

Schwertner, Harvey A., Vítek, Libor (2008). "Gilbert syndrome, UGT1A1*28 allele and cardiovascular disease risk: Possible protective effects and therapeutic applications of bilirubin". *Atherosclerosis* 198 (1): 1–11. doi: 10.1016/j. atherosclerosis.2008.01.001.

Sedlak, T. W., Saleh, M., Higginson, D. S., Paul, B. D., Juluri, K. R., Snyder, S. H. (2009)."Bilirubin and glutathione have complementary antioxidant and cytoprotective roles".Proceedings of the National Academy of Sciences 106 (13): 5171–6.Bibcode: 2009pnas.106.5171s. doi: 10.1073/pnas.0813132106. JSTOR 40455167.

Sedlak, T. W., Snyder, S. H. (2004). "Bilirubin Benefits: Cellular Protection by a Biliverdin Reductase Antioxidant Cycle". *Pediatrics* 113 (6): 1776–82. doi: 10.1542/peds. 113.6.1776.

Seyfried, H, Klicpera, M, Leithner, C, Penner, E (1976). "Bilirubin metabolism (author's transl)". Wiener klinische Wochenschrift 88 (15): 477–82.

Shu, X. (2009). "Mammalian expression of infrared fluorescent proteins engineered from a bacterial phytochrome". *Science* 324 (5928): 804–807. doi: 10.1126/science.1168683.

SI Units for Clinical Data. Archived from the original on 2013-10-11.

Stocker, R, Yamamoto, Y, McDonagh, A., Glazer, A., Ames, B. (1987). "Bilirubin is an antioxidant of possible physiological importance". *Science* 235 (4792): 1043–6. doi: 10.1126/science.3029864.

Stricker, Reto, Eberhart, Raphael, Chevailler, Marie-Christine, Quinn, Frank A., Bischof, Paul, Stricker, René (2006). "Establishment of detailed reference values for luteinizing hormone, follicle stimulating hormone, estradiol and progesterone during different phases of the menstrual cycle on the Abbott ARCHITECT® analyzer". Clinical Chemical Laboratory Medicine44 (7): 883–7. doi: 10.1515/CCLM. 2006.160.

Urinalysis: three types of examinations. Lab Tests Online (USA). Retrieved 16 August 2013.

Watson, D., Rogers, J. A. (1961). "A study of six representative methods of plasma bilirubin analysis". *Journal of Clinical Pathology* 14 (3): 271–8. doi: 10.1136/ jcp.14.3.271.

Index

α-amylase inhibitors 33, 57
α-carotene 437, 439
β-aminopropionitrile 77
β-carotene 87, 439
β-N-oxalylamino-L-alanine 78

A

Abnormalities 74
Acid esters 90, 441
Acridine derivatives 21
Actiniochrome 392
Adenochrome 392
Agriculture 154
Ajmaline 29
Albinism 483
Albinistic 380
Alcoholic glycosides 118
Alcohols 89, 440
Aldehydes 90, 441
Alfalfa 262
Alkaloids 5, 8, 15, 17, 26
Alpha-1 antitrypsin 331
Alpha-1 proteinase inhibitor 333
Amphibians 394

Amylase 339
Amylase inhibitor 345
Androgens 228
Antenna 377
Anthocyanidins 258, 410
Anthocyanins 258, 290, 379, 405, 406, 407, 411, 412, 454
Anthoxanthins 108
Anthranilic acid 26
Anthraquinone Glycosides 119
Anti-aging 458
Antibacterial 107
Antibiotics 158
Anticancer 243, 421
Anticarcinogenic 352
Anti-inflammatory 421, 422, 424
Antimetabolites 157
Antinutrients 156, 339
Antinutritional 344
Antioxidant 106, 194, 408, 422, 423, 439, 458
Antioxidant activity 284
Antitrypsin 333
Antitumor 195, 384, 444

Arabidopsis thaliana 251

Arthritis 458

Arthropods 397

Aspartic acid 26

Astringent 344

Atropine 9, 29

B

Barley 251

Beer 278

Beets 421

Berries 282, 412

Beta-aminopropionitrile 69

Betalains 378, 379, 406, 419

Bilberry 407

Bilirubin 525, 529

Bioavailability 283

Biochar 179

Biochemical 523

Biochromes 389, 390, 395, 466

Biological pigments 378

Bioluminescence 382, 388, 427

Bioluminescent 431, 432

Biomarker 470

Biomimicry 499

Biosynthesis 410

Biosynthetic pathway 258

Biphenyls 5

Blackberry 407

Blackcurrant 407

Black raspberry 407

Black rice 407

Blood clotting 460

Blueberry 407

BOAA 67

Bone health 195

Bones 290

Brain 422

By-products 7

C

Caffeine 17, 29, 279, 280

Camouflage 395, 400

Cancer 100, 107, 193, 291, 412, 457, 458

Cancer prevention 422

Cancers 438

Carcinogenesis 351, 354, 456

Carcinogenicity 284

Cardamom 175

Cardiac glycosides 120

Cardiovascular diseases 107

Carotene 377

Carotenoids 12, 85, 379, 381, 437, 439, 491

Carotenoproteins 381

Carrots 445

Caryophyllales 419

Catechins 255, 266

Cephalopods 401, 430

Chalcogen 135

Chemical composition 3

Chemotherapy 46

Cherry 407

Chili powder 175

Chloroform 17

Chlorophyll 379, 408, 447, 453, 455, 523

Chlorophyll a 377, 448

Chlorophylls 396

Chlorosis 453

Chocolate 9, 283

Cholesterol 193

Chromatophore 383, 465, 466, 471

Chromatophoric 397

Chromone glycosides 119

Chromophores 532

Chymotrypsin 95, 97, 100
Clinical significance 326
Cobalamin 526
Cocaine 17
Codeine 17, 29
Colchicine 29
Colchicine alkaloids 23
Colon cancer 193, 200
Colouration 387
Colour temperature 507
Competitive inhibition 35
Concord grape 407
Condensed tannins 238, 265, 355
Coriander 175
Corticosteroids 228
Coumarin glycosides 119
Cranberry 407
Cravings 456
Cryptoxanthin 437
Cyanidin 406
Cyanogenic glycoside 9, 10
Cyanophores 468
Cystic fibrosis 99
Cytotoxicity 198

D

Dammaranes 196
Dauricine 28
Defense 7
Delphinidin 406
Dementia 292
Deodorizing 459
Depolymerization 243
Derivatives 421
Detox 421
Detoxification 423, 458
Diabetes 292
Diadinoxanthin 491

Diaminopimelic acid 132
Diatoxanthin 491
Dietary guidelines 288
Digestibility 347
Digestion 421
Dimethyltryptamine 30
Dinoflagellates 491
Diphosphate 12
Disease 88
Dizziness 74
Dopamine 15
Drug 46

E

Eadie-Hofstee 38
Eggplant peel 407
Emetine 29
Energy 509
Enological tannins 270
Enzyme 33
Enzyme inhibitors 34
Enzymes 39, 105, 331, 411
Epicatechins 255
Epidemiological 352
Epoxides 89, 441
Ergot alkaloids 29
Erythrocytes 527
Erythrophores 466
Essential Proteins 50
Estrogen 11, 228
Ethnobotany 191
Ethnopharmacology 170
Eumelanin 477, 479, 484

F

Fatal cardiac arrhythmias 292
Fatty acid 6
Ferric chloride 264
Flavanones 109

Flavanonols 109

Flavans 109

Flavins 391

Flavonoid glycosides 120

Flavonoids 8, 9, 103, 105, 176, 266, 271

Flavonols 268

Flowers 406

Fluorescence 531

Fluoride 279

Free radicals 486

Fruits 406

Furostanes 202

Furriness 270

Fuzzy complex 50

G

Gamma-carotene 437

Garlic 175

Genetic disorders 482

Genetic manipulation 262

Ginger 175

Glare 505

Glazing 506

Glutamic acid 68

Glycosides 5, 89, 117, 410, 441

Glycosidic bond 118

Glycosylation 105, 334

Goldbeater's skin test 264

Gossypol 11, 123

Grape 267

Gymnosperms 453

H

Haemagglutinins 346

Headaches 293

Healing 456

Heart disease 292

Heavy metals 280, 294

Hematochromes 378

Hemerythins 392

Hemocyanins 392

Hemoglobin 523

Hemovanadin 392

Herbicides 47

Herbivores 3

Heteromultimeric 50

Hide-powder Method 264

Histidine 26

Homomultimeric 50

Hormones 394

Hue 465

Human nutrition 139

Hunger 456

Hybrids 6

Hydrocarbons 89, 440

Hydrolysable 238

Hydroxycinnammic Acids 271

Hyoscyamine 29

Hypokalemic paralysis 11

I

Ibogaine 30

Immune function 193

Immune system 438

Immunity 459

Immunity booster 194

Immunomodulatory 243

Immunosuppressive 384

Implications 503

Inflammation 106

Inhibitor I9 327

Inhibitors 98

Insects 400

Insomnia 459

Interference 389

Invertebrates 396

Iridophores 467

Iron 457

Irreversible inhibition 42

Isoflavones 166, 169

Isoflavonoids 110

Isolation 229

Isomerization 249

Isoquinoline 16

Isorhamnetin 271

Isoxazole derivatives 21

K

Kaempferol 271

Kidney stones 460

Kinetics 95

Km value 40

L

Laricitrin 271

Lathyrus 67

Lathyrus sativus 4

Leaves 406

Lectins 339, 346

Legumes 262, 283

Leucophores 467

Lichens 399

Light 500

Lignanas 341

Lignans 352

Linckiacyanin 381

Lineweaver–Burk 38

Lipid peroxidation 10

Lipophilic 190

Lipophilicity 204

Lizards 465

Localization 242

Lutein 439

Lycopene 437, 439

Lyochromes 391

Lysine 26

M

Maize 251, 260

Male fertility 444

Malvidin 406

Mammalian 481

Mammals 400

Marine 400

Medical research 501

Medicine 286

Melanin 382, 477, 487

Melanophores 467

Melasma 380

Mental health 423

Mescaline 15

Methylation 105

Mevalonic acid 322

Michaelis–menten equation 36

Misconceptions 55

Mixed inhibition 35

Multiprotein complex 48

Muscadine grape 407

Mutants 251

Myoglobin 378

Myricetin 271

N

Naloxone 29

Narcolepsy 73

Natural phenolic 166

Natural phenols 176

Nematodes 431

Nervous system 72

Neurological 70

Neuromelanin 477, 480

Neurotoxic 82

Nicotine 17

Nicotinic acid 26

Nitrogen cycling 242

Non-competitive inhibition 35

Non-obligate protein complex 49

Non-pharmaceutical products 4

Non-protein 139

Non-protein amino acids 82

Non-proteinogenic 127

Nutrition 265

Nuts 282

O

Obligate 49

ODAP 67

Oestrogens 342

Oleananes 196

Oligomeric ellagitannins 355

Onion 175

Ornithine 26, 45

Osteolathyrism 76

Osteotoxic amino acids 82

Oxalate 147

Oxalates 280

Oxalic acid 145

Oxalyldiaminopropionic acid 68, 78

Oxazole derivatives 21

P

Pancreatic cancer 351

Pancreatitis 460

Paprika 175

Paradox 287

Parkinson's disease, 483

P-dimethylaminocinnamaldehyde test 111

Pelargonidin 406

Peonidin 406

Pepper 175

Persimmons 282

Pesticides 47, 233

Petunidin 406

Phakellia stelliderma 384

Pharmaceutical industry 3

Phenazines 5

Phenolic 267

Phenolic acids 177, 267, 272

Phenolic compounds 272, 339, 343

Phenolic glycosides 120

Phenols 5

Phenoxazones 390

Phenylalanine 26

Pheomelanin 477, 479

Phosphorescent 429

Photo-oxidative 491

Photo-protective pigments 383

Photoprotectors 494

Photosynthesis 85, 396, 447

Photosynthetic 377

Phycocyanobilin 532

Phycoerythrobilin 532

Phycourobilin 532

Phycoviolobilin 532

Physiology 500

Phytic acid 10, 151, 155, 339, 340

Phytoestrogens 11, 165, 169, 341, 352

Phytonutrients 424

Pigment 377, 468

Ping-pong 97

Poisons 47

Pollination 378

Polyketides 6

Polymerization 247, 254

Polyphenols 175, 179, 185, 289

Polysaccharide 6

Pomegranates 280

Prebiotic amino acids 133

Precursor 217

Prenylflavonoids 166

Primary vs. secondary plant metabolism 7

Proanthocyanidin 243, 253, 262

Procyanidins 269

Prodelphinidins 269

Progestogens 228

Prostate cancer 280

Protease 326, 339

Protease inhibitor 41, 326, 341

Protein 135, 262

Protoalkaloids 16

Protochlorophyllide 453

Prymnesiophytes 491

Pseudo tannins 235

Psilocybin 30

Psychoactive drugs 30

Pterins 391

Pumpkin 445

Purine derivatives 23

Purines 391

Pyridine 16

Pyrophosphate 12

Pyrrolidine 16

Pyrrolidine derivatives 18

Q

Quantitative description 35

Quinazoline derivatives 21

R

Radiation 486

Red blood cells 458

Red cabbage 407

Red raspberry 407

Reflection 389

Reptiles 400

Resveratrol 289

Reversible inhibition 35, 38

Ribosome 6

Roots 406

S

Saponins 8, 120, 189, 192, 237, 339, 343, 354

Scattering 390

Schemochromes 389

Schiff bases 26

Sclerotins 390

Scopolamine 29

Secondary metabolites 3, 236, 339

Secondary metabolites 3

Secondary plant metabolites 7

Sensitivity 52

Sensory 401

Serine protease 97

Serotonin 15

Shinoda test 110

Sinusitis 460

Skin 486

Smoked foods 282

Sodium hydroxide test 111

Sodium-iodide 9

Solar cells 411

Soluble tannins 242

Soy 160

Specificity 33, 52

Spectrophotometry 450

Spirostanes 201

Sprouting 159

Stability 410

Stable protein complex 49

Starch blockers 345

Starchy foods 339

Stems 406

Steroidal glycosides 120

Steroidogenesis 226

Steroids 8, 25, 217

Steviol glycosides 121

Stiasny's method 264

Stigma 76

Stilbenoids 271

Sunscreen 408

Swelling 457

Syndrome 76

Synthesis 88

Syringetin 271

T

Tannins 8, 233, 266, 269

Tea 278

Terpenes 8

Terpenoids 5, 321

Tetrapyrroles 381, 523

Theaflavins 266

Thebaine 29

Thiazole derivatives 21

Thioglycosides 121

Tobacco 4

Tomatoes 445

Toxic alkaloid 4

Toxicity 125, 148, 283, 413, 528

Toxiferine 27

Transgenic plants 57

Transient 49

True alkaloids 16

Trypsin 97, 100

Trypsin inhibitor 329, 325, 335

Tryptophan 26

Tubocurarine 28

Tumor 194

Tyrosine 26

U

Uncompetitive inhibition 35

Urine 528

V

Vertigo 74

Villalstonine 27

Violet petals 407

Vitamin C 421

Vitamin D 397

Vitiligo 380

Voacamine 27

W

Wine 267, 286

Wound healing 459

X

Xanthophores 466

Xanthophyll 377

Xanthophylls 491

X-ray 51

Z

Zymogen 99